P9-EDU-793

Biochemical Adaptation

PETER W. HOCHACHKA

GEORGE N. SOMERO

Biochemical Adaptation

Princeton University Press

Princeton, New Jersey

Copyright © 1984 by Princeton University Press

Published by Princeton University Press, 41 William Street,
Princeton, New Jersey 08540
In the United Kingdom: Princeton University Press, Guildford,
Surrey

All Rights Reserved

Library of Congress Cataloging in Publication Data will be
found on the last printed page of this book

ISBN 0-691-08343-6 (cloth)
ISBN 0-691-08344-4 (paper)

This book has been composed in Lasercomp Times Roman

Clothbound editions of Princeton University Press books are
printed on acid-free paper, and binding materials are chosen for
strength and durability. Paperbacks, although satisfactory
for personal collections, are not usually suitable
for library rebinding

Printed in the United States of America by Princeton
University Press, Princeton, New Jersey

QP
82
.H63
1984 / 45,054

CONTENTS

CAMROSE LUTHERAN COLLEGE
LIBRARY

FIGURES

TABLES

PREFACE

One of the great accomplishments of biochemistry and molecular biology has been the elucidation of many of the major unifying principles and mechanisms that serve as the foundations of all living systems. Common mechanisms of energy transformation, catalysis, and the coding and processing of genetic information testify to the unity of life at the molecular level. While no one can deny these triumphs of reductionist approaches to biology, these insights into unifying principles of biochemical design in living systems seem to offer relatively few direct answers to a question of central importance to many biologists: How to account for the mechanisms underlying the immense diversity of organisms? What are the fundamental ways in which the basis biochemical structures and functions of living systems are adaptively modified to allow organisms to exploit the full range of natural environments and to maintain the radically different modes of life we see in nature?

The question of how a set of common mechanisms are extended into uncommon and diverse contexts is not new. Decades ago a similar gap existed between the fields of comparative anatomy and physiology. This gap was bridged by the concept of adaptation, and it is our belief that the concept of adaptation can be extended to the molecular level to effect a bridge between the observations of universal molecular mechanisms, on the one hand, and extreme biological diversity, on the other hand. Thus, the focus of our book is on the ways in which the ubiquitous molecular structures of organisms are modified to permit organisms to thrive in such diverse environments as the polar regions, deserts, and the deep sea, and to achieve modes of living that may involve major changes in type and quantity of nutrients available and in the oxygen that is present to support respiration.

In developing the central theme of biochemical adaptation we have selected examples for study that strike us as providing especially clear illustrations of the fundamental strategies of adaptation at the biochemical level. Our scope of treatment is not encyclopedic. Instead, we have focused on topics for which there either are numerous data, which allow a detailed analysis to be achieved, or where the basic phenomenology is so interesting that, despite a lack of large numbers of data, it seemed to us worthwhile to draw questions of potential interest to the readers' attention. Our hope is that the examples we have chosen will be exciting and will provide the reader with an impetus to examine other, less well-studied problems in biochemical adaptation.

Our indebtedness to those whose efforts have helped make this book possible is a pleasure to acknowledge. At the top of the list are the students and postdoctoral scholars who have given us the type of stimulation that has kept our enthusiasm for this writing project, and for research per se, at a high pitch. P.W.H. wishes to express particular thanks to the students and postdoctoral fellows contributing to the current fermentations in his laboratory: H. Abe, J. Ballantyne, M. A. Castellini, G. P. Dobson, J. F. Dunn, B. Emmett, R. Foreman, C. J. French, U. Hoeger, T. P. Mommsen, B. J. Murphy, W. Parkhouse, E. A. Shoubridge, and R. Suarez. Earlier students who have continued to influence the thinking and work in P.W.H.'s laboratory include J. Baldwin, H. Behrisch, J.H.A. Fields, H. Guderley, M. Guppy, T. P. Moon, T. Mustafa, T. Owen, J. Storey, and K. B. Storey. In addition, P.W.H. wishes to acknowledge his numerous colleagues around the world who have made the entire enterprise all the more exciting and who on occasion have combined the adventures of intellect with the adventures of scientific expedition.

G.N.S. wishes to express his gratitude to past and present members of the Scripps Institution of Oceanography High Pressure Zone Laboratory: L. Borowitzka, R. D. Bowlus, M. A. Castellini, B. J. Davis, K. A. Dickson, V. Donahue, J. G. Duman, H. Felbeck, S. L. French, E. Golanty, J. E. Graves, G. S. Greaney, S. C. Hand, K. H. Hoffmann, D. Kramer, G. Lopez, P. S. Low, M. S. Lowery, J. Malpica, E. Pfeiler, M. A. Powell, S. J. Roberts, J. F. Siebenaller, K. M. Sullivan, R. R. Swezey, P. J. Walsh, P. H. Yancey, and M. Yacoe. Stimulating discussions of many of the ideas in this book with Drs. J. J. Childress, M. E. Clark, and F. N. White also have been invaluable.

Neither this project nor much of our work described in it could have been achieved without support from NSERC and the Canadian Heart Foundation (P.W.H.) and the NSF and NIH (G.N.S.). Special thanks are given by P.W.H. to the Australian Department of Science and Technology for support via a Queen's Senior Fellowship.

Finally, we wish to emphasize that the task of preparing a book transcends the generation of ideas. The manipulation of manuscripts and figures, while the primary responsibility of the authors, could not have been achieved effectively without the enormous efforts of Ms. Leslie Borleski, Ms. Kathy Lingo, and Ms. Cecelia Ross. And, lastly, both authors acknowledge the emotional support of the members of the Laika family tree.

Peter W. Hochachka
Vancouver, British Columbia
George N. Somero
La Jolla, California

LIST OF ABBREVIATIONS

Common Metabolites

AMP, ADP, ATP	adenosine 5′-mono-, -di-, -triphosphate
cAMP	3′,5′-cyclic AMP
ArgP	arginine phosphate
CMP, CDP, CTP	cytidine 5′-mono-, -di-, -triphosphate
CoA	coenzyme A
Cr, CrP	creatine, creatine phosphate
DG	diglyceride
DHAP	dihydroxyacetone phosphate
DNA	deoxyribonucleic acid
2,3 DPG	2,3 diphosphoglyceric acid
FAD^+, FADH	flavin adenine dinucleotide, and its reduced form
F6P	fructose 6-phosphate
F1,6BP	fructose 1,6-bisphosphate
F2,6BP	fructose 2,6-bisphosphate
G3P	glyceraldehyde 3-phosphate
G6P	glucose 6-phosphate
GMP, GDP, GTP	guanosine 5′-mono-, -di-, -triphosphate
imid	imidazole
IMP, IDP, ITP	inosine 5′-mono-, -di-, -triphosphate
KGA	ketoglutarate
MG	monoglyceride
NAD^+, NADH	nicotinamide adenine dinucleotide, and its reduced form
$NADP^+$, NADPH	nicotinamide adenine dinucleotide phosphate, and its reduced form
P5C	pyrroline-5-carboxylate
PEP	phosphoenolpyruvate
PGA	phosphoglycerate
Pi	inorganic phosphate
PPi	inorganic pyrophosphate
TG	triglyceride
UMP, UDP, UTP	uridine 5′-mono-, -di-, -triphosphate

Common Enzymes

CPK	creatine phosphokinase
CS	citrate synthase
FBPase	fructose 1,6-bisphosphatase
α-GPDH	alphaglycerophosphate dehydrogenase

HK	hexokinase
IDH	isocitrate dehydrogenase
KGDH	ketoglutarate dehydrogenase
LDH	lactate dehydrogenase
MDH	malate dehydrogenase
ODH	octopine dehydrogenase
PDH	pyruvate dehydrogenase
PEPCK	phosphoenolpyruvate carboxykinase
PFK	phosphofructokinase
PGK	phosphoglycerate kinase
PK	pyruvate kinase

Biochemical Adaptation

CHAPTER ONE

Biochemical Adaptation: Basic Mechanisms and Strategies

The Paradigm of Adaptation

When scientists attempt to take a broad view of their field of inquiry and discern the dominant conceptual themes running through their discipline, they frequently speak of the "paradigms" of the field. Such paradigms are the world-views or conceptual frameworks within which most, if not all, of the detailed questions of investigation are phrased (Kuhn, 1970). In the chapters that follow we treat varied facets of what is probably the most encompassing and general paradigm in biology, a conceptual framework that finds expression at all levels of biological organization, ranging from the molecular level to the population level. This is the concept of "adaptation," the modification of the characteristics of organisms that facilitates an enhanced ability to survive and reproduce in a particular environment.

In the study of biology, adaptation is usually a dominant theme, whether explicitly stated or not. Often we begin our study of biology with a consideration of the vast diversity of anatomical forms present in the biosphere. In the study of fishes, for example, we note a myriad of body forms, some of which are designed for rapid cruising and capture of swiftly swimming prey, and other forms which are designed for minimal locomotory activity and a prey-capture strategy involving such passive approaches as the "float-and-wait" behaviors of many deep-sea fishes. We also find that some of these deep-living fishes have lures for attracting prey. Yet other fishes have anatomical specializations allowing both air- and water-breathing under different environmental conditions. Such highly conspicuous characteristics at the physiological, morphological, behavioral, and ecological levels of biological organization are typically the types of phenomena that constitute the literature dealing with organismal adaptations.

Our treatment of adaptations, in contrast, focuses on a different level of biological organization, namely, the biochemical attributes of organisms that are responsible for such critical capacities as the generation of adequate amounts and types of metabolic function, the transport of gases between the cells and the environment, the maintenance of a proper solute microenvironment (pH and osmotic conditions) for macromolecular function, and the abilities to exploit the particular types of energy resources available to the organism. These types of biochemical adaptations can be regarded as "interiorized" phenomena, to distinguish them from other, generally more familiar biochemical adaptations that are important in the interfacing of the organism with its environment. Such "exteriorized" biochemical adaptations include cryptic coloration, bioluminescence, chemical signaling, chemical defense, chemical predation, and antigen sharing in the molecular mimicry of certain parasites. These exteriorized biochemical adaptations appear formally analogous to the conspicuous morphological, behavioral, etc. adaptations mentioned above. In contrast, the interiorized adaptations which serve as the primary focus of our analysis generally can be discerned only when the organism is biochemically "dissected." As we emphasize throughout our analysis, the very success with which organisms have adapted their enzyme systems, membranes, respiratory pigments, etc. for function in diverse environments may lull us into thinking that the interior biochemical machineries of different species are similar, if not identical. This, as we show, is a misconception. The interiorized biochemical adaptations we examine display the same wealth of diversity found in more apparent, exterior features of organisms.

To appreciate the characteristics of this diversity, and its function in enhancing the adaptiveness of organisms for their particular environments, we must first consider the set of biochemical structures and functions that are at once absolutely essential for all living systems known and highly sensitive to perturbation by the physical and chemical features of the environment. The adaptations treated in the subsequent chapters of this volume will be found to serve the following key functions:

1. The preservation of the structural integrity of macromolecules (e.g., enzymes, contractile proteins, and nucleic acids) and macromolecular ensembles (e.g., membranes and ribosomes) for function in specific environments.
2. The provision of adequate supplies of a) the energy currency of the cell adenosine triphosphate (ATP), b) reducing power to drive biosynthetic steps, and c) metabolite intermediates for the synthesis of storage compounds (glycogen, fats, etc.) and for the synthesis of nucleic acids and proteins.

3. The maintenance of mechanisms for regulation of metabolic rates and directions of metabolic flow in response to the organisms' needs and changes in these needs as the environment varies.

These, indeed, are the common and fundamental requirements of all living systems, and they must be realized in all environments, under all conditions. How these fundamental requirements are realized by different organisms under widely varying environmental conditions is, therefore, at the heart of biochemical adaptation and is the primary focus of the chapters to follow.

Homeostasis and Adaptation

The concept of homeostasis, which can be traced back over one hundred years to the work and writings of Claude Bernard, was first formalized into a coherent theory by W. B. Cannon early in this century. The theory simply states that, in the face of external perturbations, organisms harness mechanisms for the preservation or maintenance of an almost-constant internal state. It is evident that an end result of many of the above adaptive strategies is indeed the maintenance of homeostasis. In metabolic terms, the concept requires that both the direction and rate of metabolic reactions be adaptively regulated; glucose homeostasis, for example, requires the regulation of both gluconeogenesis and glycolysis (oppositely directed pathways).

Enantiostasis and Adaptation

Although homeostasis is commonly observed in many organisms, it is also evident that in many adaptational strategies, the maintenance of homeostasis is not achieved; indeed, it may not even be attempted. The phospholipid compositions of membranes of cold- and warm-adapted species are different; the solute composition of euryhaline invertebrates depends upon the external medium; and blood and intracellular pH varies with temperature. In considering this problem, Mangum and Towle (1977) point out that function, not state, is being preserved in many of these organisms. Membrane fluidity is adjusted with temperature in order to preserve membrane-based enzyme, hormone, and transport functions; solute levels are adjusted to conserve enzyme structure, function, and regulation; blood pH is adjusted as temperature changes in order to preserve protein function. The outcome of adaptation in all such cases is not homeostasis (the same state) but is better termed enantiostatis (conserved function), a concept first formalized by Mangum and Towle (1977).

The Basic Mechanisms of Biochemical Adaptation

In analyzing the paradigm of adaptation at the biochemical level, we shall work within the framework provided by a short list of basic adaptation mechanisms. We hasten to stress at the outset of this discussion that the basic types of mechanism discussed below, and throughout this book, are not invariably clear and distinct from each other, and in some events it may be difficult to discern which particular basic mechanism is, in fact, involved in an organism's response to its environment. On balance, however, we feel that it is heuristically useful to give the reader a skeleton to flesh out in the chapters which follow, realizing that the "bones" are not disjoint, but are closely connected.

We shall frame much of the subsequent analysis in this volume in terms of the following three mechanisms or "strategies":

1. *Adjustments in the macromolecular components of the cell or body fluids.* Two distinct classes of adjustments are possible. First, the quantities (concentrations) of given types of macromolecules, e.g., enzymes, may be altered in adaptive ways. Second, new types of macromolecules, e. g., new isozyme* or allozyme forms, may be

* The terms "isozyme" and "allozyme" will appear frequently in this volume, so it is essential that the reader understand this terminology referring to enzyme variants. "Isozyme" is a generic term used to refer to different variants of a given type of enzyme. These variants can arise from two different genetic bases. First, two or more gene loci for a particular type of enzyme (or enzyme subunit) may be present in the genome of the organism. Rigorously, the expression, "multiple-gene-locus isozymes" would be preferable for use in discussing these isozymes, but we shall generally use only the word "isozyme" when discussing this type of enzyme variant. Second, in diploid organisms allelic enzyme variants may exist when a given enzyme (subunit)-coding locus is polymorphic. These allelic isozymes are commonly termed "allozymes."

Whereas the above terminology is adequate for discussing enzyme variants in a single species, interspecific comparisons necessitate additional expressions. In the discussions to follow, we will employ the expression "interspecific homologue" to refer to the "same" enzyme found in different species, where "same" generally refers to the enzyme coded by a gene locus common to all species being examined. For instance, the glycolytic enzyme lactate dehydrogenase (LDH. EC 1.1.1.27; NAD: lactate oxidoreductase) is coded by at least two, and typically three, separate gene loci in vertebrates. The LDH

added to the system and may replace previously existing macromolecules which no longer are well suited to function in the altered environment. For simplicity, one could refer to these two strategies as "quantitative" and "qualitative," respectively.

2. *Adjustments in the microenvironment within which macromolecules function.* Adaptations of this type, which again can have both quantitative and qualitative attributes (e.g., total osmotic concentration may vary as may the types of solutes used to adjust osmolarity), are to be viewed as adjustments that confer on macromolecules the proper structural and functional characteristics. Microenvironmental adaptations will be seen to play a vital and complementary role vis-à-vis macromolecular adaptations.
3. *Adjustments in the outputs of macromolecular systems, especially enzymes, without changes in the amounts or types of machinery present.* Adaptations of this class can be viewed as differential rates of use of macromolecular systems pre-existing in the cells according to the local needs in time and space of the organism for the particular type of metabolic activity. These adaptations thus involve the important phenomenon of *metabolic regulation*: the appropriate increases and decreases of enzymic activity in response to the organism's requirements for such activities as locomotion, anaerobiosis, hibernation/estivation, and growth. These regulatory phenomena will, of course, depend heavily in many cases on changes in the compositions and concentrations of low molecular weight effectors (activators and inhibitors of enzymes, for instance) in the cells.

Again, we stress that these general categories of adaptive responses are not always distinct, and in many cases an organism may respond to an environmental change via all three adaptive strategies. Within these limitations, however, the reader may find these three basic strategies a useful conceptual framework for understanding the responses made by organisms to different environments, and we thus shall consider each of these strategies in somewhat more detail below.

---

subunit found in highest concentration in locomotory muscle, especially the white muscle of fishes, is the muscle-type or "M" type. In aerobic tissues like heart, the "H" type of subunit predominates. A third subunit type is restricted to the testes or the retina. In discussing muscle-type (M_4) LDH tetramers in different species, we will be referring to subunits coded by a single type of gene locus, but the amino acid compositions of the subunits will often differ markedly from species to species.

Adaptive Changes in the Enzyme Machinery

Enzymes play two major roles, those of catalysis and the regulation of metabolic function. In viewing adaptive changes in enzyme systems, we must keep both functions clearly in mind, for in many cases the precise regulation of catalytic rates is even more crucial than the supply of a high level of catalytic power per se. To provide a basis for understanding when and how enzyme-level adaptations come into play during adaptive responses by organisms, it is appropriate to consider certain of the basic types of enzymic adaptations to be discussed later in this volume. Adjustments in the concentrations and types of enzymes present in organisms frequently are necessary for the following reasons:

1. Changes occur in metabolic demands, both in total flux rates and in the types of pathways needed, frequently as a consequence of environmental change or of developmental stage of the organism.
2. Changes in the physical environment, i.e., in temperature and hydrostatic pressure, may strongly influence enzyme function and structure.
3. Changes in the chemical environment—to the extent that these changes affect the composition and concentration of the intra- and extracellular fluids—may dictate the needs for altered quantities and types of enzymes.

The problems raised by these changes in habitat or developmental stage may necessitate both the "quantitative" and the "qualitative" changes in enzyme systems mentioned earlier. For example, a decrease in habitat temperature for an ectothermic organism (one whose body temperature is established largely, if not entirely, by the ambient temperature) may greatly reduce total metabolic flux and may alter the regulatory abilities of key enzymes governing metabolic flow. To a certain extent, these perturbations of metabolism may be offset by increasing the concentrations of enzymes in the cell. However, in other cases more of the same enzyme may not be a sufficient cure for the problem; instead, new isozyme forms may be needed to restore adequate catalytic capacity and, in particular, regulatory ability. Similar considerations apply in the case of changes in hydrostatic pressure, which also may cause alterations in catalytic rate and regulatory capacity. Alterations in the external environment's chemical composition also may create problems for enzyme function, especially in situations where these external changes, e.g., in total osmotic concentration, lead to changes in the solute composition/content of the cells or to large alterations in water activity. However, as we analyze the responses of organisms to changes in external osmotic concentration, we most commonly find that the macromolecular machinery of the cells is protected against the effects of external osmotic changes, either through preservation of the status quo of the cellular fluids, as may be achieved by regulation of ion pumping, or through judicious adjustments in the composition of the macromolecules' functional microenvironment.

Microenvironmental Adaptations

Studies of biochemical adaptation and molecular evolution have customarily placed their major emphasis on the macromolecular components of cells, the nucleic acids and proteins, and the large molecular ensembles such as membranes. Surprisingly little attention has been paid to the low molecular weight "micromolecules" which bathe the macromolecular systems and establish many of their most critical biological properties. In our analysis, we shall attempt to redress this neglect of micromolecules. Several points, many of which have been mostly overlooked in previous analyses of, e.g., osmotic regulatory strategies, will provide a focus for our analysis. One key point concerns the bases for selecting particular types of solutes for roles as intracellular osmotic effectors (osmolytes). We will suggest chemical principles that determine if a given solute, whether it be an inorganic ion or a small organic molecule, is a "fit" component of the biological solution. This analysis will lead us to another important conclusion: namely, the evolution of biological solutions, e.g., cytosols, reflects the establishment of a "hospitable" environment for macromolecules to function in. The defense of this "hospitable" environment in response to changes in the chemistry of the external environment will also be treated, and this analysis may provide new insights into the rationales for the diverse osmoregulatory strategies of both aquatic and terrestrial organisms.

The discussion of microenvironmental adaptations will also involve consideration of the lipid milieu in which many enzymes, notably membrane-associated enzymes, conduct their functions. Lipids, while not "micromolecules" in the sense employed in our analysis, share with the aqueous solution surrounding soluble enzymes many common attributes vis-à-vis the requirements for providing proteins with a "hospitable" environment for function. The pervasive changes in lipid composition noted in studies of temperature adaptation, and strongly suggested in the case of pressure adaptations as well, exhibit a logic comparable to that found in osmolyte adaptations.

In addition to an analysis of the adaptations involving osmolytes and membrane lipids, we will consider how the smallest entity of the cell, the proton, is regulated to achieve a "hospitable" pH microenvironment for enzyme function. In many respects, the selection of particular pH values and certain types of buffers to defend these values is the primordial microenvironmental problem that organisms solved at the dawn of cellular evolution. This conclusion is based on the fact that particular pH relationships are found in all organisms studied to date. We will attempt to elucidate the criteria instrumental in determining what particular pH values are optimal for cellular function, and this analysis will lead us directly to a consideration of how the optimal pH values are maintained by buffering systems and other regulatory mechanisms.

From the analysis of the diverse ways in which microenvironmental factors within the cell are adjusted to preserve satisfactory macromolecular structure and function, the reader may gain a new or enhanced appreciation for the following important principle: regulation of the microenvironments of macromolecules may achieve the required adaptation to an environmental change without there being a concomitant need for changes in the macromolecules themselves. This shifting of the evolutionary "work" onto the shoulders of micromolecules has the potentially important consequence of minimizing the requirements for amino acid substitutions in proteins, and thereby speeding the rate at which certain evolutionary processes, such as colonization of habitats differing in salinity, can take place.

Adjustments in Metabolic Output

To this point in our discussion we have emphasized that organisms can respond adaptively to environmental change by adjusting the amounts or types of macromolecules present in their cells or by controlling the milieu in which these macromolecules function, and thereby shielding them from the environmental fluctuation. In addition to these two widespread mechanisms of adaptation is a third mechanism, which is especially notable for its rapidity of response. This mechanism entails the adjustment of the rates at which macromolecules already existing in the cell conduct their functions. The signals for adjusting rates of metabolic output are varied, and may involve changes in the physical environment, e.g., in temperature or light, or a broad suite of physiological signals within the organism, e.g., hormonal and electrical events. In response to these signals, adjustments in overall metabolic rate and in the relative contributions of different, often competing, metabolic pathways are regulated:

a) in response to the variable energy demands of, and substrate sources available for, different kinds and rates of exercise, e.g., steady-state versus burst locomotion;
b) in response to changes in oxygen availability to the organism, e.g., in facultative anaerobiosis;
c) in response to starvation and migration, e.g., during salmon spawning migration;
d) in response to drastic changes in the physical environment that necessitate major metabolic adjustments, e.g., during estivation and hibernation; and
e) in response to altered hormonal conditions either routinely, as in control of glucose homeostasis, or in synchrony with reproductive events, as in the control of lactation.

This is, of course, a minimal list of conditions under which external and internal signals elicit large changes in metabolic flux and pathway participations. However, this brief list should be sufficient to impress upon the reader the crucial role that modulation of the activities of pre-existing enzymes plays in a diverse set of environmental adaptations. Common to many of these adaptive responses is the need for extremely rapid adjustment of metabolic rates, adjustments that occur far more rapidly than changes in the concentrations or isozyme forms of enzymes can be achieved. This point leads us to consider another very important general property of biochemical adaptations: the relationship between the time available to effect the adaptation and the variety of adaptive strategies accessible to the organism.

Time Courses of Biochemical Adaptations and Their Relationship to Accessible Adaptive Mechanisms

In view of the fact that biochemical adaptations may involve processes having widely different time constants, ranging from the long time periods required for genetically based changes in amino acid sequences to the rapid, split-second adjustments in the activities of pre-existing enzymes, it is clear that the time course over which a biochemical adaptation can occur will play a major role in determining how, i.e., via what mechanism(s), that adaptation is effected. The longer the time available to make the adaptive response, the larger are the numbers of combinations of adaptive strategies open to the organism. In discussing the time courses of adaptation, it is useful to employ the following categories, for they indicate the avenues of adaptive response that are possible under different time constraints.

Genetic Adaptation. When adaptations occur over many generations, all categories of adaptive strategies are open to the population. Mutations in regulatory genes may lead to alterations in the basal-level concentrations of enzymes and other molecules, i.e., to quantitative macromolecular adaptations. Amino acid substitutions in proteins may lead to new isozyme (allozyme) forms of enzymes or, in more extreme cases, to novel types of catalytic potential. Amino acid substitutions may also facilitate the development of new types of regulatory capacities for enzymes, permitting more appropriate responses to signals eliciting changes in metabolic flux rates. Genetic adaptations may lead to the capacities to produce entirely new types of molecules that give the organism an ability to colonize new habitats. A classic example of this type of adaptation is the appearance of "antifreeze" glycoproteins and polypeptides in teleost fish of high latitude seas. These antifreeze molecules allow these fishes to survive in the presence of ice, a capacity not possessed by other teleost fishes.

Acclimations/Acclimatizations. Adaptive responses that occur during the lifetime of an organism, and that occur over a sufficiently long time course to permit, e.g., the induction of new types of proteins and the restructurings of membrane phospholipids, are termed "acclimations" or "acclimatizations." The distinction between these two concepts, as defined by Prosser (1973), pertains to the setting in which the adaptations occur. Adaptations of this time course that occur in the organism's natural habitat, where a variety of environmental parameters in addition to the one parameter of interest may vary, are called acclimatizations. Laboratory-induced adaptations in response to variation in but a single environmental variable are called acclimations.

What distinguishes these adaptations from genetic adaptations is, of course, the restriction of acclimations/acclimatizations to phenotypic changes. Only the information in the organism's genome at the start of its life can be utilized to make the biochemical adaptations needed to cope with environmental change.

Immediate Adaptations. The shortest time course of biochemical adaptation is termed "immediate" to indicate that these adjustments occur too rapidly to involve alterations in gene expression or substantial restructurings of cellular systems through, e.g., biosynthetic events. Immediate adaptations frequently involve the modulation of pre-existing enzymes. These rapid adjustments in enzymic activity often appear to be the organism's first line of defense against the adverse effects of an environmental change; later, more elaborate responses may derive from changes in gene expression or, over many generations, changes in the genetic information per se.

Thus, regardless of how rapidly an environmental change occurs (and these changes must be viewed relative to the generation time of the organism), biochemical adaptation mechanisms of varying degrees of complexity—and capacity for novel change—are accessible.

Compensatory and Exploitative Adaptations

Whatever the biochemical adaptation mechanisms utilized by an organism, be they changes in the amounts or types of macromolecules, adjustments in the milieu surrounding the macromolecules, or adjustments in the rates of enzymic function, two outcomes of successful adaptations are often apparent. These outcomes are noted over all time courses of adaptation as well. Certain adaptations can be regarded as "righting" environmentally caused "wrongs" in the biochemical functions of the organism; we term these "compensatory" adaptations for this reason. Compensatory adaptations may effect homeostasis or enantiostasis. A common example of a compensatory adaptation is temperature compensation in ectothermic species, the process by which metabolic rates of ectotherms are held relatively independent of temperature. In all cases, these adaptations represent means for restoring the system to a steady-state of function or structure.

In contrast to these rather conservative adaptations are more radical biochemical changes that give organisms novel new capacities. These, in turn, may permit the organism to colonize a new habitat, to capture a new class of prey, and so forth. We term these "exploitative" adaptations since they give the organism a new ability to extract benefits from its environment. The antifreeze molecules found in polar fishes are an excellent example of what we mean by exploitative adaptations. The symbiotic relationships between sulfide-oxidizing chemoautotrophic bacteria and animals of the deep-sea hydrothermal vent communities represent another fine example of exploitative adaptations. These adaptations clearly are major components of evolutionary change, and they provide us with some of the most dramatic examples of how biochemical modifications of organisms provide the basis for exploiting new environments or new features of a given habitat.

Why Study Biochemical Adaptations?

Other than possible esthetic rewards stemming from a deeper understanding of how biological systems work, is there any compelling reason to study the various types of biochemical adaptations discussed above? Are there any insights that can be gained *only* through the study of organismal diversity at the biochemical level? We feel that affirmative answers can be given to both of these questions. Our philosophy concerning the importance of examining adaptations at the biochemical level can be summed up as follows: to fully appreciate the major "themes" running through biology, one needs to study the elaborations and variations on these themes; only through this comparative approach can the fundamental design principles of organisms be adequately understood. The following example should illustrate our point. One could spend a lifetime examining the various properties of a given type of enzyme from a single species, say a rabbit. After many decades of work a long list of the enzyme's characteristics would be assembled. Which items on this list represent fundamental properties of all enzymes, or at least all interspecific homologues of this enzyme? Which characteristics are those of "rabbitness," and are not shared by other homologues of this enzyme? Neither question can be answered without studying other variants on this protein theme. When the needed comparative work has been done, it may be clear what "really matters" to all enzymes and, in addition, what is critical for an enzyme that must function in a warm-bodied (endothermic) species such as a rabbit. We will perform just such a comparative analysis time and again throughout this volume, and by discerning which properties of biochemical systems emerge, again and again, in all interspecific variants of the system, we will gain a particularly firm understanding of what is most crucial in biochemical design.

CHAPTER TWO

Design of Cellular Metabolism

Introduction

Because they are the end results of cycles of mutation and selection, organisms are in a proper sense "designed" systems, fully analogous to products of engineering design. Although organisms are designed by adaptation while machines are designed by engineers, both design systems share the essential attributes of *function and purpose* (which are, of course, unknown in the inanimate world); both involve the trial of variants and the selection of those that work; and in both systems, origin and nature cannot be predicted from chemical and physical principles alone, any more readily than can the shape of fast-swimming fishes from knowledge, however complete, of the physics and chemistry of proteins, fats, and carbohydrates. For understanding and prediction of designed systems (whether designed by man or by adaptational processes) we need to know *the rules or principles of design in each case.* What are the constraints on each system? What are the limits of change and adjustment? Armed with such information, a ship-builder can predict the speed, stress, and strain consequences of change in hull shape, in polymer coating, or in sail position; by analogy, a biomechanic can predict the consequences to swimming speed and swimming cost of change in body shape, in fin position, or in the aspect ratio of the tail of pelagic fishes. Although design principles in metabolic and enzyme biochemistry are not as well charted as in many engineering fields, much excitement has arisen from the realization a) that nature utilizes such principles, and b) that the rules are decipherable (and, indeed, are now well appreciated in many areas of biochemistry). To ascertain just how far we have gone along the pathway to understanding the design of cell metabolism is the main goal of this chapter. This requires a fairly detailed analysis of the fundamental organization of cell metabolism (which some senior students may already have). However, the analysis is considered basic to a thorough understanding of later themes in the book on how biochemical adaptations permit organisms to possess the right types and proper amounts of metabolic activity under diverse environmental conditions.

Functional Blocks in Metabolism and How They Are Coupled

Most discussions of cellular metabolism begin with a list of the three most fundamental requirements of all (microbial, plant, and animal) cells: energy in the form of adenosine triphosphate (ATP)*, reducing power as nicotinamide adenine dinucleotide phosphate, reduced form (NADPH), and starting materials for biosynthesis. Curiously, the explanations of these needs and how they are achieved are often buried and nearly lost in the details of large metabolic maps; the loss occurs because the very complexity of such maps precludes clearly conveying the essentials at the heart of the metabolism. As emphasized by Atkinson (1977), such descriptions are in many ways comparable to schematics in electronics; as in electronics, the easiest way to explain and clarify the overall operating principles of metabolism is to use a block diagram which emphasizes functional relationships, which in total, add up to cell metabolism. Such a functional block diagram (Figure 2-1) illustrates that metabolism in animal cells can be symbolized by three functional blocks:

a) In the Catabolic Block (I), various food stuffs are oxidized to $CO_2 + H_2O$, with most of the electrons liberated in the oxidation being transferred to O_2 with the concomitant production of ATP (a process termed oxidative phosphorylation). If O_2 is absent or limiting, only partial oxidations are utilized and electrons are transferred to organic compounds, which in the reduced form accumulate as end products. The major pathways in this block are glycolysis, β-oxidation of fatty acids, pathways of amino acid degradation and the Krebs cycle, all of which not only contribute electrons and protons for the electron transport system but also generate, along with a few other reaction sequences, the carbon compounds used as precursors for all of the cell's biosynthetic processes. End products are a special problem in eukaryotic organisms (in procaryotes, they are merely lost to the medium): while some of them, like CO_2 and H_2O, are relatively harmless (even CO_2 is not totally innocuous since it raises acid-base regulation problems), others such as anaerobic end products, nitrogen waste products, and sulfur waste products are, or can be, hazardous. Thus, to avoid self-pollution, a special Block I provision in eukary-

* Many standard biochemical abbreviations, as used, for example, in *The Journal of Biological Chemistry* or in standard biochemistry textbooks, will appear throughout the text. For convenience, these abbreviations are also indexed.

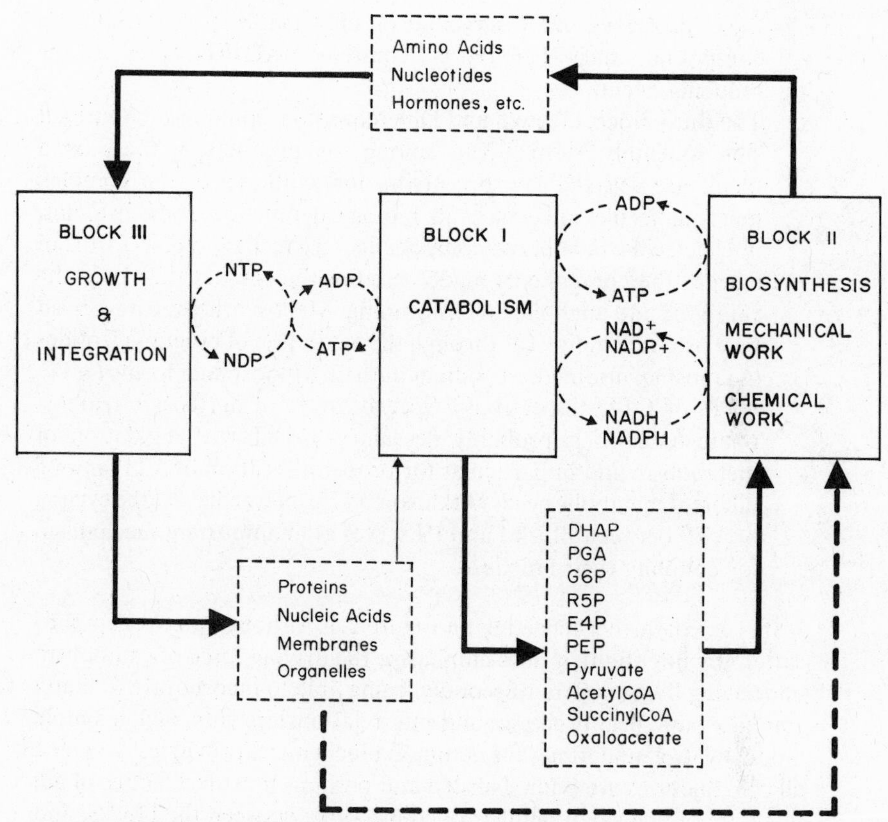

Fig. 2-1. Summary block diagram of major functional units of cellular metabolism Abbreviations: NTP, nucleotide triphosphate; NDP, nucleotide diphosphate; DHAP, dihydroxyacetone phosphate; PGA, phosphoglyceric acid; G6P, glucose-6-phosphate; R5P, ribulose-5-phosphate; E4P, erythrose-4-phosphate; PEP, phosphoenolpyruvate.

otes must be made for either excreting, detoxifying, storing, or recycling potentially detrimental end products of catabolism.

b) The Anabolic Block is chemically much more complex than the first, and basically includes the cell's major ATP-utilizing processes. In some cells, this is mainly biosynthesis, with the starting materials generated by Block I being converted to hundreds of specific cellular components. In some cells (muscle, for example), it is mainly mechanical work that is the primary sink for ATP equivalents. Still other cells are specialized for certain kinds of chemical work (e.g., ion pumping). In both the latter categories, however, at least modest biosynthetic capacities must be maintained. In all these processes, ATP serves as a universal energy-transducing compound, and where H is required, NADPH serves as the reducing agent.

c) The third block, Growth and Differentiation, could also be entitled The Assembly Shop. From among the products of the second block are selected key precursors for synthesis of the complex macromolecules on which all biological function rests: proteins, nucleic acids, membrane components, organelles, and so forth. In essence the components made here are the engines that keep the catabolic and anabolic blocks running. Most syntheses categorized here use ATP indirectly, through the mediation of other nucleotides (guanosine, uridine, or cytidine) in their triphosphate forms (GTP, UTP, or CTP, respectively). Specialization of nucleoside triphosphate function is probably necessary for efficient regulation of metabolism and in particular for proper allocation of ATP among different metabolic needs (Atkinson, 1977). Nevertheless, the cycling of ATP between Block I and III serves as an important mechanism for coupling their functions.

It is worth further reflecting on Figure 2-1. Although an oversimplification (by intention), it does emphasize the driving force of catabolism underlying living systems. Secondly, being able to incorporate so many complex biological processes and interrelationships into such a simple model by itself underlines the common biochemical principles on which all cell functions are built. Thirdly, and perhaps most instructive of all, there are only a small number of connections between the blocks, and among these, the roles of ATP and NADPH are unique: they are the only fundamental coupling agents in metabolism—the only compounds that have as their primary role the job of coupling activities between the major functional blocks. Roles for other coupling agents are essential but less broad in scope. For example, nicotinamide adenine dinucleotide, reduced form (NADH) and flavin adenine dinucleotide, reduced form ($FADH_2$) participate within the catabolic block in the transfer of electrons from substrate to oxygen in electron transfer phosphorylation (see discussion below), but are not significantly involved in coupling between major functional blocks.

An important feature of these coupling agents is that they are cycled, not consumed. When ATP and NADPH contribute to Block II functions they are converted to adenosine diphosphate (ADP) or adenosine monophosphate (AMP) and to $NADP^+$; thus they must merely be rephosphorylated or reduced by substrate oxidation in catabolic reactions in order to be re-utilized. This feature clearly distinguishes ATP and NADPH from another group of compounds which can also be viewed as coupling intermediates (Fig. 2-1): unlike ATP and NADPH, these are produced by one block and consumed by another. About ten compounds, supplied by glycolysis, the Krebs cycle, the pentose phosphate pathway, and a few anaplerotic reactions are used in Block II, mainly in biosyntheses. This group of coupling intermediates includes four sugar phosphates: triose phosphate (dihydroxyacetone phosphate (DHAP) and 3-phosphoglyceraldehyde), tetrose phosphate, pentose phosphate, and hexose phosphate; three α-keto acids: pyruvate, oxaloacetate, and 2-ketoglutarate; two activated carboxylic acids: acetylCoA and succinylCoA; and phosphoenolpyruvate. (In this list isomeric or readily interconvertible compounds are not counted). These ten compounds are consumed by Block II processes and therefore must be continuously replaced by catabolic Block I processes; unlike ATP and NADPH which are recycled, these flow through the system ultimately being deposited into more complex cellular material.

A similar but much larger group of intermediates is formed by Block II biosyntheses and serve as precursors for Block III components or functions. They, too, flow through the system, in the process coupling Block II functions to those in Block III.

The products of Block III activities (genes, proteins, membranes, organelles) are the engines that keep all three block activities going. Although they, too, flow through the system (in the sense that cellular components typically have distinct half-lives), only under unusual circumstances are they consumed in the sense that coupling intermediates are between Blocks I and II and between II and III.

Adaptation of Machinery versus Adaptation of Its Output

Biochemical adaptations can occur in all three functional blocks. In Block I such adaptations influence the way ATP and coupling intermediates are generated and at what rates; they also influence the kinds of end products formed and the means adopted for handling their accumulation so as to prevent "polluting the factory." Block II adaptations influence the need for specific biosynthetic products and/or the need for ATP in various cell work functions. Most of the first several chapters will be concerned with these kinds of adjustments.

Block III-level adjustments are fundamentally different from those in Blocks I and II. In the latter, what is usually being modulated is the rate of specific metabolic pathways or of ATP-requiring work functions (pumping, contraction, etc.). Block III adaptations, in contrast, influence the structural (and hence functional) components of cells for improving metabolic and indeed, in its broadest sense, biological performance in specific environments or conditions. For example, enzymes are adapted for function in specific salinity, temperature, and pressure regimens; isozymes of a given enzyme are kinetically adapted for function in specific metabolic fields; the actual presence or absence of terminal dehydrogenases determines the kinds of end products that can be made when O_2 becomes limiting. The structural organization of membrane-based ion pumps may determine both the direction and rate of ion translocation. All such adaptations, and many more which will be discussed in detail below, are analogous in process to factories or mechanical maintenance shops: in both it is the working parts of the system that are being adjusted. In Block III adaptations, therefore, the actual machines of cell metabolism and of cell work are the sites of modification. In contrast, in Block I and Block II adaptations, it is the outputs of the working parts that are modified. Although for convenience we will usually discuss these separately, it is evident that Block III adaptations must underlie all adaptations of Block I and II functions.

The ATP Equivalent

The pivotal position of the adenylates in Figure 2-1 needs further explanation, since it is in many ways one of the most remarkable features in the design of cell metabolism. As has been recognized for over three decades, ATP is involved in essentially every extended metabolic pathway. No other metabolites even approach the adenylates in terms of the ubiquitous roles they play in energy metabolism, in biosynthetic reactions, in contractile work, in ion pumping, and so forth. The coupling unit of exchange, defined as the ATP equivalent, is the conversion of ATP to ADP or vice versa. Since all other metabolic conversions can be chemically related to the conversion of ATP $\rightarrow$ ADP, they all can be evaluated in terms of ATP equivalents. For example, NADH oxidation to NAD^+ is priced at three ATP equivalents; $FADH_2 \rightarrow FAD^+$, at two ATP equivalents; while the oxidation of NADPH $\rightarrow$ $NADP^+$ is priced at four ATP equivalents (see discussion below).

The adenylate coupling function also means that any given metabolite can be defined in terms of ATP equivalents, either in terms of ATP equivalents obtained by oxidizing the compound (by fermenting it in anaerobes), or alternatively in terms of the number of ATP equivalents needed for synthesis of the compound. In this way, sugars and polysaccharides, amino acids and proteins, fatty acids and fats, DNA, and even growth can be evaluated in terms of ATP equivalents actually or potentially expended during their synthesis or gained during their degradation.

It is important, however, to bear in mind that this "pricing" of materials is no different than evaluating substrates in terms of energy released on hydrolysis in calories per mole. Its advantage is that it defines energetics in currency units actually used by the cell. And, equally important, it emphasizes the primary importance of the ATP equivalent in the temporary storage and transfer of chemical energy in the cell. This transfer function indeed is the chemical basis for the coupling described in Figure 2-1, so it is important to understand how it is achieved.

Role of Adenylates in Storage and Transfer of Energy

At constant temperature and pressure, the change in free energy for a chemical reaction

$$A + B \rightleftharpoons C + D$$

is given by a standard equation of elementary thermodynamics:

$$\Delta G = -RT \ln K_{eq} + RT \ln Q$$

The reaction parameter [C][D]/[A][B] is Q, where the letters in brackets refer to chemical activities (concentrations, if the solution is dilute and if ionic and other interactions are small). K_{eq} is the equilibrium constant, R is the gas constant, T is the absolute temperature, and ΔG is the change in free energy associated with the conversion of 1 mole each of A and B to C and D, with no change in concentration (strictly, activity) of any component. Combining terms on the right hand side of the above equation yields

$$\Delta G = RT \ln Q/K_{eq}$$

This equation indicates that the property Q of ΔG is of most relevance to cell metabolism, for Q is a simple measure of how far from equilibrium the system is and on which side. When $Q = K_{eq}$, $\Delta G = 0$ and there can be no net flux through the reaction considered in isolation. When $\Delta G < 0$ (i.e., $Q < K_{eq}$), the reaction will proceed in the forward direction (as written above). A reaction cannot proceed when $\Delta G > 0$; under these conditions the reaction must indeed go in the reverse direction.

The only essential requirements that must be met in order for metabolic substrates to be a good chemical fuels (i.e., sources of chemical potential energy) are the following:

1. the substrate must be thermodynamically unstable; i.e., $G \ll 0$ for its degradation;
2. it must be kinetically stable under storage conditions; and
3. there must be a feasible means for overcoming the kinetic barrier when the fuel is to be mobilized.

These conditions are necessary since metabolic substrates contain chemical potential energy only if they are connected to, but are not in equilibrium with respect to, degradative metabolic reactions. Such chemical systems furthermore must be chemically inert until their energy is needed; otherwise, they are not useful storage energy sources.

This is the situation found for the primary fuels of cell metabolism. Carbohydrates, fats, and proteins are thermodynamically unstable in the presence of oxygen, but are kinetically very stable. They are mobilized when needed through activation of appropriate enzyme systems (discussed further below) in reactions for which $\Delta G \ll 0$.

The adenylates have these same characteristics and, indeed, ATP may be usefully considered as a fuel for those reactions or metabolic processes that utilize it. However, the roles of the adenylates in transferring or providing energy for cell work processes are somewhat different. This is because a system such as the adenylates can contain chemical potential energy only when it is not at equilibrium. Since all reactions go toward equilibrium, energy cannot be stored in a system under the same conditions as are needed for transferring or providing it.

The biological option utilized is to charge energy transfer systems under one set of conditions, and utilize them under another set: in the first instance the system is moved away from the equilibrium that applies when the system is to be used in the second instance. The adenylate energy transfer system depends on just such a principle. This system is charged by phosphorylating ADP in reactions for which the equilibrium ratio [ATP]/[ADP] is high. For known reactions in which ATP is formed (e.g., pyruvate kinase), the equilibrium constants are large in the direction of ATP production; i.e., $\Delta G \ll 0$. Most ATP in aerobic cells is generated by electron transport phosphorylation and here too the equilibria must strongly favor ATP formation.

The system is discharged when ATP is consumed in energy-requiring processes; the conditions for discharging are different from those of ATP formation so that here, too, equilibrium constants are large but now in favor of ADP production. Thus reactions in which ATP donates or loses a phosphoryl group, for example,

$$\text{glucose} + \text{ATP} \xrightleftharpoons{\text{hexokinase}} \text{G6P} + \text{ADP}$$

$$\text{ATP} + \text{H}_2\text{O} \xrightleftharpoons[\text{ATPase}]{} \text{ADP} + \text{P}_i$$

and those in which ADP accepts a phosphoryl group, for example,

$$\underset{\text{1,3-diphosphoglycerate}}{\text{1,3-DPG}} + \text{ADP} \xrightleftharpoons{\text{PGK}} \underset{\text{3-phosphoglycerate}}{\text{3-PGA}} + \text{ATP}$$

$$\underset{\text{phosphoenolpyruvate}}{\text{PEP}} + \text{ADP} \xrightleftharpoons{\text{PK}} \text{pyruvate} + \text{ATP}$$

each proceeds with a large free energy drop. That is why, as the adenylate system discharges it can be harnessed to move another chemical reaction (e.g., in biosynthesis) or chemical process (in mechanical work, ion pumping, etc.) further from equilibrium than would be possible for that reaction or process considered in isolation. It is by this means that all Block II anabolic and cell work functions are coupled to ATP-generating Block I functions (Fig. 2-1). We will now proceed to examine more closely metabolic processes available to animal cells for generating ATP and other coupling materials. These functions are conventionally and conveniently divided into anaerobic and aerobic components.

Anaerobic Glycogenolysis and Glycolysis

In vertebrate cells the commonest form of anaerobic metabolism is glycolysis (with glucose as the fermentable substrate) or glycogenolysis (with glycogen as the storage substrate). This is the common fermentation pathway that is presented in detail in all introductory biochemistry textbooks; it can be formally described as a linear pathway for the fermentation of glycogen or glucose to lactate (Fig. 2-2). There are a number of hallmark characteristics of this pathway. Firstly, the enzymes of glycolysis are almost always extramitochondrial in location, sperm being the only known exception in containing intramitochondrial lactate dehydrogenase. Secondly, one oxidative step (catalyzed by glyceraldehyde-3-phosphate dehydrogenase) is balanced by one reductive step (catalyzed by lactate dehydrogenase), a redox coupling that leads to the accumulation of lactate as an anaerobic end product. It is important to emphasize that the accumulation of end product serves no other purpose than redox balance and that it is a simple consequence of an organic metabolite (pyruvate in this case) rather than oxygen serving as an electron acceptor.

Superficially, it may appear that the pathway loses efficiency by involving two ATP-utilizing reactions (at the hexokinase and phosphofructokinase steps) in preparatory phases of glycolysis. These steps

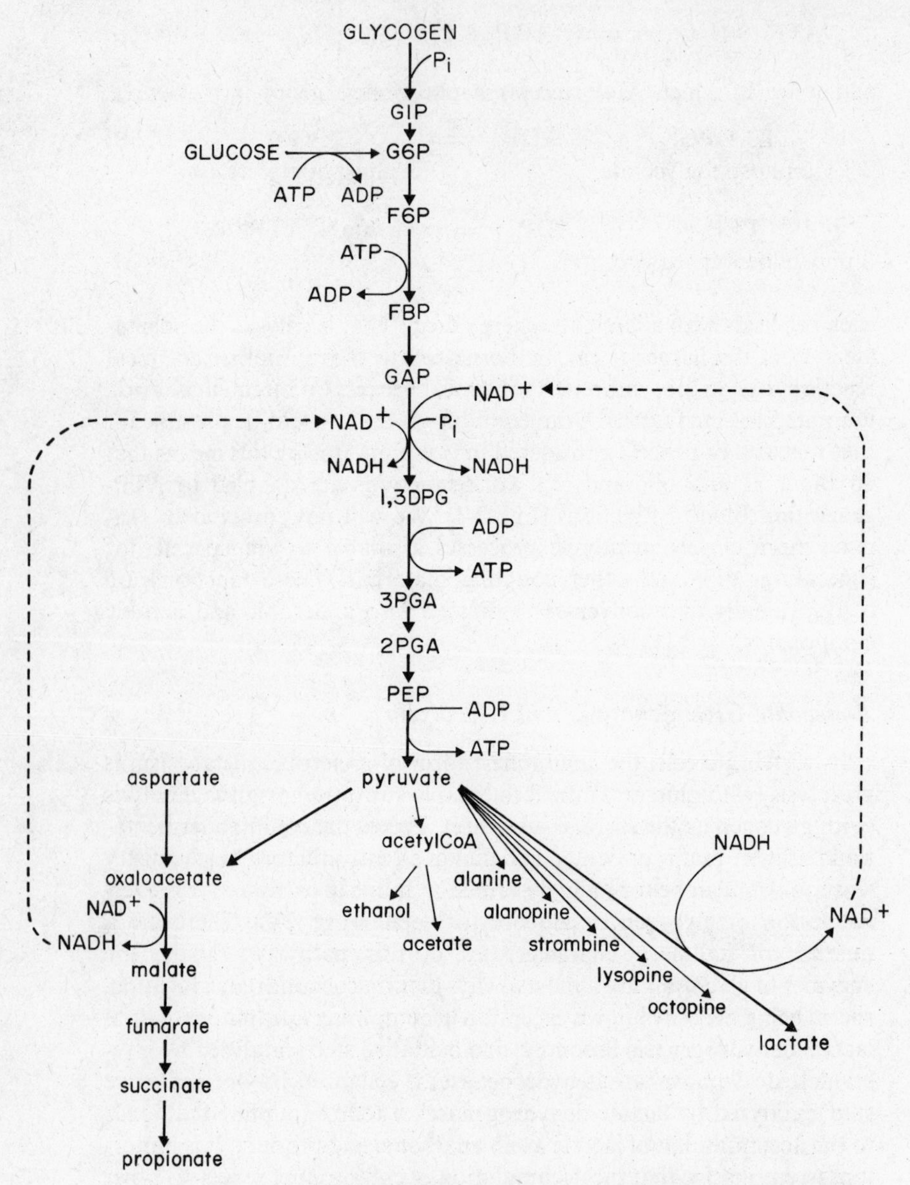

Fig. 2-2. Major fermentation pathways in animal tissues

precede the ATP-yielding reactions (phosphoglycerate kinase and pyruvate kinase), but far from being wasteful, they play a key role in allowing the cell to maintain much higher concentrations of hexose phosphate than would be possible without phosphorylation, since the phosphorylated sugars cannot pass through the plasma membrane. Thus, another trade-off seems to have occurred at this level: energy yield for controlled maintenance of higher substrate levels.

Phylogenetically the pathway of glycolysis is very old and thus has often been presented in classical biochemistry courses as being highly conservative and unchanging. Nothing could be further from the truth, and it is worth briefly emphasizing that at least several zones or functional sections of the pathway are clearly subject to change and adaptation. These occur at minimally four positions: at the level of glycogen, triose phosphate, P-enolpyruvate, and pyruvate, and each must be discussed separately since they vary in importance under differing conditions, and as we shall see in later chapters, in different species.

Glycogen as a Storage Form of Energy

Glycogen is normally stored in a granular form termed the β-particle; it is about 100–400 Å in diameter and is stored largely in the cytosol. This is the usual form of glycogen that most students will be familiar with and which appears in electron micrographs of most tissues and cells. However, glycogen can also be stored in a much larger granular form termed the α-particle. These particles can reach up to 1,000 Å in diameter and appear to be a more efficient way of packaging intracellular glycogen; that is probably why they are typically found in tissues and organs that are required to store unusual quantities of glycogen. For example, α-particles are normally found in human liver cells, but not typically in other organs of humans. On the other hand, they are a standard storage form of glycogen in the livers of goldfish and turtles, in the seal heart, or in lungfish muscle. All these tissues are characterized by storing unusual levels of glycogen, over 1 M in the case of goldfish liver (Hochachka, 1982).

In addition to β- and α-particles, glycogen can be stored in the form of glycogen bodies or glycogen-membrane complexes; so far, glycogen-membrane complexes are best known in relatively active tissues, such as heart muscle of lungfish or ocular muscle of mammals, which are also charged with unusually large quantities of glycogen and glycolytic enzymes.

Since the enzymes involved in mobilizing glycogen, and the cascade of enzymes controlling that mobilization, are all physically associated with glycogen *in situ*, it is a reasonable assumption that the storage form adopted by any given cell has an important impact upon how and when that glycogen is used. To date, very little information is available on this matter. This ignorance, however, does not extend to another highly adaptable zone of glycolysis, that occurring at the level of triose phosphate.

The Role of Alpha-Glycerophosphate Dehydrogenase

The aldolase cleavage of fructose-1,6-bisphosphate to yield glyceraldehyde-3-P and dihydroxyacetone phosphate is designed to feed the former into the terminal stages of glycolysis. However, most cells maintain significant levels of α-glycerophosphate dehydrogenase (α-GPDH) which may serve a) in the α-glycerophosphate cycle, or b) in supplying α-glycerophosphate (α-GP) for triglyceride synthesis. These are aerobic functions. Under anaerobic conditions, however, the occurrence of high activities of α-GPDH, as in flight muscle of some insects, could lead to a significant energetic drain from mainline glycolysis because of the following competition:

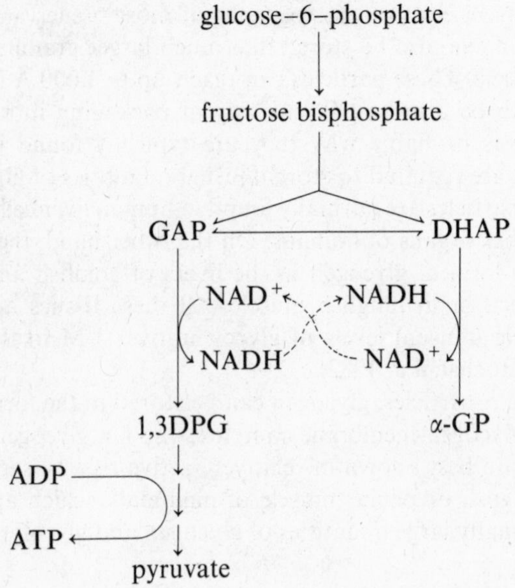

glucose–6–phosphate
↓
fructose bisphosphate

GAP ⇄ DHAP

NAD⁺ → NADH (GAP → 1,3DPG); NADH → NAD⁺ (DHAP → α-GP)

1,3DPG α-GP

ADP → ATP

pyruvate

The danger is that half the glucose carbon could flow to α-GP, half to pyruvate; since two ATPs are used in forming fructose-1,6-bisphosphate (F1,6BP), and only two would be formed in glycolysis, the net yield would be zero ATP/mole glucose fermented. Not surprisingly, the activity of α-GPDH is closely regulated in different tissues and species so as to allow its other functions while minimizing the risk of energy short-circuit (Guppy and Hochachka, 1978).

Adjustment at the Level of the P-Glycerates

The mature mammalian red cell is considered to satisfy all of its limited catabolic needs with anaerobic glycolysis, so it is particularly interesting that even in this system the pathway has been adjusted and modified to suit the specific needs of this tissue. In this case, the modification occurs at the level of 1,3-diphosphoglycerate which can either be metabolized by mainline glycolysis or be channeled into 2,3-diphosphoglycerate (2,3DPG):

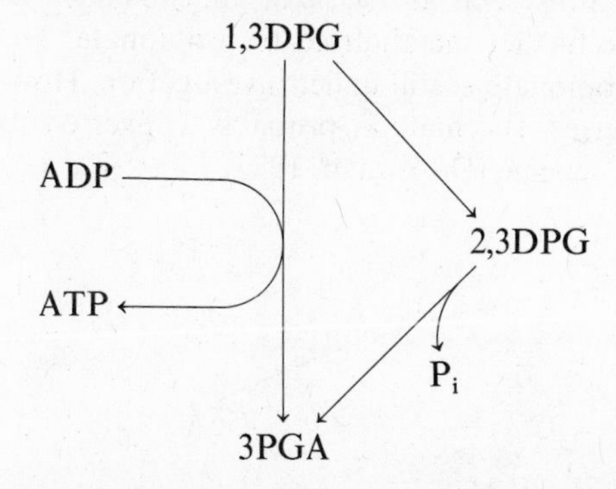

The organophosphate, 2,3DPG, is utilized in mammalian red cells as a modulator of hemoglobin binding of O_2. It occurs at high concentration (about 6 μmol/ml of whole blood) and its concentration is regulated by the amount of carbon flux that is channeled away from mainline glycolysis and by its subsequent hydrolysis to 3PGA. The latter, of course, is back on mainline glycolysis and is metabolized by conventional reactions.

It will be evident that the 2,3DPG reactions bypass phosphoglycerate kinase, an ATP-synthesizing step in glycolysis; thus, this modification in the pathway is made at energetic cost to the cell. For each mole of 2,3DPG formed per mole of glucose, the energy yield of glycolysis drops from two moles ATP to one mole ATP per mole glucose.

However, this will occur only during transition from low to higher levels of 2,3DPG; at steady state, the energy yield of glycolysis need not be affected by this bypass reaction loop.

Adjustment at the Level of P-enolpyruvate

In many invertebrate animals (helminths, annelids, marine bivalves), the classical pathway of glycolysis is critically modified at the level of PEP, allowing either carboxylation to oxaloacetate (OXA) or conversion to pyruvate (Fig. 2-2). OXA so formed is often reduced to succinate, which may accumulate as an anaerobic end product. In some cases, OXA is also formed from aspartate, and for redox reasons, may be fermented simultaneously with glucose (Hochachka, 1980). Under such conditions, glucose and aspartate carbon may accumulate as succinate. This principle of the simultaneous fermentation of carbohydrate and amino acids in fact appears over and over again in widely differing groups of animals.

In some organisms (helminths such as *Fasciola*; annelids such as *Arenicola*), succinate may be further metabolized to propionate. The pathway for formation of propionate is still under investigation. However, in helminths and bivalves, succinate is probably converted to propionate by the following scheme (De Zwaan, 1983):

$$\text{succinate} \xrightarrow{\text{CoA}} \text{succinylCoA} \longrightarrow$$

$$\text{methylmalonylCoA} \xrightarrow[\text{ADP} \rightarrow \text{ATP}]{\rightarrow CO_2} \text{propionylCoA} \xrightarrow{\rightarrow \text{CoA}} \text{propionate}$$

(CoA released in the last step is returned to the first step.)

In any event, the important conclusion to bear in mind is that glucose fermentation in these organisms proceeds in initial phases along the conventional glycolytic path; however, at the level of PEP, carbon can be diverted from that pathway to succinate or even propionate. In such a linear path (glucose → propionate) the yield of ATP/mole of glucose rises from two to as high as six (Hochachka, 1982; De Zwaan, 1983).

Adjustments at the Level of Pyruvate

Finally, a lot of modification can occur in the terminal zone of glycolysis (Fig.2-2). As already mentioned, in conventional glycolysis, pyruvate serves as a H acceptor and is reduced to lactate, the function of the reaction being to maintain an adequate oxidizing potential to keep glycolysis going. However, pyruvate can have many other fates in other organisms. These can include conversion to acetate (in some helminths), to ethanol, in goldfish and *Chironomus* larvae (Shoubridge and Hochachka, 1981), or to a host of imino acids (alanopine, strombine, lysopine, octopine) formed by the reductive condensation of pyruvate and an amino acid. The commonest amino acids used in imino acid production in invertebrate animals appear to be arginine, alanine, glycine, and lysine, although there are indications that threonine, ornithine, and proline can sometimes be utilized (Storey and Storey, 1983), in reactions that can be written as follows:

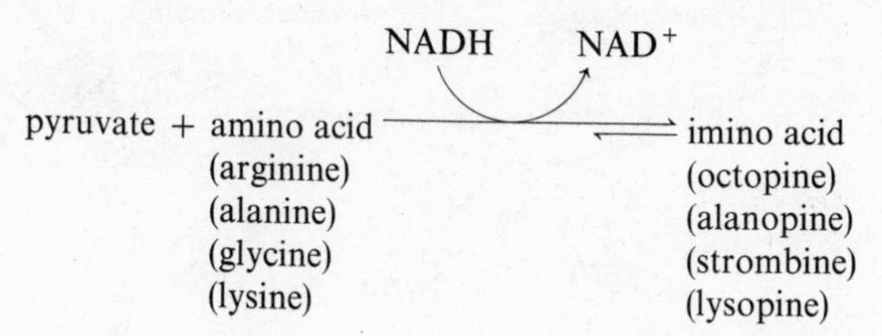

These latter reactions are of particular interest since they are a special case of the principle of coupling carbohydrate and amino acid fermentation in invertebrate animals.

Oxidative Metabolism: The Krebs Cycle as Its Hub

Among animals, it is rare for oxygen not to be an absolute requirement for life and, hence, most carbon substrates are degraded completely to CO_2 and H_2O. During the aerobic metabolism of carbohydrate, pyruvate formed in glycolysis enters the Krebs cycle reactions as acetylCoA. The enzyme catalyzing this reaction, pyruvate dehydrogenase(PDH), thus is positioned at a strategic point in metabolism, and is closely regulated by acetylCoA, NADH, and GTP inhibition plus AMP activation, and by phosphorylation/dephosphorylation mechanisms. Phosphorylation deactivates PDH, and this is enhanced by high ratios of ATP:ADP and NADH:NAD^+. Pyruvate dehydrogenase

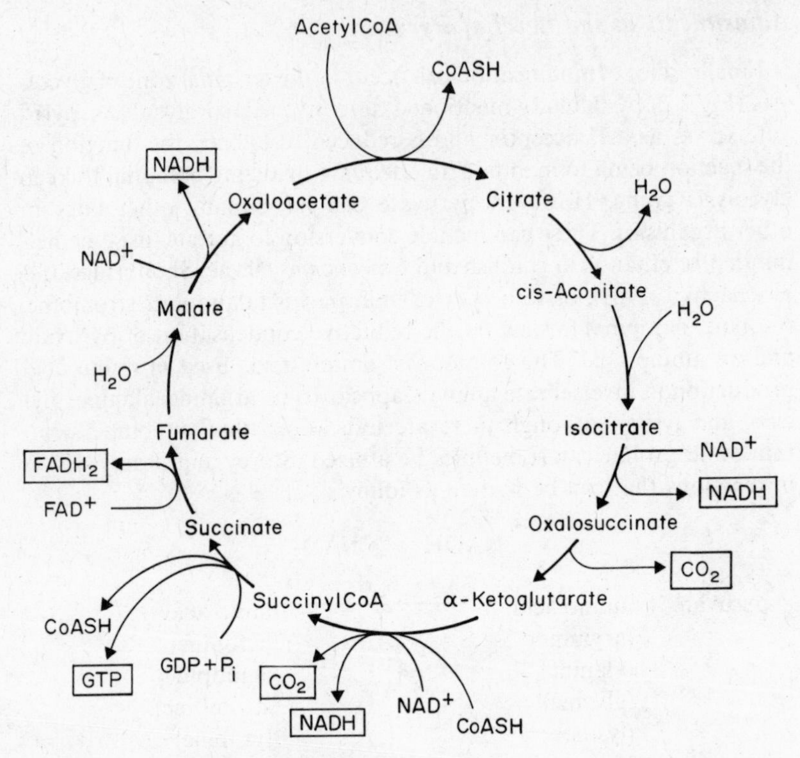

Fig. 2-3. The Krebs cycle as the hub of cellular metabolism

function sets the stage for the first span of the Krebs cycle (Fig. 2-3) which involves three component enzymes: 1) citrate synthase, 2) aconitase, and 3) isocitrate dehydrogenase. The last of these forms 2-ketoglutarate which can be viewed as the means for substrate entry into the second span of the Krebs cycle: the reconversion of 2-ketoglutarate (via succinylCoA, succinate, fumarate, and malate) to oxaloacetate. The enzyme components of this span are 2-ketoglutarate dehydrogenase, succinic thiokinase, succinic dehydrogenase, fumarase, and malate dehydrogenase (Fig. 2-3).

The reason the Krebs cycle is called the hub of metabolism is that it is the final pathway for the complete oxidation of carbohydrates, proteins, and fats. The entry mechanism for glucose-derived pyruvate has already been described. Fatty acid oxidation also leads to the formation of the acetylCoA which enters the Krebs cycle for complete oxidation at the beginning of the first span. Since glucose or fat can supply the cell with acetylCoA it is obvious that the two substrate

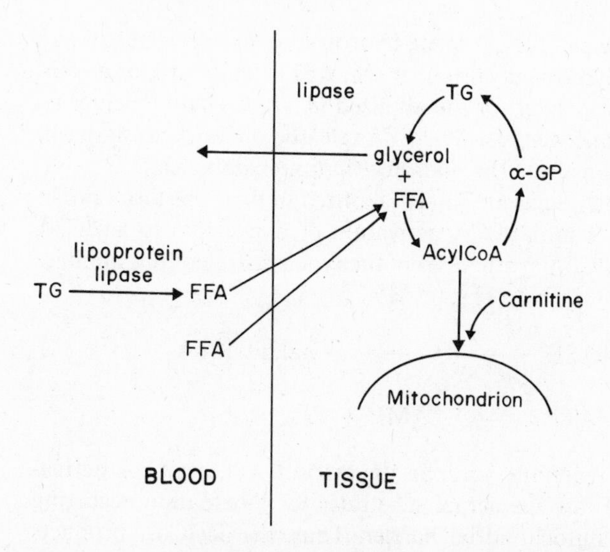

Fig. 2-4. Simplified diagram of the main sources of long-chain acylCoA for β-oxidation in mitochondria

sources could in effect compete for access to the Krebs cycle; not surprisingly there usually is an antagonistic relationship between the catabolism of glucose and fatty acid oxidation.

The Complete Oxidation of Fat is Initiated in the Cytosol

The catabolism of lipid can begin either with exogenous free fatty acids (FFA) or endogenous triacylglycerol (triglycerides) (Fig. 2-4). Although most mammalian tissues (except the brain) are able to mobilize triacylglycerol (TG), the main storage depot is located in adipose tissue which under conditions of high demand (e.g., prolonged fasting or prolonged steady-state work), supplies essentially all of the body's needs for fatty acids. As in the case of glycogen mobilization, the first enzyme initiating TG metabolism is strategically positioned in both metabolic and control terms. That is why TG lipases, particularly in adipose tissues, appear to be under stringent control (Fig. 2-4); current concepts of that control involve phosphorylation-dephosphorylation cascades similar to those found for glycogen phosphorylase; these are described in detail elsewhere (Steinberg and Khoo, 1977).

The two products of the complete hydrolysis of TG are glycerol and free fatty acids. Glycerol may be either removed from the cell for metabolism elsewhere (e.g., liver) or metabolized *in situ* (usually by entering glycolysis or glyconeogenesis). The FFA released on TG hydrolysis, on the other hand, constitute the main fuel in lipid catabolism.

Interestingly, fatty acids are not catabolized in their free form; rather their metabolism is initiated by activation or conversion to acylCoA derivatives by acylCoA synthetases of the type illustrated for palmitate activation:

$$\text{palmitate} + \text{CoASH} \xrightarrow[\text{ATP} \;\rightarrow\; \text{AMP} + \text{PP}_i]{} \text{palmitylCoA}$$

These activation reactions occur in the cytosol. Yet, acylCoA derivatives of fatty acids are the actual substrates for β-oxidation occurring within the inner mitochondrial barrier. Thus, the acylCoA products of acylCoA synthetases must be transferred into mitochondria for their further degradation.

Although it has been known for some time that acylCoA penetrates the inner mitochondrial membrane only in the presence of carnitine, the mechanism for this transfer has not been fully clarified. Recent evidence strongly favors an exchange diffusion facilitated by a translocase (Vary et al., 1981), allowing acyl-carnitine formed outside the mitochondrial matrix, by carnitine acyl-transferase I, to be transported across the inner mitochondrial membrane in exchange for carnitine. On the matrix side of the inner mitochondrial membrane, the acyl-carnitine is reconverted to carnitine and acyl-CoA by carnitine acyl-transferase II. The carnitine so formed would be available to exchange for more acyl-carnitine.

Carnitine acyl-transferases are thus seen to play a critical role in the catabolism of fatty acids. The best studied of these are the carnitine palmityl transferases, which occur in two distinguishable forms. CPT I appears to be located on the external surface of the inner membrane and catalyzes an easily reversible reaction, while CPT II appears to be located on the inner surface of the inner mitochondrial membrane and normally catalyzes the reaction in the direction of long-chain acylCoA formation:

$$\text{palmityl carnitine} + \text{CoA} \longrightarrow \text{carnitine} + \text{palmitylCoA}$$

This completes the transfer of acyl units into the mitochondria and sets the stage for β-oxidation.

The β-Oxidation Spiral

In mammalian mitochondria, fatty acid oxidation occurs almost exclusively by the β-oxidation spiral—a process in which C_2 units as acetylCoA are removed from the carboxyl end of the fatty acid. The process to thiolytic cleavage releasing acetylCoA is thought to involve the following reactions:

1. acylCoA + FAD^+ $\longrightarrow$ α-β-unsaturated acylCoA + $FADH_2$
2. α-β-unsaturated acylCoA + H_2O $\longrightarrow$ β-hydroxyacylCoA
3. β-hydroxyacylCoA + NAD^+ $\longrightarrow$ β-ketoacylCoA + NADH + H^+
4. β-ketoacylCoA + CoASH $\longrightarrow$ acylCoA (−2 carbons) + acetylCoA

The activities of the β-oxidation enzymes in rat skeletal muscle and rat liver are approximately an order of magnitude greater than that of acylCoA synthetase or carnitine acyl-transferase I. Thus, β-oxidation can readily keep pace with the production of acylCoA.

In rat liver mitochondria metabolizing palmityl carnitine, a process involving twenty-seven potential intermediates, only the acylCoA derivates that are formed by complete spirals of the β-oxidation process, i.e., C-14, C-12, C-10, C-8, are detectable but these do not accumulate in a product–precursor relationship. Rather, each represents the end of a complete passage through the β-spiral or the beginning of a new passage. Unsaturated and 3-hydroxy intermediates are only observed when NADH oxidation is inhibited (by rotenone, for example). These observations suggest that the activities of enzymes of β-oxidation are carefully regulated with respect to one another.

Recent work with intact hearts indicates that metabolic control of β-oxidation occurs at the FAD^+- and NAD^+-dependent dehydrogenase reactions. Of these two sites, primary control appears to be vested in the NAD^+-dependent (rather than the FAD^+-dependent) oxidative step (Vary et al., 1981).

Proteins and Amino Acids as Potential Fuel Sources

There are two sources of proteins and amino acids for catabolism: firstly, organisms eat proteins which are hydrolyzed to their constituent amino acids during digestion. At the same time, cellular proteins are continuously being hydrolyzed to form amino acids which mix with those derived from food as an amino acid pool, upon which through the blood vascular system all tissues may draw, each according to its needs. Under nutrient-rich conditions, nongrowing (adult) organisms maintain relatively constant amounts of protein and therefore any surplus of amino acids is roughly equivalent to that ingested; however, if food supplies are limited, endogenous protein reserves become more important as sources of carbon and energy, and in extreme situations, they may be the sole fuel source remaining to the organism.

Protein Hydrolysis

As far as we know to date, there are no special storage forms of protein specifically laid down for use as a fuel for energy metabolism. Unlike carbohydrate and fat catabolism, therefore, protein catabolic machinery can be directed at any or all classes of intracellular protein. The proteolytic enzymes initiating protein hydrolysis are largely localized in intracellular organelles called lysosomes (which also contain other hydrolytic enzymes), but some tissues (muscle, for example) are now known to contain extracellular proteases which appear to display somewhat greater substrate specificity.

In terms of the strategic positioning as the first key step in the mobilization of a major energy reserve, proteolytic enzymes are analogous to glycogen phosphorylase and triacylglycerol lipases in the mobilization of glycogen and fat, respectively. Unlike the latter two enzymes, which are under strict phosphorylation-dephosphorylation control, protease activity does not seem to be as rigorously regulated; at least to date, no comparable control mechanisms have been established. Research along these lines is moving very fast and it is possible that comparable mechanisms of control may yet be found. However, as of this writing, one pathway of proteolysis seems to reside primarily in the lysosomes and operates in large part by regulating enzyme amount (it therefore has a large time constant). Another, non-lysosomal pathway appears to be mainly directed at abnormal proteins which are degraded much more rapidly than are normal ones.

Since there are some twenty amino acids commonly released in the complete hydrolysis of proteins, the description of their subsequent metabolism inevitably becomes rather complicated, much more so than in the description of glucose and fat catabolism. Nevertheless, the temptation to skip over the subject should be resisted since the metabolism of amino acids is, to many organisms, as fundamental as the metabolism of any other substrates. Often it is the most important (or only) process available to keep the system going; therefore, it is important to consider the fate of the products of protein hydrolysis.

Deamination and Transamination

The metabolism of many amino acids begins with the removal of nitrogen, which is released as a nitrogenous end product and which is handled separately from the further catabolism of the carbon skeleton. In some organisms, waste nitrogen is formed and released as NH_4^+, about the simplest form of nitrogen handled by animal species. In mammalian organisms, the main fate of waste nitrogen is conversion to glutamine (for temporary storage or transfer to sites of further metabolism) and ultimately conversion to urea or uric acid derivatives for excretion from the body. Nitrogenous end products of catabolism are potentially harmful to the cell and mechanisms for avoiding "polluting the factory" constitute an important site of selection.

There are two common mechanisms for removing the amino group from amino acids. Firstly, the amino group can be directly removed by deamination reactions releasing free NH_4^+ and keto acids. In living systems, the metabolism of only two amino acids (serine and threonine) proceeds through deamination, forming pyruvate and 2-ketobutyrate, respectively; pyruvate enters the Krebs cycle at the pyruvate dehydrogenase (PDH) reaction while 2-ketobutyrate is converted to propionyl-CoA by a reaction analogous to PDH:

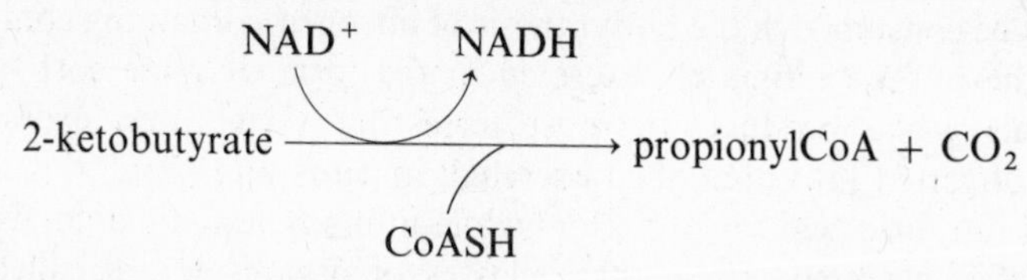

The propionylCoA can enter the Krebs cycle by conversion to succinylCoA.

Transamination reactions are much more widely used than are deaminations. In transamination reactions the amino group is transferred to a keto acid to form a new amino acid, which is usually glutamate. Glutamate can be oxidized to liberate NH_4^+ and a keto acid which can be further catabolized through the Krebs cycle.

Transaminases and Glutamate Dehydrogenase (GDH)

The oxidative deamination of glutamate is catalyzed by a mitochondrially-located enzyme termed glutamate dehydrogenase:

$$\text{glutamate} \xrightarrow[\text{NAD}^+ \;\rightarrow\; \text{NADH} + \text{H}^+]{} \text{2-ketoglutarate} + \text{NH}_4^+$$

The transaminations by which amino acids and glutamate are equilibrated are catalyzed by a number of enzymes, each specific for one or only a few amino acids. An important feature of linkage between GDH and transaminases is that the net effect of their function is equivalent to individual dehydrogenases for the other amino acids. For alanine, for example:

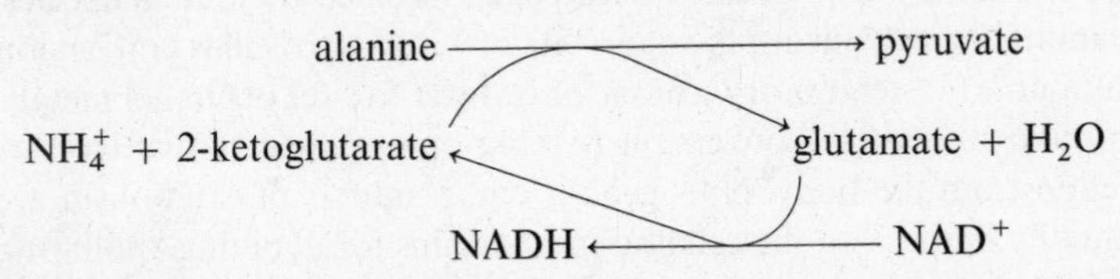

The concerted action of both enzymes achieves the same result as would a specific alanine dehydrogenase, since the overall reaction involves only the disappearance of alanine and NAD^+, with the formation of pyruvate, NADH, and NH_4^+,

Fate of Ammonia

Although some of the NH_4^+ formed in the breakdown of amino acids may be consumed in the biosynthesis of nitrogen-containing compounds, the bulk is ultimately excreted in the form of some sort of nitrogenous waste product. In most terrestrial vertebrates, excess NH_4^+ is converted into urea, then excreted; in birds and reptiles, it is converted into uric acid for excretion, while in many aquatic animals, NH_4^+ itself is excreted. These three classes of organisms are called *ureotelic*, *uricotelic*, and *ammonotelic*. In terrestrial vertebrates, urea is synthesized by the urea cycle. In this reaction scheme, one of the nitrogen atoms of the urea comes from ammonia, whereas the other nitrogen atom comes from aspartate. The carbon atom of urea is derived from CO_2. Ornithine is the carrier of these carbon and nitrogen atoms in the urea cycle (Fig. 2-5).

The overall stoichiometry of urea formation then is

$$CO_2 + NH_4^+ + 3ATP + \text{aspartate} + 2H_2O \longrightarrow$$
$$\text{urea} + 2ADP + 2P_i + AMP + PP_i + \text{fumarate}$$

In addition, pyrophosphate is rapidly hydrolyzed so, in effect, four high-energy phosphate bonds are used to synthesize one urea molecule, the energy ureotelic organisms are required to pay for detoxification of ammonia. Judging by the widespread use of this system, the cost-benefit ratio seems to be acceptably low.

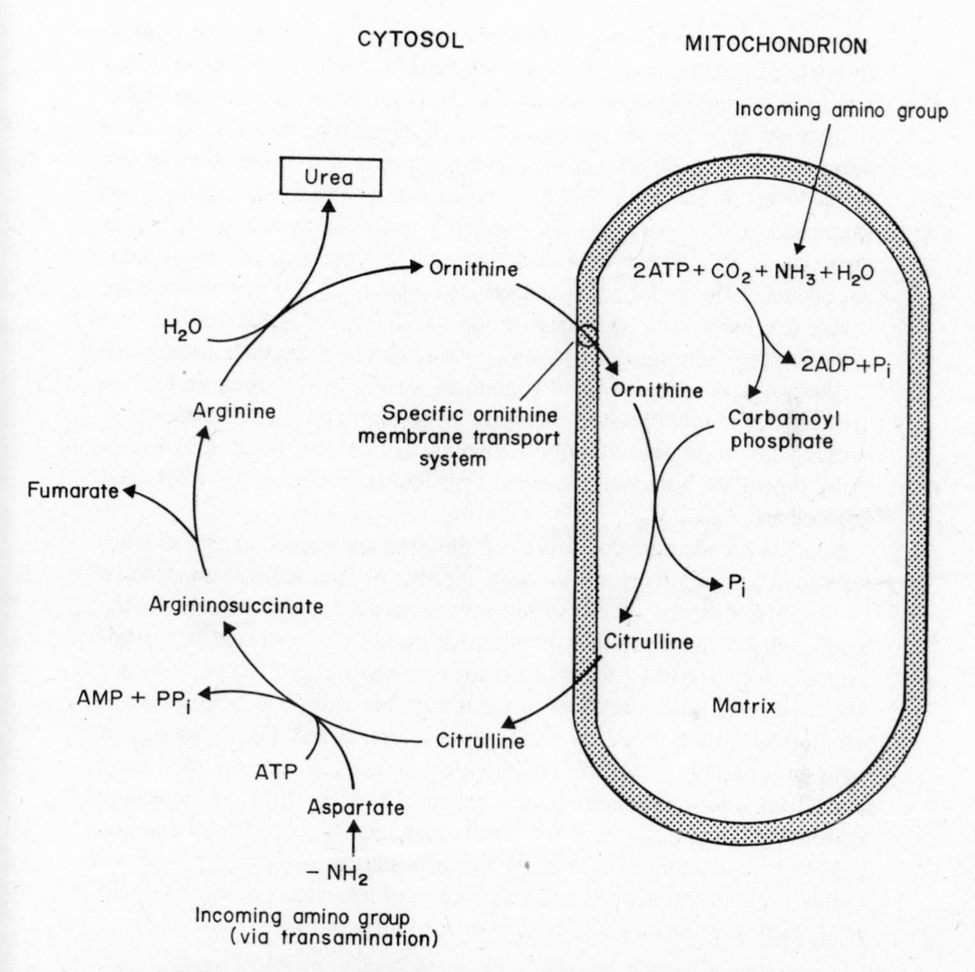

Fig. 2-5. Diagrammatic summary of the urea cycle which is compartmentalized between cytosol and mitochondria

Origin and Phylogenetic Distribution of the Urea Cycle

The urea cycle undoubtedly arose at an early stage of metazoan evolution. In some terrestrial flatworms, urea is the selected mode of nitrogenous waste excretion, and this may be an adaptation to reduced water supply, as in the case of vertebrates. Among the vertebrates, distinctly different patterns of ammonia disposal are noted in the teleosts (bony fishes) and the cartilaginous elasmobranchs. In the latter, large quantities of urea are synthesized and maintained in the body fluids for osmoregulatory purposes. Teleosts, in contrast, are primarily ammonotelic.

Interestingly, the teleost ancestors of terrestrial vertebrates quite likely were ureotelic. The closest living relative of the ancestors of terrestrial vertebrates (the South American and African lungfishes, and *Latimeria*, the coelocanth) all display a ureotelic mode of ammonia disposal. It therefore seems reasonable to assume that the immediate ancestors of the primordial amphibians possessed the enzymes of the urea cycle before the terrestrial mode of life was adopted.

Following the migration to land, evolutionary divergence took place in the ways of processing nitrogenous wastes in different vertebrate groups. Although the stem reptiles arose with ureotelic habit, most present-day reptiles (and birds) display uricotelism (uric acid excretion). However, both marsupial and placental mammals have retained ureotelism.

In all cases, the particular form of nitrogenous waste disposal characteristic of an organism represents a significant biochemical adaptation to the environment of the organism, particularly with regard to the availability of large supplies of water for flooding away excretory products. Perhaps the most vivid illustration of this pattern of evolutionary selection is found in the cases of the lungfishes and certain amphibians which are characterized by both aquatic and terrestrial life stages. In both groups, the transition from water to land is accompanied by a shift from ammonotelism to ureotelism. Thus, we find an especially dramatic example of one of the major strategies of biochemical adaptation to the environment, namely the production of new kinds of enzymes that better supply the requirements for survival in—and, indeed, exploitation of—a new environmental situation.

The Fate of the Carbon Skeletons of Amino Acids

In functional terms, the major part of amino acid degradation is the disposition of the carbon skeletons. This is the part of amino acid metabolism in which energy in the form of ATP is conserved by mitochondrial electron transfers. Unlike sugar and fat, proteins upon hydrolysis yield a wide assortment of amino acid substrates which can enter the Krebs cycle for complete oxidation at various levels. Some, like leucine, can donate carbons as acetylCoA for Krebs cycle catabolism; some, like aspartate, enter the Krebs cycle at the malate dehydrogenase step; some, like glutamate, proline, arginine, and ornithine, enter the Krebs cycle as oxoglutarate; and still others, such as the other branched-chain amino acids, can enter the Krebs cycle at the level of succinylCoA (Fig. 2-6). Thus, it is evident that the metabolism of proteins (of the free amino acid pool) in animals is rather complex and is only now being put into proper perspective.

In mammals, for example, the metabolism of the carbon skeletons begins primarily in the liver (except for the branched-chain amino acids, alanine, and glutamate, which are also metabolized by other tissues). The liver, however, has a rather low capacity for oxidation and its primary role is to convert carbon skeletons of amino acids to glucose, to fatty acids, or to ketone bodies (acetoacetate and 3-hydroxybutyrate). The result is the transformation of amino acids by liver into more standard fuels which can be utilized by other tissues and organs according to needs. Such a metabolic organization does not hold for many invertebrate and lower vertebrate species, however, wherein the capacity to directly utilize amino acids seems to be more widely retained. Whatever the sites of amino acid degradation, the mechanisms utilized for any given amino acid appear to be fairly uniform between species. These will be discussed in detail wherever necessary in later chapters.

Generation of Coupling Intermediates from the Krebs Cycle

The Krebs cycle is an amphibolic (dual-function) pathway which, as already mentioned, functions not only in catabolism but also to generate precursors for anabolic pathways. Two intermediates of the cycle in particular, 2-ketoglutarate and oxaloacetate, serve as precursors of glutamate and aspartate, respectively, to which they are converted by enzymatic transamination reactions. Citrate can also be removed from the cycle to serve as a precursor of extramitochondrial acetylCoA for fatty acid biosynthesis through the ATP-citrate lyase reaction:

$$\text{citrate} + \text{ATP} + \text{CoA} \longrightarrow \text{acetylCoA} + \text{oxaloacetate} + \text{ADP} + P_i$$

Moreover, succinylCoA can also be removed from the cycle for heme biosynthesis. Thus, in supplying critically needed intermediates for Block II anabolic functions the Krebs cycle can be drained of intermediates necessary for its functions in energy metabolism. This problem is overcome by the occurrence of anaplerotic (filling-up) reactions.

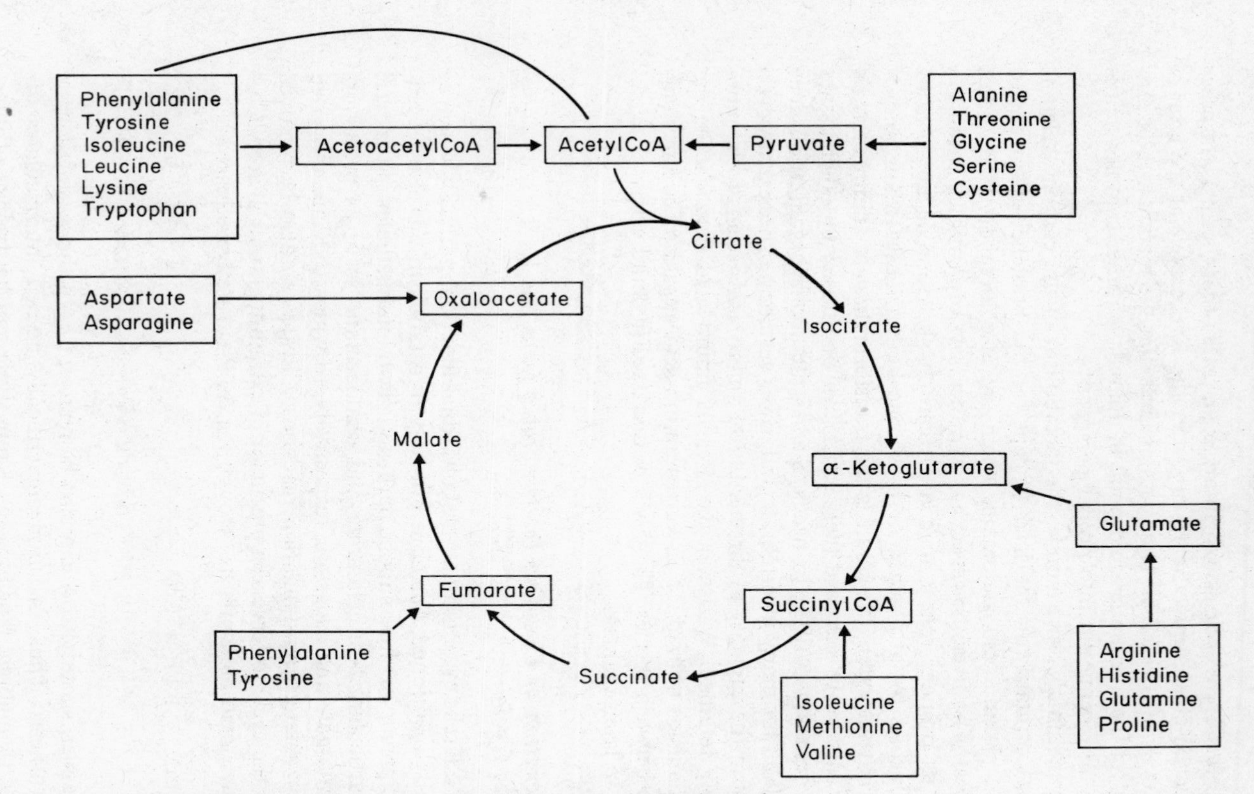

Fig. 2-6. Entry sites for the free amino acid pool into the Krebs cycle

Replenishing and Augmenting Krebs Cycle Intermediates

Anaplerotic reactions of the Krebs cycle can be defined as those that feed carbon into the Krebs cycle, usually in a single, enzyme-catalyzed step. In mammalian liver and kidney one of the most important of these is pyruvate carboxylase, a mitochondrial enzyme catalyzing

$$\text{pyruvate} + CO_2 + \text{ATP} + H_2O \xrightleftharpoons{Mn^{2+}} \text{oxaloacetate} + \text{ADP} + P_i$$

In addition to pyruvate carboxylase, a number of other reactions may also serve to replenish Krebs cycle intermediates. These include malic enzyme-catalyzed carboxylation of pyruvate

$$\text{pyruvate} + CO_2 + \text{NADPH} + H^+ \rightleftharpoons \text{L-malate} + \text{NADP}^+$$

and, particularly in invertebrate organisms, PEP carboxykinase-catalyzed carboxylation of phosphoenolpyruvate (PEP)

$$\text{PEP} + \text{GDP (or IDP)} + CO_2 \xrightleftharpoons{Mg^{2+}} \text{OXA} + \text{GTP (or ITP)}$$

Intermediates of the cycle may also be generated from aspartate and glutamate, which are converted into oxaloacetic and 2-ketoglutaric acids, respectively, by amino transferase reactions:

$$\text{glutamate} + \text{pyruvate} \rightleftharpoons \text{2-ketoglutarate} + \text{alanine}$$

$$\text{aspartate} + \text{pyruvate} \rightleftharpoons \text{oxaloacetate} + \text{alanine}$$

$NADP^+$-linked isocitrate dehydrogenase, catalyzing the carboxylation of 2-ketoglutarate, is not strictly a Krebs cycle enzyme (its NAD^+-linked analogue is). Nevertheless, it is sometimes utilized by organisms to generate NADPH required in fat biosynthesis. Under these conditions, the carbon source for the reaction is usually glutamate, which is transaminated to 2-ketoglutarate, the immediate substrate for the reaction. In the process of generating NADPH, the cell simultaneously generates isocitrate and thus helps to augment the pool of Krebs cycle intermediates.

In principle, any metabolite convertible to Krebs cycle intermediates could be utilized for replenishing or augmenting the overall pool. In addition to the above, in some animals (in the blowfly, for example, during flight initiation) proline is used as a primary mechanism for augmentation of the Krebs cycle. This augmentation proceeds by the following sequence:

proline ⟶ pyrroline-5-carboxylate (P5C) ⟶

proline dehydrogenase P5C dehydrogenase

glutamate ⟶ 2-ketoglutarate

pyruvate ⟶ alanine

The carbon of proline thus appears as 2-ketoglutarate and other Krebs cycle intermediates, while its nitrogen is deposited in alanine (Weeda et al., 1980). In mammalian muscle, branched-chain amino acids also may be used for augmenting the Krebs cycle pool (Fig. 2.6).

Sites of Decarboxylation and Electron Transfer to NAD^+ or FAD^+

Whatever the substrate sources utilized, CO_2 is released in the Krebs cycle at only two reactions, catalyzed by isocitrate and 2-ketoglutarate dehydrogenases. These are indeed the only CO_2-generating reactions in the complete degradation of fatty acids. During glucose catabolism, CO_2 is also released at the PDH reaction sequence. In amino acid catabolism, CO_2 is generated by ketocarboxylate dehydrogenases (reactions analogous to PDH) utilizing the ketocarboxylates formed from leucine, isoleucine, valine, and threonine in particular. In addition, for the complete catabolism of some important amino acids (proline, glutamate, arginine, ornithine), a decarboxylation also occurs at the level of malic enzyme, which may be either NAD^+- or $NADP^+$-linked.

In essentially all these reaction sequences, electrons and protons are transferred either to NAD^+ or FAD^+; the reoxidation of these coenzymes (involving the ultimate transfer of electrons and protons to O_2) is the process that is coupled to ATP formation. This is an important point, for it indicates that the bulk of oxidative metabolism (the pathways feeding into the Krebs cycle and the Krebs cycle per se) are not directly energy yielding or O_2 dependent. The one exception to this general statement is succinylCoA synthetase which is a substrate-level phosphorylation and which proceeds with the formation of one ATP equivalent.

Classes of Electron Transferring Enzymes

From the above, it will be evident that in animal cells there must be at least two classes of electron-transferring enzymes participating in the flow of electrons from organic substrates to O_2. These are 1) pyridine nucleotide-linked dehydrogenases, which require either NAD^+ or $NADP^+$ as coenzyme; 2) flavin nucleotide-linked dehydrogenases, which contain flavin adenine dinucleotide (FAD^+) or flavin mononucleotide (FMN) as the prosthetic group. These two kinds of electron-transferring enzymes are particularly pivotal because they *interface the metabolic reactions generating electrons with the system which transfers them to O_2*. To complete that transfer, mitochondria also contain the lipid-soluble coenzyme *Q* (ubiquinone) and the cytochrome proteins which contain iron-porphyrin prosthetic groups.

NADH and $FADH_2$ are thus the principal intermediaries between the primary degradative pathways of metabolism and the enzymes in the inner mitochondrial membrane that eventually deliver electrons and protons to O_2, forming water. The overall process, of course, is called respiration and the complex of enzymes acting as electron carriers in the membrane are called the respiratory chain, or electron transfer system (ETS).

The Respiratory Chain in Mammalian Mitochondria

The overall sequence of electron-transfer reactions in the ETS from NADH to O_2 is now relatively well known and is shown in Figure 2-7. Electrons from oxidation-reduction reactions (mainly those in the Krebs cycle) are first collected in NADH and then enter the ETS by means of the flavoprotein containing NADH dehydrogenase (FP_1 in Fig. 2-7). In contrast, some substrates are dehydrogenated by flavin-linked dehydrogenases (such as succinic dehydrogenase in the Krebs cycle or acylCoA dehydrogenase in fatty acid oxidation); from these substrates, electrons are funneled into the ETS at the level of coenzyme *Q*. The position of entry for various substrates is shown in the diagrams (Figs. 2-7 and 2-8); the point of emphasis is that the energy (ATP) captured during the oxidation of any given substrate is determined by the position of entry into the ETS. This is because electrons donated from NADH traverse all available sites of ATP formation, while those donated from $FADH_2$ do not traverse the first site of ATP conservation. A problem however arises when electrons are to be transferred from

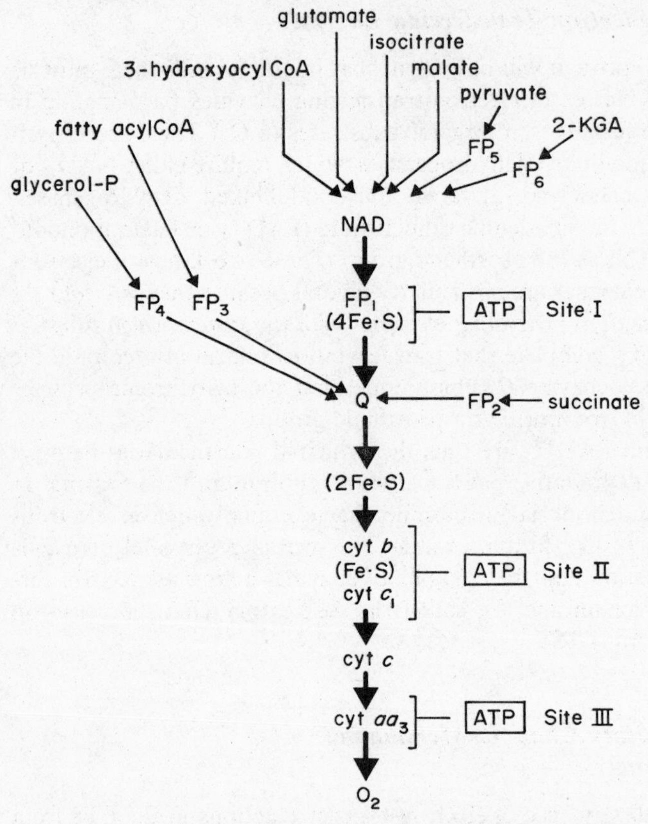

Fig. 2-7. Electron transfer system showing three ATP conservation sites and entry sites for electrons and protons from various substrates. After Lehninger (1979)

cytosolic NADH to the ETS and O_2. The problem arises because mitochondria are impermeable to NADH and it is solved by the development of hydrogen shuttling mechanisms.

The Need for Hydrogen Shuttles

Because of NADH impermeability, in all cells that generate NADH in the cytosol during aerobic glycolysis, the need arises for hydrogen shuttles to move reducing equivalents to the electron transfer system, (ETS). In mammalian tissues, two such shuttles are well documented. One is the malate-aspartate shuttle (Fig. 2-9) in which the component enzymes are a) cytosolic and mitochondrial forms of both aspartate

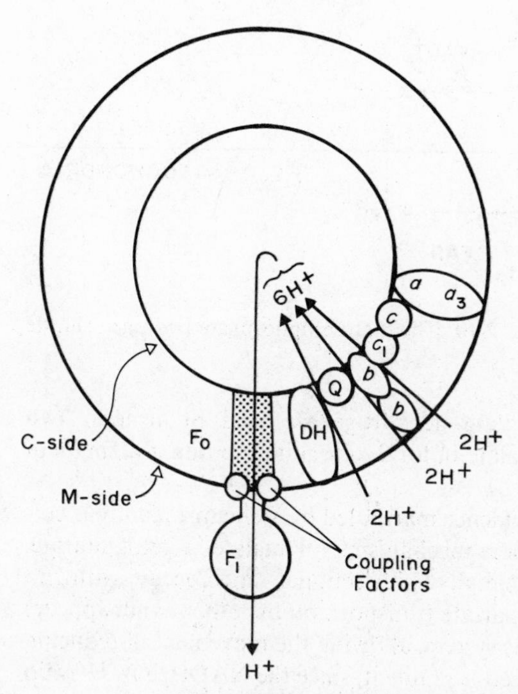

Fig. 2-8. Schematic diagram of the components of the electron–transfer chain and of the ATP–synthesizing components in a submitochondrial particle. After Alfonso and Racker (1979)

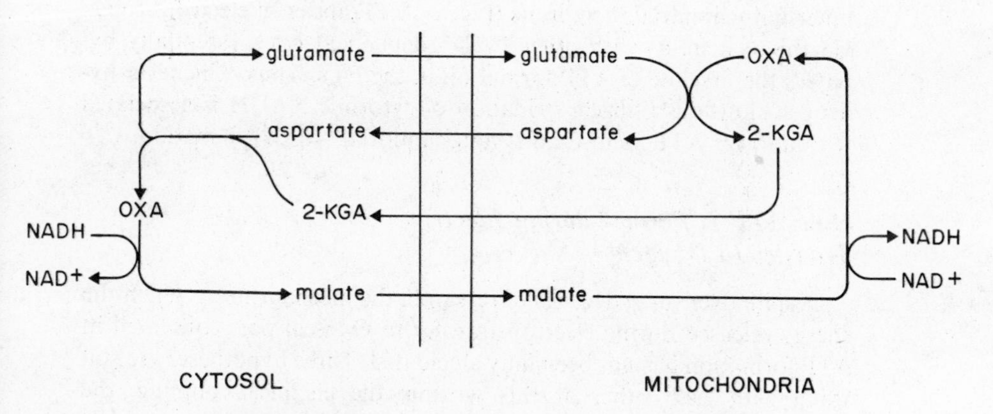

Fig. 2-9. Malate-aspartate shuttle as a mechanism for transferring reducing equivalents from cytosol to mitochondria

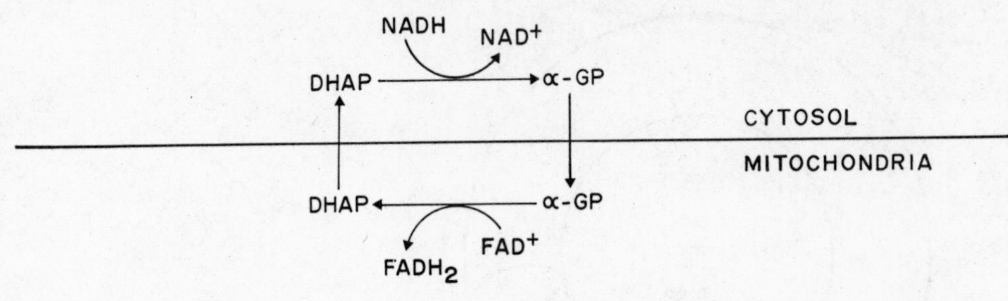

Fig. 2-10. The α-glycerophosphate hydrogen shuttle

aminotransferase and malate dehydrogenase, and b) at least two exchange mechanisms (malate in for 2-ketoglutarate out; glutamate in for aspartate out).

According to current evidence marshaled by Williamson and his colleagues (1973), the transport mechanisms for malate, 2-ketoglutarate, and glutamate are potentially bidirectional and occur with no significant energy cost; aspartate transport, on the other hand, appears to be an energy-dependent system, allowing the movement of reducing equivalents against an effective gradient, since the $NADH/NAD^+$ ratio in the mitochondria is more reduced than in the cytosol.

A second, somewhat simpler mechanism for transferring NADH-derived hydrogen into the mitochondria is the α-glycerophosphate (α-GP) cycle. In this cycle, DHAP is reduced by cytosolic NADH to α-GP, a reaction catalyzed by cytosolic α-GPDH; the α-GP is oxidized by a second FAD^+-linked enzyme, α-GP oxidase, located on the inner mitochondrial membrane (Fig. 2-10). Transfer of electrons to O_2 via this system, as with other FAD^+-dependent ones, essentially bypasses the first site of ATP formation in the ETS. Thus, whichever hydrogen shuttle is utilized, oxidation of cytosolic NADH is associated with a lower ATP yield than is mitochondrial NADH oxidation.

How ATP Is Formed during Electron Transfer to Oxygen

Despite over three decades of research, the mechanism by which the energy released during electron transfer to O_2 is in part conserved in ATP formation has not been fully elucidated. Three hypotheses are still vying with each other at this writing: the chemical-coupling, the conformational-coupling, and the chemiosmotic-coupling models. We do not propose to review the merits of each; suffice to mention that at this time, the third or chemiosmotic-coupling hypothesis seems to be gaining the most support and we will briefly describe its elements here.

To fully appreciate the main features of the chemiosmotic theory of ATP formation it is important to realize that although the overall chemistry of oxidation consists in a transfer of hydrogen, it is not necessary to transport complete hydrogen atoms at each stage in the process. In fact, according to the chemiosmotic theory, hydrogen carriers alternate with molecules that carry only electrons. The use of electron carriers is possible because protons are soluble in water and in the aqueous medium of the cell, whereas electrons are not. When a molecule that bears a whole hydrogen atom interacts with another molecule that accepts only electrons, the proton is released into solution. When an electron carrier then donates its electron to a hydrogen carrier, the hydrogen atom is reconstituted by withdrawing a proton from the medium.

In current models of mitchondrial respiration (Fig. 2-11), each pair of electrons transferred from NADH to oxygen results in the outward translocation of six protons across the membrane. The ratio is expressed in terms of pairs of electrons because the electrons appear in pairs both at the beginning of the respiratory chain (at NADH) and at the end (where they reduce an oxygen atom to water). Within the respiratory chain the electrons are transported one at a time by some carriers and in pairs by others.

As a result, energy released in electron transfer is temporarily stored in a proton gradient which inherently is formed of two components: one component is the difference in concentration or chemical activity of protons on opposite sides of the membrane. The protons tend to diffuse from a region of high concentration to a region of low concentration when a pathway through the membrane is provided.

The electric charge carried by the proton contributes the second component to the energy of the gradient. The net movement of charges across the membrane creates a difference in electric potential, and all charged particles are affected by the resulting electrostatic field. The total energy of the proton gradient is the sum of the concentration (or osmotic) component and the electric component.

Because of the difference in concentration and in electric potential a proton that has been expelled from a mitochondrion experiences a force tending to draw it back across the membrane. According to the chemiosmotic theory, proton movement in response to that force is coupled to ATP synthesis (Fig. 2-11).

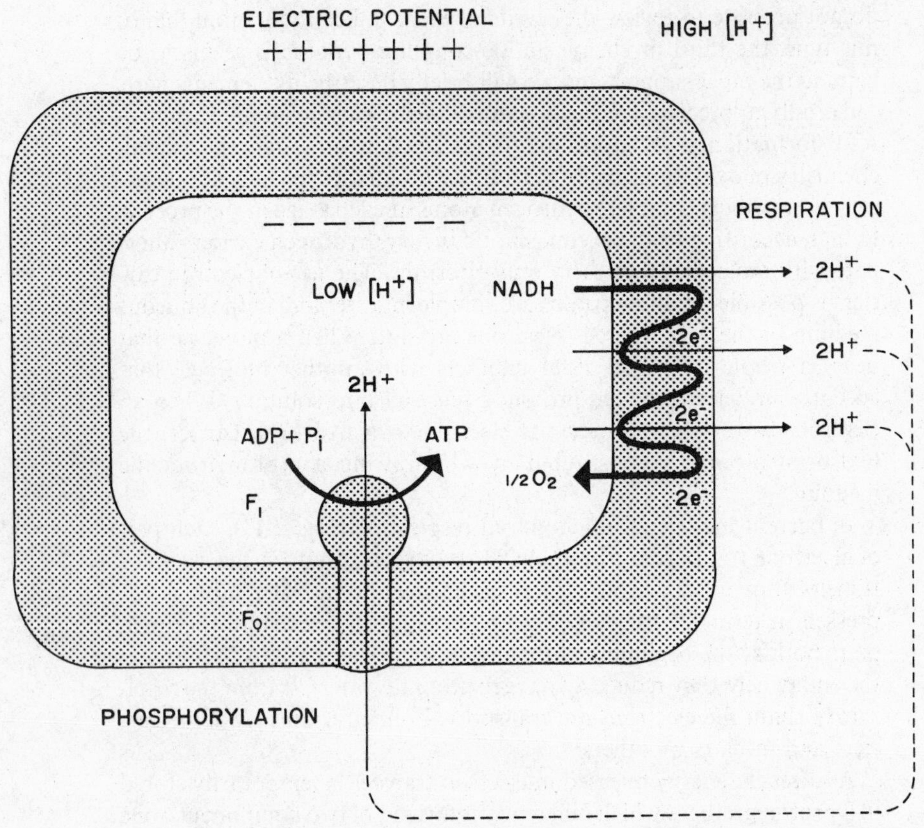

Fig. 2-11. A schematic summary of the basic events of oxidative phosphorylation as they are currently interpreted by the chemiosmotic theory

The enzymes that couple the diffusion of protons back through the membrane to the synthesis of ATP are conspicuous in electron micrographs as globular bodies protruding from the surface of the membrane. The protruding "knob," designated F_1, is a soluble protein made up of five kinds of subunits, some of which are present in multiple copies. F_1 is attached to the membrane through another protein designated F_0, which is embedded in the membrane and is thought to pass all the way through it. F_1 is readily removed from the membrane, but F_0 can be removed only when the membrane itself is destroyed with detergents. Apparently F_0 serves primarily as a channel through which protons cross the membrane to F_1. The entire system of enzymes is called the F_1-F_0 complex.

In mitochondria and in bacteria the F_1-F_0 complex is oriented with the knobs protruding into the interior matrix from the inner surface of the membrane. The evidence indicates that for every two protons passing inward through the complex, one molecule of ADP is combined with inorganic phosphate to form ATP. An important observation is that the reaction is reversible; under appropriate circumstances the F_1-F_0 complex can split ATP molecules and employ the energy made available to pump protons out of the mitochondrion or bacterial cell. Like all enzymes, the F_1-F_0 complex controls the rate of the reaction, but the direction of the reaction is determined by the balance of free energy.

Energy Balance Sheet for Oxidation of Glucose

In standard biochemistry textbooks, it is usually assumed that for each electron pair traversing the ETS from NADH to O_2 three ATP molecules can be synthesized; in this event, the P/O ratio for NADH oxidation would be 3, and 2 for $FADH_2$ oxidation. Experimentally, the value of 3 is rarely if ever found, but P/O ratios of about 2 are very commonly observed. Since earlier theories of ATP formation best accommodated P/O ratios of 3 (for NADH oxidation), it was previously assumed that the lower values obtained represented loss of energy due to poor mitochondrial preparations or experimental errors. This problem has been carefully rephrased recently, emphasizing the fact that P/O ratio for any given substrate being oxidized represents a measure of the cost of ATP synthesis in the mitochondrion plus the cost of transferring that ATP out of the mitochondrion (to where it is needed) against a chemical gradient. According to these recent studies, therefore, the P/O ratios predicted from chemiosmotic theory and in fact experimentally observed are 2 for NADH-linked substrates, and 1.3 for $FADH_2$-linked substrates. This is such a critical point in assessing a physiologically meaningful energy budget for different substrates that it is necessary to pursue the argument in some detail.

From best current estimates, the protons crossing the mitochondrial membrane have an energy of 5.3 kcal/mole, but only two protons are consumed for each ATP formed intramitochondrially. The free energy available then is 10.6 kcal/mole, which is insufficient since a maximum of about 15 kcal/mole is required for phosphorylation. In mitochondria this energetic problem is circumvented by localizing the ATP-forming reactions to the inside of the inner membrane. Because of high concentrations of ADP and P_i and low concentration of ATP in that environment, the affinity (chemical potential) of the reaction is strongly negative and effectively reduces the free energy required. Hinkle and his colleagues (e.g., Hinkle, 1981) estimate that inside the mitochondrion ATP synthesis, therefore, proceeds quite effectively with a free energy drop of only about 11 kcal/mole, which is within the range of the calculated energy available.

However, this reduction in the energy requirements of phosphorylation in mitochondria does not come without a cost: that cost is the energy needed to concentrate ADP and P_i in the mitochondrial interior and to remove ATP to the exterior as it is formed. Since the counterflow of ATP and ADP ions involves the net movement of charge across the membrane, the exchange of the two could be powered by a favorable membrane potential. Phosphate, on the other hand, is moved inwards in exchange for OH^- ions and its transport therefore is driven by the pH gradient. The combined effect of these transactions is to transport an additional proton for each ATP synthesized and transported. Thus a total of three protons is expended for each ATP appearing outside the mitochondrion and the total free energy available is 3×5.3 or 15.9 kcal/mole. That is why the stoichiometry of respiration and phosphorylation predicts a P/O ratio of 2 for NADH oxidation, and 1.3 for $FADH_2$ oxidation.

ATP Yield for Complete Glucose Oxidation

Since these estimates of P/O ratios lead to quite different yields of ATP/mole of starting substrate oxidized than usually assumed, it is useful to calculate what these are for a number of common substrates. For glucose, for example, the ATP yield/mole glucose oxidized can be shown to be 25.2 by the following calculations:

8 moles mitochondrially-generated NADH/mole glucose yields 8×2	= 16.0 ATP
2 moles mitochondrially-generated $FADH_2$/mole glucose yields 1.3×2	= 2.6 ATP
2 moles GTP (ATP equivalents) at succinic thiokinase	= 2.0 ATP
2 moles glycolytically-generated NADH/mole glucose yields 1.3×2	= 2.6 ATP
2 moles glycolytically-generated ATP/mole glucose yields	= 2.0 ATP
Total moles ATP/mole glucose fully oxidized	= 25.2 ATP
Total moles ATP/mole glucose, assuming standard P/O ratios	= 36.0 ATP

Energy Balance Sheet for Other Substrates

A similar calculation for the complete oxidation of other substrates is useful. For example, in the complete oxidation of a 16-carbon saturated fatty acid (palmitate) 91.8 moles of ATP/mole of palmitate are obtained:

8 moles NADH formed in β-oxidation 8×2	=	16.0 ATP
8 moles of $FADH_2$ formed in β-oxidation 8×1.3	=	10.4 ATP
24 moles NADH formed in the Krebs cycle 24×2	=	48.0 ATP
8 moles of $FADH_2$ formed in Krebs cycle 8×1.3	=	10.4 ATP
8 moles GTP formed at succinylCoA synthetase	=	8.0 ATP
less 1 mole ATP involved in palmitylCoA formation	=	−1.0 ATP
Total moles ATP/mole palmitate fully oxidized	=	91.8 ATP
Assuming standard P/O ratios, this value	=	129.0 ATP

Because of the uncertainty in P/O ratios, we will assume standard P/O ratios (3 for NADH-linked oxidations) because this facilitates comparisons with earlier studies. Whichever P/O ratios are used, however, it is evident that on a molar basis the yield of useful energy obtained on burning fat is substantially higher than for burning glucose. Since the starting substrates are not the same molecular weight, a molar comparison is perhaps not the best way of estimating their relative energy yields. On a weight basis, starting for purposes of calculation with ingested starch or ingested fat, the approximate ATP yields for fat are about 2.3 times greater than for carbohydrates (Table 2-1), which is very close to the caloric differences between these two substrates. The reason fat is a better substrate has been reviewed elsewhere (Hochachka and Somero, 1973); suffice it to point out at this time that fat is more reduced than glucose, and that in itself means it contains

Table 2-1 Energy Yields of Common Fuels for Metabolism

Source	moles ATP/kg fuel	moles ATP/mole O_2	kcal/gm	(RQ) CO_2/O_2
Ingested starch	216 (151)	5.8 (4.2)	4.18	1.0
Ingested fat	502 (351)	5.5 (4.0)	9.46	0.71
Ingested meat protein	215 (150)	4.8 (3.6)	4.32	0.82
Stored muscle glycogen	228 (155)	6.2 (4.4)	—	—
Stored fat	510 (357)	5.6 (4.0)	—	—

SOURCE: From McGilvary (1979) with modification, assuming P/O ratios of 3 and of 2 (given in parentheses) for NADH-linked oxidations

more electrons and protons that can be transferred to O_2 which should generate higher ATP yields.

Most amino acids, in contrast to fatty acids, are at about the same oxidation state as is glucose. For this reason, and because of common final pathways for oxidation (see above), it is not surprising that the ATP yield from amino acids is similar to that for carbohydrates. On a molar basis, this is clearly indicated for alanine, which on oxidation yields 15 moles ATP/mole alanine, exactly the same value as is obtained for pyruvate (derived from glucose). For glutamate oxidation (via malic enzyme and assuming NADPH and NADH oxidations by the ETS yield identical P/O ratios), the yield is 27 moles ATP/mole of glutamate. (Although Atkinson [1977] argues that NADPH, in terms of ATP equivalents, is worth one more than NADH, there is some uncertainty as to the exact pathway by which pyruvate is formed during the complete oxidation of glutamate or proline. Hence, to be on the conservative side, we assume that malic enzyme catalysis of malate → pyruvate is linked, for practical purposes, to NAD^+). Many lower organisms utilize proline for energy metabolism, but the energy yield from proline is difficult to ascertain since the first step in its oxidation may not be coupled to ATP formation. If it is coupled through NADH, then the yield is 30 moles ATP/mole of proline. The range of these values is consistent with the observation that the number of moles of ATP formed per kg of protein is almost identical to the yield from carbohydrate, as is the caloric yield (Table 2-1).

Adjustable Components in Mitochondrial Metabolism

As with glycolysis, the essential features of oxidative mitochondrial metabolism are considered to be extremely old in phylogenetic terms and for that reason mitochondrial metabolism is also considered to be highly conservative. Again, the most likely explanation for this view is that it is based on studies of remarkably few organisms; as our data base on organisms in different groups and different environments expands, we suspect that more and more features of mitochondrial metabolism will be shown to be adaptable. Several obvious areas of adaptation that are already well described in the literature include 1) substrate preferences, 2) overall organization, and 3) enzymatic potentials of mitochondrial metabolism in different tissues, organs, or species. In assessing the metabolic makeup of a variety of organisms the preferred substrates and entry sites into either the Krebs cycle or the ETS are obvious areas of adjustment and adaptation. The mammalian brain, for example, displays a very active oxidative metabolism, in man constituting up to 20 per cent of basal body metabolic rate; this metabolic system operates almost exclusively on glucose with pyruvate, entering into the Krebs cycle at the PDH reaction, serving as the major if not sole carbon and energy source for mitochondrial metabolism. The heart, in contrast, preferentially burns fatty acids, lactate, and glucose and even some amino acids can be utilized. Thus, the main source of carbon for mitochondrial metabolism is normally acetylCoA generated by β-oxidation, but pyruvate from either lactate or glucose can also serve to prime the Krebs cycle.

In mammalian fast-twitch glycolytic-type muscle fibers, oxidative metabolic rates are relatively low and are predominantly fueled by glucose (or glycogen); in contrast, the slow-twitch, oxidative-type muscles preferentially metabolize fatty acids, but can also utilize glucose-derived pyruvate. In tuna, the metabolic differences between white and red muscle are even more starkly exaggerated (Guppy et al., 1979). In salmon, white muscle oxidative metabolism seems primarily based on pyruvate and involves numerous amino acid interconversions, particularly during the spawning migration. While salmon red muscle is clearly geared for either fatty acid or amino acid catabolism, *when* each is used is closely timed and regulated: during early stages of migration, red muscle work is sustained by β-oxidation, while during later stages of migration, it is sustained mainly by amino acid (alanine and glutamate) catabolism (Mommsen et al., 1980).

The mitochondria of flight muscle in some insects (Hymenoptera, for example) preferentially utilize glucose-derived pyruvate and α-glycerophosphate; others (migratory locusts, for example) utilize glucose for short work periods, but depend exclusively on fat for long flights. Still others (the tsetse fly, Coprine beetles) depend upon proline *in vivo* and their mitochondria when isolated preferentially respire proline (Weeda et al., 1980).

Marine invertebrates are particularly interesting in that they maintain large intracellular amino acid pools. For example, in the cephalopods, proline is often a predominant amino acid and is utilized during activated metabolism. Isolated heart mitochondria, not surprisingly, are found to respire proline and ornithine at higher rates than all other substrates except pyruvate (Mommsen and Hochachka, 1981).

Most of the above adjustments in mitochondrial metabolism involve tissue-specific preference for entry sites into the Krebs cycle. It should be recalled, however, that this may involve automatic emphasis on different components of the ETS. For example, β-oxidation involves a different FAD^+-linked dehydrogenase than does α-glycerophosphate oxidase, which is also FAD^+-linked. The first step in proline oxidation is catalyzed by an NAD^+-dependent enzyme in insect flight muscle, but by an O_2-dependent oxidase in mammalian and squid mitochondria; if the latter is coupled to oxidative phosphorylation, it must utilize a unique component for electron transfer.

In addition to such adjustable features of the ETS, the pathways for electron transfer themselves are subject to adaptation. The most dramatic example of this is to be found in brown fat of mammals, where the thermogenic function of the tissue depends upon a fascinating "short-circuit" in the ETS leading to an increase in heat loss by the system but a decrease in ATP conservation (see Chapter 10). Similarly, the ETS composition and function are clearly modified in helminth facultative anaerobes in order to allow at least some ATP conservation by utilizing organic electron acceptors (such as fumarate) instead of O_2 (Saz, 1981).

The full significance of adjusting enzymatic potentials of oxidative metabolism, although evident in studies of ETS composition, is even more evident for Krebs cycle enzymes and the enzymes for channeling substrates into the Krebs cycle. In some cases, for example locust flight muscles, the enzymes of the Krebs cycle and β-oxidation are elevated in direct proportion to each other (Hochachka and Guppy, 1977). In others (salmon red muscle, squid heart), Krebs cycle enzyme activity and amino acid catabolizing potential are proportionately adjusted, while β-oxidation enzymes are either not proportionately adjusted (in salmon) or are hardly utilized at all (squid) (Ballantyne et al., 1981). In still other cases (bee flight muscle) the catalytic potential of the Krebs cycle is elevated in proportion to the aerobic glycolytic potential and to the α-glycerophosphate shuttling mechanism. And finally, in some cases (tuna red and white muscle), the Krebs cycle catalytic potential may have to pace the aerobic glycolytic one or the potential of β-oxidation. Whatever the combination, all such observations emphasize that control of enzymatic potential lies at the very heart of metabolic adaptation and integration in living systems. For that reason, our analysis of biochemical adaptations will begin with an exploration of how enzymes are designed for controlled metabolic functions (the topic for our next chapter).

CHAPTER THREE

Adaptation of Enzymes to Metabolic Functions

Nature of the Control Problem

Most of what we know about how enzymes work is gained from studying them *in vitro*, usually one isolated reaction at a time. But a living cell may have the potential for many thousands of reactions, so it is immediately evident that *in vivo* enzyme function must be a lot different from that observed *in vitro*. In the first place, some enzymes are designed to initiate metabolic processes, and therefore will have to be sensitive to some sort of "on-off" mechanisms for controlling their activities. Some enzymes operate at metabolic branchpoints where moment-by-moment requirements of the cell determine the direction of carbon flow through the branchpoint. These enzymes too will have to be subject to stringent *in vivo* control, so that their activities are integrated with others in the same pathway, and in different metabolic pathways as well. Still others may lack regulatory properties, and function only as highly efficient catalysts; yet, because we know different steady-state fluxes can occur, their activities must be integrated with respect to other enzymes in the pathway (including those designed to initiate the flux through the pathway). In animals, the metabolic activities of tissues and organs do not occur in a vacuum; rather they are integrated with respect to the needs of the organism as a whole (e.g., muscle work is not activated for the internal welfare of muscle, but rather for the sake of the organism as a whole). Thus the metabolic machinery of cells in a given tissue or organ must be sensitive to external signals derived from other parts of the organism and assuring that appropriate metabolic functions be turned on at appropriate rates only at metabolically appropriate times. Finally, in complex metazoan organisms, the need for different metabolic functions may vary greatly between developmental stages, between different environmental situations, and between different tissues and organs. If specific enzymes, or indeed entire metabolic pathways, are not needed at all times, then it would be advantageous that enzymes be made and maintained at appropriate activity levels. Thus *in vivo* there is the final need to control how much of any given enzyme is made when and in which tissues.

Designing the metabolic machinery, then, to allow the right intracellular reactions to work at the right rates, in the right directions, and at the right times must be one of the most fundamental levels of adaptation occurring in animals irrespective of other (environmental, physiological, or ecological) considerations. How is this control of catalytic potential (enzyme amount) and catalytic activity (enzyme function) actually achieved?

Control of Enzyme Concentration: Transcription, Translation, Assembly, and Degradation

The amount of a particular enzyme present in a cell can be regulated at several steps in the production of the enzyme and, of course, at the stage of enzyme degradation. In the metabolic control hierarchy, the most complex mechanism for regulating enzyme concentration involves the processes of gene activation and repression. In response to specific chemical signals, the *transcription* of a given DNA sequence into messenger RNA (mRNA) may be initiated or blocked, depending on whether the signal in question is an "inducer" or a "repressor," respectively. Gene-level control can lead to a) increased or decreased quantities of enzymes, b) changes in the types of enzymes which occur in the cell, and c) changes in the relative abundance of enzyme variants (or isozymes), each of which catalyzes the same given reaction but may display specific catalytic properties.

Control of enzyme concentrations at this highest point in the metabolic control hierarchy has obvious advantages and limitations. Gene activation/repression is an effective way of changing enzyme concentrations in a highly specific fashion. On this count, it is a highly versatile control mechanism. However, in eucaryotic cells, with which this book is almost exclusively concerned, gene activation is a slow process. Normally, at least hours are required for an inducing or repressing signal to exhibit its effects at the level of enzyme concentration. In contrast, environmental changes may occur within seconds or minutes and, therefore, survival may depend on biochemical adaptations that can occur at the same rapid rates.

Following transcription and the processing of newly synthesized RNA chains via removal of noncoding segments (introns) mRNA participates in reactions which terminate in the production of "nascent" polypeptide chains. This series of reactions is termed "translation," as it is these reactions that translate the information carried by the language of the genetic code into a protein or polypeptide molecule. The control of protein synthesis at the level of translation is poorly understood. Theoretically, control could be exerted at any of the several steps in the translation sequence, including a) the binding of mRNA to the 40S ribosomal subunit, b) the formation of the 80S ribosomal complex, c) the activation of amino acids, and d) the rate at which the mRNA message is read. In reality, controls at sites other than a) probably are not of sufficient specificity to be of general importance in metabolic regulation.

Following translation, the nascent polypeptide chain usually must assume a particular higher order structure before it is functional. Thus, merely having the correct amino acid sequence (primary structure) does not assure that the protein is functional. The nascent polypeptide chain must gain the proper secondary (2°), tertiary (3°), and for most enzymes, quaternary (4°) structure* before it can be of metabolic "use" to the cell. Many proteins also are covalently modified following translation; for example, disulfide bridges may form and certain side chains, notably serine residues, may be phosphorylated.

There appears to be precise regulation of the rate at which certain levels of higher order structure are attained. Whereas for 2° structure the amino acid sequence of the protein appears to bear major, if not sole, responsibility for the geometry of the molecule, 3° and 4° structures are at least partially under the control of exogenous molecules. For example, the aggregation of subunits to form the functional multimers (the 4° structure) may depend on (enzyme-catalyzed) phosphorylation of the subunits. This process, in turn, may be hormonally regulated. The equilibrium, inactive subunits $\rightleftharpoons$ active oligomer, is often influenced by substrates and cofactors of the reaction. Thus, epigenetic control of enzyme function at the level of acquisition of 3° and 4° structure may be of significance in metabolic regulation.

Lastly, many enzymes must be integrated into definite cellular structures or joined with other large molecules before they are functional. In these processes, which represent the acquisition of quintinary (5°) structure, the enzyme combines with other proteins, lipid molecules, or membranes before it takes on its *in vivo* functions. Any factor that

* *Primary structure* refers to the covalent backbone of the polypeptide chain and specifically denotes the amino acid sequence. *Secondary structure* refers to the extended or helical conformation of polypeptide chains. *Tertiary structure* refers to the manner in which the chain is bent or folded to form the compact, folded structure of globular proteins. The general term, conformation, is used to refer to the combined 2° and 3° structures. *Quaternary structure* denotes the manner in which individual chains of a protein having more than one subunit are arranged in space. Proteins possessing more than one chain are known as oligomers; their component chains are called subunits or protomers.

enhances or interferes with the assumption of these types of 5° structures will influence the concentration of *functional* enzyme present in the cell.

Once an enzyme has attained its functional state, following transcription, translation, and assembly, how does it work and how is it regulated?

Enzyme Function Is Inseparable from Enzyme Structure

Even for the simplest kinds of catalyzed reactions enzymes must bind substrates with high and closely regulated affinities; they must change conformation in such a way as to bring the reactants and participating groups on the enzyme into a reactive configuration (the transition state), and in many cases move parts of the enzyme aside to allow dissociation of products. To choose a specific example, that of M_4-LDH, about one-third of the structure of each subunit is involved in forming the binding site for NADH; once NADH is bound, a polypeptide loop closes over the coenzyme to form a final binding pocket of the enzyme, thus allowing the binding of pyruvate and the initiation of the catalytic cycle (Holbrook et al., 1975).

Saturation Kinetics

Simple enzyme-catalyzed reactions can be described in several discrete steps:

$$E + S \rightleftharpoons ES \rightleftharpoons ES^{\ddagger} \rightleftharpoons EP \rightleftharpoons E + P$$

where S is substrate, E is enzyme, ES is enzyme-substrate complex, $ES^{\ddagger}$ is the activated complex, EP is the enzyme-product complex, and P is the product. The velocity of such enzyme-catalyzed reactions displays a critical dependence upon enzyme concentration as well as upon substrate concentration. When these are varied experimentally, it is found that the initial reaction velocity (v) is directly proportional to enzyme concentration, $[E_0]$. That is as it should be, for *in vivo* it means that higher rates of metabolic reactions occur when enzyme concentrations are elevated. However, v generally follows saturation kinetics with respect to the concentration of substrate, [S] (Fig. 3-1): at low [S], v increases in linear fashion with [S], but as [S] is increased this relationship begins to break down and v increases less rapidly than [S] until at saturating [S], v reaches a limiting value termed V_{max}.

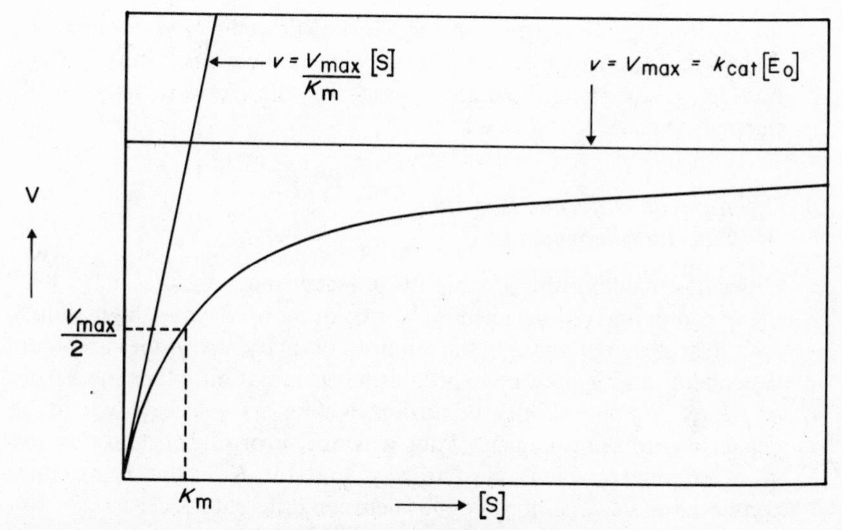

Fig. 3-1. Reaction rate plotted against substrate concentration for a reaction obeying Michaelis-Menten (or saturation) kinetics

These relationships between [S] and v are described by the Michaelis-Menten equation,

$$v = \frac{V_{max}[S]}{K_m + [S]}$$

The term K_m, the Michaelis-Menten constant, is the concentration of substrate at which $v = \frac{1}{2}V_{max}$.

In Vivo Values of [S] *and* K_m *Are Similar*

As discussed in various sections of this volume, it is widely observed (in studies of different tissues and species) that for enzymes displaying saturation kinetics, *in vivo substrate concentrations are either similar to the value of the* K_m *or are actually lower*. This is so for two reasons. In the first place, such an arrangement means that small changes in substrate concentration lead to relatively large changes in reaction velocity—and that is one way in which reaction velocities and substrate levels can, therefore, be regulated. The second reason is conceptually a mirror of the first: at any given low value of [S], a small change in the value of the K_m leads to a relatively large change in reaction velocity (Fig. 3-1). This, too, as we shall see, supplies an *in vivo*

CAMROSE LUTHERAN COLLEGE
LIBRARY

means for regulating reaction velocities. Michaelis-type enzymes, by responding to changes in [S], serve well in their basic job of transmitting carbon along the pathway, and they also serve to stabilize [S] fluctuations.

Affinities of Different Enzymes May Be Interdependent

Because metabolism is a highly branched networklike affair, with many compounds being acted on by two or more enzymes, high affinity for substrate is not enough; the affinities of different enzymes are interdependent, and it may be expected that a mutation either increasing or decreasing the affinity of an enzyme for its substrate would be deleterious to the organism. Thus it is not surprising to find, for any given enzyme or metabolic pathway, that the K_m value is carefully conserved in enzyme homologues between different species.

The Need for Isozymes

Any given metabolic reaction may be used in different metabolic "fields" depending upon the metabolic patterns and capacities of different tissues. But because the affinities of different enzymes may be interdependent, the preferred value of the K_m for any given enzyme may not be the same in each tissue. The metabolic constraints or limits on substrate affinity in each tissue may be unique. For example, some tissues may need to make a given metabolite; others may need to use it. Where the chemical reaction is the same in both tissues, it may be catalyzed by different forms of the same enzyme (isozymes), each kinetically specialized for, and locked into, its particular metabolic field. Kinetic specializations of isozymes may be especially important in terms of regulatory properties, for different isozyme forms of a given class of enzyme may need to be attuned to quite different activity-regulating signals in different tissues. We illustrate this requirement with numerous examples in subsequent chapters.

Effectiveness of Michaelis Enzymes in Stabilizing [S] Fluctuations

Metabolic homeostasis arises in large part from the stabilization of substrate/product concentrations by appropriate changes in fluxes through metabolic sequences (see Chapter 4). For example, should a substrate concentration rise, the extent of the increase is limited by the consequent increase in the rates of the enzyme reaction or reactions utilizing it. But for enzymes showing saturation kinetics, the velocity response decreases with increasing substrate concentration. That is, the effective kinetic order is less than unity and falls to zero as [S] approaches saturation. Such enzymes become less and less effective, when they might be most needed (during pulses of high [S]). Therefore they are not the most effective in holding [S]/[P] ratios steady by increase in rate. The need to overcome this deficiency of Michaelis-type enzymes, perhaps more so than any other single requirement, seems to underly further refinements in enzyme catalysis and control.

Multienzyme Complexes in Stabilizing [S]

One way of overcoming the problem is to incorporate all enzyme components of a given pathway into a single large assemblage, a kind of super-enzyme, as in the case of fatty acid synthetase, the enzymes of β-oxidation, and pyruvate dehydrogenase. In such a system at its most efficient, the product of one enzyme may be directly delivered as substrate to the next enzyme in the metabolic sequence. Although such an assemblage has other advantages (e.g., compartmentation, reduced competitiveness for common substrates with other metabolic reaction pathways, etc.) it is easy to visualize how the component enzymes could operate at vanishingly low substrate levels, where the effective kinetic order for Michaelis-type enzymes is at its highest. Nevertheless, even under these conditions the kinetic order for Michaelis-type enzymes is maximally still only unity. Enzymes with kinetic orders of greater than one would clearly be the most effective in stabilization of [S]/[P] ratios by appropriate rate changes. It is this advantage that probably best explains the origin and development of regulatory enzymes: *enzymes that bind substrate cooperatively and hence catalyze reactions of kinetic order much greater than one.* Although, as we shall see, these enzymes play other roles in metabolic regulation, their first and perhaps most fundamental contribution to homeostasis is their increased responsiveness to changes in substrate concentration.

Positive Cooperativity: An Improved Way of Stabilizing [S] *Fluctuations*

Most metabolic enzymes are composed of multiple polypeptide subunits and therefore have multiple binding sites. In enzymes showing Michaelis kinetics, these different binding sites do not significantly influence each other. For many regulatory enzymes, however, this is not the case; for these, a sigmoidal or S-shaped saturation curve is an operational hallmark. Such sigmoid saturation kinetics derive from cooperative interactions between binding sites, binding at the first site increasing affinity for substrate at the second, and so forth. This positive cooperativity, then, is the basis for sigmoidal saturation kinetics and derives directly from subunit-subunit interactions in regulatory enzymes. Positive cooperativity is also known as a positive homotropic interaction to denote that the binding of one substrate molecule facilitates binding of additional molecules of the same type. Heterotropic interactions refer to cooperative binding interactions between unlike molecules, e.g., between substrates and regulatory modulators (see below).

The original attempts to explain positive cooperativity were based on hemoglobin. Hemoglobin is composed of two pairs of chains, α and β, arranged in a symmetrical tetrahedral manner, each containing one O_2 binding site or heme group. The hemoglobin binding curve may be well explained by assuming four successive binding constants, with the affinity of the fourth being over one hundred times greater than the affinity for the first. Such an increase in affinity with increasing saturation cannot be explained by four non-interacting sites of differing affinities. If this were the case, the high affinity sites would fill first so that the partially liganded molecules would be of lower affinity than the free deoxyhaemoglobin; experimentally the reverse of course occurs. Therefore the increase in affinity with increasing saturation must be due to the sites *interacting* so that binding at the first sites filled causes an increase in affinity at unfilled sites. However, the O_2 binding sites are too far apart to interact directly, so on oxygenation there are required, and indeed observed, changes in the quaternary and tertiary structures of hemoglobin, and it is these long-range adjustments that modify binding affinities.

The kind of detailed structural information that is available for hemoglobin is only now reaching completion for regulatory enzymes showing these effects; consequently, structural explanations of sigmoidal kinetics for enzymes are thus far very general. Three widely used models are those of a) Monod, Wyman, and Changeux, b) Koshland, Nemethy, and Filmer, and c) Eigen, which are described in more detail elsewhere (Fersht, 1977). When structural studies are complete they will supply us with mechanistic explanations of positive cooperativity; functionally it is possible to get a measure of cooperativity between different binding sites by use of the Hill equation:

$$\log [v/(V_{max} - v)] = h \log [S] - \log K$$

The Hill equation (which is sometimes called the Hill plot and is usually linear) satisfactorily describes the binding of ligands to regulatory enzymes in the region of 50 percent saturation (10 to 90 percent). Outside this region the experimental curve deviates from the straight line. The value of h found from the slope in the region of 50 percent saturation is known as the *Hill constant.* It is a measure of cooperativity. The higher h is, the higher the cooperativity. At the upper limit h may equal n, the number of binding sites. If $h = 1$, there is no cooperativity; if $h > 1$, there is positive cooperativity; if $h < 1$, there is negative cooperativity.

The Symbol $(S)_{0.5}$

For enzymes showing Michaelis kinetics, we described the K_m as the substrate concentration at 50 percent of maximal velocity; the symbol $(S)_{0.5}$ is used to define the analogous substrate concentration for enzymes displaying sigmoidal saturation curves. For an enzyme catalyzing a simple first-order reaction, $(S)_{0.5}$ is identical to K_m, but the latter is not applicable for cases involving cooperative substrate binding. Another way of describing $(S)_{0.5}$ is as the substrate concentration when $\log [v/(V_{max} - v)] = 0$. As in the case of the K_m, the $(S)_{0.5}$ constant derived from kinetic studies yields useful information on enzyme-substrate affinities. And as with the K_m, *physiological substrate concentrations fluctuate in the range of* $(S)_{0.5}$ *for the regulatory enzymes utilizing them.* It is no accident that this is so, for it clearly is the outcome of adaptational tuning of enzyme-substrate affinities for allowing large amplified change in reaction velocity in response to modest changes in substrate concentration. It is this characteristic that fits these kinds of enzymes particularly well to the job of stabilization of substrate level and to the flexibility needed in rate changes from basal to maximal levels.

Negative Cooperativity and Why It Should Exist

In contrast to enzymes showing positive cooperativity in binding substrate, some enzymes exhibit negative cooperativity (negative homotropic interactions) with kinetic orders less than one, i.e., in the Hill equation for such enzymes, $h < 1$. We can easily visualize the homeostatic advantages of positive cooperativity, but why negative cooperativity? What is its functional significance? This can be readily explained using phosphofructokinase as an example.

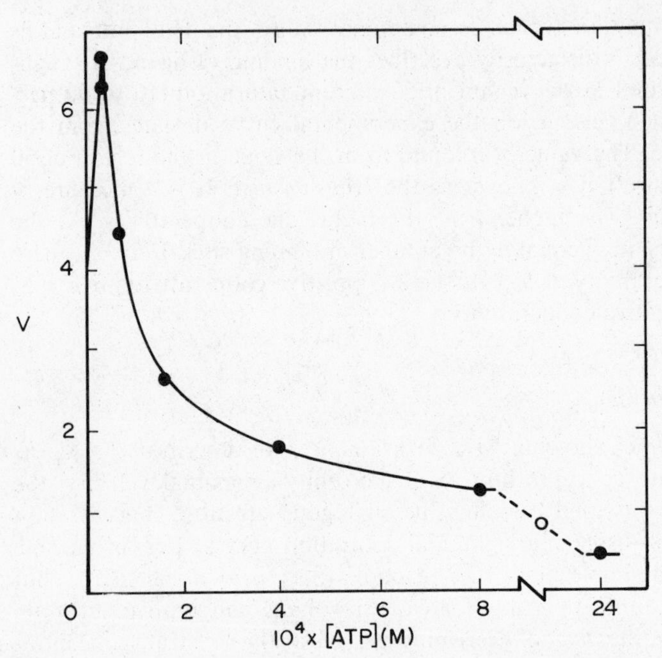

Fig. 3-2. Rate of the reaction catalyzed by yeast phosphofructokinase as a function of the concentration of ATP in the absence of AMP. Modified from Atkinson (1977)

Phosphofructokinase (PFK) catalyzes a transphosphorylation reaction

$$\text{ATP} + \text{F6P} \rightleftharpoons \text{ADP} + \text{F1,6BP}$$

As the reaction is a step in glycolysis its overall function is in the production of ATP during the oxidation or fermentation of glucose. It is, therefore, appropriate that it should be regulated, not as an ATP-utilizing reaction, but rather as a key step in an ATP-regenerating sequence, a Block I function. Not surprisingly, in the physiological range of ATP concentrations, the reaction has effectively a negative kinetic order with regard to this reactant (Fig. 3-2). Again, it is not an accident that this is so, but rather it is the outcome of highly sophisticated adaptations of the catalyst to its *in vivo* functions. These adaptations involve adjustments in the oligomeric structure of PFK allowing the development of two kinds of ATP binding sites, one catalytic and one regulatory, with appropriate adjustments in affinity at each. In the case of the catalytic site, the affinity for ATP is high so that under physiological conditions it will always be fully saturated. Interestingly, under *in vitro* conditions, ATP saturation curves are hyperbolic while those for the cosubstrate F6P are strongly sigmoidal.

On the other hand, the oligomeric structure of PFK includes regulatory subunits with ATP binding sites of lower affinity. Binding ATP at the regulatory (noncatalytic) sites causes a decrease in affinity for the cosubstrate F6P at the catalytic site. As a result, when ATP concentrations are high (or when ATP/ADP concentration ratios are high), the cell is not in need of more energy in the form of ATP and PFK activity is low; as the availability of ATP falls it serves automatically as a signal calling for activating PFK when the cell is in need of more energy.

How can one be so confident that this is the outcome of adaptation? One way to demonstrate this is to show that such catalytic and regulatory characteristics of PFK develop only when needed and are not seen in PFK homologues that serve other functions. Fortunately, at least one such example is known in the slime mold. For slime mold cells in late stages of aggregation, when the mold utilizes amino acids and proteins as carbon and energy sources, glucose is not essential for energy metabolism. Most exogenous glucose in fact is converted to glycogen. The role of glycolysis and therefore PFK here is to supply the cell with trioses and acetate units for biosyntheses. Kinetic studies of this kind of PFK indicate that none of the above characteristics are exhibited; in these cells, PFK in fact displays classical Michaelis kinetics. These are instructive observations for they clearly establish that the enzyme affinities for substrates and indeed for regulatory ligands are not fixed by chemical necessity; they clearly are sensitive to selection and adaptation (see Hochachka, 1980).

Allosterism as a Means for Controlling Enzyme Activity

To this point we have used the term "regulatory enzyme" to mean any metabolic enzyme which at least under some circumstances may display sigmoidal substrate saturation kinetics. That is fine as far as it goes. However, as already indicated in outlining negative cooperativity in PFK, such enzymes almost always are also characterized by being strongly influenced by modulators. For most regulatory enzymes, modulators are chemically unrelated metabolites which at increasing concentrations serve as signals for activation or inhibition of their target enzymes. Regulatory enzymes therefore must exhibit modulator binding sites as well as substrate binding sites. By analogy with substrate saturation, modulator binding sites may also display cooperativity, and modulator saturation curves are also often sigmoidal in nature. As with substrate saturation, the $(M)_{0.5}$ is defined as the modulator concentration for achieving half-maximal activation or inhibition of the regulatory enzyme. As with substrate binding, enzyme-modulator complexes are stabilized by weak (or noncovalent) chemical bonding, which, as we shall see later, makes this an important site of selection not only in metabolic adaptation but in adaptations to the physical environment as well.

As we have already implied, the modulator compound need not be at all chemically related to the substrate. For example, AMP is a positive modulator for glycogen phosphorylase, phosphofructokinase, and citrate synthase, among other enzymes; yet it is chemically related to the substrate (ATP) of only one of these enzymes (PFK). Moreover, the modulator binding site is spatially separated from the substrate binding site, often in a different subunit. Regulatory enzymes are therefore often termed allosteric enzymes because the effector is structurally different from the substrate and binds at its own separate site. (The term homotropic is sometimes used to indicate interactions between identical substrate binding sites in contrast to heterotropic interactions between modulator and substrate binding sites.)

Allosterism is a useful term and concept because it emphasizes the important difference between metabolite control by direct interference at the substrate binding site (e.g., competitive inhibition by product or by other substrate analogues) and metabolite control at a separate modulator binding site. On binding, allosteric modifiers influence enzyme-substrate affinity and catalytic activity by *long-range effects transmitted through the oligomeric structure of the enzyme*. Thus adaptations for such functions must adjust not only the substrate and modulator binding sites, but also the way catalytic and regulatory subunits of the enzyme interact with each other. Again, these subunit-subunit interactions are always stabilized by weak chemical bonding, which, therefore, makes them particularly sensitive sites of adaptation to the external physical and chemical environment as well as to the internal metabolic one.

How Modulators Regulate Catalytic Activity of Target Enzymes

Although, in principle, modulators can influence enzyme activity by changes in V_{max}, in $(S)_{0.5}$, or in both (examples of all three cases are known), the commonest mechanism for controlling regulatory enzymes is through changes in affinity for substrate (see Fig. 3-3, for example). This is because modulator-induced change in affinity for substrate leads

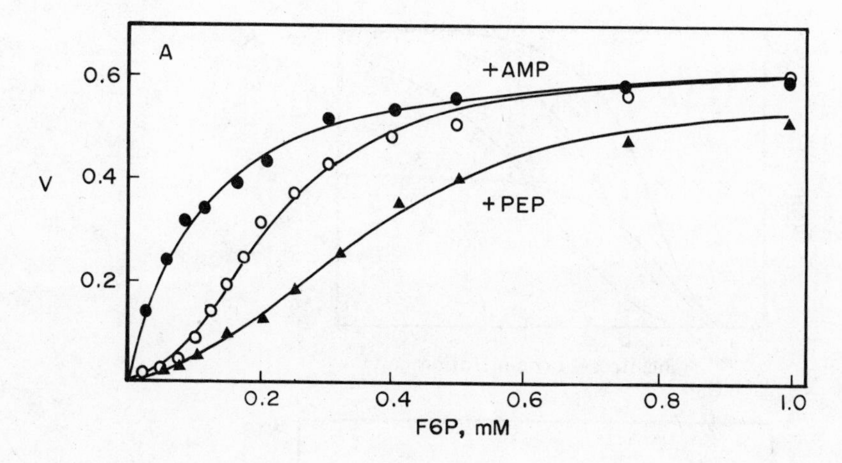

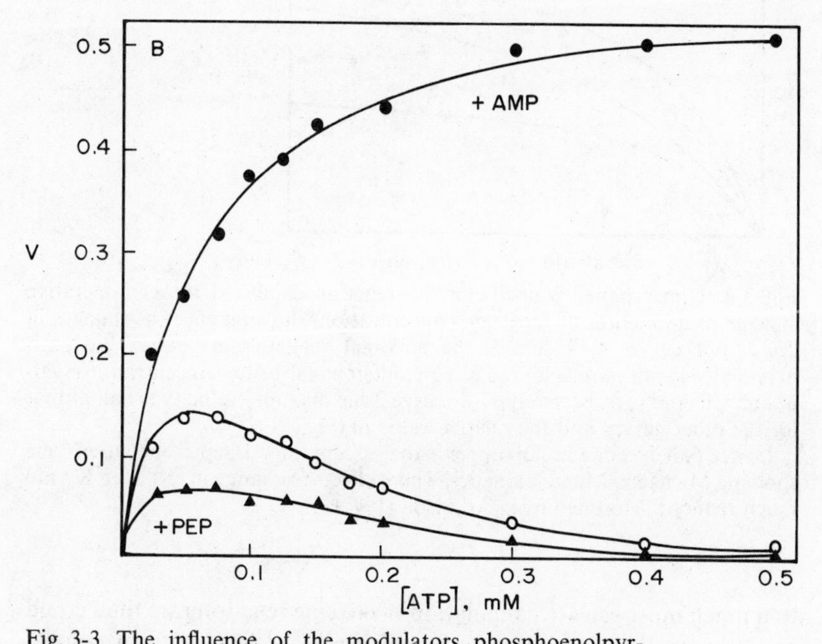

Fig. 3-3. The influence of the modulators phosphoenolpyruvate (PEP) and AMP on the regulatory kinetic behavior of *Mytilus* adductor PFK. Modified after Ebberink; see De Zwaan (1983) for further details and discussion.

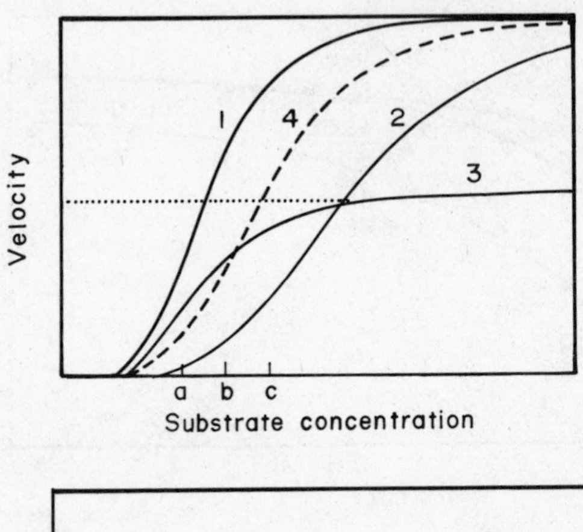

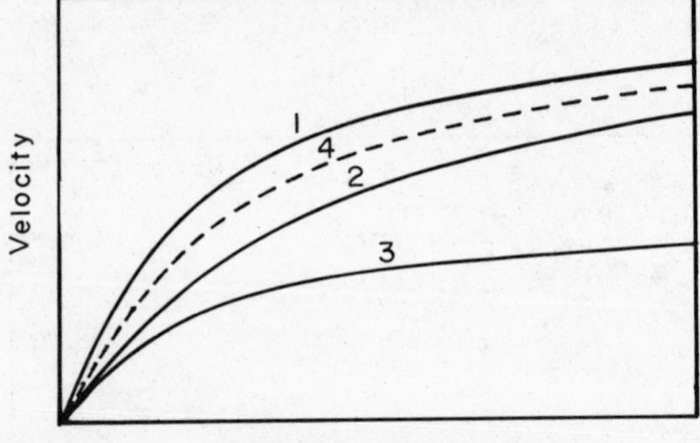

Fig. 3-4. Upper panel: Velocity of the reaction catalyzed by a cooperative enzyme as a function of substrate concentration, showing effect of changes in $(S)_{0.5}$. For curves 1, 4, and 2, the maximal velocities are equal, but half-maximal velocities are achieved at very different substrate concentrations (values of a, b, and c, respectively). For curve 3 the maximal velocity is half of that for the other curves and the relative value of $(S)_{0.5}$ is a.

Lower panel: Same as for upper panel except curves represent an enzyme showing Michaelis-Menten kinetics, where effects of change in [S] or in K_m are much reduced. Modified from Atkinson (1977)

to a much more sensitive regulation of enzyme reaction rate than could be achieved through lowering or raising V_{max}. Sensitivity of control is further facilitated by the higher-order reaction kinetics of regulatory enzymes. This is best illustrated graphically (Fig. 3-4) comparing Michaelis enzymes with regulatory ones showing strongly sigmoidal saturation curves: the same change in affinity for substrate that causes only about a 25 percent change in either direction for Michaelis-type enzymes in the case of regulatory enzymes changes the velocity from more than twice to less than one-third of control rates. Thus it is obvious that control can be much more sensitive and precise for regulatory than for nonregulatory enzymes, but the advantage of regulatory enzymes is even greater because of the way they respond to modulator signals.

This is because regulatory enzymes nearly always bind their modulators cooperatively and a twofold change in substrate affinity can be readily caused by a change of only a few percentages in modifier concentration. Thus cooperative binding increases both the extent of change in substrate affinity that results from a given change in modifier concentration and also the extent of the change in reaction velocity that results from a given affinity change. The multiplication of these two effects makes metabolite control many times more sensitive than it could be in the absence of cooperative binding of substrates and modifiers.

The Need for Bypass Enzymes

A characteristic of many (but not all) regulatory enzymes is that the reaction they catalyze proceeds with a large drop in free energy, so that there is a significant thermodynamic barrier to reversal. This problem *in vivo* is further amplified by the fact that these reactions are usually held far from equilibrium and therefore the chemical potential strongly favors predominantly unidirectional function. Yet physiological conditions may arise requiring the reverse reaction. There could be kinetic solutions to this problem. Liver glyceraldehyde-3-phosphate dehydrogenase, for example, must function in glycolysis under some conditions and in gluconeogenesis under others. One isozyme does the job in both directions and its kinetic properties have been adjusted accordingly (Smith and Velick, 1972). But this is unusual; it probably works in this case because of the redox nature of the reaction and by virtue of its being closely coupled to a kinase, phosphoglycerate kinase, which utilizes ATP during gluconeogenic function, but makes ATP during glycolysis. The more usual situation is to use a different chemical reaction for the physiologically reverse process. The metabolism of F6P and FBP supplies an excellent example.

During glycolysis, PFK, of course, catalyzes the conversion of F6P to FBP but in tissues capable of *de novo* carbohydrate synthesis, the reverse reaction is required. Neither the regulatory nor catalytic properties of PFK are suitable for *in vivo* reversal; the ratio of the forward to the back reaction under saturating conditions, for example, is about

400, and no combination of modifiers alters this situation significantly. The solution to this problem is the development of a bypass enzyme, catalyzing a completely different chemical reaction that itself is thermodynamically highly favorable and which generates F6P from F1,6BP: the enzyme fructose bisphosphatase, FBPase. This is a hydrolytic reaction (not a transphosphorylation, as in the case of PFK) and proceeds with an overall free energy change of about −4 kcal/mole:

$$\text{F1,6BP} + H_2O \rightleftharpoons \text{F6P} + P_i$$

The enzyme occurs in high activity in gluconeogenic tissues (liver and kidney), in low activity in tissues such as white muscle, which is capable of low rates of *de novo* glycogen synthesis, and not at all in tissues incapable of such biosynthesis. Numerous other examples of bypass enzymes are known, some of which will be discussed further in later chapters.

The alert reader will realize, however, that in solving the problem of kinetically and thermodynamically irreversible steps in metabolism with bypass enzymes, a new problem is created: simultaneous activity of both enzymes. The problem is one of control, for when both sets of enzymes working in opposite directions are present in the same cell compartment, their simultaneous function could set up a carbon (and often energy) short-circuit.

Organisms have invented numerous solutions to this problem. One solution is to separate the enzymes spatially. This occurs, for example, in the case of fatty acid oxidation and synthesis, the former occurring in the mitochondria, the latter in the cytosol. Similarily, in some species (trout, pigeon, rabbit, frog, lizard), the gluconeogenic bypass reactions around pyruvate kinase are located in a different cell compartment: PK is soluble, cytosolic in location, while pyruvate carboxylase and PEPCK are mitochondrial. Such a spatial separation should greatly assist in minimizing simultaneous function of oppositely directed metabolic reactions and pathways, but it is not the only mechanism used.

The commonest solution to this problem is to utilize opposing control mechanisms on the opposite partners in such a metabolic situation. In the case of the F6P–FBP interconversion, for example, at least *one positive modulator for PFK should be shared by FBPase as a negative modulator* (Fig. 3-5). That modulator turns out to be AMP, which strongly activates PFK, but is a potent inhibitor of FBPase. More recently, fructose 2,6 bisphosphate has been found to be an inhibitor of FBPase while simultaneously serving as the most powerful PFK activator thus far discovered (Uyeda et al., 1981).

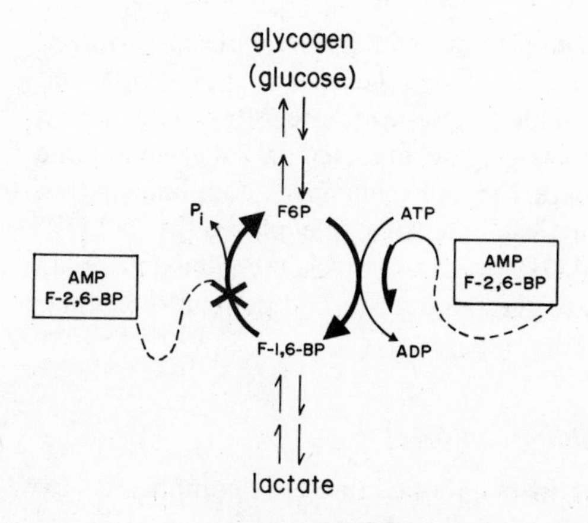

Fig. 3-5. Summary diagram of two key regulatory mechanisms in the PFK-FBPase catalyzed interconversion of F6P and Fl,-6BP. Activation shown with a dark arrow; inhibition with a dark cross.

As with other enzyme traits that are subject to adaptation, FBPase binding of AMP is also adjustable. In ectothermic organisms (or in ectothermic tissues in mammals and birds), the binding of AMP is strongly perturbed by temperature and adjustments must be made accordingly. In one group of organisms, the bees, flight muscle FBPase appears to contribute to overall thermogenesis; under some conditions, there seems to have been a selection for an energy short-circuit at this site in metabolism. That is why it is only in this group of organisms that the AMP control on the enzyme is absent (see Hochachka, 1974). The observation that AMP control of FBPase can be lost, however, is an insightful one, for it again boosts our confidence that this, like other regulatory properties of enzymes, is a product of adaptation. When it is useful, the character is selected; when it is no longer functional, it can be dropped and other mechanisms for influencing the enzyme's activity may be adopted.

Fine versus Coarse Mechanisms in Enzyme Regulation

Modulator-mediated control of regulatory enzymes supplies the cell with perhaps its most sensitive means for regulating metabolism. These are "fine tuning" mechanisms in cellular metabolism because they are so highly specific. Each signal is often appropriate in only one metabolic setting, often especially useful in only one, or at most a few, tissues.

The reaction pathways that make up cellular metabolism, however, must also be closely integrated with the general energy status and redox conditions of the cell. This coarse-level integration is achieved primarily by the adenylates, which are involved in every metabolic pathway in the cell. These are the cell's universal coupling agents, and except for a similar but less extensive role played by NAD^+/NADH and the $NADP^+$/NADPH redox couples, the adenylates play an especially important role in the integration and control of cellular metabolism

Adenylate Ratios in Metabolic Control

The role of the adenylates in metabolic control is complicated by the fact that they are all interconvertible. Firstly, ATP and ADP are interconvertible through

$$2ATP + H_2O \rightleftharpoons 2ADP + 2P_i$$

in kinase reactions, in the electron transfer system and so forth. AMP enters the picture through the adenylate kinase reaction

$$AMP + ATP \rightleftharpoons 2ADP$$

Atkinson (1977) has championed the view that in terms of metabolic control, concentration changes of individual adenylates are of lesser significance than are changes in their ratios. For convenience, and to emphasize the point, he introduced the concept of the adenylate energy storage system and energy charge: the system is fully charged when all the adenylates are present in the form of ATP, fully discharged when only AMP is present. Energy charge is formally defined as the ratio

$$\frac{[ATP] + 1/2\,[ADP]}{[ATP] + [ADP] + [AMP]}$$

and so defined is a linear measure of useful energy stored in the adenylate system.

Many regulatory enzymes possess specific binding sites for the adenylates but the manner in which they respond to change in the energy charge depends upon the pathways in which they function. If they participate in ATP-generating pathways, they tend to be inhibited by high values of energy charge, a situation analogous to a feedback negative inhibition. This is true, strikingly enough, even if the reaction itself consumes ATP, as in our PFK example above. There is no chemical necessity underlying such responses; they are true biological adaptations, expressed because they are metabolically necessary or useful.

Regulatory enzymes that function in pathways or processes utilizing ATP, on the other hand, respond to the adenylate energy charge in the opposite manner. ATP utilizing reactions or processes are most favored under energy-rich conditions in the cell and hence when energy charge is high, they should be favored. Again, the nature of the response is determined by biological adaptation, not by chemical necessity.

In the above treatment, concentration changes in P_i are ignored even if they are coincident with changes in ADP and ATP levels. Atkinson (1977) argues there are good reasons for this. In a closed system, where the sum of the phosphate occurring in the adenylates and as free P_i is constant, the concentration of free P_i clearly serves as a measure of the charge of the system. A decrease in P_i corresponds to an increase in energy charge. However, *in vivo* the system is not closed and there are many other phosphorylated compounds in quasi-equilibrium with the P_i pool associated with the adenylates. That is why the concentration of P_i cannot yield a useful measure of the energy charge.

Although that is true, it does not follow that P_i cannot play an important regulatory role in cellular metabolism. Its activating effect upon PFK was long ago noted to be exactly what would be expected and is entirely analogous to PFK activation by AMP or lowered energy charge. Whereas a role for P_i in energy charge control of cytosolic metabolism may be minor, it appears to be more involved in the regulation of electron transport system (ETS) function. One theory of respiratory control, for example, proposes that flux through the ETS is more tightly regulated by the ratio of $[ATP]/[ADP]\,[P_i]$, or the so-called phosphate potential than by energy charge, with $[P_i]$ playing a particularly pivotal role (Wilson et al., 1979).

The above fine and coarse control mechanisms allow regulatory enzymes such as PFK to accommodate changes in flux through the pathway in which they operate; sometimes, these enzymes are indeed rate-limiting. However, they appear to be rarely utilized to *initiate flux in the pathway*. The function of initiating flux is usually vested in enzymes strategically positioned at the beginning of major metabolic sequences, or sometimes at major metabolic branchpoints. Because these enzymes are pivotal in initiating flux in the pathways in which they operate, they have been termed "flux generators" by Newsholme (1978).

Flux-Generating Enzymes

There are two fundamental features of flux-generating enzymes. Firstly, such enzymes must incorporate mechanisms for maintaining themselves in either 'off' or 'on' states. And secondly, substrate supply to them cannot be limiting, i.e., they must always be fully saturated when active. If these conditions are met, the flux of carbon through the reaction will always be *directly proportional to enzyme activity* (i.e., to the concentration of enzyme in the active form): *the higher the proportion of active enzyme, the higher the flux through the reaction and into the pathway the reaction initiates.* All other enzymes along the pathway will then respond according to the chemical potential (substrate/product ratio, etc.) changes that are consequently generated.

The enzyme system for which these features are best understood is glycogen phosphorylase, which initiates glycolytic flux. It occurs in active and relatively inactive forms, and the conversion of one to the other is closely regulated. In most tissues (liver, muscle, etc.) glycogen and P_i both occur at concentrations manyfold higher than their respective K_m values. Thus, the chemical potential for the system always favors glycogenolytic function and the only additional provision needed is an activated form of glycogen phosphorylase. Turning "on" the enzyme, then, is tantamount to initiating flux into glycolysis, which is why such enzymes are termed flux generators. Because of their pivotal role and position, such enzyme systems are marvels in design of enzymes for controlled function *in vivo.* Not only are such enzymes controlled by intracellular mechanisms, they are also subject to extracellular signals, allowing control of function appropriate to the needs of the cell, the tissues, and the whole organism. Control at this level is extremely complex, involving enzyme cascades (enzymes controlling the activities of other enzymes), phosphorylation-dephosphorylation mechanisms for turning enzymes "on" and "off," and extracellular hormonal activation or inhibition mechanisms. The sheer complexity of such systems assuredly removes the possibility of their arising by chemical necessity instead of adaptational expediency: such control networks are biologically designed systems and cannot be explained as inevitable outcomes of their chemistry.

Control of Flux Generators: The Glycogen Phosphorylase Model

A simple but basic design feature of flux-generating enzymes is position, for enzymes such as glycogen phosphorylase are strategically positioned in three senses of the term. Firstly, they are positioned at the entry step into the pathway in which they operate, ideal sites for controlling functions. Secondly, not only are they metabolically located at the beginning of the pathway, but at least in muscle and probably liver they are also physically plated out on their substrate, being loosely associated with intracellular glycogen granules. Thirdly, by being subject to extracellular hormonal control, enzymes such as glycogen phosphorylase interface cellular metabolism with the metabolism and needs of the whole organism.

Although many flux-generating enzymes regulated closely by internal and extracellular signals are known, not all control aspects of all of them have been worked out. Of those that have been studied in detail, the information available for glycogen phosphorylase is most complete and so we will use it to illustrate how at least one such enzyme operates *in vivo*.

Control of Glycogen Phosphorylase Begins with Extracellular Stimuli

Organisms will often have imposed upon them external reasons for initiating glycogenolysis at specific internal sites, the classic example being the "fight or flight" syndrome. Hence, it would be advantageous if both hormonal and neuronal signals were usable for activating glycogenolysis in tissues such as muscle. The overall outlines of how this is achieved are now fairly well understood and involve cyclic AMP or calmodulin or both as special kinds of intracellular second messengers. Cyclic AMP is formed from ATP by a membrane-bound enzyme termed adenyl cyclase. This enzyme is specifically activated by various hormones. In the case of epinephrine, the hormone arrives in the blood at levels of about 10^{-9} M and becomes bound to specific receptor sites on the outer surface of cell membranes, leading to an activation of adenyl cyclase activity on the inner surface of the cell membrane. The active form of adenyl cyclase catalyzes the conversion of ATP to cyclic AMP which may attain concentrations of 10^{-6} M within the cell.

Cyclic AMP does not act directly upon glycogen phosphorylase. Rather, it binds to the regulatory subunit of protein kinase, releasing its catalytic subunit in an active form (Fig. 3-6). This kinase in turn catalyzes the phosphorylation of an enzyme whose substrate is inactive glycogen phosphorylase *b*. The latter reaction is catalyzed by phosphorylase kinase, which requires Ca^{2+} for activity. It in turn catalyzes the ATP-dependent phosphorylation of four serine residues on phosphorylase *b* to generate the active form of the enzyme, phosphorylase

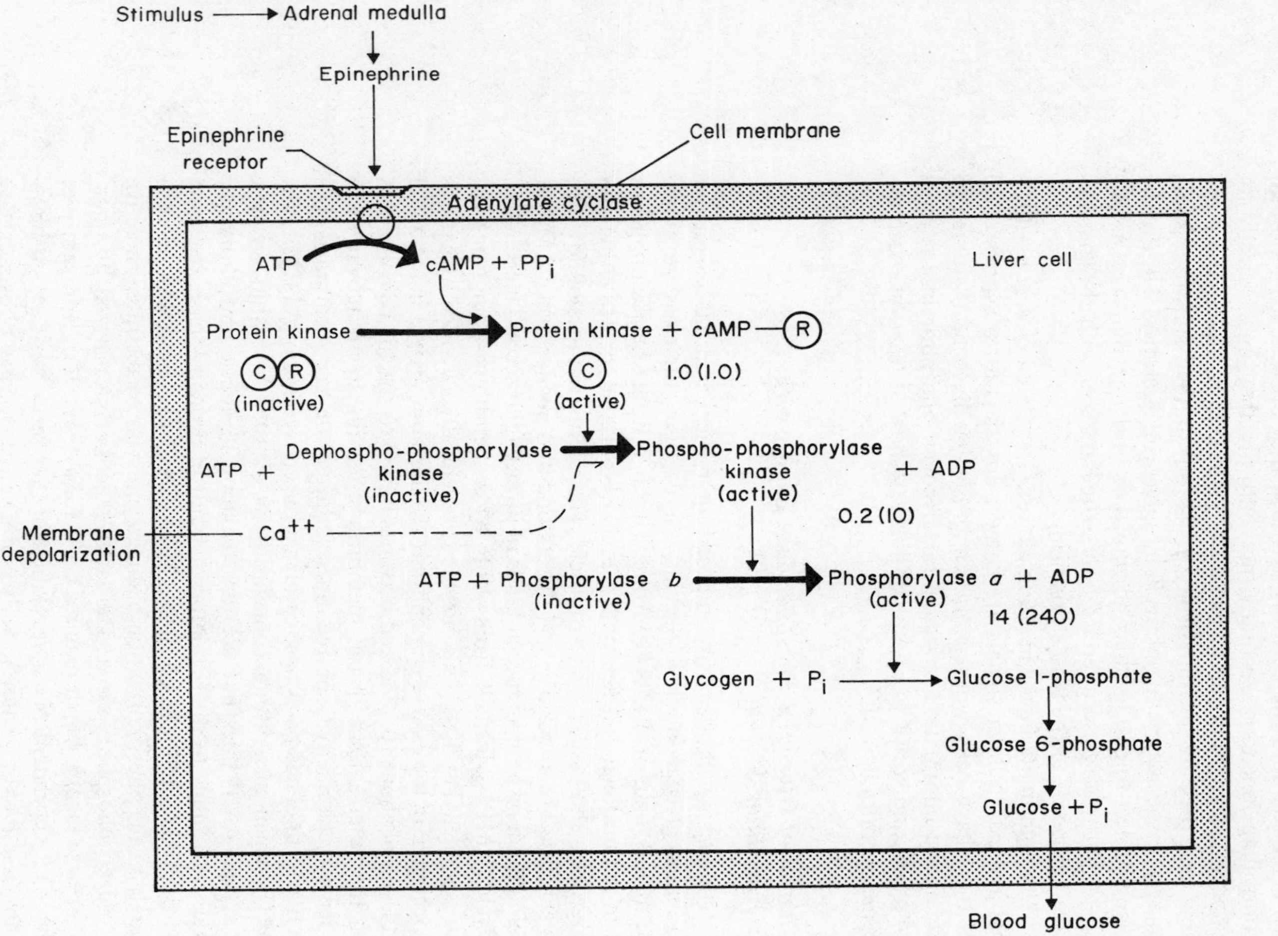

Fig. 3-6. Control cascade for activation of glycogen phosphorylase. The molar ratios of protein kinase, phosphorylase *b* kinase, and phosphorylase *a* are shown for liver and, in parentheses, for skeletal muscle. Summary diagram

a. In normal tissues, phosphorylase *a* is the form of the enzyme involved in glycogen breakdown in the presence of P_i to yield GlP. This kind of process occurs in muscle and liver, as well as in other tissues (heart, kidney, etc.). In liver, the concentration of glucose being released into the blood is in the 5 mM range. Thus an important outcome of this cascade type of control is the amplification of an incoming signal at 10^{-9} M to a product released at 5×10^{-3} M; i.e., about a 5×10^6 fold amplification!

In addition to hormonal signals, in some tissues (particularly muscle) glycogen mobilization must be responsive to neuronal activation. For some time now, this has been known to involve Ca^{2+}, but only in recent years has it been discovered that Ca^{2+} modulation of the glycogen-mobilizing cascade occurs indirectly following binding to calmodulin.

Calmodulin is a small (16,700 MW) protein highly enriched with acidic amino acids; from both binding and structural studies it is known to form four binding sites with high affinity for Ca^{2+}. There are many calmodulin-binding proteins each specific either to a given tissue or function. In muscle, calmodulin is a subunit of phosphorylase kinase. The binding of Ca^{2+} to this bound calmodulin converts this kinase into an active form and thus sets the stage for phosphorylase activation and glycogen mobilization. The cascade for neuronal activation of glycogen mobilization thus has fewer components than in the case of hormonal activation (Fig. 3-6) and activation may be more rapid as a result.

How Adaptable Are Protein Phosphorylation Systems?

Protein phosphorylation systems such as the glycogen phosphorylase one are so widespread in animal tissues that they are being postulated as an essential framework upon which other control mechanisms and indeed metabolic pathways per se are incorporated. That this basic fabric is itself subject to modification and adjustment is indicated by the different ways in which it is utilized in specific tissues and by different kinds of "signals" that feed into it. The molar ratios of protein kinase:phosphorylase kinase:phosphorylase in white muscle and liver, for example, are 1:10:240 versus 1:0.2:14, respectively. Amplification of incoming "signals" (such as epinephrine) therefore may be species- and tissues-specific and the adjustment is dependent upon controlling both the amount of each enzyme in the cascade and the isozyme type (Cohen, 1978).

It is beyond the scope of this book to review cascade control mechanisms in detail. The interested reader is referred to recent reviews (Cohen, 1978). The point of emphasis, however, is that, more so than any other feature of enzyme function discussed to this point, *the enzymes in these various cascade series are at once catalysts and controlling elements. The point of catalysis is control and control is not possible without catalysis. These are enzymes literally sitting upon the backs of other enzymes, telling them what to do, when, and at what rates.*

Although the signals feeding into protein phosphorylation systems may or may not be cyclic nucleotide dependent, may or may not be Ca^{2+} and calmodulin sensitive, may or may not be influenced directly by hormones, the actual control circuitry found in any given tissue is determined by criteria of functional utility. *These catalysts and controlling elements are, in other words, examples par excellence of enzyme adaptation to metabolic function.*

Enzymes as Editing and Proofing Mechanisms

Enzymes as controlling elements are also very well illustrated in the extraordinary fidelity with which macromolecular synthesis is achieved. DNA replication in *E. coli*, for example, is known to proceed with only one mistake in about 10^8–10^{10} nucleotides polymerized and the overall error rate in transcribing DNA and translating the message into protein is only about 1 in 10^4 amino acid residues incorporated. Fersht (1980) has emphasized that such fidelity is beyond the theoretical thermodynamic limits for simple enzymes and is only possible because of the maintenance of *editing* or *proofreading mechanisms.*

General Strategy of Editing or Proofreading

Interestingly, an analogous strategy for assuring such high fidelity of the biosynthetic machinery is found both in protein formation and DNA replication. Enzymes in both processes, in addition to the synthetic active sites involved in polymerization, *also retain secondary hydrolytic active sites which are used to break down incorrect intermediates even as they are formed. The synthetic process is thus double-checked at each step so that errors may be removed before they are permanently incorporated.*

Parenthetically, it should be mentioned that editing is distinct from repair. The double-stranded DNA molecule, because of the complementary nature of the base pairing, is capable of being repaired after it is replicated by the excision of lesions in one strand followed by patching and sealing by a DNA polymerase and ligase. Editing, by contrast, checks the fidelity of the process *before* the biosynthesis is completed. Let us consider the editing problem in greater detail in the context of protein synthesis.

The Problem of Amino Acid Selection in Protein Synthesis

Amino acids are selected for protein synthesis in a two-step reaction composed of activation of the amino acid followed by its transfer to tRNA. Both steps are catalyzed by the same enzyme (an aminoacyl-tRNA synthetase), with each amino acid (AA) having its own specific enzyme (E^{AA}) and tRNA ($tRNA^{AA}$):

$$E^{AA} + AA + ATP \longrightarrow$$

$$E^{AA} \cdot AA\text{-}AMP + PP_i$$

$$E^{AA} \cdot AA\text{-}AMP \xrightarrow{tRNA^{AA}}$$

$$AA\text{-}tRNA^{AA} + AMP + E^{AA}$$

Theoretically, two selections must be made by the enzyme: *first the correct amino acid and then the correct tRNA.* In practice, tRNA recognition is not a problem; these are large molecules with adequate distinctive structural variations. On the other hand, amino acids often are so similar in structure that there are problems in distinguishing between them. The classic difficulty is the discrimination between an amino acid and its smaller homologue. For example, in the case of alanine and glycine, the active site of the alanyl-tRNA synthetase is large enough to accomodate alanine, so it must also be able to bind the smaller glycine. Similar problems arise with other amino acids and their smaller isosteric analogues (e.g., isoleucine and valine). A paradox thus arises, for experimentally it can be shown that *aminoacyl-tRNA synthetases do make the above predicted errors in selecting for similar amino acids* (*first equation above*); *yet the overall reaction* (*sum of both equations above*) *is very precise.*

The answer to this paradox (of how the accuracy of the overall reaction can be higher than the partial reaction) is found in the occurrence of an editing mechanism. This can be nicely illustrated with a specific example, comparing how isoleucyl-tRNA synthetase complexes with isoleucyl adenylate in the "correct" reaction, and with valyl adenylate formed in the "incorrect" reaction. Whereas the addition of $tRNA^{Ile}$

Table 3-1. Error Rates and Editing

Process	Observed error rate	Predicted error rate in absence of editing	Experimental evidence for editing
Replication DNA → DNA	10^{-8}–10^{-10}	10^{-4}–10^{-5}	Yes
Protein synthesis DNA → protein (overall)	$<10^{-3}$	10^{-1}–10^{-5}	Yes
DNA → mRNA (transcription)	$<10^{-3}$	10^{-4}–10^{-5}	No
E^{AA}:AA (amino acid recognition)	$<10^{-3}$	$\leq 10^{-1}$	Yes
E^{AA}:$tRNA^{AA}$ (tRNA recognition)	$<10^{-3}$	10^{-6}–10^{-8}	No

SOURCE: From Fersht (1980), with modification

to the correct complex leads to the formation of the Ile-$rRNA^{Ile}$, its addition to the "incorrect" complex leads to the quantitative hydrolysis of valyl adenylate:

$$E^{Ile}\cdot\text{-AMP} \xrightarrow{tRNA^{Ile}}$$

$$\text{Ile-}tRNA^{Ile} + E^{Ile} + \text{AMP}$$

$$E^{Ile}\cdot\text{Val-AMP} \xrightarrow{tRNA^{Ile}}$$

$$\text{Val} + tRNA^{Ile} + E^{Ile} + \text{AMP}$$

The result of the activation and the hydrolytic reactions is that in the presence of valine and $tRNA^{Ile}$, the isoleucyl-tRNA synthetase acts as an *ATP-pyrophosphatase, wastefully and of necessity, hydrolyzing ATP*. As in the case of the protein kinases discussed above, these controlling functions too are thus seen to be somewhat energy expensive, but the compromise (some cost in ATP equivalents for fidelity of biosynthetic machinery far beyond what is achievable on thermodynamic grounds alone [Table 3-1]), is an appropriate one; as discussed by Fersht (1980), it is also made in the editing of DNA replication.

How Effective Is Editing?

A simple set of calculations indicates how necessary the above compromises actually are. A typical protein in a microorganism such as

E. coli might be taken to have a molecular weight of about 110,000, containing some 1,000 amino acid residues. If editing mechanisms were not functional, the protein would be genetically unstable since the error rate for replicating its genome would be unacceptably high. Secondly, its structure would be very loosely organized, since each molecule would contain from a few to possibly 100 incorrect amino acids. If this is multiplied by the total number of proteins present in a microbial cell ($\sim$3,000), then it is evident that there would be few, if any, viable progeny per generation and those produced would be inefficient. Thus, without editing mechanisms, the present genetic code and amino acid building blocks clearly would be unable to sustain higher forms of life.

Enzymes as Protective Elements

There is one final way in which enzymes are used more in control than in metabolism per se; namely in the detoxification or destruction of potentially noxious compounds formed either in the mainstreams of metabolism or in unwanted side reactions. The best examples of such enzymes are those charged with the catalytic breakdown of oxygen-derived radicals which may be extremely harmful to the cell. The complete reduction of a molecule of O_2 requires four electrons and in a sequential univalent process three intermediates are encountered which are too reactive to be tolerated by living systems. The formation of these intermediates—1) the superoxide anion radical, 2) hydrogen peroxide, and 3) the hydroxyl radical—proceeds as follows:

$$O_2 \xrightarrow{e^-} O_2^- \xrightarrow{e^- + 2H^+} H_2O_2 \xrightarrow{e^- + H^+} OH\cdot \xrightarrow{e^- + H^+} H_2O$$
$$\hspace{2em}\downarrow$$
$$H_2O$$

Fluxes of O_2^-, generated enzymatically or photochemically, are able to inactivate viruses, induce lipid peroxidation, damage membranes, and kill cells. Paraquat, which increases the rate of production of O_2^-, is much more toxic under aerobic than under anaerobic conditions. Indications are that O_2^- is not itself the species that causes these effects, but is the precursor of a more potent oxidant, whose generation depends on the simultaneous presence of H_2O_2 and which then attacks DNA, membrane lipids, and other essential cell components (Fridovich, 1978).

Since O_2 can be reduced in a number of ways *in vivo*, it is interesting that in the most O_2 demanding of them all, the reaction catalyzed by cytochrome oxidase, neither superoxide anion radical nor H_2O_2 is generated, because of design constraints on this enzyme (Hill, 1981). However, similar constraints are not put upon most other cellular O_2 reductions and one of the principal cellular sources of O_2^- and H_2O_2 probably is the reaction of dioxygen with reduced ubiquinone (Hill, 1981). Moreover, numerous other reactions produce substantial amounts of O_2^- *in vivo*, including such apparently innocuous steps as the conversion of hemoglobin and myoglobin to methemoglobin and metmyoglobin (see Fridovich, 1978, for other examples of reactions generating O_2-derived noxious intermediates).

The simplest way of dealing with this problem is to minimize the amount of O_2 that is available for reactions that can release O_2^-; i.e., maximize the amount utilized by cytochrome oxidase. This simplest strategy is in fact utilized (Fridovich, 1978), but of itself is insufficient for two reasons: firstly, some O_2 is always left over for reduction by reactions that generate toxic intermediates. More importantly, *cytochrome oxidase does not in itself do anything* about O_2^- and H_2O_2. The cytochrome oxidase reaction alone does not produce or use such free radicals. So it's not surprising that other ways must be found for protecting the cell against these compounds. The primary defense against O_2-derived free radicals is provided by a group of enzymes *that catalytically scavenge the intermediates of O_2 reduction*. These include a) superoxide dismutases, which catalyze the conversion of O_2^- *to hydrogen peroxide plus O_2*; b) *catalases, which convert H_2O_2 to H_2O plus* O_2; and c) *peroxidases, which reduce H_2O_2 to water using a variety of reductants* (e.g. glutathione) available to the cell. These reactions are shown below:

$$O_2^- + O_2^- + 2H^+ \xrightarrow[\text{dismutases}]{\text{Superoxide}} H_2O_2 + O_2$$

$$H_2O_2 + H_2O_2 \xrightarrow{\text{Catalases}} 2H_2O + O_2$$

$$H_2O_2 + RH_2 \xrightarrow{\text{Peroxidases}} 2H_2O + R$$

It is important to point out that the defense arsenal does not include an enzyme scavenger for hydroxyl radical. From the pathways of formation (shown above) it will be evident that efficient removal of the first two intermediates (O_2^- and H_2O_2) prevents formation of the third, $OH\cdot$. This is fortunate, for the hydroxyl radical is so extremely reactive with many substances that its specific detoxification (via enzymatic conversion) would be impossible (Fridovich, 1978).

As may well be expected, the occurrence of these enzymes appears to be adjusted according to the needs of the organism. Anaerobic organisms typically have little or no superoxide dismutase; in facultatively anaerobic micoorganisms the enzyme is induced by the presence of O_2, while in obligate aerobes, the enzyme activity is always expressed.

The catalytic properties of superoxide dismutase, and its molar concentrations, are elegantly suited for *in vivo* scavenging function. Firstly, the reaction between O_2^- and the enzyme is first order with respect to both enzyme and substrate, in contrast to the uncatalyzed reaction (which is second order with respect to substrate). Secondly, the concentration of superoxide dismutase is high (about 10^{-5} M in most tissues), substantially higher than *in vivo* O_2^- concentrations. Fridovich (1978) has calculated that these factors in themselves allow the enzyme-catalyzed reaction to proceed about 10^9 times faster than the spontaneous uncatalyzed one, and thus efficiently maintain intracellular concentrations of O_2^- at a value substantially lower than 10^{-10} M! This role of superoxide dismutase in lowering the intracellular levels of O_2^- *in vivo* is coupled with catalases and peroxidases, which together function to maintain very low levels of H_2O_2. The lower the concentrations of these two reactive intermediates, the lower the chances that they will participate in the production of even more reactive species such as $OH\cdot$ or singlet oxygen. Thus, these defensive enzymes probably exert a synergistic effect in protecting respiring cells against the consequences of "self pollution" by undesirably high fluxes of O_2^- and H_2O_2 through the cell.*

Sites and Functions under Selective Pressure

It will now be evident that, *as catalysts and controlling elements*, enzymes must be designed for a great variety of functions. At their simplest metabolic jobs, enzymes are mainly catalysts; while at their most complex metabolic functions, they can be viewed as mainly controlling or regulatory elements. Thus it is convenient to summarize this chapter by looking upon enzyme functions along a spectrum of complexity, each stage displaying its own characteristics, which serve as means for achieving specific goals. Characteristics that appear to be sensitive to selective forces would include:

1. Catalytic efficiency (high k_{cat}) as a means for rapidly transmitting substrate along the pathway in which the enzymes operate.

* An especially poignant defense against O_2-derived free radicals is reported for sea anemones, SOD enriched 100-fold to protect themselves against endosymbiotic algae (see J. Dykens and M. Schick, *Nature* 297 [1982]: 579–580).

2. Adjustable enzyme-substrate affinities (K_m values equal to, or less than, *in vivo* substrate concentration) as a means for keeping [S] low and more importantly as a means for stabilizing fluctuations in [S].
3. Isozymes as a means for developing tissue-specific metabolic fields, for putting similar or identical chemical reactions to different metabolic purpose.
4. Positive cooperativity as an improved means for stabilizing substrate concentrations.
5. Multiple-enzyme complexes as an alternative means for stabilizing and controlling substrate-level fluctuations and as an improved means for regulating flow through specific metabolic pathways.
6. Allosteric enzymes and allosteric modulators as a means for fine control of rates and direction of function of metabolic pathways.
7. Stoichiometric and allosteric sensitivity to adenylates and adenylate concentration ratios as a means for general integration of all metabolic pathways through the energy status of the cell.
8. Flux-generating enzymes as a means for initiating rate changes at strategic positions in metabolism.
9. Protein phosphorylation systems sensitive to external hormones, cyclic nucleotides, and/or Ca^{2+}-calmodulin as a means for interfacing cell and tissue metabolism with overall whole-organism metabolic needs.
10. Enzymes with dual binding sites (dual functions) allowing editing and proofing during macromolecular biosynthesis.

Designing that much function into the structure of enzymes clearly should not be underestimated. It is an incredibly complex matter and a startling far cry from the "landing platform" conceptualization of enzymes commonly proposed only twenty years ago. Although the above list is an oversimplification (by intent), it does bring focus and emphasis on the idea of enzyme machinery as the raw material for adaptation. In this chapter we have examined from the perspective of qualitative enzyme function just how nature has molded that raw material into the workings of metabolism as we know it. In subsequent chapters, we shall examine in greater detail how this raw material has to be adjusted to the solution of specific environmental or biological problems imposed upon animal metabolism.

CHAPTER FOUR

Exercise Adaptations

Fundamental Strategies

Because of a tremendous interest in exercise metabolism and physiology on the part of scientist and layman alike, a large literature has developed outlining metabolic provisions for exercise and locomotion in a variety of animals including man; these studies supply us with one of the finest frameworks in which to illustrate process and mechanism in biochemical adaptation. Physiologists tell us that, in general, the power requirements of locomotion are a function of body size, velocity of locomotion, and mode of movement. Being big and moving fast are traits that demand high power output; being small and moving slowly is energy efficient. Highest velocity short-duration burst-type performance generally requires the most effort, much more than slower, steady-state movement. Walking and running costs are higher than flying costs which in turn are higher than the costs of swimming. In the vertebrates, and at least in some invertebrate groups, three fundamental biochemical adaptations give animals tremendous flexibility in terms of how fast and how long they can move. These involve:

1. the specialization of muscle ultrastructure and contractile machinery into two or more classes of muscle cells, each with different biomechanical properties (different twitch times, different strength, and so forth); and
2. the development of high anaerobic and high aerobic metabolic potentials in different fiber types, each typically being fueled by different substrate sources (carbohydrate in anaerobic fibers; fat or fat plus carbohydrate in others). In addition, there has occurred
3. the development of finely tuned mechanisms for control of metabolic output
    a) on a short-term basis (stimuli calling for essentially instantaneous adjustments in muscle metabolism and work, hence locomotion),
    b) on a medium-term basis (stimuli such as training effects calling for more profound and stable adjustments), and
    c) on a long-term basis (stimuli involving phylogenetic time and leading to genetically fixed differences between species in the

above characteristics, and thus in performance style and capacity).

These principles are best understood for vertebrates, so we shall center our discussion mainly upon them.

Muscle Metabolism and Muscle Work

Since getting somewhere (or away from somewhere) is so primeval an animal trait it is almost considered a defining element of what an animal is, we should not be surprised to find that muscle in many animals is singly the largest tissue in the body. In vertebrates, it can constitute about three-quarters of total body mass. In man and mammals in general, muscle O_2 uptake at rest accounts for about 30 percent of basal metabolic rate (Table 4-1); during heavy work, the large absolute increase in metabolic rate observed is mainly due to activation of muscle metabolism, although heart work rate and heart energy consumption also obviously must rise (Table 4-1). Overall metabolic rates during maximum work typically increase about tenfold, but in some species, such as the horse, the increase is as high as fortyfold!

In mammals, the three main fuels available for muscle work are creatine phosphate (CrP), carbohydrate (glycogen plus glucose), and fat. The power output achievable with these different fuels, and work duration, both vary (Table 4-2). So it is not surprising, and is known

Table 4-1. Relative Oxygen Consumption of Different Organs in Man as a Function of Work Rate
(Whole body at rest = 1.00; actual value near 0.17 mmoles min^{-1} kg^{-1})

Organ	Rest	Heavy work
Skeletal muscles	0.30	6.95
Abdominal organs	0.25	0.24
Kidneys	0.07	0.07
Brain	0.20	0.20
Skin	0.02	0.08
Heart	0.11	0.40
Other	0.05	0.06
Total	1.00	8.00

SOURCE: From McGilvery (1979) with modification

Table 4-2. Estimated Maximum Power Output for Skeletal Muscle (of Man) Utilizing Different Substrates and Metabolic Pathways

Fuel/pathway	Power output μmol ATP gm wet weight^{-1}min^{-1}
Fatty acid oxidation	20.4
Glycogen oxidation	30.0
Glycogen fermentation	60.0
CrP and ATP hydrolysis	96.0–360

SOURCE: From McGilvery (1975) with modification

to all of us, that low-intensity work, sustainable for long time periods, is fueled by oxidative metabolism (burning either fat or carbohydrate); that intermediate intensities of work surpassing the power output capacities of oxidative metabolism are supported by anaerobic glycolysis; and that even higher intensity work can be sustained for short time periods by phosphogen hydrolysis.

The decline with time in the maximum power that can be generated by working muscle arises from the use of these different energy sources for contraction; and, since these are preferentially stored in different kinds of muscle fibers, it also follows that different fiber types are used for different phases of work. Before considering the mobilization of different sources of energy, it is necessary to define what we mean by different fiber types.

Skeletal Muscle Fiber Types in Vertebrates

In virtually all vertebrate animals that have been examined at least two, and usually three, fiber types are always distinguishable, which can be termed white, intermediate, and red according to their appearance (which derives mainly from their myoglobin content). In phylogenetically ancient fish groups (elasmobranchs, holosteans, chondrosteans, and some teleosts with primitive taxonomic features), the different energetic requirements of steady-state and burst-type swimming have led to a complete anatomical and functional division between fast and slow motor systems; in the dogfish, for example, steady-state swimming is entirely supported by thin lateral strips of tonically active red muscle fibers. In contrast, the bulk of the musculature is composed of phasically active white fibers which are utilized only during burst swimming. Interestingly, these distinctly separate muscles have similar patterns of innervation and electrophysiological properties to tonic (multiply innervated) fibers and the fast-twitch (white fibers with local innervation) found in other vertebrates. Thus the standard explanation for the development of white and red fibers, *that it allows a simple separation of labor*, is beautifully illustrated in these fishes.

However, this simplest of patterns is not universally found in the vertebrate kingdom. In most "modern" teleosts, for example, there are significant amounts of overlap of function, both fiber types being recruited at sustainable swimming speeds (Johnston, 1981); and, to complicate the situation further, intermediate fibers are present. In most mammalian species thus far studied, most skeletal muscles are mixtures of three different fiber types. Unlike the situation in teleosts, however, where little if any mixing of fibers occurs, in mammals skeletal muscle typically is a mosaic mixture of all three fiber types. Nevertheless, white and intermediate fibers retain fast-twitch characteristics, while red fibers typically show slow-twitch characteristics. Interestingly, in mammals most muscles are slow at birth (the fetus has no predators to escape, no prey to capture); however, growth and development in the neonate progress with the concomitant differentiation of some muscles into fast-twitch muscles, while others remain slow-twitch (Goldspink, 1977).

In addition to electrophysiological differences, white, intermediate, and red fibers differ markedly in their ultrastructure (Table 4-3). In mammals, white fibers are typically large (up to 100 Å in sarcomere diameter) but not very uniform in size; they are not well supplied with

Table 4-3. Summary of Various Ultrastructural and Metabolic Features of Different Muscle Fiber Types in Vertebrates

Fast-twitch red (intermediate)	Fast-twitch white	Slow-twitch red
High respiration rate	Low respiration rate	High respiration rate
High glycolytic rate	High glycolytic rate	Low glycolytic rate
High myoglobin levels	Low myoglobin levels	High myoglobin levels
High myosin ATPase	High myosin ATPase	Low myosin ATPase
Well-developed SR	Well-developed SR	Poorly developed SR
High buffer levels	High buffer levels	Low buffer levels

SOURCE: After Holloszy and Booth (1976)

capillaries, and contain few mitochondria; sarcoplasmic reticulum (SR) is extensively developed. Red fibers are typically surrounded by capillaries, display a reduced sarcoplasmic reticulum but high mitochondrial abundance. In addition, red fibers are usually substantially smaller (about one-quarter to one-third the sarcomere diameter). Intermediate fibers in mammals are often termed fast-twitch glycolytic-oxidative because they differ by having fairly high capacities for anaerobic and aerobic metabolism.

Myosin ATPase isozymes (Hoh, 1983) also differ in the three fiber types: white and intermediate fibers have the highest activity, while red fibers have low myosin ATPase levels.

All these kinds of data are consistent with the idea of a significant retention of the primeval division of labor noted above in primitive fishes, an interpretation supported by direct measurements: these show that both in fishes and mammals, *red fibers are preferentially recruited during work of light and moderate intensity, while white fibers are recruited either when the excitatory input into the motor neurons increases to high levels during very strenuous work or when the red fibers become fatigued* (Holloszy and Booth, 1976). The intermediate or fast-twitch red fibers combine some of the capabilities of the other two fiber types, since they have high capacities for both aerobic and anaerobic production of ATP. Thus, white muscle is physiologically specialized for short bursts of intense work (separated by long periods of recovery) while the slow red fibers are designed for prolonged work of intermediate intensity. These higher-level adjustments correlate with, and indeed critically depend upon, underlying enzymatic and metabolic adaptations.

Storage Energy Quantities and Utilization Sequence during Different Kinds of Work

One of the most important levels of exercise adaptation involves the kind of substrate that is stored and utilized. As mentioned above, the available energy stores in human skeletal muscle (Table 4-4) are high-energy phosphate compounds, glycogen, and triglycerides. Amino acids and proteins are not usually considered to be mobilized in most kinds of exercise in man but they take on much greater importance in some species (Table 4-5). In terms of useful high-energy phosphate that can be generated, ATP and CrP constitute only a minor reserve, whereas glycogen and triglycerides provide ample amounts of energy at the local level, two to three orders of magnitude more than from ATP and CrP respectively.

Table 4-4. Cellular Energy Stores in Human Muscle

Substrate	Amount[a]	Available[b] energy
ATP	25	10
CrP	75	60
Glycogen	370	14,200
Triglyceride	50	24,520
Amino acids, proteins	?	normally hardly used

SOURCE: After Howald et al. (1978)
[a] μmols/gm dry wt.
[b] μmol ~P/gm dry wt.

Table 4-5. Preferred Energy Stores in Salmon and Squid

Salmon	Squid
ATP	ATP
CrP	Arginine phosphate (ArgP)
Glycogen	Glycogen
Fat	—
Proteins	Amino acids
Amino acids	Proteins

SOURCES: After Mommsen et al. (1980); Hochachka et al. (1983)

Phosphagen-Based High-Intensity Muscle Exercise

The signal for activation of muscle work at its highest possible rate is neuronal membrane depolarization, leading to Ca^{2+} activation of myosin ATPase and thus contraction. The immediate source of energy for contraction is ATP, which is stored at highest levels in white muscles specialized for burst work. But even at these high levels, there is enough ATP to support only seconds of heavy work. In few species does that seem to be an adequate burst-work mechanism, so it is usually bolstered by two additional processes: phosphagen replenishment of the ATP supplies and anaerobic glycolysis.

In vertebrates CrP is the high-energy phosphate reservoir; in many invertebrates, arginine phosphate is utilized. CrP is most abundant in white muscles, occurring at about 30 μmoles/gm wet weight. Although stored generally in the cytosol, there is evidence that it is locally concentrated near myosin ATPase, where myosin-bound creatine phosphokinase (CPK) is found (Bessman and Geiger, 1981) and where ATP supplies must be sustained if high rates of work are to continue. Not only is the positioning of the MM-isozyme of CPK appropriate to its function, so also are some of its catalytic properties. Under highest intensity anaerobic work, the direction of net flux for this reaction is

$$\text{CrP} + \text{ADP} \longrightarrow \text{Cr} + \text{ATP}$$

Thus, the relative affinities for CrP and ADP are of particular interest. Kinetic studies of the human muscle CPK (Jacobs and Kuby, 1980) indicate that the dissociation constants of CrP from both the binary and ternary complexes are high (72 and 32 mM, respectively), implying that under normal *in vivo* conditions [CrP] is not saturating and the enzyme is maximally responsive to [CrP] changes. Although the enzyme shows a low affinity for CrP, its affinity for ADP is high (dissociation constants of ADP for the binary and ternary complex, respectively, are about 200 μM and 60 μM), thus making it more competitive for ADP than are phosphoglycerate kinase (PGK) or pyruvate kinase (PK); this means that under highest intensity work conditions, when ADP concentrations may rise to values above the enzyme-CrP affinity for ADP, the enzyme may become saturated with this substrate and its activity would then be largely determined by changing concentrations of CrP.

Finally, since the total adenylate pool in muscle is only 6 μmol/gm, even if all available ADP were converted to ATP, the reaction

$$\text{CrP} + \text{ADP} \longrightarrow \text{ATP} + \text{Cr}$$

could lead to a [CrP] change of substantially less than the adenylate pool size; yet nearly a total depletion of CrP (about 30 μmol/gm) is commonly observed during highest intensity muscle work, and this same occurs for arginine phosphate in many invertebrates (Hochachka et al., 1983). As with the paradox of lactate accumulation (much more being found than could be accounted for by the adenylate pool), this is only possible because the above reaction is coupled with (in fact is initiated by) myosin ATPase. As with glycolysis (see below), the significance of this interaction is usually overlooked in discussion of anaerobic muscle work.

Whereas the total amount of creatine phosphate stored in muscle determines how much work can be sustained by this mechanism, the *rate* of energy production (in terms of μmols ATP hydrolyzed/min.) depends upon the amount of CPK available. That is why white fiber types typically have higher levels of this enzyme than red muscle fibers and why species that are particularly good at burst work have higher levels of CPK or of arginine phosphokinase (APK) than do their more sluggish relatives.

Although phosphagen-driven muscle work may be adequate for quick-strike predators, most animals that need to work at elevated rates for longer than five to ten seconds must utilize an endogenously driven backup system to augment the ATP initially being produced by phosphagen hydrolysis. In both vertebrate and invertebrate locomotion, that backing system is always some form of anaerobic glycolysis.

Elevated Glycogen Stores and Elevated Enzyme Levels in White Muscle

Anaerobic glycolysis behaves as a closed system during burst work, so the total amount of work it can support depends upon the available glycogen while the rate of energy generation depends upon the amount of glycolytic enzyme machinery. Thus, the simplest way to elevate glycolytic capacity is to store more glycogen and maintain higher levels of glycolytic enzymes. Nature in fact relies upon this strategy, which is why in mammals glycogen is stored at higher levels in white muscle than in red. Similarly, the levels of glycolytic enzymes are higher in white muscle than in any other tissue or organ in the vertebrate body (see Table 4-6). That is also why the rate of lactate

Table 4-6. Adaptation of Levels of Citrate Synthase and LDH in White (WM) and Red Muscle (RM) in Selected Fishes

	Citrate synthase		Lactate dehydrogenase[a]		
	WM	RM	WM	RM	Ratio
Inactive species					
Aruana	1.0	—	760	—	
Pirarucu	1.7	3.3	260	263	1.0
Hoplias	2.0	3.7	576	419	1.4
Hoplerythrinus	1.3	1.4	1,060	810	1.3
Active species					
Salmon	8.8	68.0	5,500	300	18.3
Tuna	2–7	21.0	5,500	515	10.7

SOURCES: From Hochachka et al. (1978); Guppy et al. (1979); Mommsen et al. (1980).

[a] Enzyme activities expressed as μmols substrate converted to product per min. per gram wet weight of muscle at 25°C.

production and the total amount that can be accumulated are both higher in this tissue than in any other. McGilvery (1975) and Newsholme (1978) have argued that, in well-trained sprint runners, leg muscle glycolysis may be activated by as much as two thousandfold; that is, an increase in absolute flux of carbon (glycogen → lactate) of over three orders of magnitude! This characteristic depends not only on the amount, but also on the kind of enzymes present, particularly on their regulatory properties. Most critical of these, in terms of initiating changes in glycolytic rate, is the initial flux generator: muscle glycogen phosphorylase.

Cascade Control of Muscle Glycogen Phosphorylase

In mammals, glycogen phosphorylase occurs in at least three isozyme forms, one of which is characteristic of skeletal and cardiac muscle (Cohen, 1978). Two classes of signals, endocrine and neuronal, act indirectly to activate glycogen phosphorolysis. Various endocrines (epinephrine being one of the most potent) serve to activate membrane-bound adenyl cyclase which initiates a cascade control system. Most of this cascade, including phosphorylase per se, is bound to glycogen particles in muscle, a physical positioning that probably facilitates activation of the system (Cohen, 1978).

The neuronal mechanism for activating glycogen breakdown involves membrane depolarization and Ca^{2+} release from the sarcoplasmic reticulum (SR). Ca^{2+} control cuts in at the level of inactive phosphorylase *b* kinase by binding to its smallest subunit: calmodulin. Calmodulin in muscle, unlike in other tissues, is incorporated into the integral oligomeric structure of phosphorylase *b* kinase; as in calmodulin from other tissues, it has two Ca^{2+} binding sites. The binding of Ca^{2+} causes a large (about fifteenfold) increase in affinity for substrate which under physiological conditions leads to an ATP-dependent phosphorylation of *b* kinase and thus its activation. Phosphorylase *b* kinase in turn leads to the polymerization of phosphorylase *b* to phosphorylase *a* (Fig. 4-1). In this cascade system in liver, the degree of amplification of the incoming epinephrine signal is about 10^6. The molar ratio (Cohen, 1978) of the cascade enzymes in muscle (1:10:240), compared to 1:0.25:14 in liver, indicates that the amplification capacity is higher in the former than in the latter, and seemingly adequate to account for the 10^3 change in glycolytic flux (Fig. 4-1).

In its capacity to function over such large absolute ranges of flux the pathway of glycogenolysis is unique in metabolism; no other pathway ever achieves the same degree of activation. All intermediates

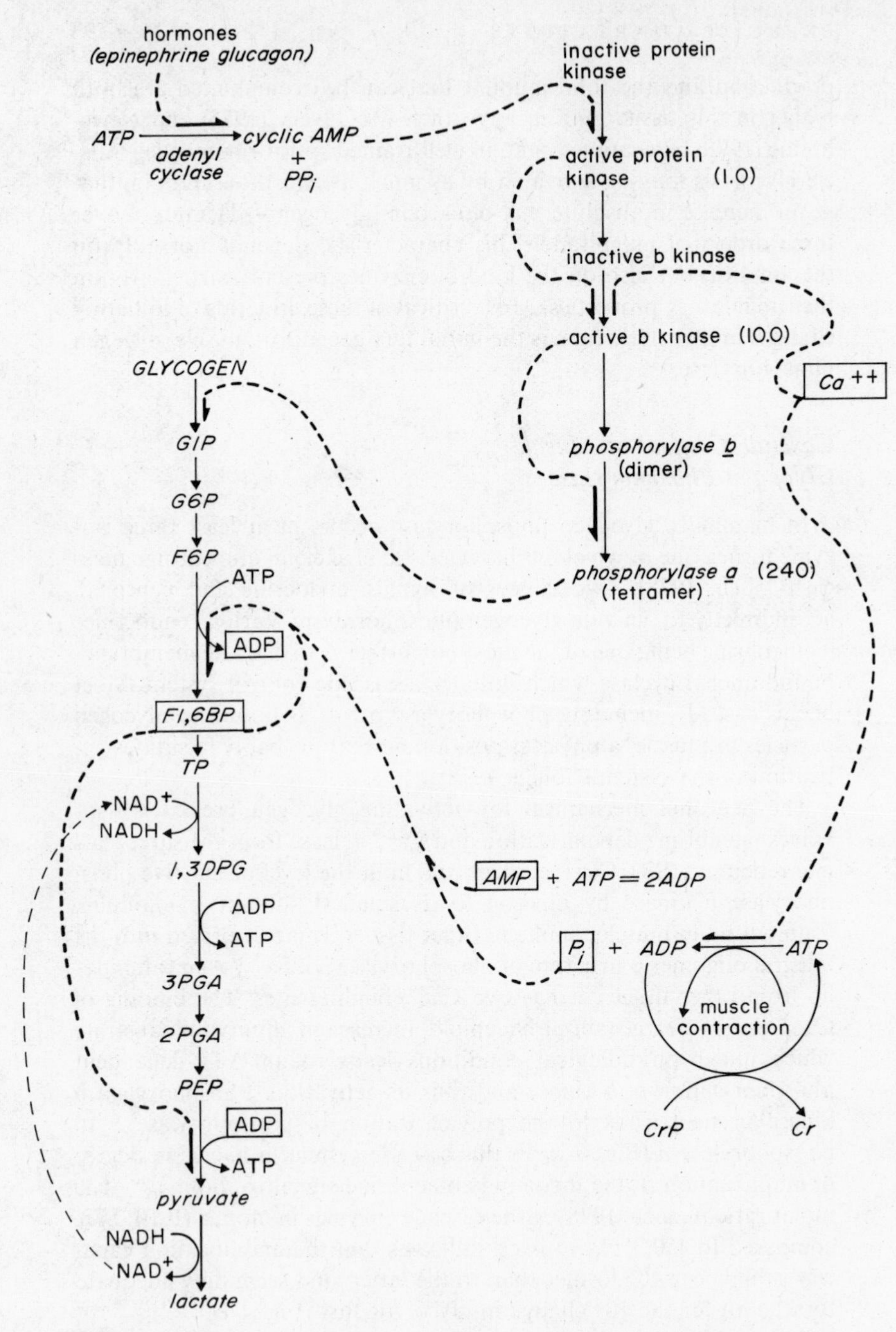

Fig. 4-1. Regulatory properties of glycogen fermentation to lactate in vertebrate muscles. Molar concentration ratio of cascade enzymes given in parentheses.

along the pathway have been monitored through these kinds of transitions and although many if not all of these increase in concentration during activated flux, none change nearly as much as can the actual flux through the system (Hintz et al., 1982). That is why amplification mechanisms must exist for the large activation of catalytic machinery in muscle during burst work. And that is why strategic significance is attached to the cascade control of glycogen phosphorylase, for it would otherwise be difficult to understand how flux changes of 10^3 could be achieved concomitant with less than tenfold changes in the levels of key pathway intermediates. But even if it is probably the most important single mechanism for turning on muscle glycolysis, the glycogen phosphorylase system in turn must be integrated with at least one additional control site in the pathway: the step catalyzed by phosphofructokinase (PFK).

Regulatory Functions of Muscle Phosphofructokinase

Phosphofructokinase (PFK) catalyzes the first committed step in glycolysis and as such has long been recognized as an important locus of control. As in the case of phosphorylase, muscle PFK occurs as a tissue-specific isozyme form, displaying unique catalytic and regulatory properties that tailor it for burst-type work in muscle. In all mammals (Tsai et al., 1975) and most other animals that have been studied in this regard (see Storey and Hochachka, 1974a), PFK control is based upon its interactions with both substrates, F6P and ATP: the F6P saturation curve is typically sigmoidal (at physiological pH) with the $S_{(0.5)}$ in the 0.1 mM range, similar to physiological levels of F6P. The saturation curve for ATP, in sharp contrast, is hyperbolic and through physiological concentrations of ATP (above 1 mM), the enzyme displays a kinetic order less than one. This is as it should be for an enzyme in a pathway whose function is to make ATP: the lower the *in vivo* level of ATP, the less inhibited by ATP the enzyme becomes, the greater its catalytic potential.

In principle, these two key functional characteristics could go a long way toward appropriate control of PFK: during burst work, for example, F6P levels rise as flux is activated, and increasing F6P availability would serve to increase its rate of utilization by PFK (positive cooperativity) while reversing ATP substrate inhibition. At the same time, falling ATP levels simultaneously lead to reduced substrate (ATP) inhibition and to an increased affinity for F6P, in effect increasing reaction velocity. These are synergistic regulatory properties that help

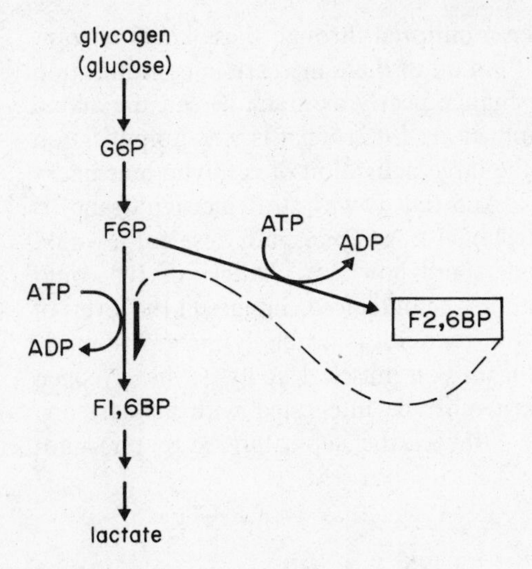

Fig. 4-2. Role of F2,6BP in the control of muscle phosphofructokinase

in integrating PFK function with that of glycogen phosphorylase and, indeed, with myosin ATPase. But built upon this control foundation are other inputs that add even more versatility to the control of this locus in glycolysis: the most important of these are activation by fructose-2,6-bisphosphate (F2,6BP), F1,6BP, AMP, ADP, P_i, and NH_4^+, as well as inhibition by citrate and fatty acylCoA.

Much of the older information is reviewed by Tsai et al. (1975). In metabolic terms an exciting new development in our understanding of how PFK may be regulated *in vivo* is the discovery of F2,6BP as the most potent activator of PFK thus far known (see Furuya and Uyeda, 1980). The modulator is formed from F6P by the action of a specific enzyme and has several critical effects on PFK: a) it increases enzyme-F6P affinity substantially, while reducing the strength of F6P site-site cooperative interactions (in effect, saturation curves for F6P are moved to the left and become less sigmoidal); b) it increases the affinity for the cosubstrate, ATP; c) it reverses inhibition by high levels of ATP; d) it reverses inhibition by citrate; and e) it is synergistic in its effects with other potent positive modulators, in particular AMP and P_i (see Uyeda et al., 1981). Recent metabolic measurements indicate that [F2,6BP] rises by two-to-fourfold during insulin or epinephrine activation of glycolysis in perfused rat hind limb muscle, thus providing *in situ* support for a novel control situation (Fig. 4-2).

ADP and F1,6BP effects on muscle PFK can be viewed as adding an autocatalytic component to PFK control: these are the immediate products of the reaction and they in effect stimulate their own further rate of formation. At least in part for this reason, PFK catalytic rate is not a linear function of time, but increases exponentially with time, so making an important contribution to the flare-up of glycolysis that is characteristic of burst work.

Role of ADP in Integrating PFK and Glycolytic ATP-Forming Reactions

There is another critical role for ADP in glycolytic control: as a substrate for phosphoglycerate kinase (PGK) and pyruvate kinase (PK), it serves to integrate these activities with PFK catalysis. Muscle PGK is a high-activity enzyme with kinetic properties suited for glycolytic function. In particular, it displays very high affinity for 1,3 diphosphoglycerate (micromolar range) and under most (especially working) conditions, the enzyme should be saturated with this substrate. Its affinity for ADP, however, is not so high and the apparent Michaelis constant clearly falls in the physiological range (Krietsch and Bücher, 1970). Thus, PGK activity *in vivo* is probably paced by ADP availability.

The situation at the PK locus is more complex. In lower vertebrates and invertebrates, F1,6BP is a feedforward activator of PK (Fig. 4-1) and serves as a coupling signal between PFK and PK, a mechanism potentiated as glycolysis proceeds and pH drops (see Hochachka, 1980; Storey and Hochachka, 1974b). In mammalian muscle PK, this control is lost and PFK-PK integration depends more upon ADP (being the product of one, and the substrate of the other). The kinetic properties of muscle PK (particularly its marked positive cooperativity in binding of substrates) admirably suits this isozyme for glycolytic function in muscle. The dissociation constant for ADP binding to the free enzyme is lowered from a nonphysiological range (>2 mM ADP) down to about 0.2–0.3 mM by addition of PEP; i.e., the binding of PEP leads to a very large (about tenfold) increase in the affinity of the enzyme for its cosubstrate, ADP (Dann and Britton, 1978). In metabolic terms, this means that two signals are needed to integrate muscle PK and muscle PFK activities: the first one is PEP, formed (via several reaction steps) from F1,6BP, a product of PFK reaction; the second is ADP, supplied directly by the PFK reaction. The first (PEP) is needed to make PK competitive for the second (ADP).

Role of ADP in Coupling Anaerobic Glycolysis with ATPases

The end products of anaerobic glycolysis (considered at about pH 8 to avoid the problem of H^+ stoichiometry (Hochachka and Mommsen, 1983) are lactate, ATP, and water:

$$\text{glycogen (glucosyl)} + 3\text{ADP}^{3-} + 3\text{P}_i^{2-} \longrightarrow 2\ \text{lactate}^{1-} + 3\text{ATP}^{4-} + 2\text{H}_2\text{O}$$

The cytosolic adenylate pool size in muscle of vertebrates ranges between 3–8 μmol/gm, (Beis and Newsholme, 1975), so even if all endogenous ADP were phosphorylated to ATP, maximum levels of lactate produced by this reaction pathway as written would be two-thirds the adenylate pool size or a maximum of about 6 μmol lactate/gm muscle (in fact it would have to be less, since ADP can never account for the total adenylate pool). In contrast, *in vivo*, mammalian muscle can readily accumulate 20–40 μmols lactate/gm; tuna white muscle can accumulate over 100 μmols/gm (Guppy et al., 1979); diving turtles can achieve up to 200 μmols/gm (Ultsch and Jackson, 1982). That is, more lactate is observed than can be produced by the above fermentation pathway as written. This apparent paradox does not mean that the glycogen → lactate conversion is dissociated from ADP phosphorylation but rather that glycolysis and myosin ATPase are coupled *in vivo* and that the continuous hydrolysis of ATP serves as a mechanism for *generating ADP and driving the above glycolytic reaction to the right.* This aspect of glycolytic regulation is little discussed in reviews of this field, although it is clear from reconstituted glycolytic systems that flux can be effectively activated by addition of ATPases (Wu and Davis, 1981).

Maintenance of Redox Balance during Anaerobic Glycolysis

In skeletal muscles, the concentration of glycogen is high (in the range of 100 μmol glucosyl-glycogen/gm), much if not all of which can be converted to lactate during anaerobic glycolysis. The pool size of NADH plus NAD^+, however, is small, about 1 μmol/gm. So for anaerobic glycolysis to be prolonged there is of course the need for continuously reoxidizing NADH formed at the glyceraldehyde-3-phosphate dehydrogenase (GAPDH) step. Usually, it is explained that this is achieved by a 1:1 functional coupling between the GAPDH oxidative step and the LDH-catalyzed reductive step, lactate accumulating as the process continues. In addition, however, malate dehydrogenase (MDH) is kinetically well suited for function during early stages of glycolytic activation (because of a very high affinity for both NADH and oxaloacetate). Its key carbon substrate (oxaloacetate) is derived from aspartate via transamination catalyzed by a high-activity enzyme, aspartate aminotransferase. In invertebrate muscles, aspartate is stored at quite high levels (>10 μmol/gm), but in vertebrate muscles, only modest amounts of aspartate are stored, so this system cannot sustain glycolytic redox for very long; long enough, however, to build up pyruvate levels and make LDH ever more competititve (Hochachka, 1980). It is presumed that as LDH begins to take on its usually assigned role in redox regulation, the accumulation of H^+ concomitant with lactate serves to activate further lactate formation by decreasing the K_m for pyruvate. The latter is particularly important, for it indicates an increasing LDH affinity for its carbon substrate as pyruvate availability is rising, in extreme cases (tuna, for example) reaching levels as high as 1 μmol/gm (Guppy et al., 1979). These functions of LDH in anaerobic muscle work appear to be facilitated by the occurrence of LDH in isozymic form. The importance of having M_4 LDH during anaerobic muscle work is particularly graphically illustrated in a recently described myopathy: individuals lacking M_4 LDH in skeletal muscle are found to display abnormally low capacities for lactate production and drastically reduced capacity for anaerobic muscle work (Kanno et al., 1980).

Muscle Buffering Power and Anaerobic Work Capacity

The end products of phosphagen-based work are creatine and P_i, and work proceeds with a modest net proton consumption: at physiological pH, about 0.1 μmol H^+ are consumed per μmol ATP cycled through the CPK and myosin ATPase reactions (Hochachka and Mommsen, 1983). The end products of anaerobic work supported by glycogen fermentation in mammals are lactate anions, water, and protons; about 0.7 μmols H^+ are formed per μmol ATP cycled through the glycolytic and ATPase reactions. Thus, during prolonged anaerobic work, muscle pH must gradually decrease; yet the process cannot be left totally unbridled. It, too, must be regulated. That is why muscles with high myosin ATPase activities coupled with either a high glycolytic capacity or with at least a high anoxia tolerance have elevated buffering capacities.

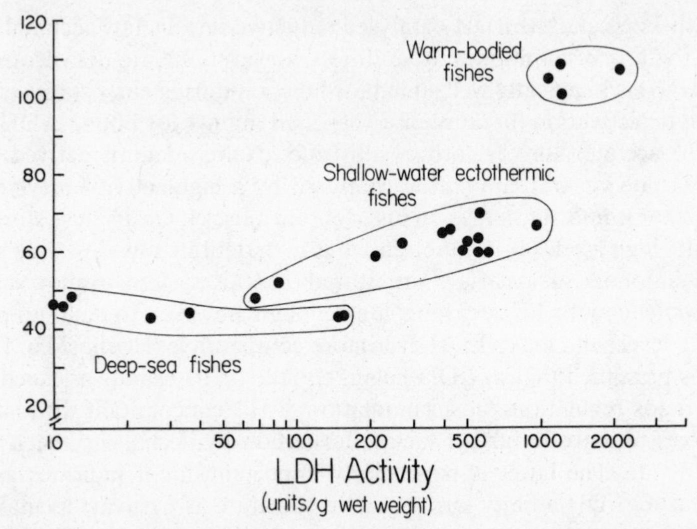

Fig. 4-3. Buffering capacities (β) in white skeletal muscle of different marine teleost fishes plotted against LDH activity of the muscles. The fishes are grouped into three sets: warm-bodied fishes (tunas, albacore), shallow-water ectothermic fishes, and deep-sea species. Buffering capacity is measured as μ-moles of base needed to titrate the pH of a one gram sample of muscle by 1 pH unit between pH values of approximately 6 and 7. LDH activities are in units of activity per gram fresh weight of muscle at a measurement temperature of 10°C. Figure modified after Castellini and Somero (1981)

The buffering capacity due to nonbicarbonate buffers is measured in "slykes" (named after the famous respiratory physiologist, Van Slyke) which have the units of μmoles of base required to titrate the pH of one gram wet weight of tissue by one pH unit usually chosen to be between pH 6 and pH 7 (Castellini and Somero, 1981). The total buffering capacity in slykes is abbreviated as β. For a group of fishes, examined by Castellini and Somero (1981), large interspecific differences are found in LDH activity and in β. The highest values for both occur in warm-bodied fishes (tunas); lowest values are found in deep-sea species. (Fig. 4-3). Shallow-living ectothermic fishes yield intermediate values.

The interspecific differences in β and LDH activity can be readily explained by the differences in locomotory capacities and feeding strategies among these fishes (Sullivan and Somero, 1980). At one extreme lie the warm-bodied fishes, which have the highest LDH activities and buffering capacities of all animals studied. These fishes are remarkable for their swimming speeds and capacities to produce lactate during high-speed performance. Deep-sea fishes, in contrast, usually are weak swimmers, and commonly display "float and wait" feeding strategies. High-speed locomotion is not usual in most deep-sea fishes and, consequently, one does not find in their white muscles large capacities for generating lactate or high β values. The shallow-living ectothermic fishes studied show some variation in both LDH activity and β, and a trend is apparent between these two variables.

In keeping with the functional and compositional differences discussed above between white and red muscle, the β values of fish red muscle are much lower than those for white muscle. For instance, even in the case of the warm-bodied fishes, values for red muscle fall in the range of only 50 slykes. Clearly, the buffering capacity of a tissue is closely regulated to fit the tissue's capacities for generating protons during metabolic activity. This relationship holds among animals besides fishes (see Somero, 1981, for further literature) and, as discussed below, is noted for both burst-type locomotion and locomotion that entails relatively low-level, but very long-term anaerobic metabolism.

Among mammals the buffering capacity of locomotory muscles is extremely high in most marine (diving) mammals and in burst-locomotory species like the rabbit (Castellini and Somero, 1981). It is instructive to examine in detail the relationships between anaerobic function and muscle buffering capacity and Mb content in diving mammals. In the marine mammal which has the greatest capacity for breath-hold diving, the Weddell seal, lactate generally does not rise in the blood until after twenty minutes of submersion (Kooyman et al., 1980). During the initial twenty or so minutes of the dive, muscle function remains aerobic, and the high content of Mb in the swimming muscle must play a major role in facilitating prolonged aerobic ability. After approximately twenty minutes of diving, lactate concentrations generally rise. After this point the high buffering capacity of the Weddell's muscle may be critical for continued swimming. It should be noted that muscle LDH activities in diving mammals are not in general higher than those of common terrestrial mammals (cow, pig, dog; Castellini et al., 1981). The abilities of diving mammals to support prolonged muscle function in the absence of oxygen derive from an elevated ability to deal with proton accumulation, not from an increased capacity to produce these end products on a per unit time basis. The correlation between β and [Mb] in mammals, notably diving species, is not paradoxical when analyzed in the above context, for the high [Mb] is seen to extend the period of aerobic function, while the high β is appreciated as an adaptation which comes into play after the aerobic capacity conferred by high [Mb] is exhausted.

Similar relationships between buffering and burst work capacities are developed in invertebrates. In squids, for example, buffering capacity of mantle muscle is higher than in the fin muscle. In several gastropod species highest buffering capacity is known to occur in the powerful pedal retractor. In crustaceans, β of tail flexor muscles is higher than in leg muscles, while between-species studies show that higher β values in tail flexor muscle correlate with the relative size of the muscle, its total arginine-arginine phosphate pool, and the activities per gram of arginine kinases and lactate dehydrogenases (England and Baldwin, pers. comm.) In these organisms as well, then, muscle buffering power seems tailored to anaerobic work capacity.

Adaptive Strategies for Burst Work: Overview

In discussing classes of adaptational responses (Chapter 1), we emphasized that organisms could call upon a combination of three basic mechanisms:

1. adaptation of enzyme machinery,
2. adaptation of the enzyme milieu, and
3. adaptation of metabolic output.

It will be evident that, at least in vertebrate organisms, burst work adaptations involve all three of the above. Enzyme machinery is adjusted both in amount and kind, with white muscle fibers containing the highest activities of glycolytic enzymes found anywhere in the organism. What is more, essentially every step in the pathway is catalyzed by a tissue-specific isozyme displaying catalytic and regulatory properties suited for function during burst work.

The cellular milieu in which muscle glycolytic enzymes operate is adapted in at least three ways. Firstly, large quantities of glycogen (about 100 μmol glucosyl units/gm) are stored, usually as β-granules, upon which glycogen phosphorylase and most of the enzymes of its control cascade are "plated out." Secondly, the elevated glycolytic potential of white muscle is matched by a similar elevation in imidazole-based buffering power. And thirdly, modest changes in pH per se as well as in the ratio of $[H^+]/[OH^-]$ necessarily develop during burst work and even these have been harnessed to facilitate or to regulate function of key enzymes in glycolysis.

The third mechanism, involving adaptation of metabolic output of the system, must be viewed as more than just the sum of enzyme adjustments plus adjustments in cellular milieu. This is because many of the control features of muscle glycolysis are not formally a part of the pathway. The most critical of these is the control cascade initiated by hormone activation of adenyl cyclase, but others include the calmodulin-based Ca^{2+} activation, as well as bypass steps such FBPase, and facilitatory functions such as those of MDH in early redox regulation. Their summed effects, however, raise the potential for a controlled two thousandfold activation of anaerobic glycolysis during burst work in man. This is an unique feature, in no way representative of metabolic control and metabolic pathways in general; it has developed in vertebrate muscles as a backup to CrP priming of anaerobically generated ATP for burst work. And because the potential metabolic output and its control vary both between species and between tissues in a way that correlates with their burst work capacities, we can confidently conclude that these characteristics are products of selection and adaptation. Yet even at their best, they can sustain high work rates for only a short time, two to three minutes in a man, five to ten minutes in the tuna, to mention two examples. For exercise that must be sustained for longer periods, the organism must finally harness oxidative energy metabolism, and the question now arises of how aerobic metabolism is adapted for this purpose.

Elevated Oxidative Capacity in Red and Intermediate Muscles

Just as in the case of glycolysis, where anaerobic power output depends upon the catalytic potential of glycolytic enzymes, so in the case of aerobic sustained performance the maximum *power output depends upon the catalytic capacity of oxidative enzymes*, largely in this case in red and intermediate fiber types. That is why across essentially all vertebrate groups, red muscle typically contains substantially higher levels of oxidative enzymes than does white muscle (Table 4-6). This difference between red and white muscle of course goes hand in hand with differences in mitochondrial abundance and ultrastructure (discussed above). The biochemical adaptation strategy, therefore, is one we have seen before, merely maintaining higher amounts of what otherwise might be termed standard metabolic machinery.

What Determines Total Amount of Energy Produced in Aerobic Work?

A critical difference between anaerobic and aerobic performance arises when we consider the total amount of power that can be generated. In anaerobic metabolism, because it is in effect a closed system, the total amount of power that can be released is determined by the endogenous stores of substrate (mainly glycogen). Aerobically working muscles, in constrast, remain open systems: they are continuously perfused with blood and are thus in communication with central depot supplies of carbon and energy. The total amount of work they can generate is influenced in part by endogenous substrate supply and in part by substrate supply in central depots. The two main endogenous substrates available for red muscle metabolism are glycogen and fat; both occur in red and intermediate-type muscles, but fat is especially abundant. In addition, however, the total amount of power that can be generated is also influenced by the availability of substrates in other depots, fat in adipose tissue and glycogen in liver.

Relationship between Glucose and Fat Metabolism

For some time now a reciprocal relationship between glucose and fat metabolism has been well known, and it is widely accepted that when fat metabolism is activated, glucose oxidation is reduced. In part, the explanation for this reciprocal control lies in the effect of citrate on phosphofructokinase (PFK): during activated fatty acid oxidation, citrate levels are known to increase to levels where they are inhibitory to PFK (Tsai et al., 1975). Although a number of modulators are known (AMP, NH_4^+, F2,6BP, etc.) which can reverse citrate inhibition, this idea has become entrenched in the literature, and there is widespread belief that simultaneous metabolism of both substrates is the exception rather than the rule. That is why it may come to the reader as somewhat of a surprise to realize that high and continuing rates of aerobic work are best supported by a mixed and simultaneous catabolism of glucose and fat.

Power Output from Carbohydrate and Fat for Substained Exercise

In man, the available carbohydrate stores are adequate to sustain work at near maximum rates for only twenty to thirty minutes; the amount of lipid, on the other hand, is adequate to sustain fairly high work rates for several days. To appreciate why a mixed glucose and fat metabolism is a useful strategy it is important to realize that the rate of energy generation (and hence the intensity of work) sustainable on fat alone is only about half that sustainable on carbohydrate or a mixture of the two (Table 4-2). Why this should be so is not entirely clear, for in general it is argued that on a weight basis fat is a more efficient fuel. Three possible factors may be involved:

1. Glycogen is more efficient than fat in terms of *energy yield per mole of* O_2 *consumed*, about 6.2 moles ATP being formed/mole O_2 during glycogen catabolism compared to 5.6 moles ATP/mole O_2 during fat oxidation.
2. When working muscle uses free fatty acids from the blood (rather than endogenous triglyceride) the rate of utilization varies with concentration, which is low and which may make the process *substrate limited.* In man, free fatty acids in blood occur at about 0.5 microequivalents/ml, and even though these levels rise during exercise, they never become saturating for uptake by muscle, possibly because they are normally transferred in blood in an albumin-bound form.
3. The β-oxidation spiral feeds acetylCoA units into the Krebs cycle, whose activity is elevated mainly by increasing the steady-state levels of its intermediates (see below). During extremes of exercise, when all muscle and body glycogen (glucose) reserves are reduced, muscle may not be able to sustain as high levels of Krebs cycle intermediates; these are normally augmented from amino acids and from glycolytic intermediates. In effect, complete transition to fat as the sole source of carbon and energy in working muscle may simultaneously remove an important means for priming the Krebs cycle. This in turn would lead to a reduced rate of ATP synthesis and thus allow only a reduced rate of power output.

The Dual-Substrate Compromise

For a combination of the above reasons, glycogen would appear to be the best fuel for highest intensity aerobic work but because of limited supply, the duration of this process also is limited. Not surprisingly, therefore, if it can, nature strikes a compromise, so that at least in man at high work rates, sustainable for two to three hours or more, *both glycogen and fat are utilized.* RQ values at such time are about 0.9, indicating that about 65 percent of the O_2 consumed is accounted for by glycogen oxidation, while the rest is due mainly to fatty acid metabolism. Actually, this ratio is probably continuously decreasing with duration of performance, indicating that the longer the aerobic work period, the greater the contribution of fat metabolism to overall energy production.

Since the reserves of glycogen are lower than of fat (Table 4-4), this high level of aerobic work is normally limited by the supply of glycogen. In fact, at maximum aerobic work rates in man, dependence on muscle glycogen is essentially absolute, there being a close correlation between duration of exercise to exhaustion and muscle glycogen concentration at the beginning of exercise (Holloszy and Booth, 1976).

If nature cannot strike the above dual-substrate compromise, and it cannot when muscle glycogen reserves are fully depleted, then it must rely much more exclusively upon fat metabolism. This is indeed the usual strategy and a good long-term solution, but as already mentioned its cost is a reduction in the level of power output to about 60 percent or less of the aerobic maximum sustainable on glycogen. What is more, when muscle glycogen is gone, muscle utilization of exogenous free fatty acid and endogenous triglyceride must occur at the same time as blood glucose levels are conserved for the brain and other tissues (e.g., kidney medulla), which rely predominantly or exclusively upon glucose for normal metabolism. It is at such time that the above antagonistic interactions between fat and glucose metabolism become particularly important.

Thus transition from rest to high-intensity work rates for periods of several hours actually involves the carefully regulated utilization initially of two substrates, followed by a complete switch to fat-based metabolism in muscle, and with concomitant conservation of enough glucose to satisfy minimal body needs. Such transitions are explainable not by simple adjustments in the amounts of enzymes, but rather in the kinds of enzymes retained. As in the case of the glycolytic system described above, the tuning up of metabolic machinery for aerobic work not only involves gross adjustments in the amounts of oxidative muscle fibers and oxidative enzymatic capacity, but also finer level adjustments in how that capacity is used; that is, in the control of muscle metabolism and its integration with the metabolic activities of other tissues.

Flux-Generating Steps in Muscle Fat Metabolism

It will be evident that β-oxidation in muscle metabolism can draw upon two input sources, endogenous triglyceride and exogenous free fatty acids. Interestingly, although endogenous triglyceride utilization in avian, insect, and fish red muscles was well appreciated early on, it was not until 1971 that its utilization was clearly established in human skeletal muscle. In man, during the first thirty minutes of long-lasting exercise, intramuscular triglyceride contributes about half the total caloric expenditure, but the contribution of free fatty acids gradually increases as duration of work increases. In runners covering a distance of 100 km (in a mean time of 8 hr. 49 min.!) there occurs about a 7 percent drop in intracellular lipid content which is adequate to account for about 25 percent of the total caloric requirements; in such long work periods, the remaining 75 percent of the caloric requirements are supplied by free fatty acids derived from adipose tissue stores (Howald et al., 1978). Although muscle triglyceride lipase (TG lipase) has not been analyzed in detail, its homologue in cardiac muscle has been. In common with the adipocyte enzyme, this muscle TG lipase displays the two necessary requirements for function in flux initiation; namely, the enzyme occurs in essentially "on" or "off" states, and substrate supply is saturating. Thus, starting with triglyceride, flux is directly proportional to the amount of TG lipase in its active form.

The flux-generating step for muscle oxidation of exogenous free fatty acids, however, is not so clear. It is widely accepted that the rate of free fatty acid oxidation by muscle is determined by substrate availability and substrate demand. Whereas metabolic controls may be exerted at the level of acylCoA formation and on entry into the Krebs cycle, no such controls have been found for β-oxidation per se, which indeed seems to be primarily regulated by the redox state of mitochondria (Vary et al., 1981). Moreover, none of the known control mechanisms express the essential requirements for flux initiation, which is why Newsholme (1978) suggests that the complete pathway for fatty acid metabolism in exercising muscle actually begins with triglyceride in adipose tissue. In that event, the flux-generating step for *muscle* fatty acid oxidation is *adipocyte* TG lipase.

Control of Fat Mobilization in Adipose Tissue

Because in adipose tissue the process of lipolysis occurs simultaneously with that of esterification (triglyceride is broken down to fatty acids which are reactivated and re-esterified to form triglyceride), the rate at which triglyceride is mobilized must be determined by the difference between the two processes. Control of lipolysis resides primarily in a hormone-sensitive TG lipase, with secondary metabolite effects, while control of esterification seems to be largely under metabolite control.

The control of hormone-sensitive lipase is formally analogous to that already described for glycogen phosphorylase. As in the latter case, TG lipase occupies a strategic position in being a first and rate-limiting step in fat catabolism, and in being interconvertible from an inactive *b* form to an active, phosphorylated *a* form (Steinberg and Khoo, 1977).

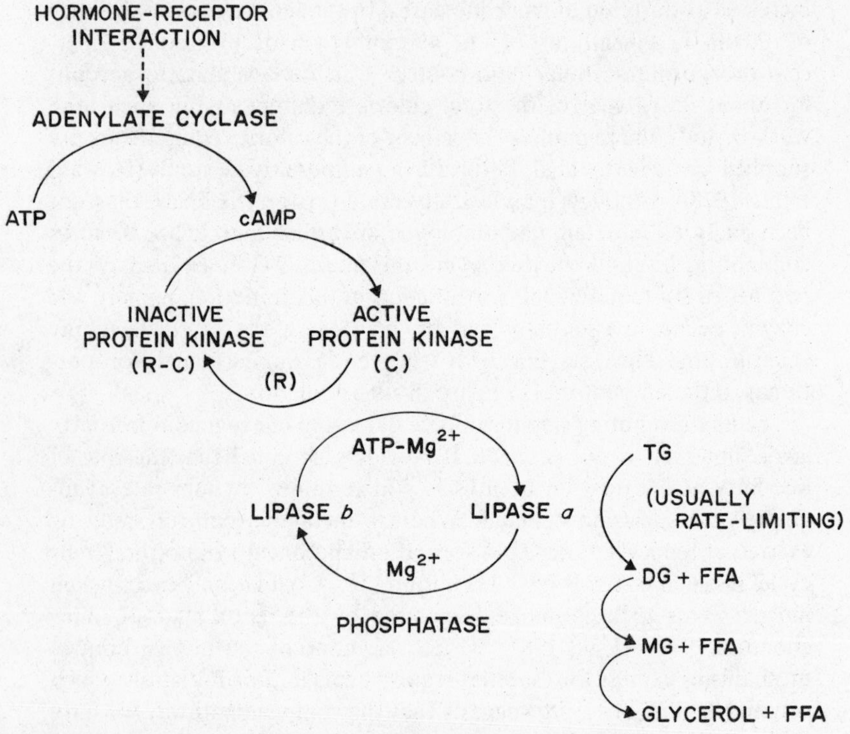

Fig. 4-4. Enzyme cascade in the control of lipase-catalyzed hydrolysis of tri-, di-, and monoglycerides. See Steinberg and Khoo (1977)

During exercise, increases in blood levels of lipolytic hormones (adrenaline, glucagon, and nonadrenaline) together with a drop in the level of insulin, lead to an activation of membrane-bound adenyl cyclase, which initiates a control cascade similar to the one described for glycogen phosphorylase. This system has been best worked out in rat adipose tissue, but studies of human and avian adipose and even of heart muscle indicate analagous control features for hormone-sensitive lipases. As shown in Figure 4-4, TG lipase is but one of at least three acyl hydrolases that are involved in the complete lipolysis of storage fat, and it is interesting to note that recent studies are consistent with a similar protein-kinase dependent activation of diglyceride (DG) and monoglyceride (MG) hydrolases. However, the absolute catalytic activities of the DG and MG hydrolases are much higher than those of TG lipase; consequently, even after protein kinase activation, the hydrolysis of the first ester bond probably remains the rate-limiting step in fat mobilization (Steinberg and Khoo, 1977).

Thus, through the concerted action of lipolysis-stimulating hormones, antilipolytic hormones (insulin), and metabolites (fatty acids, acyl CoA, CoASH, etc.), the stage is set a) for the controlled mobilization of the major body stores of fat (adipose tissue triglycerides), b) for their delivery in the form of albumin-bound fatty acids to muscle, and c) for the activation of β-oxidation in muscle. The latter process feeds acetylCoA units into the Krebs cycle, so we must finally turn to the question of how this latter process is controlled.

Krebs Cycle Control Options

We have pointed out that in tuning up anaerobic metabolism for support of muscle burst work, organisms adjust both the amount and the isozyme type of enzyme present in muscle tissue. Tissue-specific isozyme arrays, then, are an important part of the adaptational arsenal that makes muscle glycolysis exceptionally suitable for supporting burst-type work. In principle, the same two mechanistic strategies could be utilized in Krebs cycle adaptation for muscle work, so it is instructive that this qualitative strategy is rarely if ever utilized. Although as we have seen the levels of Krebs cycle enzymes in red muscle are high, sometimes an order of magnitude higher than in white muscle, for example (Guppy et al., 1979) *these typically do not occur in tissue-specific isozymic form.* That is, the catalytic and regulatory properties of Krebs cycle enzymes in red muscle are the same as they are in any other tissue, which means it is the unique *compositional and content changes of muscle metabolites during transition from rest to exercise that must be at the heart of Krebs cycle regulation in muscle.*

Control of the Krebs Cycle

Singly, the most important feature of the Krebs cycle is that it breaks down acetyl units without any change in levels of cycle intermediates. That is, *the process is cyclic and catalytic*, with no net accumulation or depletion of intermediates. If for a moment one views the Krebs cycle in isolation, one can imagine it as a super-enzyme catalyzing the breakdown of acetylCoA. Although of course that is an oversimplified view, it is useful because it clearly emphasizes that the simplest way of increasing acetylCoA oxidation is merely to increase the spinning rate of the Krebs cycle without any concomitant changes in cycle intermediates. This kind of control has two fundamental prerequisites. The first is that at least one (or more) enzymes in the process occur in "on" and "off" (low versus high activity) states. Within the cycle, three such enzymes are well known: *citrate synthase (CS) and NAD^+-linked isocitrate dehydrogenase (IDH) in the first span of the cycle; 2-ketoglutarate dehydrogenase (KGDH) in the second span.* CS is under close regulation by the adenylates and CoASH. IDH is regulated by the adenylates, ADP being essentially an absolute allosteric requirement for catalysis, while 2-ketoglutarate dehydrogenase is product inhibited by NADH. Thus any changes in the mitochondrial redox state, in CoASH availability, or in the concentration ratios of ATP, ADP, AMP, and P_i, as may be expected during transition from rest to work in muscle, may lead to concerted activation of all these enzymes, providing nothing else is limiting. Put another way, if substrate supply for any one of these enzymes were saturating under resting conditions, flux through the Krebs cycle could be activated by simply increasing spinning rate (i.e., increasing the amounts of catalytically active enzymes at these key control sites). This condition, *that these enzymes not be limited by substrate supply*, in fact is the second prerequisite for this simplest kind of control mechanism, and unfortunately it cannot be met in muscle at rest. From the best evidence available, at least one of these enzymes (CS) is undoubtedly limited by vanishingly low concentrations of oxaloacetate; moreover, it is likely that IDH and KGDH are also limited by isocitrate and 2-ketoglutarate availability. That is why, in order to increase the spinning rate of the Krebs cycle it is also necessary to augment the pool size of the cycle intermediates.

Augmentation of Krebs Cycle Intermediates

There are a variety of pathways by which intermediates can either be bled from, or fed into, the Krebs cycle (see Chapter 2). In the powerfully energetic flight muscles of Hymenoptera, pyruvate carboxylase channels pyruvate from mainline aerobic glycolysis into the mitochondrial oxaloacetate (OXA) pool from which carbon is spread throughout the cycle intermediates. In the flight muscles of the blowfly and the tsetse fly, and the mantle muscle of squid, proline serves as both a substrate for complete oxidation and as a means for augmenting the Krebs cycle (Mommsen and Hochachka, 1981). Whereas in principle these could work to a modest extent in mammalian muscle, they are far less important quantitatively than are aspartate and P-enolpyruvate (PEP). Aspartate is mobilized through the coupled operation of two amino transferases, forming oxaloacetate for the Krebs cycle (Fig. 4-5); alanine

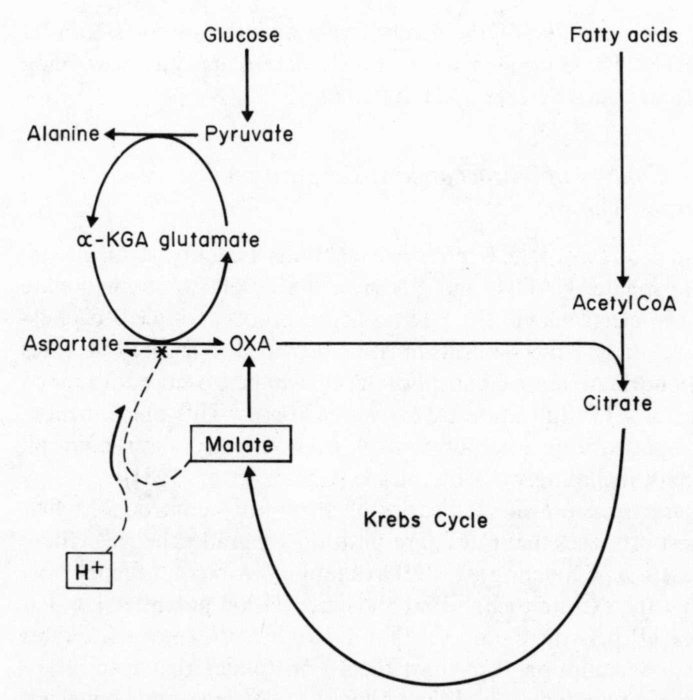

Fig. 4-5. Favored direction of aspartate aminotransferase function during Krebs cycle augmentation. The aspartate affinity of the mitochondrial enzyme is about fivefold higher than that of the cytoplasmic isozyme, and it therefore competes effectively for intramitochondrial aspartate. Its K_m for oxaloacetate is about tenfold higher than intramitochondrial oxaloacetate concentrations, a factor that also favors function in the direction of oxaloacetate production. Moreover, malate concentrations are increasing during Krebs cycle activation, and malate serves as a potent inhibitor of the backward reaction. A low pH further potentiates the malate control at this locus. Modified after Hochachka and Storey (1975)

accumulates and in cases such as cardiac muscle, where this is the main means for augmenting the Krebs cycle, the alanine formed equals the summed increase in levels of all cycle intermediates. Skeletal muscle, in addition to this mechanism, contains significant levels of PEP carboxykinase, which is capable of channeling carbon from mainline glycolysis for purposes of Krebs cycle augmentation during activated fat catabolism. Thus, tissue-specific and work-specific effects of metabolites serve to adjust Krebs cycle spinning rates according to moment-by-moment changes in need. In principle, very similar mechanisms integrate the activities of the Krebs cycle and the electron-transfer system (ETS). These mechanisms in ETS function are only now being clarified, and hence deserve a careful analysis.

Acceptor Control of Mitochondrial Respiration and Phosphorylation

The end products of the Krebs cycle reactions are CO_2, H_2O, guanosine triphosphate, NADH, and $FADH_2$; the latter two both donate protons and electrons via the ETS to O_2, a process coupled to phosphorylation. *In vivo* mechanisms must be provided to assure that rates of mitochondrial oxidative phosphorylation can be closely coordinated with rates of ATP utilization by myosin ATPases. This phenomenon, termed respiratory or acceptor control, is empirically observed in all aerobic cells, including working muscle (Chance et al., 1981).

There are three prevalent theories of respiratory control. The first and earliest proposes that rates of respiration are graded between State 3 and State 4 as a function of ADP availability. A second theory postulates that the extramitochondrial phosphorylation potential, $[ATP]/([ADP] \times [P_i])$, is the parameter that determines the immediate rates of oxygen consumption. The third theory postulates that respiratory control is simply a function of the [ATP]/[ADP] ratio and somewhat independent of $[P_i]$ (see Jacobus et al., 1982 for literature).

Although the matter has been controversial, recent data suggest that free [ADP] in tissues is much lower than estimated from the values of [ADP] measured in cellular acid extracts, thus undermining both the second and third of the above theories of respiratory control. Since the new values for [ADP] fall in ranges where only low rates of mitochondrial respiration are usually measured, they create a paradox as to how mitochondria actively respire in tissues under the presumably inhibitory conditions of high phosphorylation potentials or high [ATP]/[ADP] ratios. This paradox appears to have been resolved in recent studies by Jacobus et al., (1982) who experimentally managed to separate two critical parameters: 1) availability of ADP to the F_1 ATP synthetase and 2) the exogenous [ATP]/[ADP] ratio. Under such conditions, State 3 respiration rates correlate with absolute [ADP] and the most plausible *explanation of respiratory control is the availability of ADP to the ATP synthetase* (and possibly the kinetics of its transport into the mitochondria by the adenine nucleotide translocase). It is particularly interesting that apparent K_m values for ADP of mitochondria from heart muscle are in the 15 μM range if assessed by exogenous ADP. In contrast, apparent K_m values for ADP of liver mitochondria are about 50–70 μM (Jacobus et al., 1982; McGilvary, 1975). This difference, which may arise from structural differences between liver and muscle mitochondria or from isozymic forms of ATP synthetase (see Hochachka et al., 1983), is even greater if assessed by endogenous ADP, utilizing the cytosolic and mitochondrial creatine phosphokinase reactions.

Energy Transport Functions of Creatine Phosphate

It is now widely appreciated that CrP and creatine phosphokinase (CPK) play important roles in aerobic energy metabolism of heart and muscle (Bessman and Geiger, 1981). As mentioned above, the MM-type CPK is found in the cytosol (the so-called "soluble" form of the enzyme), but is also localized within the thick filaments and on the sarcoplasmic reticulum. The CPK_m isozyme, on the other hand, is bound to the exterior aspect of the inner mitochondrial membrane. For over a decade, it has been considered plausible that the mitochondrial isozyme participates in the production of CrP in concert with oxidative phosphorylation, supplying CrP for cytoplasmic CPK to generate locally ATP required by the excitation-contraction-relaxation cycle (Fig. 4-6). Much of the evidence for this scheme is reviewed by Bessman and Geiger (1981). At this point suffice to mention that one of the most conclusive pieces of evidence supporting an energy transport function for CrP derives from tracer experiments showing that oxidative phosphorylation can supply ATP to cytoplasmic CPK without first mixing with the extramitochondrial pool of ATP. This result indeed is predictable from the shuttling models proposed (Fig. 4-6), but is otherwise hard to explain.

Role of Mitochondrial Creatine Phosphokinase

Since the function of CPK_m may be the delivery of ADP via the adenylate translocase to F_1 ATP synthetase, the *in vivo* direction of net flux is

$$\text{Creatine (Cr)} + \text{ATP} \longrightarrow \text{CrP} + \text{ADP}$$

Kinetic studies of CPK_m *in situ* (i.e., in purified heart muscle mitochondria) indicate apparent K_m values for Cr and ATP of 5 mM and 0.7 mM respectively, each being about an order of magnitude *higher* than the K_m values for CrP and ADP (Saks et al., 1975). This suggests that CPK_m is usually saturated with ATP, while [Cr] may or may not be

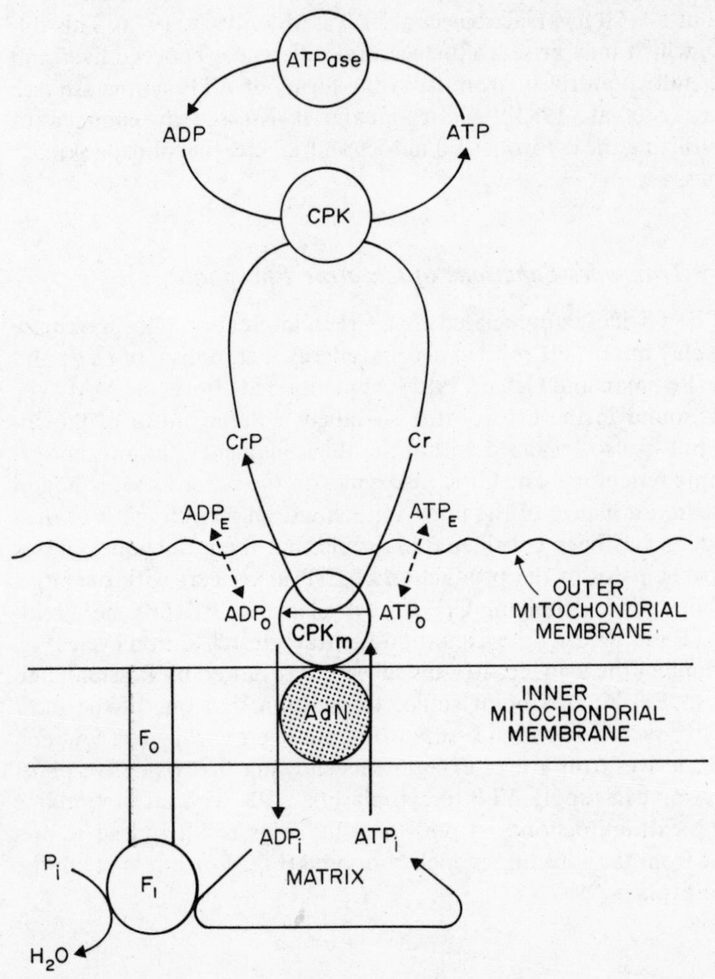

Fig. 4-6. Roles of cytosolic CPK and CPK_m in shuttling high energy phosphate from sites of ATP formation to sites of ATP utilization. After Moreadith and Jacobus (1982) and Bessman and Geiger (1981)

saturating. Equally important, the kinetic properties show that some provision must be made for rapidly removing the products of the above reaction, ADP and CrP; for, given the above relative affinities, *product inhibition is inevitable and would thus preclude efficient CPK_m function in the direction written.* This kinetic problem apparently is avoided by CPK_m being strategically positioned so as to supply a preferred pathway for ADP from CPK_m to the adenine dinucleotide translocase. To demonstrate this role for CPK_m, Moreadith and Jacobus (1982) initiated State 3 respiration either with exogenous ADP (readily inhibitable with a competitive inhibitor, atractyloside) or with ATP in the presence of 20 mM creatine (i.e. utilizing mitochondrial CPK to generate ADP endogenously). In the latter case, four- to fivefold higher amounts of inhibitor were required to block respiration, even though the actual [ADP] sparking respiration was *less* than in the control experiments. The possibility that this endogenous effect with heart mitochondria was simply explainable by intermembrane ADP generation was ruled out by parallel studies using liver mitochondria and nucleoside diphosphokinase for endogenous ADP generation (CPK_m was not used because liver mitochondria do not contain this isozyme); State 3 respiration in this case showed normal behavior. Taken together, these data therefore suggest that in heart muscle mitochondria, the ADP generated by CPK_m is formed in such close proximity to the adenine nucleotide translocase that atractyloside inhibition of respiration is effectively overcome; i.e., *there is a functional coupling between the two systems allowing a preferential access route for CPK-generated ADP to the translocase* (Fig. 4-6). Aside from obvious functional implications, these experiments allow more accurate estimates of ADP affinity and establish a) that the apparent K_m values for State 3 respiration of heart muscle mitochondria are as low as 2 μM ADP, values close to the 0.5–1.0 μM values observed by direct measurements of F_1-ADP binding, and b) that the apparent affinity of heart muscle F_1 ATP synthetase for ADP is not three- to fourfold, but rather may be some thirtyfold, higher than in the case of the liver homologue.

Metabolic Significance of High Mitochondrial ADP Affinities

We have argued elsewhere (Hochachka et al., 1983) that if these data are generally applicable, two important implications arise. Firstly, in the case of muscle or heart cells working under aerobic conditions, the high apparent affinity for ADP assures that mitochondrial respiration can respond sensitively to even modest changes in free [ADP]. A second implication of the very low $K_{m(ADP)}$ values plus the key role played by ADP in respiratory control is that *in vivo* heart and muscle mitochondria probably operate *at below saturating levels of this key substrate.* Otherwise mitochondrial respiration could not respond to change in [ADP]. As with other metabolic enzymes, small changes in substrate (ADP) availability thus can cause large changes in flux.

Conversely, since F_1 demonstrates catalytic properties typically found for regulatory enzymes, small changes in the ADP affinity of F_1 ATP synthetase may lead to large changes in catalytic rate. In this view, *acceptor control of mitochondrial respiration revolves around changes in ADP availability or changes in F_1 ATP synthetase affinity for ADP, or both.*

ADP Control of Mitochondrial Respiration during Exercise and Training

To understand better the above mechanisms in the context of exercise metabolism, three series of critical observations must be recalled:

1. Sustained muscle work, supported by aerobic metabolism, is associated with only modest drops in [ATP], about 0.2 μmol/gm, and therefore even smaller increases in available ADP (since a significant fraction of the latter is protein-bound). As expected from CrP shuttling models (Fig. 4-6), the cytosolic ADP "pulse" is reflected in a [creatine] change to which heart and muscle mitochondria appear to respond via CPK_m (Mahler, 1980).
2. During endurance training, the oxidative capacity of skeletal muscle approximately doubles even in a relatively glycolytic muscle such as the gastrocnemius. In principle, this end result could be achieved in two ways: by adjustments in enzyme amount or by adjustments in isozyme kind. Currently, it appears that although both mechanisms are utilized, the dependence upon isozyme adjustments is minimal. Most enzyme adaptations of muscle mitochondria are known to involve change in activity/gm tissue, but isozymic forms are either not known (e.g., most components of the ETS) or have not been looked during training (e.g., cytochrome oxidase). Included in this category of adjustment are citrate synthase and other Krebs cycle enzymes, various ETS enzymes in a constant proportion manner, and F_1 ATP synthetase (see Hochachka et al., 1983, for literature in this area).
3. Interestingly, many comparable enzyme adaptations occur during high altitude adaptation, which at the level of muscle metabolism apparently involve some adjustments that are similar to those in endurance training. In this case, in addition to enzyme content and composition adjustments, the apparent $K_{m(ADP)}$ of mitochondria also may be reduced, although mechanisms for this effect are thus far unknown.
4. In species genetically adapted for sustained exercise, the activity levels of oxidative enzymes are always substantially higher than in

homologous muscles of more sedentary relatives (for examples, see Marsh, 1981 and Table 4-6).

When we integrate all these observations with the new developments in acceptor control of mitochondrial metabolism several unexpected and previously overlooked features of metabolism during exercise arise (Hochachka et al., 1983). Firstly, for a given level of work, muscle mitochondria from endurance trained (and high altitude adapted) individuals sustain about half the flux rate/gram compared to mitochondria from muscle of untrained individuals (because they have about twice the enzyme capacity); exactly the same considerations apply for between-species comparisons, muscle mitochondria from species adapted for prolonged aerobic work (e.g., migrant birds) sustaining *lower* flux rates/gm muscle for a given work level than mitochondria from species adapted for short-term work. Secondly, the ADP saturation curves for mitochondria from high altitude adapted (and from trained?) individuals are left-shifted (i.e., display lower K_m values for ADP). Thirdly (Fig. 4-7), because of the first two features above, for a given level of work, muscle mitochondria from trained, (or high altitude adapted) individuals and from species adapted for long-term aerobic work *operate at substantially lower ADP concentrations ranges and therefore can respond more sensitively to change in [ADP]*. And fourthly, the mitochondrial catalytic potentials (*amounts* of oxidative enzymes/gm muscle) are elevated during adaptations to strenuous conditions such as endurance performance or high altitude not only to elevate maximum flux rates and power output attainable, but also to assure effective operation at low [ADP]; i.e., to maintain and indeed improve control of mitochondrial respiration by assuring that these mitochondria are responsive to even the smallest, micromolar pulses of [ADP].

In addition, increasing ADP affinity while lowering the concentration range over which F_1 operates should buffer mitochondrial function against falling O_2 concentrations, which are to be expected during strenuous muscle work or during chronic hypoxia (as at high altitude). Wilson et al. (1979) point out that adjustments to falling O_2 concentrations involve increasing the [ADP], the [ADP]/[ATP] ratio, and the reduction state of cytochrome *c*, so as to maintain unaltered rates of ATP synthesis. This is one mechanism proposed to explain why mitochondrial respiration can be so apparently insensitive to changing O_2 concentrations (down to less than 1 μM levels). What this means is that at any given level of work, mitochondria from trained individuals should be able to maintain ATP synthesis rates independent of O_2 down to lower [O_2] than in untrained individuals, and this can be viewed as a final benefit of the above enzyme adaptations.

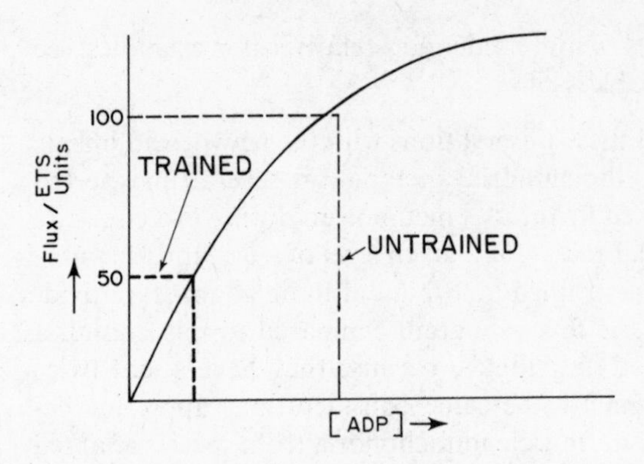

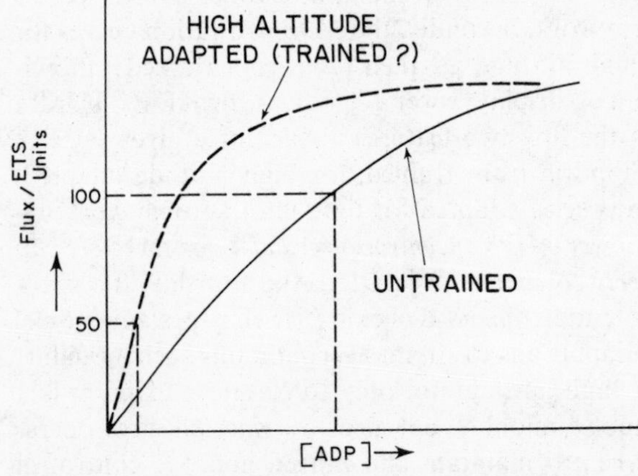

Fig. 4-7. Graphic representation of hypothetical ADP saturation curves for muscle mitochondria from trained versus untrained individuals, showing that for a given level of work, the former operate at lower [ADP]. A drop in the apparent K_m for ADP left-shifts these curves and implies function at even lower [ADP] ranges

Adaptive Strategies for Sustained Work: Overview

To return to our theme, it will now be clear that all three basic adaptational strategies (adaptation of enzyme machinery, enzyme milieu, and metabolic output) are utilized in adjusting skeletal muscle (particularly red and intermediate fibers) for sustained performance. The main adjustments at the enzyme level appear to involve simple increases in the levels of enzyme activities per gram of muscle. This does not hold for all catabolic enzymes, but rather applies mainly to enzymes in fat catabolism, the Krebs cycle, and the electron transport system. Thus the potential aerobic power output is elevated in red and intermediate muscles, but unlike the situation in glycolysis, where essentially every enzyme step in the pathway is catalyzed by a muscle-specific isozyme or isozyme array, the enzymes of fat mobilization, fatty acid activation, β-oxidation, and the Krebs cycle (particularly the latter two for which the best data are available) do not occur in isozymic form. Thus, transition from rest to maximum work rates requires adjustments at other levels, namely, in cellular milieu and in the way metabolic output is regulated.

The milieu in which muscle enzymes operate during aerobic exercise is adjusted in at least three ways. Firstly, large quantities of intracellular triglyceride are retained in red muscle. Secondly, creatine phosphate, although at lower levels than in white muscle, plays a rather critical role as a shuttle between sites of ATP formation (in the mitochondria) and sites of ATP utilization (at the myosin ATPase). And thirdly, CPK_m functions so as to facilitate the flow of ADP from cytosol to mitochondrial sites of oxidative phosphorylation.

The third and final level of adaptation of muscle metabolism for aerobic exercise occurs in the way in which energy generation is controlled. During early stages of highest-intensity aerobic work, at least in man, glycogen and fat are *simultaneously* mobilized in order to obtain high and sustainable rates of energy generation. As endogenous glycogen stores are depleted, the importance of fat as a carbon and energy source rises, until in long-duration aerobic exercise it is the sole fuel being utilized. This ultimate transition to fat and fat alone as a fuel occurs, however, at an energetic cost, since fat can only sustain about 60 percent the rate of energy generation that is possible with glycogen or with glycogen plus fat. Also, in such prolonged aerobic work, endogenous triglyceride gradually becomes depleted and blood free fatty acids take on an increasingly important role as substrates for oxidative metabolism.

Control of the transition from rest to work, at least for fat metabolism, resides at the level of TG lipase, for which the two key features of a flux generator are realized: firstly, the enzyme is fully saturated with substrate (intracellular triglyceride is abundant in red muscle), and secondly, the enzyme occurs in "on" and "off" states. Thus flux becomes proportional to the amount of enzyme in the active form.

For the utilization of blood free fatty acids by aerobically working muscle, the flux-generating step does not appear to lie within muscle metabolism per se, for at no step between fatty acid uptake through β- oxidation to delivery of acetylCoA to the Krebs cycle are the conditions for a flux generator realized. The effective flux generator for fatty acid metabolism in *muscle* during aerobic exercise therefore is probably TG lipase in *adipose* tissue. Its activity is known to be well integrated with the needs of the working muscles through the balanced effects of lipolytic hormones (epinephrine, glucagon), antilipolytic hormones (insulin), and metabolites (which also control the physiologically reverse process of fatty acid esterification). As in the control of glycogen phosphorylase, adenyl cyclase plays a pivotal role in interfacing triglyceride mobilization in adipocytes with the overall metabolic requirements of the organism per se. Through its protein kinase activation TG, DG, and MG lipases are phosphorylated and so converted to their catalytically more active forms, leading ultimately to a release of free fatty acids into the blood when muscle demands are rising. At the level of working muscle, the rate of free fatty acid metabolism is roughly balanced between substrate supply and substrate demand. At a finer level of control, the process is probably regulated at the level of acylCoA formation, at transfer into the mitochondria, or at the level of β-oxidation per se. Finally, the activity of the Krebs cycle and of oxidative phosphorylation must also be regulated to pace that of β-oxidation. The first is achieved by turning on mechanisms for augmenting the pool of Krebs cycle intermediates, which in concert with modulator effects on key regulatory enzymes lead to increased rate of spinning of the cycle. Oxidative phosphorylation, on the other hand, seems to be controlled either by change in ADP availability or by change in ATP synthetase affinity for ADP, or by both.

Summed together, these adjustments underpin sustained aerobic performance. Although the degree of activation of aerobic power output is less than that for phosphagen-based or glycolysis-based performance, it is nevertheless impressive. In insects, particularly in those which warm up their flight muscles before takeoff, flight metabolism may require a several-hundredfold increase in rates of O_2 consumption. In man, and most vertebrates, the figure is less, usually about tenfold (Weibel and Taylor, 1981) although in some mammalian species it can reach fortyfold (Thomas and Fregin, 1981). As with anaerobic glycolysis, this magnitude of swing from basal to activated rates of metabolism is in no way representative of metabolic control and metabolic pathways in general; rather, it has developed in different species as a specific kind of adaptational response to specific kinds of aerobic locomotory needs. The response is a good and long-term strategy that covers the majority of exercise problems the majority of time encountered by the majority of organisms, so much so that the literature on exercise adaptations in mammals usually ignores a third kind of need that often develops in nature: the need to be able to perform for longer periods than are allowed by available glycogen (glucose) and fat supplies. In a number of situations, these reserves cannot be recharged by rest and nutrition: the long-range overland migrations of caribou, the upstream spawning migrations of anadromous fishes (such as salmon), the overseas nonstop migrations of birds, and so forth. For this kind of performance, yet another source of energy must be harnessed: body protein reserves (see Mommsen et al., 1980). Because in man and other mammals these are not normally mobilized during exercise, the literature in this area has developed in another context (starvation and migration).

Competitive Interplay between Different Energy-Yielding Pathways

Since different ATP-replenishing processes are selected for support of specific kinds of work, we are still faced with the question: what determines which ATP-yielding pathways are utilized during muscle work of varying intensity? Although many of the above-discussed regulatory mechanisms may play a role in selecting which fuels and which pathways are utilized, one mechanism which may be most pivotal involves a functional competition for ADP. A *modus operandi* for the system is given by Hochachka et al. (1983) and runs as follows:

During work of highest power output (up to 6 μmol ATP $gm^{-1} \cdot sec^{-1}$), myosin ATPases are activated to approximately their physiological maximum (Table 4-2). Since the myosin isozymes of white muscle display higher turnover numbers than do their homologues in any other fiber types, this may involve their specific activation. Yet in spite of the near maximum activation of myosin ATPases, their catalytic potential apparently does not exceed that of CPK, whose activity level responds to changing [adenylate]. During early stages of phosphagen-powered work, neither a decrease nor an increase [ATP] can be easily detected, probably because its percentage change is small compared to the total (5 μmol/gm) pool size available. In contrast, the change in concentration of free ADP is more readily estimated (see Gadian et al., 1981) presumably because its fractional increase is larger compared to its starting pool size. The latter is particularly important, for CPK (MM isozyme), is a high-activity enzyme and displays in the presence of CrP an avid affinity for ADP; *by outcompeting PGK and PK for ADP, a preferential mobilization of CrP is assured at the appropriate time.* Since ADP may gradually become saturating for CPK

(MM) at such times, this process can continue until phosphagen is depleted, *but at decreasing rates*, because CPK flux becomes proportional to falling [CrP]. In fact, this probably is why the highest power output achievable by CrP hydrolysis cannot be sustained for more than a few seconds.

During sustained muscle work (power output of about 0.25–0.5 μmol ATP $gm^{-1}sec^{-1}$) myosin ATPases are activated to only about one-sixth to one-tenth of their physiological maximum (Table 4-2). Changes in [ATP] and [ADP] at this level of work are rather modest; nevertheless, [ATP] changes usually are observed often in the range of about 0.2–0.5 μmol ATP/gm, and the changes in [ADP] are accordingly even smaller. As at highest work rates, the MM isozyme of CPK is catalytically and kinetically tailored to outcompete glycolysis for small amounts of ADP; as a result, the change in [ADP] is expressed as an increase in [Cr], which serves to activate CPK_m (i.e., to complete the CrP-Cr shuttle). In essence, the CrP-Cr shuttle, which under highest power output conditions may be blocked by O_2 lack, is now allowed to operate and it serves to transmit a small [ADP] signal from cytosol to F_1 ATP synthetases in the mitochondria. Because of the very high ADP affinity of F_1 ATP synthetases, this assures that *oxidative phosphorylation is the only pathway that can respond to the modest* (*μmolar*) *changes in* [*ADP*] *that may be expected during sustained work of low power output.*

In both cases above, the overall system appears to be set up so as to give cytosolic CPK preferential access to the ADP liberated by myosin ATPase hydrolysis of ATP. At highest power output, CPK (MM) per se in effect precludes significant glycolytic contribution, while under aerobic conditions, the CrP-Cr shuttle serves as a means for minimizing glycolytic consumption of ADP, and thus helps to assure that anaerobic glycolysis is not utilized inappropriately. When anaerobic glycolysis *is* appropriate during work of intermediate power output (about 1 μmol ATP $gm^{-1}sec^{-1}$) and when CrP supplies are depleted, myosin ATPases are activated to only about one-third of their physiological maximum rate (Table 4-2). Falling [ATP] and rising [ADP] are always reported when phosphagen supplies are depleted and these changes are larger than during higher or lower power output periods (Gadian et al. 1981). Under these conditions, dropping [ATP] may be large enough to deinhibit PFK, allowing glycolytic carbon flow to increase PEP availability; at this point, and as dropping [CrP] becomes limiting to CPK, *muscle PK becomes ever more competitive for ADP, thus completing the preferential glycolytic activation and mobilization of glycogen at intermediate power outputs.*

Models in science are useful for making predictions and for explaining earlier and sometimes perplexing data; the above model, accounting for fuel and pathway function at appropriate work rates, displays several such characteristics (Hochachka et al., 1983):

1. Rates of glycolysis should be higher in the presence of an active ATPase, because of the cycling of ADP between the latter and the former, as in fact has been experimentally observed.
2. The MM form of CPK can outcompete glycolysis for ADP in the 0.1 μmol/gm range through all [CrP] changes (i.e., until CrP is depleted). This prediction is based on kinetic properties of CPK (MM isozyme) and PK (muscle isozyme), and empirically is often observed, which is why, under highest work rates, phosphagen depletion may be largely completed *prior* to any significant glycolytic activity. An important enzyme basis for this observation is that PK is not an effective competitor for ADP until PEP is available. And even then, it has a lower affinity for ADP then does CPK. That is why, while accepting usually emphasized glycolytic control mechanisms (focused on PFK and glycogen phosphorylase) the above model places *additional emphasis on ADP regulation of the lower half of the glycolytic path.* In fact, this model would predict that in reconstituted glycolytic systems, exogenous addition of CPK and CrP would strongly *inhibit* glycolysis by outcompeting PK for limiting ADP.
3. At [ADP] of about 0.1 μmol/gm, exogenous CPK and CrP should potently inhibit mitochondrial respiration, which indeed has been observed.
4. Another prediction is that at about 0.1 μmol ADP/gm, glycolysis should potently inhibit mitochondrial respiration (if CrP is low or CPK is absent). This too has been experimentally observed. Moreover, such inhibition a) is dependent upon a competition for ADP and b) is *abolished* if PGK and PK are deleted, as would be expected.
5. Although experimentally it is already known that the competition for ADP between glycolysis and mitochondrial metabolism can lead to respiratory inhibition, the ADP concentration dependence of this process has not been clarified. If this model is correct, the above inhibitory interaction should be abolished at low [ADP] (in the μmolar range), where only mitochondrial metabolism appears competitive. Moreover, the [ADP] dependence of these interactions should be different in muscle and liver mitochondria because of the higher ADP affinities of the former.

6. This model also suggests that it should be possible to show CrP depletion or glycolytic activation *before* O_2 supplies of working muscle are depleted; the latter has in fact been observed but never properly explained.
7. Finally, this interpretative model supplies a biological rationale and explanation for why, of the three pathways for generating ATP, mitochondrial metabolism displays by far the highest ADP affinity: this assures the organism that whenever O_2 availability is adequate, oxidative metabolism will outcompete anaerobic pathways for limiting ADP, thus gaining efficiency advantages while avoiding the undesirable problems of fermentation.

Contrasts and Similarity, Anaerobic versus Aerobic Metabolic Performance

Before proceeding further, it may be instructive to compare briefly the properties of metabolism supporting anaerobic performance to those in aerobic work. There are at least six fundamental differences between the two systems:

1. Anaerobic energy generation is activated in response to short-term locomotory needs; aerobic energy generation, in contrast, is used to support long-term performance.
2. Anaerobic metabolism behaves as a closed, self-contained system, while aerobic work and metabolism require an open system with functional exchange and communication channels between muscle and the rest of the body.
3. The power output (rate of ATP turnover) in both depends upon the maximum catalytic potential of enzymes in anaerobic and aerobic metabolism, respectively, but the actual power output of the former exceeds that of the latter by two- to fourfold. However, this advantage can be maintained for only a short time.
4. Total power output in anaerobic performance is determined by the amount of endogenous substrate, while in aerobic performance it is determined by the amount of endogeneous substrate *plus* exogenous substrate that can be delivered to working muscle from liver and adipose tissue.
5. In white muscle burst work, creatine phosphate serves primarily as a $\sim$P donor, while in aerobically working muscle, it serves as a $\sim$P shuttle between mitochondria and muscle.
6. Burst-working white muscle typically relies upon only one carbon fuel (glycogen) and one metabolic pathway for its catabolism, while

O_2-based work in red muscle is preferentially fueled by glycogen *plus* fat; in prolonged work, however, aerobic muscle work is sustained by fat metabolism alone. In either event, several metabolic pathways must be simultaneously regulated.

Scaling of Glycolytic and Oxidative Enzyme Capacities

The important conclusion emphasized above that white muscle relies on endogenous substrates (phosphogen and glycogen) during burst function while the long-term function of red muscle can be strongly reliant on oxygen and substrates transported into the muscle during work finds an interesting manifestation in size-dependent scaling of enzymic activities in skeletal muscle. A wide spectrum of physiological processes display what is termed a "scaling" relationship with body mass (see Schmidt-Nielsen, 1979). Processes associated with aerobic performance typically scale with an exponent less than one; that is, larger individuals display a lower intensity of the process per unit mass than smaller individuals. The well-known "shrew to elephant" curve of oxygen consumption per unit mass is perhaps the most familiar aspect of the scaling of aerobic processes. Whereas an elephant obviously consumes more oxygen than a small mammal like a shrew, per unit mass the shrew far outstrips the larger mammals.

Even though metabolic scaling relationships have long been appreciated in terms of oxygen consumption rates, scant attention has been paid to the scaling of anaerobically powered metabolism. Thus, when Somero and Childress (1980) examined the activities of two glycolytic enzymes, PK and LDH, in white skeletal muscle of thirteen species of teleost fishes, it came as somewhat of a surprise that larger-sized individuals exhibited significantly *higher* PK and LDH activities per unit mass of muscle than smaller specimens. In contrast, the Krebs cycle enzyme citrate synthase scaled according to the expected relationships for aerobically poised systems (i.e., lower activities per gram in larger individuals). These relationships are illustrated in Figure 4-8, along with PK, LDH, and CS data for brain.

The three distinct size-versus-enzymic activity patterns reflect important characteristics of locomotory tissue design, and show very clearly how differently aerobic and anaerobic pathways may scale with body size. The decrease in CS activity per gram of white muscle in larger-sized specimens of a species may reflect limitations on oxygen supply in larger individuals (see Schmidt-Nielsen, 1979, for a discussion of possible mechanistic bases of aerobic scaling). Aerobic scaling may

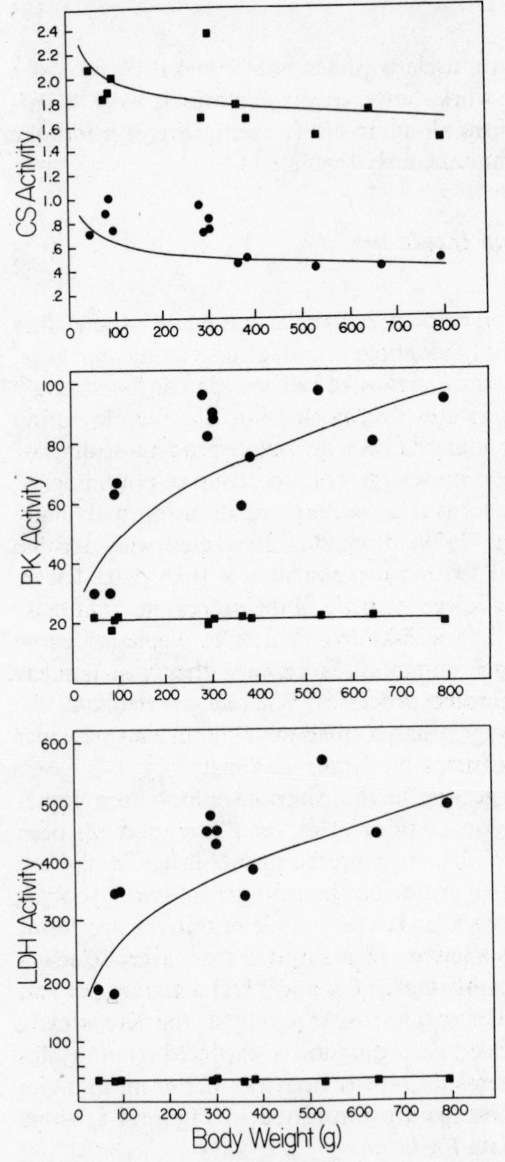

Fig. 4-8. Scaling of LDH, PK, and CS activities in muscle (circles) and brain (squares) of *Paralabrax clathratus* and *P. nebulifer* (data for both species are combined). Enzymic activities are expressed as international units of activity per gram wet weight of tissue, at a measurement temperature of 10°C. Figure modified after Somero and Childress (1980) *Physiological Zoology* 53:322-337 by permission of the University of Chicago Press. © 1980 by the University of Chicago

involve surface-to-volume aspects of organismal design which influence the capacity of the respiratory and circulatory systems to supply working muscle with fuel and oxygen; however, hypotheses in this area remain controversial. In the case of PK and LDH scaling, we are concerned with a system which, as emphasized above, is not dependent on exogenous supplies of oxygen and substrate during short-term burst function. Thus, we must try to account for the PK and LDH scaling patterns in terms of the design of muscle for burst function during short periods of activity.

The key to understanding the increases in glycolytic enzymic activity per unit mass in larger-sized individuals of a species involves the important fact that burst swimming abilities, as measured on the relative scale of body lengths traversed per unit time, are virtually the same in small and large fishes of a species (see Somero and Childress, 1980 for literature). In contrast, the aerobic swimming abilities of larger fish, as measured on the relative scale of body lengths per unit time, are lower than those of smaller fish. The latter fact seems consistent with the reduced CS activity per gram of muscle in larger fish. Can we account for the rise in PK and LDH activity per gram muscle in terms of power requirements for maintaining the same burst swimming performance in different-sized fishes of a species? Somero and Childress (1980) have calculated that the increases in PK and LDH activity per gram muscle are precisely the scaling patterns needed to overcome the increase in drag that occurs as the size of the fish increases. The glycolytic power of fish white muscle therefore varies with body size in such a way as to supply the additional power required to overcome length-dependent increases in hydrodynamic drag. The importance of this scaling adaptation probably relates to predator-prey relationships, where burst swimming abilities are often of critical importance (see Somero and Childress, 1980).

The scaling patterns noted in brain merit contrasting with the patterns noted in white muscle (Fig. 4-8). PK and LDH activities are virtually constant in brain, regardless of body size or species (Somero and Childress, 1980; Sullivan and Somero, 1980). Thus the scaling of LDH and PK activities found in white muscle appears to be associated with locomotory function per se, not with some ubiquitous constraints operative in all tissues. CS activity of brain exhibits the classical scaling relationship noted for aerobic processes.

In an independent study of ten mammalian species varying in body mass by about 10^6, Emmett also observed that glycolytic enzyme activities in skeletal muscle scale directly with body mass, while oxidative enzymes scale inversely. The catalytic activities of oxidative enzymes in fact scale with a slope similar to, or less than, that observed for maximum O_2 uptake of the whole organism, and are therefore consistent with the observation that the aerobic scope for activity of large mammals is greater than for small mammals (Weibel and Taylor, 1981). The scaling of glycolytic enzyme potentials, in contrast, cannot be similarily rationalized, since there is no information available on the scaling of glycolytically powered muscle work in mammals (Emmett and Hochachka, 1981).

Exercise Adaptations in Invertebrates: Insects as Invertebrate Champion Athletes

Many of the mechanisms discussed above for vertebrates (mainly mammals) may very well be applicable to most invertebrate groups. However, there are now some well-known and rather novel metabolic situations in invertebrates that are interesting in their own right and may better illuminate basic metabolic and adaptational principles. Because of their economic importance, insects supply a particularly rich source of data from which to draw examples of exercise adaptations. Such adaptations have been reviewed in detail elsewhere (Sacktor, 1976), so we will here merely mention some of the overriding strategies and mechanisms utilized.

Although insect skeletal muscle is differentiated into at least two functional types (flight and leg muscle) showing significant functional differences, the associated electrophysiological and structural specializations that so clearly separate red and white muscle types in vertebrates are lacking. Thus the energetic demands of different kinds of performance (burst versus steady state) must be met by enzymatic and regulatory specializations. That indeed is observed, and it is commonly reported that leg muscles (used for jumping) depend much more strongly on phosphagen (arginine phosphate) hydrolysis and anaerobic glycolysis than do flight muscles. In fact, the flight muscles of some insects are so highly adapted for sustained aerobic work that their metabolism is essentially obligatorily aerobic. Flight muscle lactate dehydrogenase in such species is often completely deleted from metabolism and cytosolic redox during aerobic metabolism is balanced by α-glycerophosphate dehydrogenase, which typically occurs in high activity and is kinetically well adapted for function in the α-glycerophosphate cycle (Chapter 2). Whereas anaerobic capacity is greatly reduced, enzymes in aerobic activity occur at very high levels, thus assuring potentially high rates of energy generation during flight.

Substrates used for firing flight metabolism vary in different groups. Insects with short-term flight (or foraging) needs often utilize carbohydrate, blood trehalose, and glycogen (either endogenous or derived from exogenous depots in the fat body). Bees and wasps fall into this category. These groups in addition display an interesting biochemical adaptation at the level of Krebs cycle augmentation: the enzyme pyruvate carboxylase, although not typically active in muscle tissues, occurs in high activity in flight muscle of bees. It shows an absolute allosteric requirement for acetylCoA and is thought to function in augmenting oxaloacetate supplies (and thus the pool of Krebs cycle intermediates as a whole). It is an elegantly designed augmentation mechanism, since these animals utilize nectar (sugar) as a sole carbon and energy source; because of the built-in acetylCoA requirement (Fig. 4-9), pyruvate carboxylase activity is turned on to the degree oxaloacetate is needed for CS-catalyzed condensation with glucose-derived acetylCoA!

In contrast to *Hymenoptera*, insects that are adapted for longer-term flight either utilize a mixture of carbohydrate and fat (locusts during the first hour or so of flight) or rely solely upon fat (locusts during migration flights, moths, etc). To our knowledge, however, the relative power outputs of these fuels in insects have not been ascertained.

Some insect flyers display an interesting dependence upon proline during activated flight muscle metabolism. Included here are organisms like the blow fly, some beetles, the locust, and the tsetse fly, all of which store high concentrations of proline in their flight muscles. Proline can serve two functions: its simplest role is in augmentation of Krebs cycle intermediates, in which case there need not be any net oxidation of the proline carbon skeleton. In some cases (e.g., Colorado potato beetle), proline enters the Krebs cycle as 2-ketoglutarate and is partially oxidized via the Krebs cycle and malic enzyme, forming pyruvate which accumulates as alanine (see Hochachka et al., 1983 for literature). The adaptational significance of using proline for these functions is not entirely clear. One possibility is that because of its neutrality it is a "compatible" solute and may be stored at high concentrations. Another is that it is energetically advantageous, for even in simple Krebs cycle augmentation it contributes to the energy yield. Furthermore, if a part of the proline pool can be fully oxidized, the energy yield rises even further, since the complete oxidation of proline is potentially capable of yielding 30 moles ATP/mole amino acid (see Chapter 2). For whatever reasons, proline is a critical substrate for many insects, which in this regard show some interesting parallels with cephalopods, particularly squids.

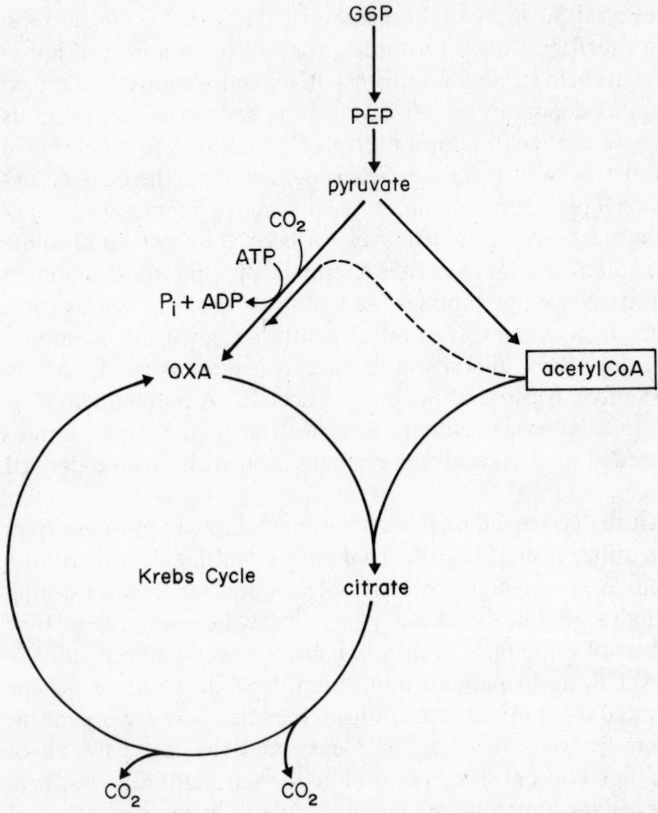

Fig 4-9. Krebs cycle augmentation mechanism in bee flight muscle. The pyruvate branchpoint in bee flight muscle is more complicated than in mammalian muscle because of the presence of an active pyruvate carboxylase. The activity of pyruvate carboxylase shows an absolute allosteric requirement for acetylCoA; hence it is believed that during flight metabolism acetylCoA determines the degree to which pyruvate carboxylase augments the pool of oxaloacetate (OXA) plus other Krebs cycle intermediates. After Hochachka and Guppy (1977)

Squids as Invertebrate Athletic Champions of the Sea

If insects, of all invertebrate animals on land, are capable of highest metabolic rates during sustained work, squids are their metabolic counterparts in the sea. Metabolic rates of fast-swimming squids are high and similar to those of fast-swimming fishes such as salmon. The mantle musculature of squids seems to be differentiated into a kind of doughnut structure (Mommsen et al., 1981), with outer and inner muscle layers displaying relatively high oxidative capacities, the middle muscle layer displaying higher ratios of anaerobic/oxidative enzyme activity, an organization rather reminiscent of that noted above in vertebrate skeletal muscle. Substrate sources for cephalopod metabolism initially presented scientists with a mystery; this is because many squids are not only powerful short-term swimmers, but also undertake long spawning migrations. On the eastern seaboard of North America, for example, the squid *Illex* feeds and grows inshore, but for spawning migrates distances of thousands of kilometers. By analogy with mammals, it was natural to anticipate that the short-term burst type jet propulsions of squid would be powered by anaerobic glycolysis, while the long migrations would be powered by fat. The problem was that early workers found little or no glycogen or fat in the mantle muscle (which supplies the jet power for swimming). The absence of a vigorous fat metabolism has in fact been confirmed, but the absence of glycogen has been refuted.

The problem arises because the isolation of polysaccharide from cephalopod muscles required some special precautions (thus far for reasons that are not understood). When these are taken, values for stored carbohydrate are in reasonable vertebrate ranges (over 0.1 M in terms of glycosyl units/gm muscle). This indeed is consistent with the high levels of glycolytic enzymes found in squid muscle (see Hochachka and Fields, 1982).

Aerobic performance in squids is another matter however; what energy sources are utilized for jetting that can be sustained for prolonged migrations? Although a lot of work remains before this question can be fully answered, current evidence implicates protein plus a large free amino acid pool (about 300 milliequivalents/kg) found in muscle. Of the amino acids present, proline seems particularly important.

It is stored at high concentrations; in *Illex* mantle muscle, for example, it is in the 20 μmol/gm range, and it can be vigorously oxidized (Mommsen and Hochachka, 1981). In their dependence on amino acids and proteins for long-range migrations, then, squids may be similar to salmon during their upstream spawning migrations. The latter, however, seem to display a much higher dependence on alanine and a much lower one on proline than do the squids (Mommsen et al., 1980). As in insects, the phosphagen of cephalopod muscles is arginine phosphate, which is formed by arginine phosphokinase (APK) catalyzing the reaction

$$\text{arginine} + \text{ATP} \longrightarrow \text{arginine phosphate} + \text{ADP}$$

If we assume similar *in vivo* [ADP] in cephalopods as in vertebrates, and a similar equilibrium constant for APK, it appears that free arginine occurs at low levels and that the bulk of the arginine plus arginine phosphate pool is represented by the phosphagen. The higher values of arginine reported may arise from arginine phosphate breakdown. Since overall concentrations are very high (20–50 μmol/gm range), it would be expected that if the phosphagen were used the same way as in CrP in vertebrate white muscle, the cell would be flooded with high levels of free arginine as a consequence of burst work. This may be avoided in cephalopods by "dumping" arginine into a harmless, near neutral compound, octopine. This is achieved by coupling arginine to the terminal step in glycolysis in a reaction catalyzed by octopine dehydrogenase:

$$\text{pyruvate} + \text{arginine} \xrightarrow[\text{NADH} \;\rightarrow\; \text{NAD}^+]{} \text{octopine}$$

In squid mantle muscle, this enzyme occurs in high activity, while LDH is almost fully depleted; the ratio of ODH/LDH is usually about 100:1 (see Storey and Storey, 1983). The twin problems of recycling octopine and of integrating octopine metabolism with that of arginine are only currently being explored; nevertheless, it already seems clear that ODH is advantageous in tissues like squid mantle, where during high-intensity short bursts of jetting, the hydrolysis of a large pool of arginine phosphate can proceed without the concomitant flooding of the cell with arginine, a highly basic and thus a potentially disruptive solute (Somero and Bowlus, 1983).

Roles of Phosphagen and Multiple Terminal Dehydrogenases in Bivalves and Gastropods

At the other end of the activity spectrum in the Phylum Mollusca are much more sluggish animals, the true slugs, bivalves, and lamellibranchs. It is interesting to note that, in a number of instances where adequate data are available, the predominant energy source seems to be arginine phosphate (De Zwaan, 1983). In some cases, anaerobic glycolysis is activated mainly in recovery, presumably for recharging phosphagen reserves, while in others, recovery seems mediated primarily by oxidative metabolism. In still others, recovery may depend upon an interesting interplay between two pathways of glycolysis.

In this context it is worth mentioning that many of these molluscs contain octopine and lactate dehydrogenases (in different ratios, depending upon tissue and species) *plus* one or more other imino-forming dehydrogenases. Of these, alanopine dehydrogenase, the first such ODH-analogue discovered (Fields et al., 1980), and strombine dehydrogenase are particularly common. These dehydrogenases catalyze the reductive condensation of pyruvate with alanine and glycine, respectively, forming alanopine or strombine:

$$\text{keto acid} + \text{amino acid} \xrightarrow[\text{NADH} \curvearrowright \text{NAD}^+]{} \text{imino acid}$$

Just why bivalves and gastropods frequently retain more than one terminal glycolytic dehydrogenase in the same tissue or organ, and how the dehydrogenases partition hydrogen between themselves are essential but thus far poorly understood problems in the metabolic regulation of these fascinating animals. One possibility is that they work at different times for different reasons. During strenous jumping in *Strombus* for example, ODH is utilized as a terminal dehydrogenase once arginine phosphate supplies are depleted. Fatigue correlates with octopine accumulation and octopine product-inhibition of ODH. During *recovery* on the other hand, strombine accumulates, indicating that glycogen fermentation to strombine (Figure 4-10) is being utilized as an ATP source for recharging phosphagen reserves (J. Baldwin, pers. comm.). For a poorly perfused muscle with a dearth of mitochondria, the use of the glycogen → strombine fermentation is an elegant solution to the energy and redox needs of the cell in recovery. Glycogen → octopine fermentation, on the other hand, would not work since octopine oxidation is a requisite for arginine reconversion to arginine phosphate.

Training for Anaerobic Exercise: Possible Sites of Adaptation

In the above overall outline of how the metabolic machinery works in exercise, we have primarily emphasized adaptational qualities on two (short-term and phylogenetic) time scales. Now we wish to consider intermediate time-courses of adaptation and to outline those sites and functions which in principle might be further modified during training; these may then be compared to training effects that are actually demonstrable in practice. A minimal list of such theoretically adjustable aspects of anaerobic metabolism in muscle must include:

1. increasing size of muscles utilized for the exercise (i.e., hypertrophy),
2. fiber-type replacement,
3. increasing levels of endogenous substrates (CrP and glycogen),

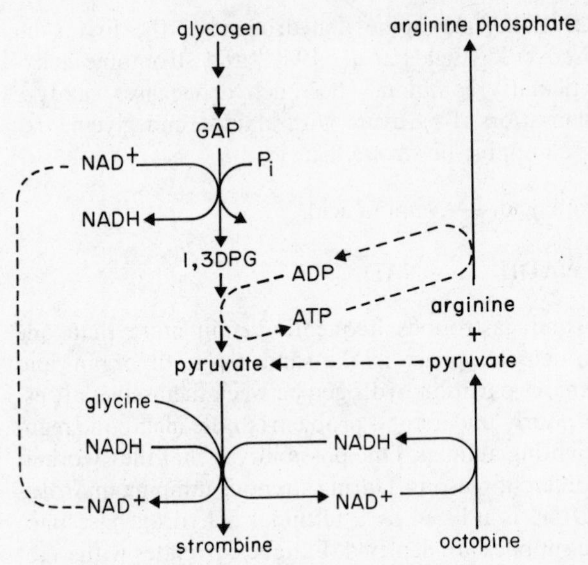

Fig. 4-10. Possible use during recovery from strenuous anoxic work of glycogen-strombine fermentation (as a source of ATP) and the ODH back-reaction (as a source of arginine) for recharging arginine phosphate reserves. If the system were redox-coupled as shown, the use of two sources of NADH and pyruvate would lead to a stoichiometry of 4 moles strombine per glucosyl unit fermented plus octopine oxidized

4. increasing levels of key enzymes in anaerobic metabolism and its control,
5. adjustments in isozyme type at specific loci in metabolism, and
6. improving buffering capacity of working muscles.

Work-Induced Hypertrophy of Muscle

The simplest adaptive response to prolonged if periodic burst-work demands on muscle, *simply to make more of what is already present, i.e., to hypertrophy,* is in fact so obviously utilized it has become a part of society's folklore. All of us will be familiar with the image of the ninety-eight-pound weakling reading with avid interest appealing advertisements on body-building and sexual conquest. This popular lore seems to have a scientific basis because the widely desired hypertrophy of muscle cells and their concomitant increase in strength in fact is an *adaptational response mainly to repeated bouts of anaerobic work* and is either not elicited or is more modestly expressed in endurance training. Although widely recognized, hypertrophy is poorly understood and underlying mechanisms are still largely unclear with only a few general features being well established (Goldberg et al., 1978):

a) Hypertrophy is stimulated by exercise per se, for exercise can act independently of anabolic hormones (growth hormone, insulin, glucocorticoids) and even takes precedence over endocrine signals for muscle depletion (e.g., during starvation).
b) The process involves an activation of RNA and protein biosynthesis and even some DNA synthesis, although it is still unclear if any cell division per se is involved. Inhibition of RNA or protein synthesis predictably blocks hypertrophy.
c) The increase in muscle mass represents primarily an increase in tissue protein, as a result of increased rates of synthesis and decreased rates of degradation. Increase in muscle weight, for example, is directly proportional to the increase in ^{3}H-labeled amino acid incorporation into proteins. Interestingly, even before increased RNA and protein synthesis can be demonstrated, stimulated muscle shows an increased ability to accumulate specific amino acids from the blood, and this indeed is one of the earliest biochemical events signaling that the tissue will hypertrophy.
d) Several lines of evidence suggest that some biochemical consequence of exercise per se or the contractile process per se turns on amino acid uptake. (For example, muscle contracting isometrically consistently takes up amino acids more rapidly than muscle shortening against no load).
e) The increase in muscle mass during such work-stimulated growth represents mainly an increase in fiber size (Gollnick et al., 1981).
f) The increase in muscle mass represents an increase in soluble and myofibrillar proteins, as well as in collagen. However, the soluble proteins tend to be augmented disproportionately, which may reflect changes in contractile properties (reduced contractile speed, for example) as well as changes in metabolic capacities. Such changes could also be mediated by interconversions between slow-twitch oxidative and fast-twitch glycolytic fibers, so it is important to assess this possibility as well.

Adjusting the Proportion of Fast-Twitch Glycolytic Fibers

Because of major biochemical differences between red, intermediate, and white fiber types, increasing the percentage of white fibers may represent another possible way of adjusting the anerobic power of a muscle. This theoretical possibility is raised by electrical stimulation and cross-innervation studies in animals which show that fibers *can* be converted one into the other (see Whalen et al., 1981). What is more, when one considers a longer (phylogenetic) time scale, one of the commonest adaptations for performance style is in the type and amount of muscle fiber retained in working muscles. This is best illustrated in teleosts, where species that rely mainly or solely upon burst-type swimming maintain a large white muscle mass but reduced red muscle; furthermore, the red muscle that is present takes on many of the biochemical properties of white muscle (reduced capillarity, reduced mitochondrial abundance, reduced levels of oxidative enzymes, and so forth). Similar differences are also well illustrated in the muscles of burst-flying birds, such as grouse and partridges. Even in man there is evidence for such gene-fixed differences between individuals. In athletes that are particularly good at anaerobic exercises (weight-lifters, for example), the proportion of white fibers is typically higher than in individuals particularly good at sustained, aerobic work.

Earlier studies assessing this possibility in training programs had difficulty in demonstrating that such effects might actually occur during, or in response to, the exercise regime, although there was some evidence for the interconversion of intermediate and white fiber types. More recently, however, the use of immunohistochemical techniques for distinguishing fast myosin (FM) and slow myosin (SM) have shown that intermediate fiber types *contain both* myosin isozymes, and moreover that the proportions of these can change with physical training. However, no such changes can be measured for the SM and FM isozymes in red and white fiber types (see Lutz et al., 1979; Holloszy et al., 1978). Thus, as repeatedly observed, the potential for adjusting the percentage fiber composition of mammalian muscle by exercise training programs is limited (probably to some conversion of intermediate to white fibers during anaerobic training).

Effect of Anaerobic Exercise Regimes on Glycogen and Creatine Phosphate Supplies

If the proportion of white fibers in working muscle is not increased during training for anaerobic work, to what extent might the anaerobic potential of muscle be improved simply by elevating the stored supplies of creatine phosphate and glycogen? Again, this mechanism appears to be utilized in long-term phylogenetic adaptation, particularly in the case of glycogen. It turns out that it also is utilized to some extent during training programs. In sprint-trained rats, for example, muscle creatine phosphate levels are higher than in controls particularly when the training is coupled with altitude stress. Such an elevation in creatine phosphate level does not typically occur in man, but even if the concentration does not rise, the hypertrophy of skeletal muscle that accompanies anaerobic training in effect increases the total *amount* of phosphagen available during burst work; what this means is that the *total amount of energy extractable from phosphagen is proportionately elevated.* Exactly the same considerations apply for glycogen, which also occurs in higher concentrations in muscles of athletes trained for anaerobic exercises.

Effect of Anaerobic Exercise Regimes on Enzyme Levels

In animals that are specialized for burst-type locomotion not only is the proportion of white type fibers high, so also are the levels of enzymes in adenylate, creatine phosphate, and glycogen metabolism. In tuna white muscle, for example, which can sustain one of the highest rates of burst swimming thus far analyzed by biologists, the activities of myokinase, CPK, and of various glycolytic enzymes are unusually high; lactate dehydrogenase, for example, occurs at over 5,000 units/gm (assayed at 25°C), which contributes to potentially very high rates of glycolysis (Guppy et al., 1979). Thus, it is not surprising to find the same mechanism being utilized during anaerobic exercise regimes in animals and man; particularly good data are available for CPK, myokinase, and LDH. The effects are presumably quite general, and indicate that enzyme concentrations stabilize after training at about 120–150 percent of pretraining levels (Holloszy and Booth, 1976). In addition, it must be remembered that the total content (total catalytic power) of enzymes in anaerobic metabolism is also elevated by simple hypertrophy of muscle during training. Thus the power output of muscle following such training will be elevated by an amount determined by a combination of a) the degree of hypertrophy, and b) the percentage change in concentration of enzymes in anaerobic metabolism.

Effects of Anaerobic Exercise Regimes on Isozyme Composition

As discussed above, the nature of anaerobic metabolism is also strongly determined by the isozyme arrays present at each step in the anaerobic glycolysis of vertebrate muscle, so the question arises as to whether or not training can lead to adjustments in isozyme arrays utilized. The answer appears to be affirmative, but it must be cautioned that only a few studies are available. One of the best involves LDH, which is composed of five isozyme types; these are tetramers formed from the random association of H and M subunits to generate H_4, H_3M_1, H_2M_2, H_1M_3, and M_4-type LDH. In man, during sprint training, both the total LDH activity and the proportion of M_4 LDH increase and this appears to be the case in horses as well (Guy and Snow, 1977). Whether or not such isozyme adjustments during anaerobic exercise regimes turns out to be general remains uncertain and in need of further study.

Effect of Anaerobic Exercise Regime on Muscle Buffering Capacity

The production of large amounts of lactate plus protons during burst work indicates the need for intracellular buffers. In vertebrate muscles, intracellular buffering is dominated by histidine imidazole groups. These may either occur as free histidine (in some teleosts), as histidine-containing dipeptides, or as protein-bound histidine residues. In teleosts, for example, the concentration of free histidine is highest in powerful burst-swimmers; tuna white muscle contains nearly 100 μmols free histidine/gm tissue and can accumulate lactate during intense anaerobic glycogenolysis to about the same levels (Abe, 1981). In mammals, about half the total histidine in muscle is protein-bound; the rest is present in the form of dipeptides, carnosine, anserine, and ophidine; concentrations can get up to about 50 μmols/gm, making these the dominant solutes in the cytosol. In general, the buffering capacity of muscle is proportional to its glycolytic activity (Castellini and Somero, 1981), and so it is reasonable that during training for anaerobic exercises the buffering capacity of muscle is appropriately adjusted (Parkhouse et al., 1982).

Training for Aerobic Exercise: Possible Sites of Adaptation

Just as in our analysis of anaerobic training, it is perhaps useful to begin our discussion of aerobic training by considering possible sites and functions in aerobic work which theoretically may be adjustable; these then will supply us with a framework for analyzing aerobic training effects that are actually demonstrable in practice. A list of such theoretically adjustable determinants of aerobic performance must include:

1. increasing size of muscles utilized (i.e., hypertrophy),
2. adjusting proportions of red, intermediate, and white fiber types,
3. increasing levels of endogenous substrates (triglyceride and glycogen),
4. increasing levels of key enzymes in pathways contributing to aerobic oxidation of substrates,
5. increasing number (and possibly enzyme organization) of mitochondria, and
6. reducing activity levels of enzymes in anaerobic metabolism in concert with elevating aerobic metabolic potential.

In addition, because aerobically working muscles remain open systems, we may expect important adaptations occurring at higher levels of organization: in cooperative metabolic interactions between liver and muscle and between adipose tissue and muscle, in particular. These adaptations may involve

1. simple adjustments in glycogen and triglyceride storage depots,
2. adjustments in the capacity of liver and adipose tissue to release glucose and fatty acids, respectively, during aerobic work,
3. adjustments in the capacity of muscle to take up blood glucose and blood free fatty acids, either for immediate catabolism or for deposition as endogenous storage (as muscle glycogen or muscle triglyceride), and
4. adjustments in the capacity to deliver O_2 to working muscle and in its capacity to take up O_2 from blood.

These are the kinds of adjustments that would appear to be theoretically possible, given what we have learned above about metabolic support for aerobic work. How do such theoretical expectations match up with fact? It turns out not as well as in our analysis of anaerobic training effects, since the first two possibilities actually appear not to be utilized.

Hypertrophy Is Not a Characteristic Response to Aerobic Training

One of the most striking differences between aerobic and anaerobic training is that two key characteristics of the latter (*hypertrophy* plus an increase in *strength*) are not evident in the former. In rodents, for example, training on motor-driven exercise treadmills for several weeks leads to large increases in endurance and in respiratory capacity, but no indication of muscle hypertrophy. Nor under such training regimes is there any increase in muscle strength (Holloszy and Booth, 1976). Thus this potentially useful adaptive response does not seem to be a part of the arsenal of training adaptations utilized by mammals.

White Fibers Are Not Converted to Red during Aerobic Training

Because between-species comparisons in teleosts, birds, and mammals consistently show a correlation between the proportion of red fibers in working muscles and aerobic performance preferences (Marsh, 1981), it is surprising that again this obvious theoretical possibility is not typically utilized during aerobic training in mammals. On the contrary, some of the differences between red and white muscle fibers are accentuated (see below) in aerobic training, implying there are advantages to retaining different fiber types. These advantages may arise from the way in which substrates are utilized (by working red and white muscles) and from enzymatic adjustments occurring in each.

Aerobic Training Effects on Endogenous Substrate Supplies

Since the power output of muscle is highest when it utilizes either glycogen or glycogen plus fat in combination, it is instructive that endogenous depots of *both* these substrate sources are elevated following aerobic training regimes (Holloszy and Booth, 1976). This means the total amount of energy that can be extracted from endogenous sources is elevated by an exactly proportionate amount. Similarly, high levels of endogenous substrates in red muscles are observed in species well adapted for sustained aerobic performance.

Aerobic Training Effects on Enzymes in Aerobic Metabolism

As we emphasized above, the aerobic metabolic potential of muscle depends upon the integrated activities of several metabolic pathways (unlike the single pathway of glycolysis that is used in anaerobic work). So it is to be expected and is indeed observed (Holloszy and Booth, 1976) that increased respiratory capacity of muscle depends upon elevated activities of enzymes

1. in the activation, transport, and oxidation of long chain fatty acids,
2. in the oxidation of ketone bodies,
3. in the Krebs cycle,
4. in the mitochondrial electron transfer system,
5. in Krebs cycle augmentation pathways,
6. in hydrogen shuttling pathways, and
7. even in the initiation of blood glucose metabolism, catalyzed by hexokinase.

Typically, these elevations occur in all three fiber types, but they are most pronounced in red muscle, least in white muscle. Whereas the total amount of power that a muscle can generate is determined by the substrate that is available to it (from endogenous or exogenous sources), the rate of energy generation of course depends upon the catalytic power of its catabolic machinery; the observed training effects on level of key enzymes in aerobic metabolism and upon substrate supply therefore imply that both the total power output *and* its rate of generation are elevated during aerobic training. Again, similar adjustments are apparently genetically fixed in phylogenetic adaptations for sustained performance.

Aerobic Training Effects on Mitochondrial Abundance and Composition

The above changes in enzyme levels with training appear to result from an increase in enzyme protein concentration (rather than from other possible mechanisms such as altered catalytic efficiency). For those enzymes in aerobic metabolism that are mitochondrially located, ultrastructural studies of muscle in men and rodents indicate elevations in *both size and number* of mitochondria, leading to an increase in total mitochondrial protein. In addition to making more of what is already present, adaptation at this level also leads to alterations in enzymic composition and organization of mitochondria. An expression of such reorganization of the mitochondria is gained by comparing quantitatively the degree of training-induced change (Holloszy and Booth, 1976). Thus, some enzymes increase in activity a lot, some a little, while still others are essentially unchanged in activity, when expressed on a muscle weight basis. Analogous genetically fixed elevations in mitochondrial abundance are observed in between-species comparisons of animals varying greatly in endurance performance capacities (Hochachka et al., 1978; Marsh, 1981).

Aerobic Training Effects on Glycolytic Enzymes

Since essentially the same pathway may be utilized in both aerobic and anaerobic glycolysis, it is interesting to inquire into the effects of aerobic training on glycolytic enzyme levels in muscles. For one enzyme, hexokinase, the situation is quite clear: it displays an elevation in activity in all muscle fiber types following aerobic training. Levels nearly double in intermediate fibers, increase by about 1.5-fold in red muscles, and increase by about 30 percent in white muscle. The data on other glycolytic enzymes are less clear. Red muscles show a slight increase in levels of glycogen phosphorylase, pyruvate kinase, and lactate dehydrogenase, while in intermediate muscles these same enzymes decrease in activity by about 20 percent. Similarly, there are reported adjustments in the ratios of M_4 and H_4 type lactate dehydrogenases. In white muscle, there occurs a small (15 percent) decrease in lactate dehydrogenase activity following aerobic training.

In our view such modest changes in glycolytic potential are consistent with the idea that these enzyme levels are conserved largely for emergency (i.e., anaerobic) needs periodically imposed on the system. Since flux through anaerobic glycolysis is much higher than through aerobic glycolysis, enzyme levels must be tailored to the former, not the latter. As a result, under aerobic conditions, the catalytic potential of glycolytic enzymes should be well in excess of needs. The one exception to this, understandably enough, is hexokinase. During anaerobic conditions, the main carbon source being fermented is glycogen, but during sustained aerobic work, the importance of glucose rises as a function of duration of performance. In performance intense and long enough to fully deplete glycogen in working muscle in an hour, liver-derived glucose accounts initially for about 10 percent of the total energy being utilized but for about 20 percent by the end of the work bout. (Hultman, 1978). Since hexokinase acts as a control site in aerobic glycolysis, its elevation would facilitate such utilization of glucose, so it is not surprising that aerobic training leads to higher activity levels in muscle. Indeed, it would be more surprising if this did not occur.

Aerobic Training Effects on Liver Glycogen and Glucose

The increasing dependence of muscle metabolism upon liver-derived glucose during sustained performance emphasizes the importance of intertissue metabolic interactions during exercise. The increased glucose production from the liver (with increasing work load or work duration) is initially due primarily to activated liver glycogenolysis. At later stages, as liver glycogen becomes depleted, the continued release of glucose depends upon activated gluconeogenesis (mediated by insulin and glucagon). About two- to tenfold gluconeogenic activation can occur during exercise, but to our knowledge the effect of training on this process has not been ascertained (Hultman, 1978). Although several precursors for *de novo* glucose formation in the liver under these conditions are utilized (including alanine, glycerol, pyruvate, and lactate), of these lactate is quantitatively by far the most important. This critical and provocative observation begs the question of the origin of such lactate, which again seems to remain unanswered by researchers in this field. Be that as it may, it is when this process becomes critical that fat becomes the main carbon and energy source for continued aerobic exercise. It is important to ask if the mobilization of fat is similarly influenced by aerobic training regimes.

Aerobic Training Effects on Triglyceride Metabolism

In both rodents and men at least, several aspects of triglyceride metabolism are now known to be influenced by training. Capacities in muscles for fatty acid oxidation, for triglyceride formation by esterification of α-glycerophosphate, and for triglyceride deposition all are elevated following training (Holloszy and Booth, 1976). Interestingly, even the capacity to take up triglyceride from the blood is elevated. This capacity is represented biochemically by the membrane-bound enzyme lipoprotein lipase, which hydrolyzes triglyceride in chylomicrons in the blood. Recent studies show that lipoprotein lipase activity is increased threefold in intermediate muscle fibers, twofold in red and white fibers of rats following a long aerobic training program. A metabolically critical consequence of improved triglyceride and fatty acid uptake and metabolism by muscle following aerobic training is that it may cause a reduction in rates of glucose oxidation. This indeed is considered an important reason why trained animals deplete their liver glycogen reserves more slowly than untrained animals during extended, submaximal exercise. Although there are no direct measurements on rates of fatty acid release from adipose tissue to blood, the slight elevation in blood free fatty acids during extended exercise (when fatty acid oxidation rates are also elevated) is clearly consistent with increased flux from adipocytes to muscle; however, the effect of aerobic training on this process has not been assessed.

Aerobic Training Effects on O_2 Delivery and O_2 Extraction

To this point in our discussion of aerobic training, we have focussed our attention on carbon fuels for working metabolism. For completion of the discussion it is useful to remind the reader that the delivery of O_2 to working muscles, and the extraction of O_2 by working muscles, must clearly be closely integrated with the metabolic adjustments described above. That indeed is the case, and it is of course well known that aerobic training induces an adaptive increase in the maximum cardiac output and thus in the total amount and total rate of O_2 delivered to muscle. Although in theory the delivery of O_2 to muscles could also be increased by preferential perfusion (i.e., redistribution of cardiac output), this mechanism does not appear to be utilized. So the question arises: does the increase in cardiac output during aerobic exercise proceed in step with the overall increase in maximum O_2 uptake rates? The answer turns out to be negative.

Although there is a great deal of individual variation, on average the increase in maximum cardiac output appears to account for only about 50 percent of the rise in maximum O_2 uptake rates occurring in response to training. The other 50 percent or so of the increase is accounted for by improved extraction of O_2 by the working muscles, which is reflected in larger AV PO_2 differences and lower O_2 tensions in venous blood (see Holloszy and Booth, 1976). How improved O_2 extraction is achieved, however, is not well understood.

Addendum

Recent studies of metabolite cycling rates during exercise have generated some unexpected insights; one of these is that a large fraction of carbohydrate-derived CO_2 in sustained muscle work passes through lactate. Presumably lactate is formed in fast-twitch glycolytic fibers at about the same rates as it is utilized in more oxidative fibers, the net effect being the simultaneous harnessing of both fiber types during sustained performance. Both mechanical and metabolic advantages may accrue (see C. M. Donovan and G. A. Brooks, Endurance training affects lactate clearance, not lactate production, *Amer. J. Physiol.* 244 [1983]: E83–E92).

CHAPTER FIVE

Limiting Oxygen Availability

Primordial Organisms Probably Were Anaerobes

Extracting the maximal amount of chemical bond energy from a reduced organic molecule, be it a carbohydrate, lipid, or protein, typically involves degradation to CO_2 and H_2O. To achieve this complete combustion, molecular oxygen (O_2) usually must be available to the organism. That is why for most present-day organisms, O_2-based combustion of foodstuffs is the sine qua non of metabolic efficiency. Yet it was not always so. The best available evidence indicates that early stages in the development of our planet occurred under highly reducing conditions (Crick, 1981). If this view is correct, the planet was initially devoid of molecular oxygen and therefore was first "colonized" by prokaryotic anaerobes. The introduction and gradual accumulation of O_2 by photosynthesis is therefore considered a major turning point in the development of life on the planet (Wald, 1964); indeed, the photosynthetic release of O_2 as a metabolic end product may have represented a devastating pollution to the anaerobic world which was already thriving (Lovelock, 1979). Hyperoxic conditions can be severely debilitating even to modern-day organisms, probably due to the formation of reactive free radicals at concentrations that surpass the cell's defense mechanisms (see Chapter 3). The recent demonstrations that molecular oxygen may not be randomly dispersed throughout the cell by molecular diffusion, but rather may be restricted to (transfer) channels leading to mitochondria, may represent a remnant of the primordial O_2 sensitivity of internal metabolic machinery (Longmuir et. al., 1979). This, too, may be the phylogenetic explanation of the occurrence of enzymes such as LDH_k, an unique isozyme of LDH whose catalytic activity is so O_2-sensitive the enzyme's catalytic activity can only be expressed under hypoxic or anoxic conditions; the "saturation curve" for LDH_k inactivation by O_2 shows a half-saturation value that in fact is rather similar to the P_{50} for hemoglobin (Anderson et al., 1983). In the case of LDH_k, the evidence indicates that the enzyme is inactivated by molecular oxygen per se and not by O_2-derived free radicals. Although this kind of O_2-sensitive enzyme may merely represent an historical artifact (a telling indicator of an anaerobic origin to life on this planet), it is alternatively possible that enzymes such as LDH_k take on special metabolic functions during hypoxia or anoxia; in such event, the so-called O_2 inactivation would be more properly viewed as a mechanism for turning off those functions during normal O_2-rich states. Whichever interpretation turns out to be correct, it appears that the advantages of molecular oxygen are obtained at some risk to the functional integrity of the internal metabolic machinery.

As far as we know today, this risk is modest by comparison with the risks that would be taken if organisms were to develop absolute dependence upon O_2 with no anaerobic capacities at all. This is because O_2 supply is variable. O_2 limiting conditions may arise for any organism for a variety of environmental or biological reasons. Momentary shortfalls of O_2 delivery to a tissue may preclude a reliance on oxidative metabolism. Periodic depletion of environmental O_2 supplies (in muds, small ponds, or self-contained burrows) would be extremely hazardous to organisms with an absolute O_2 dependence. In more extreme cases, O_2 may be absent throughout the life of the organism and the simple equation: organic foodstuff + O_2 = CO_2 + H_2O may never pertain. Because the advantages of utilizing O_2 are great, but its availability at all times at all places in adequate amounts is not assured, most organisms have retained some capacity for anaerobiosis. Anaerobic metabolism typically is selected for function in one of two modes. In Chapter 4 we discussed the use of anaerobic glycolysis and phosphagen hydrolysis for purposes of achieving high rates of ATP turnover (the high power mode of anaerobic metabolism). This strategy, although useful for high power output, is critically time-constrained, being useful for only seconds to minutes of anoxia. That is why using anaerobic metabolism in a low power output mode (often coupled with improved yield of ATP/mole of substrate fermented) is more commonly observed in organisms when the urgent need is to survive O_2 lack on a sustained basis. In this chapter, we examine the variety of mechanisms used by prokaryotic and eukaryotic organisms to cope with a prolonged lack of O_2 as terminal electron acceptor. We will find a number of solutions to the O_2 deficiency problem, ranging from O_2 substitutes like nitrate (NO_3^-), which is used in a process termed "dissimilatory nitrate reduction," to novel fermentative end products which allow a greater extraction of energy from a foodstuff molecule than is possible in standard fermentation pathways. We consider first the "oxygen substitutes" utilized by certain bacteria, for these metabolic inventions are virtually on a par in their efficiencies with the oxygen-based combustion processes.

True Fermentations versus Anaerobic and Aerobic Respiration

In metabolic terms, all strategies for dealing with the problem of O_2 deficiency revolve around replacement functions that in some way compensate for the loss of molecular oxygen. The best of these, at least in energetic terms, are found in denitrifying bacteria. The key metabolic feature of these microorganisms is that when oxygen is absent they are able to respire nitrate instead. The oxidation of substrates such as glucose with either oxygen or nitrate proceeds with a similar, large free energy change and thus both processes are thermodynamically extremely favorable:

$$\text{glucose} + 6O_2 \longrightarrow 6CO_2 + 6H_2O \qquad \Delta G'_0 = -686 \text{ kcal}$$

$$\text{glucose} + 4.8NO_3^- + 4.8H^+ \longrightarrow 6CO_2 + 2.4N_2 + 8.4H_2O \qquad \Delta G'_0 = -638 \text{ kcal}$$

Denitrifying bacteria have taken advantage of the similarities between these two processes, developing means to respire nitrate rather than oxygen whenever the latter is unavailable. In this process, nitrate is reduced to N_2, a metabolic end product formally analogous to CO_2 and even less harmful to living cells. Such denitrifiers encompass quite a large group of microorganisms including many bacilli and pseudomonads. In addition, a number of bacteria are able to perform a nitrate-nitrite respiration in which nitrate is reduced to nitrite:

$$\text{glucose} + 12NO_3^- \longrightarrow 6CO_2 + 6H_2O + 12NO_2^- \qquad \Delta G'_0 = -422 \text{ kcal}$$

The nitrite formed is either excreted directly or reduced by non-ATP-yielding reactions to ammonia. The enzyme machinery for both processes, nitrate/nitrite respiration and denitrification, is formed only under anaerobic conditions or conditions of low oxygen tension. In fact, the activities of the enzymes involved in dissimilatory nitrate reduction are strongly inhibited by oxygen. Thus, denitrification and nitrate/nitrite respiration take place only when oxygen is absent or available in insufficient amounts.

Like oxygen respiration, denitrification allows a complete oxidation of the organic substrate to CO_2 and H_2O. For instance, when *Bacillus licheniformis* grows with glucose and nitrate under anaerobic conditions, the substrate is degraded via glycolysis and the Krebs cycle, while $NADH_2$ and $FADH_2$ serve as electron donors for the respiratory

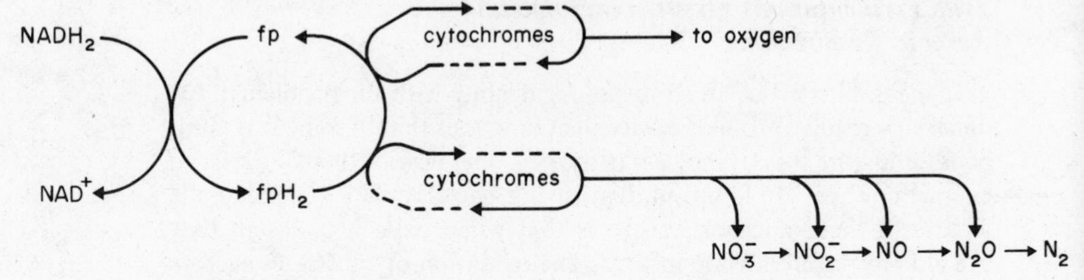

Fig. 5-1. Mutually exclusive processes of nitrate- and O_2-based respiration in denitrifiers

chain. Nitrate, however, does not simply replace oxygen; special types of cytochromes and membrane-bound enzyme systems are utilized, which systematically reduce nitrate to nitrite and further to nitrogen in at least four distinguishable steps (Fig. 5-1).

It is now evident that at least two, and probably more, of the four possible reductive steps are coupled to ATP formation in denitrifying bacteria. As may be expected from thermodynamic consideration, this crucial observation implies an ATP yield/mole of glucose similar to that for normal oxidative metabolism (see Gottschalk, 1979, for literature in this area).

From these considerations, it is clear that the three most fundamental features of oxygen-based respiration are also expressed in nitrate-based respiration:

1. in both, the free energy drop of glucose oxidation is large and negative and the process, therefore, is thermodynamically very favorable;
2. in both, the process leads to the complete degradation of glucose to CO_2 and H_2O without the concomitant accumulation of large amounts of partially catabolized anaerobic end products; and
3. in both, the process is relatively efficient in terms of ATP yield/mole of carbon substrate because of a tight-coupling between electron transfer and phosphorylation.

That is why anaerobic respiration, based on nitrate as a terminal electron acceptor, is more similar to oxygen-based (aerobic) respiration than it is to fermentation and why it must by definition be clearly distinguished from the latter. The great pioneer in this area, Louis Pasteur, first and simply defined fermentation as life in the absence of oxygen.

But today, a century after his pathbreaking work, fermentations are more precisely defined as *those metabolic processes that occur in the dark and do not involve respiratory chains with either oxygen or nitrate as terminal electron acceptors.*

The Nature of Fermentative Pathways

In true fermentation, the free energy drop between substrate (say glucose) and anaerobic end products is always modest by comparison with respiration, because fermentation is never based on electron transfer chains coupled to phosphorylation. Rather, true fermentations depend upon a variety of oxidation-reduction reactions involving organic compounds, CO_2, molecular hydrogen, or sulfur compounds. All these reactions are inefficient in terms of energy yield (moles ATP/mole substrate fermented), and, therefore, the mass of cells obtainable/mole of substrate is much smaller than with respiratory-dependent species.

Microbes capable of carrying out fermentations are classified as either facultative or obligate anaerobes. Facultative anaerobes, such as the enterobacteria, utilize O_2 if and when it is present, but if it is absent, they carry out fermentative metabolism. In contrast, obligate anaerobes are unable to synthesize the components of electron transport systems; consequently, they cannot grow as aerobes. Moreover, many of the obligate anaerobes cannot even tolerate oxygen and perish in air; these organisms are referred to as strict anaerobes.

The standard hallmark of fermentation is the accumulation of partially metabolized anaerobic end products. This is an inefficient metabolic strategy because a lot of potential chemical energy is still retained in the end products being formed; moreover, these are often noxious and therefore at high levels may well hinder further microbial growth or metabolism. Anaerobic end products are so distinguishing that true bacterial fermentations are often classified according to the main end product formed: alcohol, lactate, propionate, butyrate, mixed acid, acetate, methane, and sulfide fermentations are commonly found among anaerobic bacteria.

There are two fermentative processes that at first appear to be quite similar to oxygen and nitrate-dependent respirations: the reduction of CO_2 to methane and of sulfate to sulfide. However, on closer examination, it is clear that they bear little resemblance to the process of denitrification. In the first place, the reduction of CO_2 and of sulfate is carried out by strict anaerobes whereas nitrate reduction is carried out by aerobes only if oxygen is unavailable. Equally important, nitrate respirers contain a true respiratory chain; sulfate and CO_2 reducers do not. Furthermore, the energetics of these processes are very different. Whereas the free energy changes of O_2 and nitrate reduction are about the same, the values are much lower for CO_2 and sulfate reduction. In fact, the values are so low that the formation of one ATP per H_2 or NADH oxidized cannot be expected. Consequently, not all the reduction steps in methane and sulfide formation can be coupled to ATP synthesis. Only the reduction of one or two intermediates may yield ATP by electron transport phosphorylation, and the ATP gain is therefore small, as is typical of fermentative reactions.

Design Rules for Bacterial Fermentative Metabolism

The design rules for fermentative metabolism in bacteria are few in number and are shared by all species. Firstly, the fermentation process always involves the partial oxidation of substrate, although there is a tremendous diversity in choice of substrate. Almost any organic compound can be fermented by some microorganism somewhere. Secondly, the oxidation reaction or reactions must always be balanced by subsequent reductive reactions in order to allow sustained function; organic compounds usually serve as electron and proton acceptors in the reductive reactions leading to the formation of organic anaerobic end products. The end products typically accumulate to some extent and are released to the outside. Thirdly, because the free energy changes associated with substrate converion to end products are always modest, the ATP yield/mole of substrate fermented is always relatively low. One to two moles ATP/mole substrate fermented is not unusual. Fourthly, some fermentative reactions must be retained not for energy purposes per se but for the generation of key metabolite intermediates which are required for biosyntheses and growth; these may be directly related to anaerobic energy-producing pathways or may be unrelated to them; in the latter case, different substrates may be fermented to satisfy these different needs. Finally, for a unicellular system, it is reasonable and economical not to synthesize all the time all of the enzymes it is able to make but to make only those that are needed under specific and current physiological conditions. That means, the mere presence or absence of specific enzymes in microorganisms is of critical importance and must be very closely regulated. As has been well known in microbiology for over two decades now, this regulation is achieved by enzyme induction and repression. The former is usually used in regulating levels of catabolic enzymes, while enzyme repression

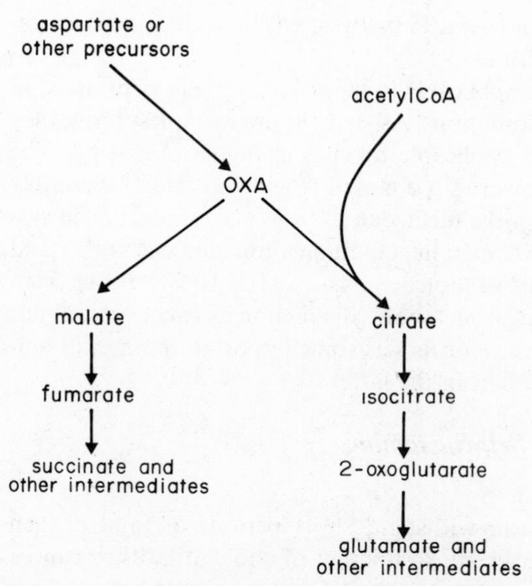

Fig. 5-2. Metabolic functions during O_2 lack of Krebs cycle reactions in facultatively anaerobic bacteria and yeast

is usually the mechanism exploited for control of anabolic pathways (Gottschalk, 1979).

Enzyme induction is so much associated with catabolic enzymes, enzyme repression with anabolic ones, it is necessary to emphasize that the levels of enzymes in central pathways must also be regulated. This is because the requirement for coupling intermediates changes with varying O_2 and substrate availability, varying growth rate, and varying metabolic rate. In facultatively anaerobic bacteria, for example, a change from aerobic to anaerobic environments removes the need for a complete Krebs cycle; not surprisingly, anaerobic *E. coli* cells stop making 2-ketoglutarate dehydrogenase, the enzyme connecting the first and second spans of the Krebs cycle. Other enzymes of the Krebs cycle, in contrast, are still synthesized, and, it turns out, for good reasons (Fig. 5-2). The first span of the Krebs cycle is still required for glutamate synthesis (which in turn is needed for various biosyntheses). Citrate synthase, aconitase, and isocitrate dehydrogenase therefore are maintained even under anaerobic conditions, but their levels are accordingly reduced. The second span of the Krebs cycle (succinate → OXA) under anaerobic conditions works backwards (OXA → succinate) forming ATP at the level of fumarate reduction to succinate, which accumulates as an anaerobic end product; succinate may in turn be used for other biosyntheses. Thus these enzymes, too, must be retained during anoxic periods.

These design rules are applicable in theory to any cells considered in isolation under anoxic conditions. As such, in their simplest form, they are prerequisite for, and applicable to, cells in higher organisms. The anoxic microbial cell, however, is a closed system and must be entirely self-sufficient; in contrast, the metazoan even in anoxia is an open system. Tissues and organs remain in communication with each other and their metabolic functions cannot be considered in isolation one from the other. This is a crucial metabolic distinction between unicellular and multicellular organisms for its sets some important needs and limits during anoxia adaptation in the latter.

Animal Anaerobes as Self-Sustaining Life-Support Systems

Among animals, all can withstand short periods of total oxygen lack, but only some species are so tolerent of anoxia that they can be considered to be facultative anaerobes. Obligate anaerobes rarely if ever are recorded, although such may infrequently have developed amongst the helminths. The most outstanding anaerobic abilities, in terms of *duration* of anoxia tolerance, are expressed by invertebrate groups (parasitic helminths, burrowing annelids, intertidal bivalves), but even amongst the vertebrates there are to be found species with surprising tolerance to anoxia. The common goldfish can survive at least several days of total O_2 lack, while some freshwater turtles (the snapping turtles, for example) at low temperature are able to withstand up to months of anoxia. Cut off from their normal sources of O_2, these organisms in effect become, for shorter or longer duration, multiple-component, self-sustaining life-support systems.

Although the various components of any life-support system can be put together differently, it is the centerpiece of them all that imposes strict metabolic limits upon animal anaerobes. That centerpiece is a metabolic organization supplying provisions for three critical needs and functions:

1. Provision must be made for suitable storage substrate(s) and for their regulated utilization, so that internal stores are not depleted and energy needs of various tissues and organs during anoxia can be matched by substrate availability.
2. Provision must be made to store, recycle, or minimize the production of noxious waste products which in high concentration could limit duration of anoxia tolerance.

3. Some provision must be made for re-establishing metabolic homeostasis in recovery.

These provisions are made for the life-support system as a whole, not only for its component parts. In fact, as we shall see below, interactions between the component parts are often critically involved in satisfying the needs of the system as a whole. That is why the number of cellular, metabolic options and mechanisms for dealing with anoxia are much reduced compared to microorganisms; they can, however, be equally effective, and understanding how is a major goal of this chapter.

Provision I: Matching Substrate Availability with Energy Needs

Glycogen in a Central Depot

For over a half century it has been well appreciated that carbohydrate in general, and glycogen in particular, is the primary fuel for anaerobic metabolism in all animals. Although some glycogen is stored in most tissues, in vertebrates most is stored in the liver, which supplies the bulk body needs for glucose and, therefore, can be viewed as a central depot of this storage substrate. Because of this special role, the primary carbon and energy source utilized during periods of severe O_2 deprivation in good anaerobes like the goldfish (Table 5-1) is undoubtedly liver glycogen, which is mobilized as blood glucose. Similar data are not available for many invertebrate groups, but those, such

Table 5-1. Glycogen Levels in Liver of Several Vertebrates Varying in Anoxia Tolerance

	Glycogen (μmol glucose/gm)
Anoxia-tolerant species	
Goldfish	1,300
Turtle	860
Anoxia-intolerant species	
Trout	235
Sunfish	185
Rat	210
Mouse	220

SOURCE: See Hochachka (1982) and Van den Thrillart (1982) for references

as *Mytilus*, which have been analyzed show a similar organization; in this case the central depots appear to be laid down in the hepatopancreas and the mantle, which during energy-rich conditions can accumulate glycogen to 50 percent of dry weight (De Zwaan, 1983). As nutritive conditions change through the year, the glycogen content of these central depots is drastically altered, probably in large part to buffer glucose changes in other tissues.

If central depot glycogen is to prime fermentative metabolism of the whole organism during anoxic stress, it follows that, in general, the more anoxia-tolerant the species, the higher the levels of glycogen in this depot should be. Because glycogen levels in central depots can be influenced by other factors (in particular by the nutritional state of the animal), easily comparable data are hard to come by. However, from values for vertebrates (Table 5-1), the most hypoxia-tolerant ones, the goldfish and turtle, clearly stand out from all others. In both these animals, liver glycogen (as glucosyl units) is stored at about 1M concentrations, or even higher, and is undoubtedly utilized during anoxia. The metabolic problem here is the same as in anaerobic bacteria: how to manage with the energetic inefficiency of fermentation. The physiological problem, however, is unique to animals under anaerobiosis: *how to partition out a diminishing resource (central depot glycogen) in the face of a high demand being made upon it be all tissues and organs in the body.*

Typically, in animal cells burning carbohydrate, a depletion of O_2 leads to a large increase in glucose consumption (the so-called Pasteur effect) because of the energetic inefficiency of anaerobic glycolysis. In oxidative metabolism, the yield of ATP is about 38 moles ATP/mole glucosyl-glycogen oxidized (Chapter 2), but in glycogen → lactate fermentation it is only 3 moles. That is why glycogen and glucose consumption rates by all tissues must rise by twelve- to thirteenfold or more *if, during anoxia, demand for ATP remains unchanged*, a condition that should lead to rapid depletion of central depot supplies and to declining blood glucose levels. Surprisingly, this is not observed during anoxic stress in facultative anaerobes. Liver glycogen stores in anoxic goldfish, even after extreme periods (four days) of anoxia, are by no means fully depleted (Shoubridge and Hochachka, 1981). In mussels and oysters, glycogen depots are only partially depleted even after days of anoxia (De Zwaan, 1983). Thus it is evident that even if central depot glycogen reserves are mobilized during anoxic stress, the problem of the massive and rapid glucose depletion that might be expected is in fact avoided; the question is how. It turns out this problem is avoided by at least three mechanisms:

1. *by storing more endogenous glycogen in other tissues,*
2. *by utilizing more efficient fermentations,*
3. *by reductions in metabolic rates.*

Each of these strategies deserves separate consideration since they vary in importance in different species.

Glycogen at Noncentral Depots

Perhaps the simplest way of reducing dependence upon central depot glycogen is to maintain elevated levels of glycogen in other tissues as well; these depots could then also be mobilized during anoxia independently or concurrently with blood glucose utilization. Not surprisingly, glycogen levels in most tissues that have been examined in the best of animal anaerobes are notably high (see Hochachka, 1982). Tissues such as gill and muscle in *Mytilus* can store up to about 200 μmoles glycosyl-glycogen/gm tissue. Turtle hearts store glycogen at 250 μmoles/gm, higher than any other species of vertebrates thus far studied. Glycogen levels in white and red muscle of the goldfish are also high, one-tenth and one-third, respectively, of the astronomical levels found in the liver. It is safe to conclude that this pattern prevails generally in most if not all effective animal anaerobes.

Not only is the amount of glycogen clearly subject to adjustment, so also is the manner in which it is stored. In vertebrates, when glycogen levels are low, glycogen is stored as β-particles; when glycogen levels are greatly elevated, as in goldfish liver, glycogen is stored as β-particles but also as large (1000 Å diameter) α-particles and as glycogen bodies.

Such adjustments in glycogen form and content in many tissues obviously aid in circumventing the glucose depletion problem during O_2 lack. However, this mechanism on its own cannot account for known durations of anoxia. In the anoxic oyster heart, for example, neither the amount of glycogen nor its utilization rate could possibly sustain the organ for the weeks-long anoxia periods it is able to withstand. Since the overall problem stems from the massive amount of glycogen (glucose) needed for an energetically inefficient metabolism, one obvious solution is to develop and utilize more efficient anaerobic pathways (in terms of ATP yield).

Energetically Improved Fermentations

It is among the most capable of invertebrate anaerobes, the helminths and the marine bivalves, that we find the best examples of alternative fermentation pathways. Many of these have been reviewed several times in recent years (De Zwaan, 1983), so only a brief summary will be considered here. Current concepts view the organization of anaerobic metabolism as a series of linear, and loosely linked, pathways. The most important of these, aside from classical glucose → lactate fermentation (yielding 2 moles ATP/mole glucose) are:

1. glucose → octopine, lysopine, alanopine, or strombine, energy yield 2 moles ATP/mole glucose;
2. glucose → succinate, energy yield 4 moles ATP/mole glucose;
3. glucose → propionate, energy yield 6 moles ATP/mole glucose;
4. aspartate → succinate, energy yield 1 mole ATP/mole aspartate;
5. aspartate → propionate, energy yield 2 moles ATP/mole aspartate;
6. glutamate → succinate, energy yield 1 mole ATP/mole glutamate;
7. glutamate → propionate, energy yield 2 moles ATP/mole glutamate;
8. branched chain amino acids → volatile fatty acids, energy yield 1 mole ATP/mole substrate;
9. glucose → acetate, energy yield 4 moles ATP/mole glucose;
10. $CH_2O + SO_4^{2-} + H^+ \rightarrow H_2S + HS^- + H_2O + CO_2$, energy yield 6 moles ATP/mole glucose.

Pathways 1–4 are known in various bivalve molluscs; 2, 3, 8, and 9 are often utilized by helminths; 5, 6, and 7 while theoretically possible in bivalve molluscs, do not appear to be utilized to any significant extent (De Zwaan, 1983; Hochachka et al., 1983). Pathway 10, or sulfate oxidation of organic substrates, is well known to occur in the highly reduced layers of benthic silt, but the distribution of this activity between bacteria and lower invertebrate animals is not yet clarified. Since both groups of organisms coexist in what has been termed a "sulfur-sulfide biome" and since both bacteria and at least some invertebrates (which may house sulfur bacteria symbionts: Felbeck et al., 1983) have the enzymes required for sulfate and sulfide metabolism, we tentatively assume that both contribute to the observed sulfate reduction in benthic sediments (Powell et al., 1980).

Except for systems using sulfate as an electron acceptor to facilitate the oxidation of organic substrates, the potential for which is only now being investigated, there is reasonable agreement concerning the pathways by which most of the above end products are formed. There are a few problem areas, however.

For example, in many invertebrates (parasitic helminths, polychaetes, sipunculids, bivalves, cephalopods, gastropods, and probably many other groups as well), PEP functions at a distinct metabolic

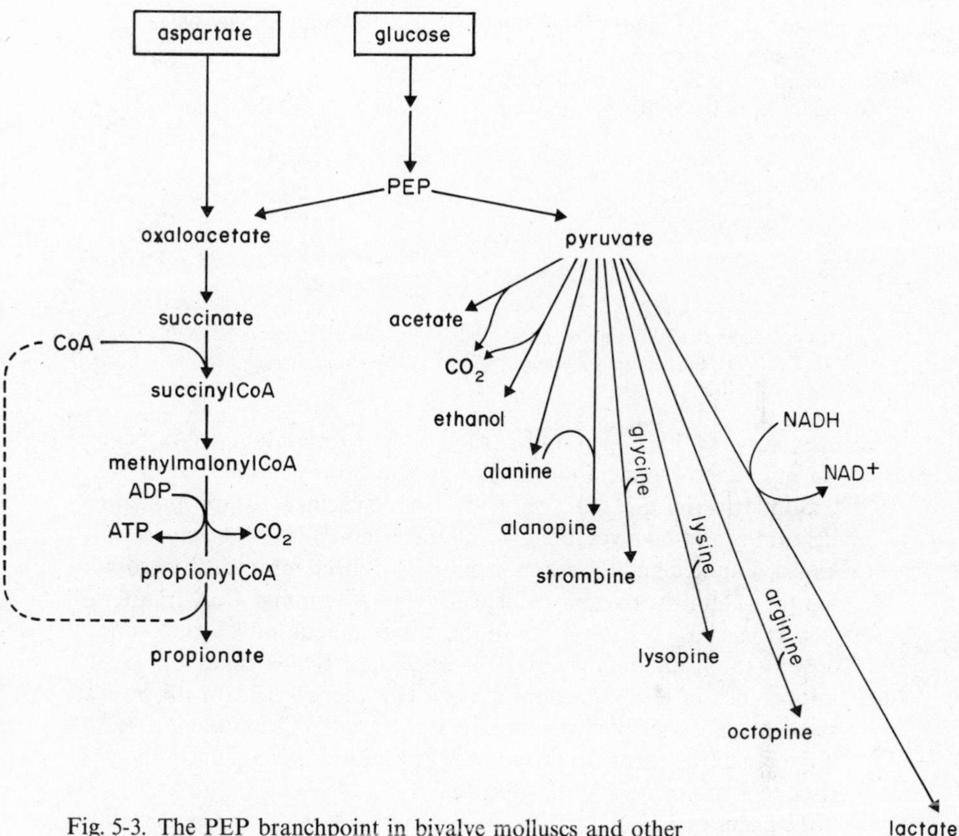

Fig. 5-3. The PEP branchpoint in bivalve molluscs and other facultatively anaerobic invertebrates

branchpoint (Fig. 5-3). PEP carboxykinase (PEPCK) catalyzes carboxylation of PEP to form oxaloacetate, from which ultimately succinate and propionate can be formed. Under other circumstances and in other species (particularly molluscs), PEP is converted to pyruvate, which, depending upon species, or tissue, can be converted into one of several kinds of end products. Some of the literature in this area has been reviewed; other developments, aimed at trying to understand which end products predominate under what conditions, are in progress. It is not necessary to redescribe them here. It is, however, important to emphasize that end products such as octopine, alanopine, strombine, and lysopine and the enzymes forming them are formally analogous to, and serve the same functions as, lactate and lactate dehydrogenase in us:

Table 5-2. Phosphorylation Coupled to the Formation of Succinate

Added malate	Uncoupler	Succinate formed	ATP formed	ATP/succinate
5	0	3.3	1.2	0.35
5	0	3.5	1.4	0.41
5	0	3.7	1.8	0.50
5	1.5	3.9	0.3	0.09
5	1.5	3.7	0.2	0.07

SOURCE: Schöttler (1977), using tubifex preparations under anoxic conditions. See Saz (1981) for helminth literature. Metabolites given in μmoles.

they serve as hydrogen and carbon sinks to maintain redox balance during anaerobic metabolism (Storey and Storey, 1983).

Similarly, in the reaction path from succinate to propionate in helminths and bivalves, there is still uncertainty as to the way in which CoA is handled and the reaction actually utilized for releasing propionate; in helminths current evidence favors a terminal CoA-transferase step (Fig. 5-3). Whatever its route, the fermentation of glucose (glycogen) to propionate in bivalves such as *Mytilus* takes on greater importance in long-duration anoxia. The energy yield to the level of succinate is considered maximally at 4 moles ATP/mole glucose and if the reaction scheme to propionate in Figure 5-3 is assumed, the yield rises to 6 moles ATP/mole glucose.

The sites of ATP formation involve substrate-level phosphorylations (PGK and PEPCK) to the level of fumarate, plus an electron-transfer-based ATP synthesis at the level of furmarate reductase (Table 5-2). The final site of ATP formation in propionate fermentation occurs at the carboxylase reaction (Fig. 5-3).

The general pathways for volatile fatty acid (VFA) formation seem to involve three phases:

CoASH CO_2

branched-chain amino acid ⟶ ketocarboxylate ⟶ acylCoA

NAD^+ NADH

acylCoA ⟶ volatile fatty acid

ADP + P_i ATP

with the last thiokinase step potentially capable of generating 1 mole of ATP/mole of VFA generated. These reactions are not major contributors to ATP yield in most animal anaerobes, but in *Fasciola* they are, probably primarily through being linked to glucose fermentation to propionate. In this situation, overall energy yield in these species can theoretically reach 8 moles ATP/mole glucose plus 2 moles leucine or other branched-chain amino acid utilized.

In oyster heart (Collicutt and Hochachka, 1977), in *Mytilus*, and other bivalves as well (De Zwaan, 1983), aspartate occurs at high (about 15 μmol/gm) concentrations in the normoxic state. In anoxia the bulk of the aspartate is fermented to succinate, a process that is particularly important during early stages of anoxia. At this time, it is probably redox coupled to glycolysis. Thus the energy yield increases by 1 mole ATP/mole aspartate fermented to succinate, or by 2 moles ATP/mole aspartate if the fermentation continues to the level of propionate, and the maximum total energy yield becomes 8 moles ATP/mole of glucose plus aspartate fermented to propionate.

From the above discussion, it is evident that anaerobic metabolism in many animals can be far more versatile than commonly observed in higher vertebrates and that this raises the possibility of utilizing energetically more efficient fermentations. Estimating how much more efficient than anaerobic glycolysis is complicated by the observation that different combinations of pathways can be utilized at different times in anoxia (De Zwann, 1983). But even if a combination of pathways usually is used, the ATP yield can nevertheless be elevated two- to fourfold; any animal anaerobes utilizing such fermentations therefore automatically reduce by a factor of two to four their anaerobic needs for glucose. Although impressive, this factor is still a long way from the order-of-magnitude difference between anaerobic glycolysis and oxidative glucose metabolism. That leaves a final, interesting alternative: *to minimize the anaerobic demands for glucose, effective animal anaerobes may sustain a depression of metabolic rate and thus of demand for ATP during anoxia.*

Reducing Anaerobic Demands for Glucose By Depressing Metabolism

Among vertebrate anaerobes, the goldfish and diving turtle are perhaps the best studied and thus supply us with the most comprehensive data (the diving turtle is discussed in Chapter 5). The goldfish at low temperature is well known to be capable of surviving several days of total anoxia. The actual duration seems to vary depending upon the condition of the fish, the season, and especially, ambient temperature, but survival for several days to several weeks is reported. Liver glycogen levels for winter acclimatized goldfish are about 1,300 μmoles glucosyl units/gm tissue. For a 100-gram goldfish, with about 6 grams of liver, this means about 7,800 μmoles glycogen (glucose) are available in the central depot. The resting metabolic rate of the goldfish at 4°C is about 0.05 μmoles ATP gm^{-1} min^{-1}, which is equivalent to about 7,200 μmoles ATP 100-gm $fish^{-1}$ day^{-1}. If this metabolic rate is to be maintained, the anoxic animal could survive on liver glycogen-derived glucose for only 2.2 days. If the resting normoxic metabolic rate were depressed by fivefold, this glucose supply could keep the goldfish going for about eleven days of anoxia; both this metabolic rate and this degree of anoxia tolerance are about right for this species, as indicated by direct calorimetry under anoxic conditions (see Shoubridge and Hochachka, 1981, for further discussion and literature).

A similar strategy of reducing anaerobic demands for glucose by reducing metabolic rate is also utilized by bivalve molluscs. The most reliable information comes from studies of *Mytilus*. From direct respiratory measurements (De Zwaan and Wijsman, 1976), the oxidative metabolic rate of *Mytilus* under normoxic conditions is known to be about 20 μmoles ATP gm wet $weight^{-1}$ day^{-1}, shell weight being excluded. Peak concentrations of glycogen in mantle and liver are about 50 percent of dry weight of the tissue, or about 550 μmoles glucosyl unit/gm, which for a 2-gram animal can be taken as the size of the central depot. If energetic efficiency in *Mytilus* anaerobic metabolism is assumed to be the same as in classical glycolysis, a normoxic metabolic rate during anoxia could be maintained by glucose from these two tissues for only three days. If we assume all glucose is fermented to succinate, the ATP yield doubles, and anoxia could be sustained for about six days. If we assume all glucose is fermented to propionate, the ATP yield rises again and the animal could survive nine days of anoxia. Under all three conditions, a Pasteur effect would have to occur: in the first case, glucose consumption would have to rise by twelvefold; in the second, by sixfold; in the third case, by fourfold. In contrast, a Pasteur effect is not actually demonstrable in *Mytilus* on aerobic/anaerobic transition; nor is it evident in a number of other molluscs studied. The reason why it is not observed is because the anoxic bivalve sustains a marked reduction in metabolic rate. From measurements of succinate accumulation, phosphoarginine depletion, and ATP depletion, the rate of ATP turnover during anoxia is estimated at only one-twentieth the normoxic metabolic rate (De Zwann and Wijsman, 1976). Thus even at the normal glycolytic efficiency level, *Mytilus* can be sustained by mantle and hepatopancreas glycogen for about sixty days! If all glucose were channeled into succinate, it could survive 120 days, while if all glucose flowed to propionate, it could survive 150 days, assuming nothing else were limiting.

These are highly instructive calculations for they indicate that organisms such as *Mytilus* have a metabolic strategy and organization *that literally frees them from the uncertainties of environmental O_2 supply.* They also explain why animals may be selective in the anaerobic metabolic pathways utilized in different organs and tissues and why quantitatively differing amounts and kinds of end products may accumulate at different stages of anoxia. And thirdly, these calculations also well illustrate the relative effectiveness of depressing metabolism versus increasing fermentative efficiency in reducing the overall anaerobic demands for glucose: *in Mytilus, anoxia duration can be extended from three → nine days by increasing energetic efficiency, but from three → sixty days by metabolic depression.*

A combination of both mechanisms is so effective in parceling out storage glucose carbon that the outstanding anoxia tolerance of these organisms could hardly be limited by substrate supply. Anoxia tolerance could, however, be determined by accumulating too much end product. This problem is so serious that a means for its solution is considered a fundamental design feature for any anaerobic life-support system.

Provision II: Taking the Edge off the End Products Problem

The Problem and Possible Solutions

The accumulation of anaerobic end products to very high levels is normally associated with perturbations in pH (the commonest situation), with drastic osmotic perturbations, or with metabolic derangements through end product inhibitory effects. Early in phylogenetic time, there would also have been the need to preclude the possibility of uncontrolled side reactions which might be favored by high concentrations of end products. Of these potential problems, that of acidification is the most serious. The nature of this problem is widely misunderstood, so it must be closely examined.

The Proton as a Metabolic Intermediate

From the discussion in Chapter 2 on ATP synthesis, it should be evident that the proton is one of the most fundamental intermediates of cellular metabolism. What may come to the reader as a surprise is just how vast are the quantities of H^+ ions that are turned over during normal aerobic metabolism. A simple calculation for a 70 kg man, with a resting metabolic rate of about 700 mmoles O_2/hour, shows that about 150 gms of H^+ ions are produced per day! The main sink for this huge amount of H^+ is oxidative phosphorylation for there is a close balance between rates of H^+ production and H^+ removal during ATP resynthesis and reoxidation of components of the electron transfer system. Because of this balance, pH remains unperturbed during aerobic metabolism (Krebs et al., 1975; Vaghy, 1979).

This closely balanced system dissipates during anaerobic metabolism, but there is controversy as to how and why the breakdown occurs. This misunderstanding probably arises mainly from the belief that glycolysis yields lactic *acid*, a fairly strong acid, which then dissociates into lactate anions and H^+ ions. This oversimplified statement is in fact rather misleading. Glycolysis per se represents the cleavage of glucose (or glycogen-derived glucosyl-unit) into two lactate anions with the concomitant production of ATP from adenosine disphosphate (ADP) and inorganic phosphate (P_i). If the overall reaction is considered in isolation and at high pH (>pH 8.0), it will become evident that H^+ ions neither need be accumulated nor depleted:

$$\text{glucose} + 2ADP^{3-} + 2HPO_4^{2-} \longrightarrow 2\ \text{lactate}^{1-} + 2H_2O + 2ATP^{4-}$$

The equation as written is balanced with regard to all constituents and charge. Only at low pH (<pH 6.0), well below physiological range, would glycolysis lead to the classical proton stiochiometry:

$$\text{glucose} + 2HADP^{2-} + 2H_2PO_4^{-} \longrightarrow 2\ \text{lactate}^{1-} + 2HATP^{3-} + 2H_2O + 2H^+$$

Although the above equations illustrate important points, such simplest of situations do not occur in living systems largely because of the influence of three factors: magnesium ions (Mg^{2+}), pH, and the substrate source (glucose or glycogen) being fermented (Table 5-3).

Mg^{2+} *Influences the Proton Stoichiometry of Glycolysis*

Because pH, free Mg^{2+} levels, and availability of glucose versus glycogen are tissue-specific parameters, it turns out not to be possible to write a single equation for glycolysis in all tissues or even for the

Table 5-3.

a) Proton Stoichiometry of Anaerobic Glycolysis

pH	No Mg^{2+} H^+/glucose	4.4 mM Mg^{2+} H^+/glucose
6.8	0.93	1.15
7.4	0.33	0.40
8.0	0.09	0.11

b) H^+ Stiochiometry for Glycogen → Lactate Fermentation[a]

pH	H^+ released per glucose	H^+ consumed per glucose
6.8	0.72	—
7.4	—	0.40
8.0	—	0.84

SOURCE: From Hochachka and Mommsen (1983).

[a] Fermentation assumed at 4.4 mM Mg^{2+}

same tissue under differing conditions. This can be easily illustrated by considering anoxic vertebrate muscle. From recent evidence, it appear that free Mg^{2+} levels in muscle are in the 3–4.4 mM range, so through *in vivo* pH ranges, essentially all ATP is in the complexed form; ADP is only partially complexed while P_i binds Mg^{2+} so weakly it can be assumed to be mainly uncomplexed. What this means is that the true adenylate reactants in glycolysis are strongly determined by Mg^{2+} concentrations (Table 5-3a,b) and since free Mg^{2+} levels are species- and tissue-specific (0.25 mM levels are found in red blood cells, for example) the overall equations for glycolysis are necessarily very cell-specific. Because the pK values for MgATP and free ATP are different, moles H^+ formed/mole of glucose fermented differ, and at any given pH are greater if the ATP is fully complexed than if ATP is uncomplexed (Table 5-3a,b).

pH and Proton Stoichiometry of Glycolysis and Glycogenolysis

We have already implied that the proton stoichiometry of glycolysis is pH dependent. The effect of pH change on glycolytic H^+ production also can be illustrated more precisely if we extend the example of anoxic vertebrate muscle fermenting glucose, assuming Mg^{2+} levels remain constant. Such analysis shows that *the lower the pH, the greater the amount of H^+ formed/mole of glucose fermented* (Table 5-3a). The same pattern prevails if glycogen is the starting fermentable substrate. However, the stoichiometry is quantitatively different because a H^+ producing step, that catalyzed by hexokinase, is not utilized. In this case (Table 5-3b), at pH values only slightly above neutrality, *glycogen* *fermentation to lactate proceeds with the consumption (not the production) of H^+* (at pH 7.4, for example, 0.3 moles of H^+ are consumed per mole glucosyl unit fermented). A close examination of the reaction pathways leading to alanopine, strombine, or octopine indicates that, in terms of proton production, they are essentially equivalent to mammalian glycolysis. Where bivalve anaerobiosis differs most from that in vertebrates is in its capacity for, and utilization of, pathways of glucose fermentation either to succinate or to propionate, so it is interesting to examine their proton stoichiometry.

H^+ Balance during Succinate and Propionate Formation

Again, assuming similar and unchanging free Mg^{2+} concentrations as above, the overall equations for glucose fermentation to succinate and to propionate show some striking characteristics of these pathways (Tables 5-4, 5-5). During propionate fermentation, for example, the reaction proceeds with *the consumption of H^+* ions; with glucose as substrate at near neutral pH, about 1 mole H^+ is consumed/mole of glucose (Table 5-5). At higher pH values, which may be expected in ectotherms at low temperatures, the molar uptake of H^+ is even higher. Glycogen fermentation to propionate also consumes H^+ but in sub-

Table 5-4. H^+ Stoichiometry for Succinate Fermentation

Glucose as fermentable substrate:

pH	H^+ released per glucose	H^+ consumed per glucose
6.8	0.77	
7.4		1.07
8.0		1.75

Glycogen as fermentable substrate:

pH	H^+ consumed per glucosyl
7.4	1.87

SOURCE: From Hochachka and Mommsen (1983)

Table 5-5. H^+ Stoichiometry for Propionate Fermentation

Glucose as fermentable substrate:

pH	H^+ released per glucose	H^+ consumed per glucose
6.8		0.55
7.4		2.80
8.0		3.67

Glycogen as fermentable substrate:

pH	H^+ consumed per glucosyl
7.4	3.60

SOURCE: From Hochachka and Mommsen (1983)

stantially higher amounts than when glucose is fermented (again because hexokinase is not a part of the pathway). Only when *the pH falls below about pH 6.5 does the glycogen → propionate reaction yield a net production of* H^+ (Table 5-5).

From these considerations, it is evident that fermentation reaction pathways per se do not necessarily lead to net proton accumulation; indeed, some pathways actually consume protons. Why, then, is an association between end product accumulation and acidosis so commonly reported? If the observed H^+ ions are not formed primarily in fermentation reactions, from where do they arise? The answer is from ATP hydrolysis.

ATP Hydrolysis and H^+ *Production*

During anaerobiosis, the fate of fermentatively generated ATP is hydrolysis for the support of various cell work functions. The hydrolysis of ATP is not determined by whether the ATP is derived from aerobic or anaerobic metabolism; in both cases it proceeds with the release of ADP, P_i, and H^+ (Table 5-6).

The two main intracellular factors determining the amount of H^+ released during ATP hydrolysis again are Mg^{2+} and pH. For mammalian muscle at constant Mg^{2+} levels, the equations (Table 5-6) show an interesting dependence upon pH: *the lower the pH, the fewer the moles of* H^+ *formed/ATP hydrolyzed. The effect of pH on* H^+ *production during ATP hydrolysis is almost exactly opposite to that observed for the glycolytic production of* H^+ (Figure 5-4). This is fundamental both in mechanistic and metabolic terms, for as we have seen in Chapter 4, *in vivo* the two processes (glycolytic generation of ATP; ATPase catalyzed hydrolysis of ATP) are, to a lesser or greater extent, coupled; neither enzyme pathway proceeds in a vacuum.

Table 5-6. H^+ Stoichiometry for ATP Hydrolysis

pH	H^+ released/ATP
6.8	0.425
7.4	0.800
8.0	0.945

NOTE: Assuming 4.4 mM Mg^{2+} and dissociation constants given in Hochachka and Mommsen (1983)

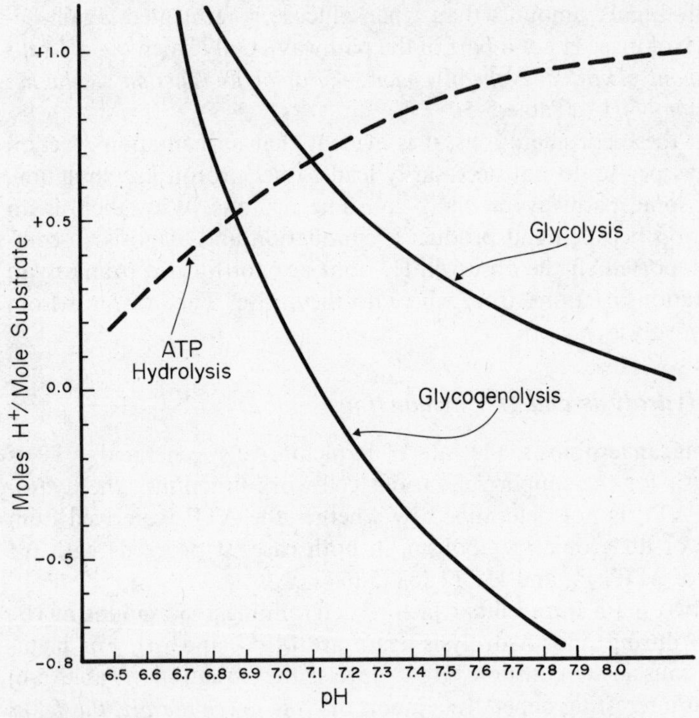

Fig. 5-4. Effect of pH on proton stoichiometry of ATP hydrolysis and of glycogen/glucose fermentation. The same curves are valid for lactate, alanopine, strombine, or octopine formation. Based on equations in Hochachka and Mommsen (1983)

Effect of Coupling of Fermentation Pathways and Cell ATPase on H^+ Stoichiometry

As emphasized by Atkinson (1977) and in Chapter 2, the coupling of ATP-forming and ATP-utilizing reactions is a center piece of aerobic and of anaerobic metabolism. The adenylates are the basis for the coupling, since ATP is one of the products of glycolysis or of oxidative phosphorylation at the same time as it is the key substrate for cell ATPases. This coupling in anaerobic glycolysis is different from that observed in aerobic metabolism. In the latter, all the products of ATP hydrolysis (ADP, P_i, and H^+) are reutilized during oxidative phosphorylation (Vaghy, 1979). In the former, by contrast, although ADP and P_i are stoichiometrically reutilized during glycolytic replenishment

Table 5-7. Fermentations Coupled to ATPases

Pathway	pH	Moles H^+/mole glucosyl unit	Moles H^+ from hydrolysis of ATP formed in fermentation[a]	Sum: moles H^+/ mole glucosyl + *moles H^+ from hydrolysis of ATP* formed in fermentation
Glucose → lactate	6.8	1.15	0.85 (2)	2.00
(alanopine)	7.4	0.40	1.60 (2)	2.00
(octopine)	8.0	0.11	1.89 (2)	2.00
Glycogen → lactate	6.8	0.72	1.28 (3)	2.00
(alanopine)	7.4	−0.40	2.40 (3)	2.00
(octopine)	8.0	−0.84	2.84 (3)	2.00
Glucose → propionate	7.4	−2.80	4.80 (6)	2.00
Glucose → succinate	7.4	−1.07	3.20 (4)	2.13

[a] Numbers in parentheses refer to assumed moles ATP formed/mole unit glucosyl during the fermentation process. From Hochachka and Mommsen (1983)

of ATP, H^+ ions are not reutilized if glucose is the fermentable substrate, and are only partially reutilized if glycogen is being fermented above about pH 7.2 (see Table 5-3b). Because of the opposite pH dependencies of H^+ production by glycolysis and by ATP hydrolysis, however, the total number of moles of H^+ generated (moles H^+/mole of glucose or glucosyl unit fermented *plus* moles H^+ released during ATPase-catalyzed hydrolysis of the ATP formed during the fermentation) *is always the same*: *two* (*Table 5-7*). *For the overall system (glycolysis plus ATPase), the process can be described by one equation with the proton stoichiometry being independent of pH and Mg^{2+} levels:*

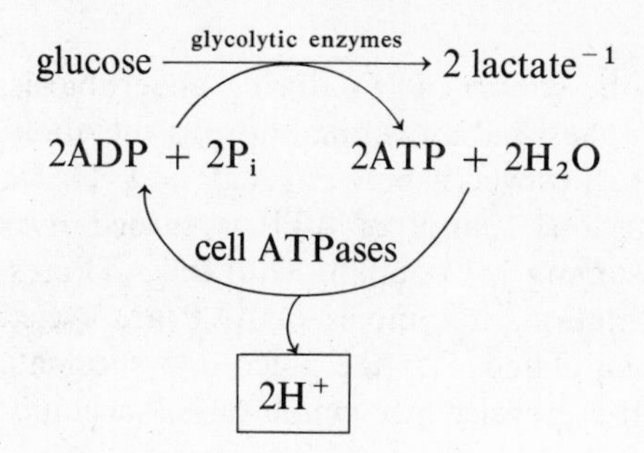

With glycogen as the substrate fermented, the summed proton stoichiometry is the same: even if some H^+ ions are consumed during the fermentation per se, 3 rather than 2 moles ATP/mole glucosyl unit are generated and when this greater amount of ATP is hydrolyzed by cell ATPases the net yield is still $2H^+$/glucosyl unit (Table 5-7).

Perhaps a more surprising conclusion from analyzing other animal fermentation pathways is that *exactly the same stoichiometry of net H^+ production prevails.* If pH and Mg^{++} levels are the same at sites of ATP formation as at ATPase sites, this means that the fermentation of either glucose or glycogen a) to alanopine, strombine, or octopine, or b) to succinate or propionate proceeds with a variable utilization or production of protons associated with ATP replenishment (Tables 5-3–5-5), yet when this process is coupled to the hydrolysis of the ATP formed in the fermentation (Table 5-7) *one equation always satisfactorily describes the net stoichiometry of proton production*:

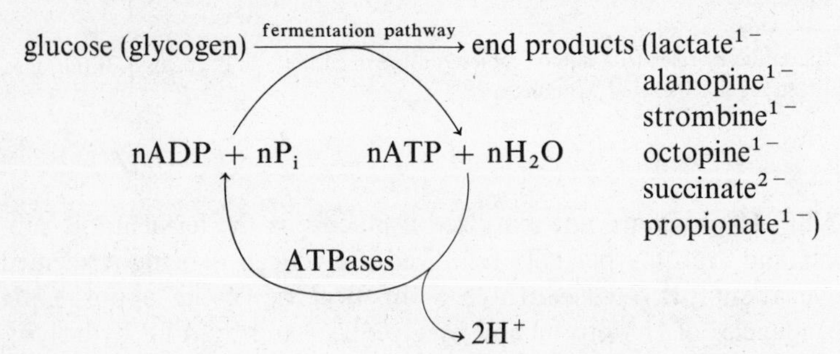

That is why in all animals, sustained anoxia results in a net accumulation of H^+ ions and why end product accumulation and acidosis are usually correlated.

ATP Cycled/H^+ Produced Varies with Different Metabolic Pathways

Because of confusion as to the source of H^+ during anaerobiosis, the usual H^+ stoichiometry emphasized is the one above; a metabolically more relevant relationship, however, is between ATP and H^+: for the glucose → lactate fermentation, 1 μmol of ATP is turned over per μmol H^+ accumulated, assuming 1:1 coupling with cell ATPases. For glycogen → lactate fermentation, 1.5 μmoles of ATP are cycled through per μmole of H^+ accumulated. For the glucose → succinate fermentation, 2 μmoles of ATP are cycled per μmole of H^+ accumulated, while for the glucose → propionate fermentation, the value increases to 3. This insight is important in adaptational terms for it identifies a principal advantage gained through the utilization of succinate or propionate fermentations by "good" invertebrate anaerobes: *during anoxia, up to three times more ATP can be turned over per mole of* H^+ *accumulated* than in the case of mammalian tissues relying solely upon glycolysis.

That is a considerable advantage, yet it does not and cannot fully overcome the problem of gradual acidification in anoxia; so other mechanisms must also be harnessed. Analysis of the metabolism of facultative animal anaerobes indicates at least four such additional mechanisms for minimizing the problem of excessive acidification during anoxia:

1. The simplest solution is to minimize the rate of end product accumulation by depression of metabolic rate during anoxia.
2. Another possibility is to tolerate high accumulations by maintaining high buffering capacities of blood and tissues.
3. The severity of the problem can be reduced by "detoxifying" anaerobic end products through their further metabolism at sites remote from sites of production and/or through their excretion.
4. The problem can also be circumvented by utilizing H^+-consuming reaction pathways.

Again, each strategy should be discussed separately as different combinations are often used by different organisms.

Reducing End Product Formation by Reducing Demands for ATP

It is important to realize at once that by depressing demands for ATP during anoxia, an organism not only reduces the need for pouring glucose into inefficient fermentative reactions, it also automatically reduces the rate of formation of anaerobic end products. Both outcomes are advantageous and related. But there is a critical time constraint on this strategy: when an organism ferments 7,800 μmoles of glucose it will generate 15,600 μmoles of lactate, whether the process occurs in three days or three months. This strategy is therefore only useful in submaximal anoxic episodes or if the excretion of H^+ can be dissociated from the further metabolism of lactate, as seems to be the case in muscles of frogs, rats, and teleosts (Benade and Heisler, 1978).

Tolerating End Product Accumulation

Another simple means for dealing with periodic massive accumulations of anaerobic end products such as lactate is simply to tolerate the simultaneous H^+ production by improving buffering capacity, as in the high bicarbonate reserves in diving turtles. Similarly adjustments in buffering capacity are probably also utilized by bivalves, cephalopods, and crustaceans (Baldwin, pers. comm.). In fact there is a good correlation between tissue glycolytic capacity and tissue buffering capacity in vertebrate animals (Castellini and Somero, 1981), implying a general strategy widely spread across phylogenetic boundaries.

In terms of intracellular enzyme function another way to tolerate high accumulations of end product is to design enzymes whose function is either actually favored, or at least unaffected, by acidification. That perhaps is why in good anaerobes such as goldfish and turtles, a key regulatory enzyme such as pyruvate kinase (see Hochachka and Storey, 1975) displays distinctly acid pH optima while the homologous enzymes in terrestrial animals have much higher pH requirements. Conditions therefore may be appropriately more favorable for pyruvate kinase function when O_2 is lacking and the organism is depending more and more upon anaerobic glycolysis. Lactate dehydrogenase supplies another example of an enzyme which in turtles has been adjusted to function adequately even if lactate levels greatly increase (Hochachka and Storey, 1975).

Minimizing Lactate Accumulation by Further Metabolism

In the above strategies for dealing with high concentrations of lactate, the animal ultimately must tolerate the consequences of accumulation. A priori a more effective strategy would be to convert the end product into less harmful metabolites; i.e., to "detoxify" it. At this writing, the best vertebrate example of this mechanism is found in the anoxic goldfish. The goldfish, like most vertebrates, ferments glycogen (glucose) to lactate during anoxia; essentially all tissues display relatively high activities of lactate dehydrogenase and thus the capacity to form lactate. However, for over two decades what has impressed students of this problem is not how much lactate, but rather how little, is accumulated during extended anoxia in the goldfish. Although intriguing to numerous workers over the last fifteen to twenty years, this problem has been clarified only recently. The key observation is that ^{14}C-lactate is partially oxidized in anoxic goldfish; whereas this is

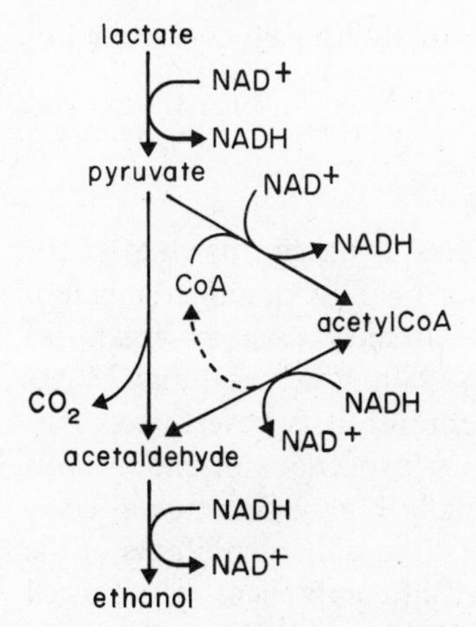

Fig. 5-5. Possible routes from pyruvate to ethanol in anoxic goldfish. After Shoubridge and Hochachka (1981) and Van den Thillart (1982)

also true for ^{14}C-glucose, lactate is the preferred substrate for this process, and is oxidized at nearly ten times higher rates (Shoubridge and Hochachka, 1981). There is still uncertainty about the reaction path generating CO_2 from lactate (Fig. 5-5), although it evidently involves acetaldehyde as an intermediate, which explains why carboxyl-labeled lactate yields $^{14}CO_2$ at high rates, but no carbon from methyl-labeled lactate appears in CO_2 (Shoubridge and Hochachka, 1981; Van den Thillart, 1982). The subsequent fate of acetaldehyde is reduction to ethanol, a reaction catalyzed by ethanol dehydrogenase and occurring in skeletal muscle. In winter-acclimatized goldfish, ethanol dehydrogenase in white and red muscles occurs at about 30 and 90 μmoles product formed gm^{-1} min^{-1} at 25°C, but the enzyme is not measurable in extracts of other tissues (brain, heart, liver). Unlike lactate in other anaerobes, ethanol does not build up to high levels, typical concentrations reaching only about 4 μmoles/gm in tissues (4 μmoles/ml in blood). This is because ethanol is easily lost to the outside water.

The metabolic mechanisms used by anoxic goldfish may not be general among anoxia-tolerant organisms. But at least several hypoxia-adapted teleosts are known to maintain unusually high RQ values, with rates of CO_2 production exceeding rates of O_2 uptake under hypoxic conditions by 1.5- to 2-fold. Similarly, true metabolic CO_2 production under anoxic conditions has been observed in tissue preparations (Shoubridge and Hochachka, 1981). Thus, it may turn out that comparable mechanisms for the further metabolism of lactate are utilized by other species as well.

Detoxification of Hydrogen Sulfide

In some environments, H_2S accumulates as an end product of the anaerobic metabolism of either bacteria or possibly of animal metabolism as well. H_2S is considered to be extremely toxic to organisms even at low concentrations (Powell et al., 1980). Whereas it may be less disruptive to metabolism in lower vertebrates, it is nevertheless true that organisms routinely exposed to either exogenous or endogenous sources of H_2S are faced with a potentially highly detrimental compound. In at least five meiofaunal species (the only inhabitants of the "sulfur-sulfide biome" thus far studied at all comprehensively, Powell et al., 1980), the strategy for dealing with this end product is exactly analogous to the goldfish's handling of lactate: any disrupting effects of H_2S are minimized by its further metabolism to other less harmful compounds. The available routes are:

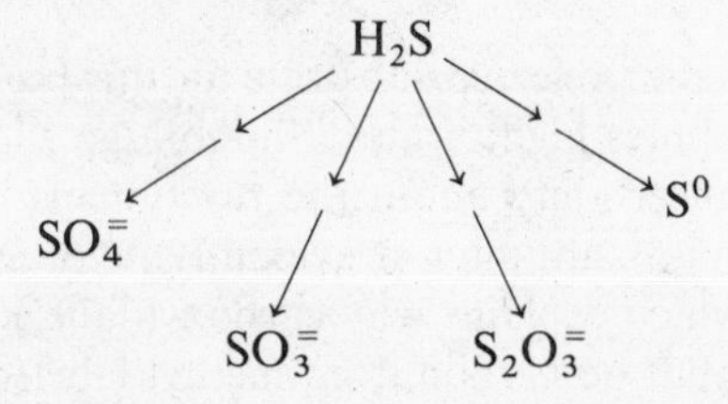

It turns out all four routes are possible alternatives, the actual one being utilized probably depending upon O_2 availability (Powell et al., 1980). As we shall see in our chapter on the deep sea, some organisms have harnessed these metabolic interconversions for other purposes (using H_2S as an energy source for the net fixation of CO_2), and some deep-sea species possess additional mechanisms for prevention of poisoning by H_2S.

Utilization of H^+-Consuming Reactions

To appreciate properly the use of H^+-consuming reactions as a means for minimizing the acidification problem, it needs to be emphasized that important consequences stem from the way in which anaerobic ATP-generating pathways are coupled to cell ATPase functions. At one extreme (very tight coupling), acidification during sustained anoxia is independent of the kind of fermentation pathways utilized: 2 moles H^+/mole of glucosyl unit are formed by them all. Since the fermentative and the ATPase functions are spatially separate (catalyzed by different enzyme pathways), *temporally independent operation would represent the other extreme situation of uncoupled function.* In anoxic muscle, for example, this situation might lead to a depletion of ATP (by myosin ATPase during anaerobic work) prior to activation of glycolysis. In nature, neither extreme is typically observed: activation of fermentative reactions and of ATPases are fairly closely coordinated, so that increased work (increased ATP turnover) usually occurs with minimal changes in ATP concentrations, both in vertebrate and invertebrate systems. However, during drastic rise in ATP requirements the rates of ATP utilization exceed rates of production and ATP levels fall substantially. In man at extreme work levels, muscle ATP concentrations fall from initial levels of about 5 μmol/gm to about 2.5 μmol/gm; in salmonids forced to swim at maximum rates for maximum (about forty-five minute) time periods, ATP concentrations in white muscle fall by a surprisingly large amount, from about 9 to 2 μmol/gm (Mommsen and Dobson, pers. comm.). All such cases represent a certain amount of uncoupling between ATPase function and fermentative ATP replenishment, so proton production rates must become unusually high. In the case of salmonids, the drop in ATP levels represents a potential increase in H^+ concentrations from about 0.1 nmol/gm to 7 μmol/gm: in the absence of protective buffering mechanisms, this would represent several orders of magnitude increase in H^+ levels.

Under these conditions, there may be selective advantage to retaining H^+-consuming processes, such as

$$\text{AMP} \xrightarrow[\text{AMP deaminase}]{} \text{IMP} + NH_3 \xrightarrow{H^+} NH_4^+$$

However, this reaction sequence depletes the adenylate pool while contributing nothing to the energy status of the anoxic cell. Thus where the urgent need is to absorb or consume protons, the best solution is to *utilize fermentative pathways that consume H^+* during ATP replenishment: in vertebrates, glycogen → lactate fermentation; in invertebrates, glycogen → succinate or glycogen → propionate. In both cases, two advantages acrue over simple glucose → lactate fermentation. Firstly, the molar yield of ATP/mole glucosyl unit is in each case higher. Secondly, these pathways at least partially redress the overproduction of H^+ by ATP hydrolysis. This advantage seems so obvious that a biologist might wonder why fermentation pathways leading to no net change in H^+ balance in the coupled system have not been developed. It turns out that they have: in the fermentation of glucose to ethanol (used by many microorganisms and by some animals) and in the fermentation of glucose to butanol (used by some microorganisms.)

H^+ Balance during Alcohol Fermentation

As pointed out elsewhere (Hochachka and Mommsen, 1983), the overall system of alcohol fermentation coupled with cell ATPases neither produces nor consumes protons:

$$\text{glucose} \xrightarrow{\text{fermentation pathway}} 2\,\text{ethanol} + 2CO_2$$

$$2H^+ + 2ADP + 2P_i \qquad 2ATP + 2H_2O$$

$$\text{ATPases}$$

If CO_2 is hydrated, its production may of course become equivalent to the production of $2H^+$/mole of glucose used up, in effect negating the advantage of the system. That is why organisms utilizing this scheme must also provide *a means for removing CO_2*. In the anoxic goldfish the probable solution is CO_2 removal across the gill into the external water which always serves as a large CO_2 sink for aquatic animals. Similar conditions for CO_2 release apply for *Chironomus* larvae and various endoparasites which are known to produce ethanol during O_2 limitation.

In the case of butanol formation in microorganisms, the pathway becomes activated only at acidic pH when CO_2 hydration would not be favored and CO_2 along with another gas, H_2, again can be removed to the outside medium.

The advantages of these pathways come with a cost; namely, loss to the medium of a carbon source (the end product) which could be utilized for oxidative energy metabolism during aerobic recovery. Nevertheless, it is clear that nature has invented fermentation pathways that consume the protons released during ATP hydrolysis.

The paradox is that such pathways are rarely used by "good" animal anaerobes; only the goldfish and a few invertebrate species are currently known to employ this strategy of anoxia adaptation. Why should it not be more generally used? One possibility, which is in need of careful exploration, is that in most animal species there may be useful, but thus far largely unexplored, functions for net proton production during O_2 limiting periods. Four such possible functions are discussed elsewhere (Hochachka and Mommsen, 1983). In this context, the most important of these may be to promote end product efflux from anaerobic tissues into blood and to promote their subsequent conversion back to glucose. These processes become particularly important during recovery from anoxic episodes.

Provision III: Re-establishing Metabolic Homeostasis

Nature of the Problem

The third major metabolic provision that must be incorporated into the metabolic organization of facultative anaerobes is a means for the re-establishment of metabolic homeostasis when the period of anoxia ends. There are two related components to this problem. Firstly, there is the problem of clearing the pool of lactate or other anaerobic end products formed during anoxia; and secondly, the central and peripheral storage depots of glycogen must be recharged. Again, it is convenient to consider these separately.

Clearance of Anaerobic End Products in Invertebrates

In invertebrate anaerobes, the problem of re-establishing metabolic homeostasis following periods of anoxia is not well worked out. For some end products, such as ethanol, acetate, or propionate, the problem is in a sense solved before it arises (since these may be transferred into the medium). For others, such as succinate and alanine, the problem of clearance is presumably relatively simple, since these metabolites are components of mainline pathways of oxidative metabolism. In cephalopods, for example, alanine is vigorously oxidized *in vivo*. The problem therefore really arises when we consider the fate of "true" glycolytic end products such as lactate, alanopine, strombine, lysopine, and octopine. The best, in fact the only, information that is available concerns the clearance of octopine.

In cephalopods, the main tissue site of octopine formation during hypoxia combined with exercise is mantle muscle (Storey and Storey, 1983); the metabolic situation here is as follows:

arginine-phosphate ⟶ arginine ⟶ octopine
glycogen (glucose) → → → pyruvate ↗

At least when hypoxic stress is coupled with exercise, the time course of events has been considered relevant, for arginine accumulation precedes by a distinctly measurable time the accumulation of octopine; indeed, the published data suggest that arginine accumulates to a larger extent than does octopine. However, this conclusion cannot be relied upon, since arginine phosphate is very unstable and most previous estimates of its level are probably much too low; that is, arginine estimates are probably too high (see Hochachka et al., 1983). The problem therefore may be somewhat more complex than simply inquiring about the fate of octopine; the fate of arginine is clearly a closely related issue.

Blood arginine levels are always low (in the μmolar range) and do not seem to change during recovery; thus the bulk of arginine clearance must occur *in situ*, and clearly must involve major reconversion to arginine phosphate. But what about the fate of octopine?

Unlike the handling of lactate in vertebrates, which occurs both endogenously and exogenously at other tissues in the body (discussed in Chapters 4 and 6), the bulk of octopine appears to be remetabolized *in situ*. The highest estimated fraction of octopine plus arginine that may be transferred in *Sepia* (see Storey and Storey, 1983) from muscle to other tissues is about one-sixth. Blood concentrations of octopine during recovery as a result rise by about eightfold but absolute concentrations remain fairly low, implying a rapid circulation time and rapid metabolism elsewhere in the body. Experiments with ^{14}C octopine confirm rapid uptake and conversion in several tissues (heart, brain, kidney, gill, among others.)

Modified Cori Cycle in the Cephalopods?

The fate of octopine taken up by different tissues in fact is not fully quantitated for any species thus far studied; however, the major alternatives seem to be clear and consistently support the operation of a modified Cori cycle. The first span involves uptake and further metabolism initiated by the ODH back reaction and occurring in the kidney (and perhaps other tissues as well):

$$\text{octopine} \xrightarrow[\text{NAD}^+ \;\rightarrow\; \text{NADH}]{} \text{pyruvate} + \text{arginine}$$

Pyruvate may be partially oxidized and partially converted to glucose for delivery back to muscle, thus completing what in vertebrates is called the Cori cycle.

The metabolism of arginine, in contrast, may be more complex. One possibility is cleavage via arginase to ornithine and urea, with the ornithine being ultimately convertible to glutamate or proline:

$$\text{arginine} \xrightarrow{\nearrow \text{urea}} \text{ornithine} \longrightarrow$$

$$\text{glutamaldehyde} \xrightarrow[\text{NAD}^+ \searrow \text{NADH}]{} \text{glutamate} \xrightarrow{\text{NADH} \curvearrowright \text{NAD}^+} \text{P5C} \xrightarrow{\text{NADH} \curvearrowright \text{NAD}^+} \text{proline}$$

Either proline or glutamate could then be transferred back to muscle. Some arginine could of course be fully oxidized via glutamate entry into the Krebs cycle. Alternatively, from the Krebs cycle pool, arginine-derived carbon skeletons could enter the complex and large free amino acid pool of molluscan tissues. Of these alternatives, the first seems most important (Hochachka and Fields, 1983).

This process then of cycling arginine carbon skeletons in glutamate or proline is interesting since it supplies a direct link between arginine (which through its role as a phosphagen and as a substrate for ODH is pivotal in anaerobic metabolism) and proline (which is utilized during activated *oxidative* metabolism). This implies that arginine, glutamate, and proline are all interconvertible (Hochachka and Fields, 1983), and that an overall process which is analogous to the Cori cycle in vertebrates may be occurring in cephalopods (Storey and Storey, 1983).

Lactate Clearance Process in Vertebrates

As in the invertebrates, re-establishing metabolic homeostasis following anoxia in vertebrate organisms also involves the twin problems of clearing lactate and recharging glycogen reserves. Because the situation here has been much more extensively studied, we can begin by summarizing the known processes that in sum regain pre-anoxic conditions. These include

1. Lactate oxidation *in situ* at sites of formation.
2. Lactate reconversion to glycogen *in situ.*
3. Lactate transport into the blood for oxidation elsewhere.
4. Lactate transport into blood for reconversion to glycogen in liver and kidney.
5. Lactate conversion through pyruvate to alanine (and subsequent metabolism along with the amino acid pool).

All these processes are known to occur to some extent in vertebrates, and what most research has been recently directed toward is quantifying each of them. The data for mammals (the rat in particular) nicely illustrate the way lactate is partitioned between these various processes. For the perfused hind limb, estimates are available on lactate oxidation *in situ* (7 percent), lactate reconversion to glycogen in muscle (45 percent), and lactate entering the Cori cycle (20 percent). How this partitioning is achieved has only recently been clarified and is worth discussing in some detail.

Recharging Peripheral and Central Glycogen Depots

In terms of bulk, skeletal muscle is the largest single tissue mass in the vertebrate body and during anoxic stresses of various sorts it is usually the largest source of lactate. During recovery, about 7 percent is directly oxidized *in situ* (McLane and Holloszy, 1979). However, lactate reconversion to glycogen is not easily demonstrable. In part for this reason, and because gluconeogenic enzymes are also difficult to assay in muscle, it was concluded about fifteen years ago that lactate reconversion to glucose (or glycogen) cannot occur in muscle, but only in liver and kidney. This idea was of course consistent with the high gluconeogenic capacity of the latter two organs. Although gluconeogenesis in overall effect is reversed glycolysis, in detail the process is different because a number of glycolytic enzymes are essentially irreversible under physiological conditions. Thus, to say the liver and kidney have high gluconeogenic capacities is tantamount to saying they possess specific bypass enzymes for getting around the one-way glycolytic steps. Four such are usually mentioned: pyruvate carboxylase and PEPCK, as a means for bypassing the pyruvate kinase reaction; FBPase as a means for bypassing PFK; and glucose-6-phosphatase, as a means for bypassing hexokinase and forming free glucose. For the synthesis of glycogen, the latter of course is not requisite, but two other enzymes (UDP glucose pyrophosphorylase and glycogen synthase) serve as an effective bypass of glycogen phosphorylase.

The picture emerging, therefore, is the one that most of us are familiar with in standard biochemistry textbooks, according to which the primary way of recharging peripheral (muscle) glycogen reserves is via glucose formed in the liver and kidney. But within two years of rejecting the possibility of *in situ* resynthesis from lactate, it was also being reported that PEPCK and malic enzyme are present in vertebrate muscles; these, along with malate dehydrogenase, supply another route around the pyruvate kinase reaction:

$$\text{pyruvate} + CO_2 \longrightarrow \text{malate} \longrightarrow \text{oxaloacetate} \longrightarrow \text{PEP}$$

Newsholme and his colleagues also confirmed the occurrence of fructose bisphosphatase, another indicator of glycogenic capacity in both frog muscle and in mixed mammalian muscles. These findings served as points of departure for the later demonstration of fairly rapid conversion of lactate to glycogen in respiring frog sartorius. More recently, several other studies have appeared (see McLane and Holloszy, 1979); from all of them it is now clear on balance that a large fraction of the lactate formed in fast-twitch white and fast-twitch red muscles during anaerobic glycolysis may be reconverted to glycogen *in situ*. Slow-twitch red fibers in mammals are unable to make this conversion mainly because of the absence of key gluconeogenic enzymes, fructose bisphosphatase in particular. The same is true of the heart. In fishes, recent studies show that the bulk of lactate formed in white muscle does not equilibrate with blood and it is a reasonable guess, since requisite enzymes are present (Guppy et al., 1979), that here, too, much of it is reconverted to glycogen *in situ*.

However, even if an important fate of muscle lactate is reconversion to glycogen *in situ*, a significant fraction (>20 percent) is washed out into the blood and is thus metabolized elsewhere. It is in this context that a physiological function for H^+ accumulation in anoxia may be identified. For many decades now it has been widely assumed that lactate exchange between body compartments is achieved by simple diffusion. Recently, in contrast, it has become apparent that lactate transfer is usually carrier-mediated. In the kidney, it appears to be linked to Na^+ transport, but in all other tissues examined (several tumor lines, red blood cells, liver, and cardiac muscle), the carrier-mediated transfer of lactate appears to be an antiport system, lactate anions exchanging for OH^- ions (which is why it is found to be highly pH-sensitive). Acidification of tissues during anoxia would then create conditions suitable for lactate efflux during recovery; acidification of the blood would in turn create conditions suitable for lactate uptake and further metabolism at the liver (see Hochachka and Mommsen, 1983 for literature). For years it has been known that an important fate of lactate is reconversion in liver to glycogen and glucose; glucose derived from this central depot in turn is used to replenish glycogen depots everywhere else in the body.

In addition to serving as a glucose precursor, lactate also is an excellent oxidative substrate for various organs and tissues in the vertebrate body and therefore an additionally significant fate of lactate during recovery from anoxia is complete oxidation. At elevated blood concentrations, skeletal muscle, lung, heart, and even brain are all fully capable of utilizing lactate as a carbon and energy source, often in preference to glucose. These tissues then complete the process of lactate clearance during recovery from anoxia.

The Anoxic Life Support System in Perspective

In summary, this chapter has concentrated upon the view of organisms in anoxia as self-sustaining life-support systems. The chief carbon and energy source for such systems in all animals is glycogen, usually stored in central depots for the eventuality of anoxic stress; however, other carbon sources can be utilized to augment energy yield from glucose fermentation. The most important thus far identified are free amino acids (particularly aspartate and branched-chain amino acids). Two distinct problems arise from using anaerobic energy metabolism: it is in all cases much less efficient (in terms of ATP yield/mole of substrate mobilized) than oxidative metabolism; also, most fermentation pathways when coupled with cell ATPases lead to accumulation of H^+ and other end products. Thus, two major problems in the design of an anaerobic life-support system are firstly, the conservation of glycogen (glucose) and secondly, the handling of potentially noxious end products. Massive depletion of carbohydrate from central stores is minimized by 1) storing more glycogen in peripheral depots, 2) utilizing more efficient fermentations, and 3) depressing metabolic rates during anoxia. The end products problem is minimized by 1) utilizing fermentation pathways which allow more ATP to be turned over per mole of H^+ accumulated than in classical glycolysis, 2) tolerating proton accumulation by improved tissue buffering capacity, 3) minimizing end product accumulation by recycling it for further metabolism or excretion, 4) utilizing H^+-consuming reaction pathways, and 5) depressing metabolic rates during anoxia. It is instructive that of all these contributing mechanisms, metabolic depression mechanisms yield by far the most effective protection against O_2 lack. *That is probably why facultatively anaerobic animals* (bivalves, goldfish) *all seem capable of potent metabolic arrests during anoxia.* To be effective when O_2 supplies are limiting or totally lacking, metabolic life-support systems therefore must be able to switch down; how this is achieved mechanistically is not yet well understood, although it is represented by the absence of a Pasteur effect, or, indeed, a reversed Pasteur effect.

A third key provision needed is a means for effective re-establishment of metabolic homeostasis when the anoxic episode ends. Unfortunately, there is little information on this matter in invertebrate anaerobes, and this represents an obvious area for future research. In vertebrates, the situation is somewhat better understood. The two parts of the problem (clearing lactate and regenerating glucose reserves) are of course related. Lactate accumulated during anoxia in recovery can be 1) reconverted to glycogen at sites of formation, or 2) removed to the blood for metabolism elsewhere. In the liver and kidney its main metabolic fate is gluconeogenesis, while in most other tissues including the brain, heart, and lung, it serves as an excellent substrate for oxidative metabolism, often being utilized in preference to glucose, an organization that speeds up re-establishment of metabolic homeostasis.

Addendum

Particularly novel adaptations for surviving freezing in several terrestrial frogs has recently been reported, with lactate accumulating as an anaerobic end product but with glucose accumulating as a cryoprotectant (see K. B. Storey and J. M. Storey, *Experientia* [1984]: in press).

CHAPTER SIX

Metabolic Adaptations to Diving

Problems and Strategies

We can think of no more vivid an example of a self-sustaining life-support system than an aquatic vertebrate when it dives, whether it be a lungfish in lakes of East Africa, a pirarucu in the Amazon, a cormorant in an American lake, or a seal, porpoise, or whale at sea. As with any of the systems discussed in Chapter 5, diving animals must have "on board" all materials required during the dive and mechanisms for their regulated utilization so that needs and supplies can be matched. Any deleterious end products formed during diving must be stored, cycled, tolerated, or excreted so that they have minimal impact on diving duration or on diving activities. And at the end of diving, mechanisms must be available for restoring the original state. These standard provisions were introduced earlier as features of metabolic life-support systems designed for sustaining complete anoxia. As such, they are fully applicable to diving animals in extreme situations when all tissues and organs become totally anoxic (e.g., long-term diving in turtles). But most diving animals do not go totally anoxic during diving; they time diving activities and duration to rates of O_2 and substrate depletion so that the limit conditions considered above do not strictly apply. Our goal in this chapter is to unravel how the above design rules are adapted for metabolism during diving and recovery sequences in aquatic vertebrates.

In lower animals dealing with low O_2 problems, wholesale adjustments in enzymatic organization of cells and tissues are clearly evident (De Zwann, 1983). So at the outset it is important to emphasize that this is not a part of the arsenal of adaptational strategies utilized by diving vertebrates: from available data, it is evident that neither enzyme levels nor metabolic organization correlate in any simple way with the diving habit. Such differences as have been observed are related not to the diving habit per se, but, it is now believed, to other characteristics, such as performance capacity, swimming style, swimming speed, substrate preferences, and so forth. That is, diving animals do not differ from nondiving ones by virtue of major qualitative reorganization of enzymatic machinery but rather in the way that machinery is used (Castellini et al., 1981). How, then, is it used?

Turtles and Lungfishes Rely on Liver Glycogen during Submergence

Let us first consider lower vertebrates, where the situation is exemplified by two well-studied groups, the aquatic turtles and lungfishes. In both, submergence under severe hypoxic or anoxic conditions leads to active mobilization of large glycogen stores in the liver. Blood glucose levels rise at the expense of liver glycogen, and the glucose concentration gradient between liver and blood remains favorable for glucose export (Fig. 6-1). Interestingly, there seems to be a built-in safety factor in both species, because in both, even after maximum duration submergence, the central reserves of glycogen are not fully depleted. Thus, a massive glucose depletion by energetically inefficient fermentation is somehow avoided. At least three mechanisms are employed to conserve liver carbohydrate reserves, the same three we discussed for animal anaerobes:

1. the animal relies upon additional glycogen stores in other tissues;
2. it utilizes more efficient fermentation pathways; and
3. it reduces metabolic rates and thus demands for substrate and O_2.

The first is particularly well displayed by aquatic turtles. For example, turtle hearts contain higher levels of glycogen than found in any other species of vertebrates thus far studied, about 250 μmol/gm heart. In the African lungfish heart glycogen levels are also high (Dunn et al., 1983). In addition, glycogen is stored as large, 1,000 Å diameter α-particles and, in lungfishes, even as glycogen membrane complexes termed glycogen bodies. Similar comparisons of glycogen levels in the turtle brain indicate elevations by about an order of magnitude relative to nondiving vertebrates (see Hochachka, 1982, for references).

But even if such adjustments help, available heart glycogen stores are largely consumed within only a few hours of hypoxic episodes that in nature could last for many weeks, and maybe months; brain glycogen levels in turtles are about one-tenth as high and would be depleted even faster. So simply relying on noncentral depots of glycogen is insufficient.

Facultative anaerobic invertebrates, when facing this problem, reduce demands for liver carbohydrate by utilizing energetically more efficient fermentations. This process does occur to some extent (succinate, for example, accumulates in hypoxic tissues in vertebrates in general, and in turtles in particular). However, alternative fermentation pathways are not very active in vertebrates and their contribution to the anaerobic energy budget is considered quantitatively unimportant (see

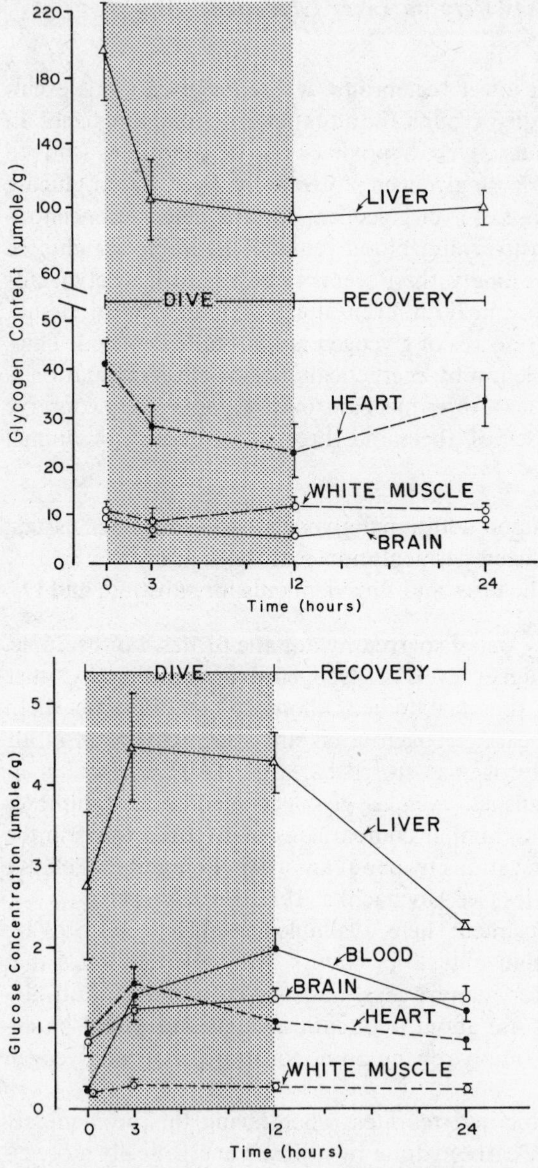

Fig. 6-1. Upper panel: Tissue glycogen concentration changes (in μmoles glucosyl unit/gm wet weight of tissue) during twelve hours of submergence and recovery in the African lungfish

Lower panel: Tissue and blood glucose concentration changes during twelve hours of submergence and recovery in the African lungfish. Modified after Dunn et al. (1983)

Hochachka, 1980; Hochachka and Murphy, 1979, for discussion of other possible functions for these processes in vertebrate anaerobes). They cannot and do not aid in minimizing the animal's anaerobic demands for glucose. To extend diving duration, therefore, aquatic turtles and lungfish are left with a final alternative: *they can minimize large glycolytic demands for glucose by depressing metabolic rate and thus demand for ATP by anoxic hypoperfused tissues.* In its quantitative impact, this alternative turns out to be by far the most important. In lungfish, the metabolic depression appears to be localized mainly in muscle (Dunn et al., 1983), but in turtles, it appears more general.

Reduction in Metabolic Rate during Diving in Turtles

Three lines of evidence bear upon this matter in turtles: a) direct calorimetry during diving, b) indirect evidence obtained from O_2 debt analysis, and c) estimates of metabolic rates of isolated heart and brain preparations. Direct calorimetry of diving turtles, first performed over a decade ago, supplies convincing evidence that during diving the overall metabolic rate of the animal drops to about 15 percent of normoxic rates. This older observation has been more recently confirmed by O_2 debt analysis. The argument here is simply that postdiving O_2 debt repayment should be exactly equivalent to the calculated O_2 debt incurred during diving if normoxic metabolic rate is sustained throughout the diving period. Experimentally, it is in fact found that the incurred debt is only about one-sixth of that expected, implying that during diving the animal's metabolic rate had been depressed down to about one-sixth of normal, a value in good agreement with the calorimetric one (Caligiuri et al., 1981).

Studies with isolated preparations of heart and brain indicate extremely low metabolic rates compared to the same organs in mammals, and are thus in general agreement with the above whole-animal studies. The data on the turtle brain nicely illustrate the point. From isolated preparations it can be shown that the metabolic rate (μmoles ATP turned over $gm^{-1}min^{-1}$) is about one-tenth the level of mouse brain when compared at 37°C, but if the turtle brain metabolic rate at 4°C (its biological temperature during diving) is compared to the mouse brain at 37°C (its normal biological temperature), the latter is 250 times the former! That is, the turtle brain metabolizes at a slow rate at all times, but especially so during diving when it lets brain temperature fall to the low ambient temperature of lake bottoms. Equally important, although the brain electron transfer system (ETS) becomes totally reduced when O_2 is absent (Lutz et al., 1980), as would of course be expected, no Pasteur effect can be demonstrated in *in vitro* preparations (Robin et al., 1979). No Pasteur effect by definition means that during anoxia brain metabolism is reduced by about an order of magnitude.

From these kinds of studies it is readily evident that metabolic depression is an advantageous option and strategy for extending diving duration in turtles. But by how much can it extend the anoxic diving duration?

Metabolic Depression Vastly Extends Turtle Anoxia Tolerance

The answer can be readily illustrated with a set of calculations. In the turtle, liver glycogen levels in the normoxic state are about 880 μmoles/gm (see Hochachka, 1982). A 100-gm turtle (excluding the shell) has about 8.6 gm liver, and thus about 7,600 μmoles of glycogen (glucose). The resting metabolic rate of turtles at 3°C and PaO_2 of about 70 mm Hg can be estimated at 0.036 μmoles ATP $gm^{-1}min^{-1}$, assuming a Q_{10} for respiratory metabolism of about 2; that is, a rate equivalent to 5,180 μmoles ATP 100 gm $turtle^{-1}day^{-1}$. At this metabolic rate, liver glycogen stores are adequate to maintain a 100-gm anoxic turtle for about 2.9 days. Other body stores of glycogen may extend this time period but in tissues such as the heart, glycogen stores are largely depleted after a few hours of anoxia. In short-duration diving (for a few hours) at 24°C, the metabolic rate drops to about 15 percent of predive levels. If these results are extrapolated to the situation at 3°C, then the anoxia tolerance of the turtle can be extended by six- to sevenfold; i.e., to about nineteen days. But even this is not enough.

In northern extensions of their range, for example in Ontario, Canada, a number of aquatic turtle species survive a five- to six-month winter period (October through March) at the bottoms of small lakes and ponds. The snapping turtle in particular is known to be partly buried in mud and some individuals remain there for the entire "diving" period. That kind of field observation led Ultsch and Jackson (1982) to maintain turtles in anoxic waters under laboratory conditions and at 3°C for six months! It is unlikely that turtles augment anaerobic glycolysis with amino acid fermentation since no changes in free amino acid profiles are observed during three- to four-week diving periods; nor do succinate and alanine, two anaerobic end products that would be expected from such a metabolism, accumulate to any large extend during diving (B. Emmett, pers. comm.). Thus, the outstanding duration of anoxia of the diving turtle would not be possible without a metabolic depression, down to about one-sixtieth of the resting normoxic metabolic rate of 0.036 μmoles ATP $gm^{-1}min^{-1}$. The same estimate is obtained if the calculation is made on the basis of lactate accumulation during prolonged anoxia (Ultsch and Jackson, 1982; Hochachka, 1982), which supports the contention that during anoxic diving turtles operate primarily on anaerobic glycolysis.

A less dramatic but similar picture emerges from simulated diving studies of the African lungfish (Dunn et al., 1983). In these and other good animal anaerobes (Hochachka and Dunn, 1984), by far the most effective means for extending anoxia duration involves a reduction in rates of ATP turnover. The aquatic turtle in particular appears so well adapted for sustaining O_2 lack one wonders if these are special cases among diving animals. The turtle is cold-blooded, after all, and its strategies for long-duration diving might be quite different from those utilized by diving mammals and birds. Moreover, one could argue that the overwintering of air-breathing aquatic turtles is not really diving: it is hibernation. Although either way it is a stunning achievement, yet in the context of this chapter what applies to the turtle may not apply to the seal or dolphin. We must therefore inquire as to whether aquatic mammals and birds sustain reductions in metabolic rates during diving as well.

Physiological Diving Reflexes Imply Reductions in Metabolic Rate

In one sense, the answer to the quandry of ATP turnover during diving is easy and is generated by the physiologist. Four decades ago, Scholander (1940) first demonstrated that experimental diving in marine mammals is accompanied by bradycardia and peripheral vasoconstriction, and this observation has been made repeatedly in many different species since that time (Butler and Jones, 1982). Whatever else these data may imply, they clearly suggest that heart work rates and therefore heart metabolic rates are reduced during diving, and the same may occur in many tissues whose normal work functions may be perfusion-dependent. In fact, we shall see that the overall metabolic depression can be quite large, but to show this quantitatively requires a careful and detailed analysis of the diving mammal. The most comprehensive metabolic and physiological data available are for the Weddell seal, which we will use to illustrate this point.

Although the Weddell seal is one of the most capable of mammalian divers, it is interesting that its physiological responses to simulated diving are qualitatively similar to those seen in numerous marine mammals and birds. Although the onset of bradycardia is rapid, the percentage drop in heart rate from about 60 to 15 beats/min. is not at all unusual. Cardiac output falls from about 40 to 6 L/min., but a mean blood pressure of about 120 torr is maintained through simulated diving, indicating the activation of extensive peripheral vasoconstriction. Quantitative measurements on the selective distribution of blood flow indicate that only the tissues of the central nervous system receive an unchanging blood flow through diving and recovery sequences. Essentially all other organs and tissues (except for the adrenals) sustain markedly reduced blood flow, down to one-sixth to one-twentieth of normal. Even coronary blood flow decreases to about one-sixth the resting values, coincident with a reduced work load on the heart. Similarly, pulmonary arterial blood flow drops by the same factor in step with a decrease in cardiac output (Zapol et al., 1979).

Biochemical Representation of the Diving Response

When the above physiological adjustments are brought into play they are represented biochemically by changes in blood gases, in metabolite profiles, and in substrate turnover rates. In natural voluntary diving, these changes are not yet fully clarified, so for simplicity, we will base our analysis primarily upon simulated or laboratory diving, when the diving reflexes are probably maximally activated. In seals, as in other marine mammals, fat is the preferred fuel for normoxic metabolism and the turnover of plasma fatty acids is reasonably rapid. In the gray seal, for which data are available, free fatty acid (FFA) turnover rate is about 10 μmoles $kg^{-1}min^{-1}$ with a half-life for plasma palmitate of about four to five minutes (Castellini and Hochachka, 1982). During diving, the turnover rate of FFA (assayed with ^{14}C-palmitate) drops drastically and is so low it cannot be quantified by usual techniques. In such experimental conditions in the Weddell seal, a decrease in blood glucose concentration is consistently seen during short- and long-term dives in samples of whole blood taken either from the pulmonary artery or the aorta. Concomitantly, blood lactate concentrations increase in arterial blood usually from less than 1 μmol/ml in predive states to over 3 μmol/ml at the end of the diving period. The lactate appearing in the blood at this time is but a small fraction of the total, most of which is retained at sites of formation during diving but is rapidly washed out in the recovery, when lactate pools in blood and peripheral tissues equilibrate (Hochachka et al., 1977).

These findings taken together indicate that the physiological adjustments to diving convert the diving animal into well-perfused and poorly perfused regions, each with changing substrate preferences through the diving-recovery sequence. Although for no species are these fuel preferences fully clarified, if we can answer the question of how much energy is required by each region during diving, we automatically will expose the role of metabolic depression mechanisms in the process.

Brain Energy Requirements during Diving

As in other vertebrates, brain metabolism in seals is supported mainly by blood glucose. Measurements of glucose uptake by the brain during simulated diving indicate that glucose uptake is essentially unchanged from resting control states, which is expected from perfusion measurements showing no change in blood flow to the brain during diving. Interestingly, lactate release adds up to about 20 to 25 percent of the cerebral glucose uptake under both states (Murphy et al., 1980). Because of the size of the brain and the blood volume relative to body mass, the brain fractional contribution to overall energy metabolism is low, about 1 percent or less (!) of total whole-organism metabolism, while in man it is at least twenty-five times higher and contributes minimally 15 percent to total basal metabolism (Table 6-1).

Table 6-1. Size and Metabolism of the Brain in the Weddell Seal during Experimental Diving Compared to Man at Rest

	Seal (450 kg)	Man (70 kg)
Brain weight, kg	0.5	1.4
Brain weight as % of body weight	0.1%	2.0%
Brain metabolic rate[a] mmoles O_2/1.2 hr.	48	151
Whole-organism basal metabolic rate mmoles O_2/1.2 hr.	5,697	960
Brain metabolic rate as % of total	0.8%	15.0%

SOURCE: From Hochachka (1981)

[a] Calculated from glucose uptake rates assuming 80 percent oxidation (Murphy et al, 1980). A time unit of 1.2 hours is chosen as this is the known maximum duration diving time for the Weddell seal.

Table 6-2. Brain Utilization of Blood Glucose in the Weddell Seal during Experimental Diving Compared to Man at Rest

	Seal (60 L Blood)	Man (5.6 L Blood)
Total blood glucose pool size, mmoles	300.0	28.0
Brain uptake rate, mmoles/1.2 hr.	10.8	25.2
% of total blood glucose used by brain/1.2 hr.	3.6%	90.0%

SOURCES: From Hochachka (1981); Murphy et al. (1980)

An informative comparison of brain metabolism in the Weddell seal and man can be made from data on glucose uptake since it demonstrates the role of scaling blood volume. Although normal whole-body glucose levels in both species are similar (about 5 μmol/ml), the blood volume in the Weddell seal is amplified relative to body size, and its absolute volume is eleven to twelve times larger than in man. With only this adjustment, the total blood glucose pool in the seal is some elevenfold larger than in man. A simple calculation shows that as a result the Weddell seal brain consumes per 1.2 hours only 3.6 percent of the total blood glucose, compared to 90 percent utilization per 1.2 hours by the human brain (Table 6-2). Maximum diving time for the Weddell seal is 1.2 hours, which clearly could not be determined by the brain's depletion of blood O_2 or glucose supplies. Moreover, this modest glucose utilization could not account for the large drop in blood glucose concentration observed during diving.

Lung Energy Requirements during Diving

To obtain similar information on the lung during diving, it is important to realize that during diving the main carbon and energy source probably is blood lactate. AV concentration gradients across the seal lung indicate a significant uptake of lactate whose major fate appears to be oxidation. Although glucose can also be utilized, *in vitro* tissue slice experiments establish that lactate is preferentially taken up and oxidized at two to ten times higher rates, depending upon the concentration ratio of glucose/lactate. From both *in vivo* and *in vitro* isolated tissue studies, a similar rate of metabolism by the lung is obtained. As in man, this represents only about 1 percent of the whole-organism metabolic rate (Table 6-3).

Table 6-3. Lung Metabolic Rate in the Weddell Seal during Experimental Diving Compared to Man at Rest

	Seal (450 kg)	Man (70 kg)
Lung weight, kg	4.0	0.5
Lung weight as % of body weight	0.9%	0.7%
Lung metabolic rate mmoles/1.2hr	72.0	12.6
Lung metabolic rate as % of total	1.2%	1.3%

SOURCES: From Hochachka (1981); Murphy et al. (1980) with modification

Heart Energy Requirements during Diving

To date, direct metabolite or O_2 gradients across the heart of the Weddell seal are unavailable, so our estimates of its energy requirements have to be indirect. The situation is somewhat complicated in the Weddell seal because the heart stores exceptionally high amounts of glycogen and displays very high levels of lactate dehydrogenase, both indicative of a high glycolytic capacity. However, isozyme distribution is relatively normal for mammalian heart, being dominated by heart-type subunits, kinetically bifunctional, with a high lactate-scavenging potential. From measurements of blood flow, cardiac output, and arterial pressure, it can be shown that coronary flow per unit of cardiac work is essentially unchanged during diving. That is, heart work during diving in the seal remains supported, primarily and probably solely, by oxidative metabolism. Whereas oxidative metabolism in the mammalian heart may be fired by a variety of substances (glucose, fatty acids, lactate), lactate is known to be preferentially utilized whenever concentrations rise above normal (Drake, 1982). This is precisely the situation developing through the diving period which is why, we assume, the Weddell seal uses its high levels of lactate dehydrogenase for lactate oxidation. Under simulated diving in the harbor seal, however, the heart may turn to net lactate production in later stages (Kjekshus et al., 1982). Whatever the substrate sources, heart metabolic rates of the Weddell seal can be calculated from work rates and obviously depend critically upon how the diving response is used. If an extreme bradycardia is activated immediately upon diving, the O_2 uptake may fall to values below 100 mmoles/1.2 hours. If the bradycardia is activated in a stepwise fashion, down to about 25 beats/min. for the first part of the dive, then further down to about 10 for the rest of the dive, heart metabolic rates are equivalent to about 150–300 mmoles O_2/1.2 hours. The

Table 6-4. Metabolic Rates of Weddell Seal Heart at Varying Work Loads

Conditions	mmoles O_2 consumed[a]	Fraction consumed of total 1,000 mmoles of blood O_2 stores
No bradycardia (HR 60; SV 568.5 for 1.2 hr.)	993	99%
First 0.3 hours:[b]		
HR 25, SV 568.5	130	
Last 0.9 hour:		
HR 15; SV 366	148	
Total in 1.2 hour	278	27.8%
First 0.3 hour:[b]		
HR 25, SV 568.5	130	
Last 0.9 hour:		
HR 7.5; SV 366	74	
Total in 1.2 hour	204	20.4%
HR 10; SV 366 for 1.2 hour[c]	132	13.2%
HR 7.5; SV 366 for 1.2 hour[c]	80.2	8.9%

[a] Metabolic rate calculated from work assuming 10 percent efficiency. For example, work of left ventricle, at 122 mm Hg blood pressure, is 122 × 1330 (conversion factor to ergs) × stroke volume (SV) × heart rate (HR) × time (min.), given in ergs/unit time. Heart functional parameters from Zapol et al. (1979). Right ventricle, working against a lower blood pressure (30 mm Hg), uses about one-quarter the energy of the left ventricle

[b] Conditions that appear to be most similar to those expected in long-duration diving at sea (Kooyman and Campbell, 1972; Kooyman et al., 1980)

[c] Conditions during experimental diving of long duration (Zapol et al., 1979; Liggins et al., 1980; Zapol, unpubl. data)

values for this, or a similar reverse sequence, represent 2–6 percent of the whole-organism metabolic rate (see Table 6-4).

Brain, Lung, and Heart Utilization of "On Board" Oxygen Stores

The above calculations indicate that summed together the metabolic rates of the three central organs represent a modest fraction of the whole-organism metabolic rate. But that may be the wrong way to look at the problem, for during diving the seal has only a limited O_2 storage capacity and it is this that is priming oxidative metabolism. Could these "on board" oxygen stores become limiting due to uptake by the brain, lung, and heart? Again, the answer is negative. From previous studies, we know that the total blood O_2 stores in the Weddell seal equal about 1,000 mmoles; another 500 mmoles are myoglobin-bound in muscle. Thus, in a maximum (1.2 hour) dive, the seal brain would use up only 3–4 percent of the blood reserves of O_2. Similarly, because of upward scaling of the blood volume, the seal heart utilizes during 1.2 hours of diving only about 15 to 20 percent of the available O_2, while the lung utilizes about 7 percent. Summed together, brain, heart, and lung metabolic rates utilize in 1.2 hours only about 25 percent of the total blood O_2 reserves (Hochachka, 1981; see also Tables 6-1, 6-2, and 6-4).

Two crucial conclusions arise. Firstly, during maximum duration (1.2 hour) diving, the brain, heart, and lung rates of O_2 and substrate depletion cannot determine the time limits observed. And secondly, *about 95 percent and 75 percent of the immediately available blood glucose and O_2 supplies, respectively, are "spared" for other organs and tissues during maximum duration diving.* How long then, could the seal be sustained by strictly oxidative metabolism? It turns out the answer depends upon how the diving response is used.

Maximum Aerobic Diving Times

Knowing the O_2 uptake rates of the central organs allows easy calculation of the metabolic rate of the rest of the body since the two values must add up to the metabolic rate of the whole organism (about 5,697 mmoles O_2 1.2 hr^{-1} 450 kg^{-1}. It is thus an easy matter to show (Table 6-5) that with *all systems fully aerobic* maximum diving time is approximately twenty minutes. It is instructive that maximum aerobic diving is relatively independent of heart work; with *no bradycardia at all,* maximum aerobic diving is only three to four minutes less than with a profound drop in heart work (down to 7.5 beats/min.). This is because heart metabolic rates even at normal work rates are fairly small fractions of whole-organism metabolic rates. An important consequence of all systems being fully aerobic (in the initial twenty minutes or so) is that *any further diving would require the heart, lung, and brain all to function entirely by anaerobic glycolysis.* In most mammals, neither the heart nor the brain can sustain anything like normal function strictly by anaerobic glycolysis; hence, there is a serious risk involved in diving without activation of the diving responses. Nevertheless, field data

Table 6-5. Maximum Aerobic Diving Times in 450 kg Weddell Seal

Assumed heart rate	mmoles O_2/min by H, L, & B	Max. dive time: all systems aerobic	O_2 left at end of dive[b]
60	15.3[a]	19 min.	0
25	10.1	21 min.	0
7.5	2.6	22.6 min.	0

SOURCES: From Zapol et al. (1979); Liggins et al., (1980); Kooyman and Campbell (1972)
[a] Metabolic rate of heart, lung, and brain (H, L, & B) calculated as in Hochachka (1981); metabolic rate of rest of the body taken as 63.8 mmoles O_2/min. from Kooyman et al. (1980)
[b] Any further diving would require *H, L, & B functions to be totally anaerobic*

show that such dives, termed feeding dives, may be utilized (Kooyman et al., 1980), because on the benefit side of the compromise, the seal is able to dive almost immediately again (i.e., recovery time from aerobic diving is extremely short).

Partitioning (Aerobic versus Anaerobic) Function of Diving Response

Whereas the diving response does little if anything to extend the duration of strictly aerobic diving, it is the foundation of longer-term or exploratory diving. Another simple calculation (Table 6-6) shows that the way in which the diving response is utilized *influences both the maximum length of time the peripheral tissues can operate strictly aerobically and how long anaerobically.* Two important insights arise from these calculations: firstly, the maximum time the noncentral tissues can be fully oxidative again approaches twenty minutes (they must be fully anaerobic for the last fifty-five to sixty minutes in maximum duration diving); secondly, *in this case enough O_2 is spared to keep the heart, lung and brain oxidative for the full 1.2 hours. That is, with but a small (about five-minute sacrifice) in maximum duration of aerobic metabolism by noncentral tissues, the diving response can extend total dive time from about twenty-two minutes to seventy-two minutes, without risk of the central organs going anoxic.* (That is the benefit side of using the diving response; the cost side is the length of recovery which we will discuss further below).

Finally, in this context, it is important to emphasize that these estimates of maximum duration of aerobic metabolism during diving assume that the metabolic rates of different tissues and organs are similar to rates under resting, normoxic conditions and are almost

Table 6-6. Partitioning (Aerobic versus Anaerobic) Functions of Diving Response

Assumed heart rate[a]	% of total blood O_2 used by H, L, & B in 1.2 hours	Time rest of body can be fully aerobic	Time rest of body must be anaerobic[b]	Total maximum dive time
25	48.5	12 min.	60 min.	72 min.
25 → 15	38.6	14 min.	58 min.	72 min.
25 → 7.5	31.2	15 min.	57 min.	72 min.
10	24.0	16 min.	56 min.	72 min.
7.5	18.8	17 min.	55 min.	72 min.

SOURCES: From Kooyman and Campbell (1972); Zapol et al. (1979); Liggins et al. (1980)
[a] In condition (2) above, initial heart rate is taken as 25; stepdown to 15 assumed at 0.3 hours into dive. In condition (3), stepdown to 7.5 beats/min. assumed at 0.3 hours into dive. Heart rates of below 10 observed during prolonged dives under laboratory conditions (Liggins et al., 1980, for examples)
[b] During these time periods, enough O_2 is conserved for the heart, lung, and brain (H, L, & B) *to allow a fully oxidative metabolism*

identical to those calculated by Kooyman et al. (1980). Both are also close to observed aerobic time limits of natural diving at sea when the Weddell seal must activate muscle metabolism for swimming. To do this for twenty-minute time periods at sea means that the aerobic metabolic rates of the rest of the body must drop to allow for muscle metabolic rates to rise. Otherwise, muscle metabolism would have to go anaerobic or diving time would have to be reduced. Or both. This issue is central to understanding diving metabolism, so we must examine it in somewhat more detail.

Nature of Muscle Metabolism during Feeding and Exploratory Diving

Of the two diving modes identified in the field (Kooyman et al., 1980), the energy needs for feeding dives (less than twenty-minute duration) can be most readily estimated because the storage supplies of O_2 are known. Assuming 135 kg of muscle and that 75 percent of available O_2 is partitioned for muscle work, a twenty-minute dive for a 450 kg seal would be supported by a muscle metabolism consuming 1,125 mmoles O_2 (equivalent to turning over about 6,750 mmoles ATP). If all its muscle mass were used for swimming (most unlikely!), energy utilization rate would be only 2.5 μmol ATP gm^{-1} min^{-1}, which is only about $\frac{1}{8}$–$\frac{1}{12}$ the maximum for mammalian muscle (see Chapter 4). If only 50 percent of the muscle mass is used to power swimming, the rate of work is about 5 μmol ATP gm^{-1} min^{-1} or a value about one-third that in other mammals. Since the scope for aerobic metabolism in diving seals is only three- to fourfold (see Hochachka, 1981), compared to tenfold for other mammals, this lower value is in fact in the right range.

During longer-term exploratory diving, the situation is more complex and can be illustrated by assuming that the seal must extend the first twenty minutes of diving through an additional twenty-minute anaerobic period. If the seal had to swim at the same speed as before, muscle work would now have to be supported by the anaerobic production of 6,750 mmoles ATP (equivalent to an accumulation of lactate to over 100 μmol gm^{-1} in the working muscle mass). If the dive were extended to maximum duration following the initial twenty-minute aerobic period, lactate accumulation in working muscle would surpass 300 mM!

Even when the above values are reduced by three- to fourfold (for reasons explained below), they still are much higher than ever observed. That is why we conclude that, as in other animals, anaerobic muscle work in the seal is used largely as a high-power output system for short-term work, but it is *not* used for the extended work typifying exploratory diving. Neither on land nor in the sea are marathon efforts powered by anaerobic glycolysis. We suggest that muscle work at this time, as in feeding dives, is therefore sustained by an energetically efficient aerobic metabolism, but because "on board" supplies of O_2 are limited, diving time can only be extended by compromising swimming speed. By reducing muscle work rates to one-third to one-fourth of those used in feeding dives, seals automatically may extend by three- to fourfold the *time* they are able to swim aerobically. The ATP turnover rates needed for these swim speeds are only about 1.5 μmol ATP gm^{-1} min^{-1}, compared to values of about 5 μmol ATP gm^{-1} min^{-1} in feeding dives and about 0.3 μmol ATP gm^{-1} min^{-1} for resting mammalian muscle (Chapter 4). The O_2 supplies available "on board" are adequate for this rate of aerobic muscle work to continue for up to about seventy minutes (i.e., close to that observed in nature).

In general, we would predict that the larger the fraction of "on board" O_2 used for muscle work, the shorter the dive or the greater the depression of aerobic metabolism of hypoperfused tissues, to a theoretical limit of zero O_2 uptake by them. This brings us full circle back to the question of a Pasteur effect. By not activating such a mechanism, turtles are able to depress ATP turnover rates. The question arising is whether or not the same process can operate in seals.

Depression of ATP Turnover during O_2 *limited Diving*

The answer to the above question is probably affirmative, but to demonstrate it we need to know the metabolic rate of hypoperfused tissues once these are in anaerobic phases of diving. As a first approximation, these rates can be obtained from lactate levels, assuming the above 25 percent/75 percent partitioning of "on-board" O_2 supplies. Following near maximum (about 50 min.) simulated dives with muscles quiescent, lactate levels quickly rise to about 15 μmoles/ml blood (Hochachka et al., 1977). If this representative value is assumed to be fully equilibrated with tissues of origin, it means that close to 30 mmoles of lactate kg are formed in peripheral tissues during maximum duration diving. Since the hypoperfused tissues remain anaerobic for about an hour (Table 6-6), it means a rate of lactate production of about 13.5 moles/450 kg seal, or an ATP turnover rate of 13.5 to 20 moles ATP 450 $kg^{-1}hr^{-1}$, compared to normoxic ATP turnover rates of about 34; that is, during near maximum duration simulated diving, *the ATP turnover rate is reduced to about one-half to two-thirds of normal.* Since the above estimate is on the conservative side, the depression in ATP turnover rate may be even greater, although it does not appear to be as great as in aquatic turtles.

Size and Diving Duration

Although intuitively it may be evident that all of the above diving-time estimates may be strongly influenced by body mass, a quantitative analysis of this relationship is not available for diving mammals. But it is for diving birds. Jones and his coworkers have shown that the relation between "on board" O_2 supplies and body mass, expressed in the allometric form, is

$$\text{total } O_2 \text{ stores } \alpha \text{ body mass}^{1.131}$$

Aerobic metabolic rate, on the other hand, varies with body mass to the 0.723 power. Therefore,

$$\text{maximum aerobic diving time} \quad \alpha \quad \frac{\text{body mass}^{1.131}}{\text{body mass}^{0.723}} = \text{body mass}^{0.4}$$

Thus if mass increases twenty times, maximum aerobic diving time also increases, but by only about fivefold (D. R. Jones, pers. comm.).

In forced dives in ducks, unlike the situation in Weddell seals, most of the "on board" O_2 supplies are used by the heart and brain. The mass of these two organs combined varies with body mass to the 0.65 power, while their metabolic rates vary with body mass to the 0.73 power. Hence, in forced diving (aerobic plus anaerobic components),

$$\text{maximum diving time (aerobic + anaerobic modes)} \quad \alpha \quad \frac{\text{body mass}^{1.131}}{(\text{body mass}^{0.65})^{0.73}} = \text{body mass}^{0.65}$$

For a twentyfold increase in body mass, therefore, maximum duration diving (aerobic plus anaerobic modes) increases by nearly tenfold, in contrast to the fivefold increase in the strictly aerobic mode of diving. That is, being big extends diving time in both aerobic and anaerobic modes, but the scaling effect confers a greater advantage for the anaerobic mode than for the aerobic one. Since glycolytic enzyme activities of working muscle scale to about a 1.1–1.2 power, compared to the 0.73 power of aerobic enzyme activities and metabolism (Emmett and Hochachka, 1981), such advantages of large size may be even greater than implied from the allometric relationships used by Jones and his coworkers in their calculations.

End Product Problems: Tolerating Accumulation as the Major Solution

As far as is known to date, the central organs during even maximum duration diving in mammals and birds remain supported mainly by an oxidative metabolism whose end products (CO_2 and H_2O) are relatively harmless. The hypoperfused periphery, however, accumulates lactate, pyruvate, succinate, alanine, and glutamine (see Murphy et al., 1982), but only lactate is found in high concentration. Of the various theoretical possibilities for dealing with the end products problem (accumulation, slow removal and metabolism at sites remote from those of formation, or excretion), only the first is clearly important, the second is utilized to a minor extent, and the third is not utilized at all. Since the pathfinding work of Scholander (1940) four decades ago, we have known that most of the lactate formed in hypoperfused tissues during experimental diving is simply accumulated at sites of formation, equilibration of blood and tissue lactate pools being prevented by vasoconstriction. Hence the main solution to end product effects involves maintaining mechanisms in tissues such as muscle for tolerating accumulation.

Although postdiving levels of blood lactate can be greatly elevated, during actual diving in mammals and birds blood lactate levels increase only modestly (Hochachka and Murphy, 1979). In this regard, turtles are very different, for during extreme diving, they may accumulate lactate in the blood to about 200 μmol/ml, truly astonishing levels which are far higher than seen in any other group of vertebrates! The levels are so high that in addition to serious acid-base problems, this amount of lactate ion requires similar amounts of counter ions. At least during simulated overwintering dives, calcium is released into the blood and serves to balance charge, but it is not known in what form the calcium occurs (Ultsch and Jackson, 1982). In shorter dives, Robin and his colleagues (see Caligiuri et al., 1981) find that the "cation gap problem" is solved by protein buffering.

Two additional biochemical adjustments for dealing with such high levels of lactate seem to have been developed in turtles. Firstly, the buffering capacities of blood and tissues seem appropriately adjusted upwards through maintenance of large bicarbonate reserves. Secondly, at least in the case of two regulatory enzymes in turtle heart, namely phosphofructokinase and pyruvate kinase, the pH optimum is much lower than for homologues in higher vertebrates (see Hochachka and Storey, 1975). This might be viewed as an adaptation for function in either a lower pH range or in a wider, and more fluctuating, intracellular pH, which in turtles seems to vary from about 7.4 down to at least 6.8 during diving (anoxic) stress.

The only biochemical adjustments thus far known for tolerating lactate accumulation in diving marine mammals do not appear unique to these groups and involve the buffering capacity of the tissue. Particularly in muscle, a good correlation is obtained between buffering capacity and lactate dehydrogenase activity (Castellini and Somero, 1981), implying the first (buffering) is adjusted to meet the demands of the second (acidification during anaerobiosis).

End Product Problems: Recycling Lactate as a Minor Solution

Although peripheral tissues and organs are vasoconstricted during diving, their blood flow is not reduced to zero and thus during diving there is a release of lactate carbon and a gradual accumulation of lactate in the blood. This accumulation would be greater if it were not for lactate oxidation at various sites in the body such as the lung. As already mentioned, AV gradients across the lung in the Weddell seal indicate lactate uptake during diving when blood lactate concentrations typically rise to about 3 μmol/ml. *In vivo* experiments, monitoring ^{14}C-lactate oxidation on a single circulatory pass through the lung, demonstrate that shortly after injection into the pulmonary artery, when the ^{14}C-lactate and absolute lactate concentrations are still decreasing in the pulmonary circulation, $^{14}CO_2$ is already appearing in aortic blood; this CO_2 can only be derived from lung metabolism since the lung is the only tissue (other than blood) to have received ^{14}C-lactate. These experiments, plus others with isolated lung slices showing higher rates of lactate oxidation than of glucose oxidation under a variety of conditions, firmly establish that the lung in seals, as in other mammals (Wolfe et al., 1979), clearly favors lactate over glucose as a substrate for oxidative metabolism. Other tissues and organs such as heart (Drake, 1982) or working red muscle may assist the lung in this process. Such lactate recycling has also been demonstrated during routine, free-diving of the Weddell seal at sea (Kooyman et al., 1980), supplying an independent field-study confirmation of the above laboratory studies.

There is a mildly disturbing question remaining: since recycling processes are clearly the best long-term solutions to problems of end product accumulation, and are extensively utilized in lower animals, why are they not more emphasized during diving in aquatic vertebrates?

Two answers are evident. Firstly, recycling lactate requires maintaining a relatively open circulation, but vasoconstriction is part and parcel of the diving response. Thus, the physiological requirements of the diving response are at variance with, and preclude, major recycling of end products like lactate.

If the contrary requirements for O_2 redistribution versus lactate recycling constitute the main reason for not emphasizing this process in diving animals, it is not the only one. A second reason is that many, if not most tissues, probably sustain significant reductions in metabolic rate during diving, as already discussed above. There is therefore little need for more substrate for oxidative metabolism at this time. But end products such as lactate still retain a lot of chemical potential energy which it is appropriate to conserve for subsequent utilization under more appropriate conditions. Those conditions are in fact realized during recovery from diving, when glucose and glycogen homeostasis must be regained and when there is a large need for good gluconeogenic substrates. It is therefore a most sensible arrangement to greatly increase lactate availability at this time, when reperfusion is explosively re-established.

Regaining Homeostasis following Diving

In many ways, the recovery from diving is as interesting in metabolism terms as the diving sequence itself. To put the matter in context, it is important to remember that within the first moments of ending the dive, heart rate and cardiac output both return to normal and within the first minute of recovery may about double prediving levels (M. Snyder, pers. comm.). At the same time, tissues which through diving have been in a hypoperfused state are now perfused at normal, or even above normal rates due to vasodilation (Zapol et al., 1979). End products which have accumulated during the diving period are now washed out into the general circulation allowing equilibration of tissue and blood pools of these metabolites. Experimentally, the best way of illustrating these processes is to monitor blood metabolites in the recovery period; when this is done, there are characteristic large pulses of end products spilling out into the blood. The main metabolites released are the same as those accumulating during diving: pyruvate, lactate, succinate, alanine, and glutamine (Murphy et al., 1982). The twin problems in re-establishing metabolic homeostasis are therefore a) the clearance of these amplified pools of end products, and b) the recharging of glycogen and glucose reserves to prediving levels. Although these are related problems, it is convenient to consider them separately, in part because different tissues contribute differently to the processes.

Clearance of Lactate, Pyruvate, Alanine, and Glutamine

We have already reviewed in Chapter 4 the main ways of returning lactate to prediving levels; namely, a) *in situ* oxidation, b) oxidation elsewhere in the body, c) *in situ* reconversion to glycogen, or d) entrance into the Cori cycle for reconversion to glucose and glycogen in the liver and kidneys. In theory pyruvate clearance could follow the same pathways, and since it has never been investigated in diving animals, we will assume it is in fact metabolically equivalent to lactate and will not further discuss it in this context.

Of the above possibilities, (a) and (c) have not been experimentally examined, (b) has been considered for three organs, the brain, lung, and heart, and (d) has been experimentally quantified in the harbor seal (Davis, 1983).

Recharging Glucose and Glycogen Stores: The Critical Role of Time

As in other mammals clearing high lactate pulses, in diving animals an important fate of lactate is uptake by the liver and kidneys for reconversion to glucose and glycogen. At least in seals, this process is probably augmented by alanine and glutamine which are also released into the circulation after diving. All three are potentially good gluconeogenic percursors, but judging from relative levels in the blood during recovery, lactate contributions to glucose resynthesis are about two to four times greater than from alanine and glutamine taken together.

The time course of recovery is interesting in that it usually takes at least as long as the dive and usually longer to regain fully blood glucose homeostasis. This process is under control of glucagon, which is elevated in early stages of recovery and promotes gluconeogenesis. The combined effect of glucagon (on stimulating flow of lactate, alanine, and glutamine carbon into glucose) and of elevated catecholamines (on stimulating glycogenolysis) typically leads to a modest overshoot in blood glucose levels during the recovery period (Murphy et al., 1980; Robin et al., 1981).

The fate of alanine and glutamine nitrogen until now has not been assessed, but by analogy with other mammals it is safe to assume it is mainly deposited into urea in the liver; in fact the main function of alanine and glutamine in mammals is to carry waste nitrogen to the liver for deposition into urea. Whether or not this leads to momentary elevations of blood urea during recovery, however, is not yet known.*

* A longstanding dream of many marine biologists—the comprehensive analysis of blood biochemistry and heart rates of marine mammals monitored during natural diving at sea—was recently realized in studies of the Weddell seal. This achievement was made possible by designing and building a dedicated microcomputer and peristaltic pump for remote measurements of ECG and for sampling arterial blood. The computer could command withdrawal of blood samples at programmable combinations of diving times, diving depths, and heart rates; it could communicate with a laboratory computer via fiber optics for program adjustments with time, thus allowing samples to be drawn in early, middle, and late periods of diving as well as during recovery. Detailed ECG analyses (Hill et al., *Fed. Proc. ABSTRACTS* 42 [1983]) indicated the use of graded bradycardia (with lowest heart rates during long exploratory dives) and a direct relationship between heart rate and vertical swimming speeds during diving. Of the potential substrates analyzed (plasma triglycerides, free fatty acids, amino acids, and glucose) only plasma glucose was found to decrease in diving. Although the amino acid pool was not perturbed, plasma urea levels increased in both diving modes. Plasma lactate pools were augmented during and following exploratory, but not short-term feeding dives. The data were taken to support the concept of diving responses being plastic and graded according to the needs of each specific dive (R. D. Hill, W. M. Zapol, G. C. Liggins, R. C. Schneider, A. H. Schuette, and P. W. Hochachka, Metabolite biochemistry of the Weddell seal diving at sea, *Amer, J. Physiol.* [1984], in press).

Why Most Diving Is Aerobic

The time-consuming nature of these multiple metabolic processes in recovery has an immense impact on diving frequency, since a diving mammal pushed to its diving duration limit requires at least the same length of time and usually longer for full recovery of metabolic homeostatis. It is a costly and inefficient strategy, therefore, for an animal to activate the anaerobic components of its metabolic arsenal every time it dives. As Kooyman and his colleagues emphasize (1980), during foraging for food the Weddell seal could spend a much greater fraction of its time underwater if most of its dives were within the time limits of strictly aerobic capacities. Not surprisingly, it is the latter strategy that is utilized by the Weddell seal most of the time, for field studies show that about 97 percent of natural dives at sea are less than about twenty minutes long; i.e., the animal appears to time its dives so as to allow most systems to be supported by oxidative metabolism most of the time. On the other hand, exploratory dives or dives in which emergencies develop appear to be of longest duration when the full metabolic potentials of the animal are utilized to greatly extend the territory and time covered during diving. The compromise, efficiency for duration, becomes a reasonable trade-off only because it facilitates future exploitation of the new environment being explored.

Metabolic plasticity appears to be the key to success: by adjustments of tissue metabolic needs and enzymatic capacities, the diving marine mammal as an isolated, self-contained life-support system with novel metabolic rules for survival, enters an environment in which it cannot live indefinitely. Yet for excursions of impressive duration and range, it can successfully explore and exploit the sea.

CHAPTER SEVEN

Off-Switches in Metabolism: From Anhydrobiosis to Hibernation

When and Why Animals Turn Down Metabolic Rate

There are numerous examples in nature of animals entering dormant or semidormant states for variable time periods. For example, for surviving the total dehydration that may occur when *Artemia* encounter completely desiccated conditions, this small crustacean enters a period of encystment and metabolic arrest, which can last indefinitely. Such anhydrobiosis, although dramatic, is not unique to this particular organism. Another complex, multicellular animal that can totally dehydrate without ill effects is the larva of a chironomid fly (*Polypedilum vanderplanki*). This larva lives in shallow and exposed rock pools in Nigeria and Uganda. At the beginning of the rainy season the pools may fill and dry several times, and they may occasionally be filled with water for short periods during the dry season. When the pools dry up the larvae also dry up and remain dry until the next rains.

The degree to which the larvae dry up depends upon humidity. At a relative humidity of 60 percent their moisture content is 8 percent and it falls below 3 percent at <1 percent relative humidity. The larvae can rehydrate in water within about an hour, at which time their normal moisture content of 80 to 90 percent is restored and they may actually be feeding again (Hinton, 1968). The larvae are thus adapted to an environment which is subject to flooding alternating with periods of extreme drought. A similar capacity is known for rotifers, tardigrades, and nematodes, and is probably relatively widespread among lower invertebrates.

The function of such anhydrobiological states is to survive in places that periodically suffer a complete lack of water; during such times, there is no possibility of obtaining food in the usual sense of that term and it is therefore advantageous to turn down metabolism maximally. In these "dry biological systems" metabolic rates can fall to zero, thus completely removing a requirement for nutrition.

Metabolic rates of zero, by definition, represent the extreme case. In most other organisms, periodically depending upon dormant states, metabolism is not switched off completely; rather it is turned down usually one to two orders of magnitude below basal or standard metabolic rates, sometimes in concert with enzymatic reorganization. Although the conditions for metabolic arrest vary, reduced availability of water or food (or both) forms a common stimulus. The African and South American lungfishes, for example, are famous for their capacities to estivate during dry seasons when they periodically become stranded in areas which are drying out. The desert snail is able to survive years of no rain by withdrawing into, and closing off, its shell in what is analogous to estivation in the lungfishes. In both cases, estivation is incompatible with feeding and a depression of metabolic rate below standard levels becomes a necessary biochemical strategy.

For small desert mammals, entry into a dormant phase often is determined by a combination of heat, water, and food problems. Cooling in hot and dry periods of the year requires large amounts of evaporative water, which cannot easily be replaced by drinking. Moreover the high metabolic rates of small mammals are not consistent with simply starving until the environmental situation should improve. All aspects of such environmental problems are well resolved by entering a dormant state—estivation—in which the metabolic requirements of the organisms are reduced far below normal.

For small mammals and birds living in temperate climates, the stimulus for dormancy may be a combination of limited food and cold. In this case, maintenance of a high body temperature requires a continuous input of food; by the same token, a lack of food may be incompatible with maintaining a high metabolic rate (i.e, a high body temperature). Thus, daily torpor, which is observed in small temperate bats, small birds (e.g., hummingbirds), and some small terrestrial mammals, becomes an excellent and widely practiced solution to the problem.

The adverse conditions not only of a cold night but also of winter can be at least partially avoided by entering a dormant phase, a process that in insects is termed diapause; in mammals and birds, it is termed hibernation (or winter sleep). Diapause can occur at any phase in the life history of the insect—embryo, larva, pupa, or even adult—but it usually occurs only once in the life history of any species. Whenever it occurs, diapause is characterized by a profound metabolic arrest, allowing survival through the stress period on endogenous reserves.

Hibernation, too, may take on several patterns. As displayed by small mammals (about 10 kg or less), hibernation is a form of dormancy in which the animal shows a profound drop in body temperature, in metabolic, breathing, and heart rates, and in activity. One group of hibernators maintain a low body temperature for about a week at a time, then arouse for about a day to normothermic conditions. Because these animals do not eat during such arousals, they are referred to as nonfeeding hibernators. The woodchuck, which does not store food in its hibernaculum in preparation for hibernation, is a good example of this form of hibernation.

Another form of "winter sleep" is found in numerous other small feeding hibernators. In contrast to the above, these mammals store large quantities of food in anticipation of hibernation. They remain dormant for five-to-ten day periods, then, like the woodchuck, they arouse between these bouts for about a day. Unlike the woodchuck, they feed quite voraciously during arousal and thus return to another bout of dormancy in a postabsorptive state.

Yet another type of "winter sleep" is found in large mammals that go into dormancy for as long as seven months, yet do not drop their body temperature to a large extent; nor do they eat, defecate, or urinate during this time. This pattern is well known for three species of bears and, by some biologists, is considered the most highly specialized of all mammalian hibernation patterns.

All of the dormancy states described above (and there are many more) display a simple biological characteristic in common: in each case, *the organism ceases its consumption of external nutrients concomitant with a metabolic depression of variable but controlled degree.* The period of spontaneous fasting sometimes is relatively short; in other cases, it is extended for a significant fraction of the organism's life cycle. This general characteristic (of hypophagia) explains a commonly held idea that metabolic processes occurring in spontaneous fasting are somehow similar to those in dormancy states, and indeed may be better displayed (in slow motion, as it were) than in normal metabolic states. This view turns out to be simplistic and basically inadequate. The metabolic characteristics of various dormancy states are in fact very different from those found in spontaneous fasting and serve different functions. Because these functions (and their underlying metabolic mechanisms) are often unique to specific dormancy state, we shall discuss each of them separately, emphasizing similarities and contrasts as we proceed.

Anhydrobiosis and Anhydrobiotic Organisms

So well established is the biological principle that water is essential for life that it may come as a surprise to some readers to learn of numerous exceptions to this principle; all of these, however, involve inactive metabolic states. The ability to survive the loss of all cellular water (save perhaps that bound very tightly to macromolecules) without irreversible damage is actually quite widespread, being expressed in nearly all the major taxa. Keilin (1959) introduced the term cryptobiosis ("hidden life"), but later workers prefer the term anhydrobiosis ("life without water") to describe the phenomenon.

Two readily distinguishable groups of organisms exhibit anhydrobiosis: 1) organisms capable of anhydrobiosis only in their early developmental stages—these include the seeds of plants, bacterial and fungal spores, the eggs and early embryos of certain crustaceans, and the larvae of certain insects; 2) organisms capable of anhydrobiosis during any stage of their life histories—these include certain protozoans, rotifers, nematodes, and tardigrades (Crowe, 1971). Despite this fundamental difference between the two groups, from currently available information it appears that they share many similar biochemical mechanisms allowing entrance into, and emergence from, the anhydrobiotic state. We shall examine in some detail two well-studied systems, the brine shrimp (*Artemia*), which falls into the first category, and nematodes, which fall into the second category.

General Morphological Features of Artemia Cysts

The *Artemia* cyst is composed of an embryo in the early gastrula stage, surrounded by a noncellular shell. The embryo contains around 4,000 cells, which appear to be essentially undifferentiated. The cells of the *Artemia* embryo display no apparent morphological features that mark them as "desiccation-adapted." Nonetheless, the embryos are capable of tolerating essentially complete water loss while remaining viable. Restoration of cellular water leads to a rapid reactivation of metabolic and developmental processes.

Water Content and Metabolic States of Artemia Cysts

In addition to providing an excellent example of a desiccation-tolerant system for experimental study, the cysts of *Artemia* provide a vehicle for examining the ubiquitous properties of cellular water in living systems. Of particular importance is the information provided by studies of *Artemia* on the classes of cellular water and the role of each class of water in stabilizing macromolecular structure and governing metabolic flux.

Table 7-1. Hydration-dependence of Cellular Metabolism in *Artemia* Cysts

Cyst hydration[a] (gH_2O/g cysts)	Metabolic events initiated	
0 to 0.1	None observed	ametabolic domain
0.1	Decrease in ATP concentration	
0.1 to 0.3 ± 0.05	No additional events observed	
0.3 ± 0.05	Metabolism involving several amino acids, Krebs cycle and related intermediates, short-chain aliphatic acids, pyrimidine nucleotides, slight decrease in glycogen concentration	domain of restricted metabolism
0.3 to 0.6 ± 0.07	No additional events observed	
0.6 ± 0.07	Cellular respiration, carbohydrate synthesis, mobilization of trehalose, net increase in ATP, major changes in the free amino acid pool, hydrolysis of yolk protein, RNA and protein synthesis, resumption of embryonic development	domain of conventional metabolism
0.6 to 1.4	No additional metabolic events observed	

SOURCE: Modified from Clegg (1981)
[a] Maximum hydration achieved by these cysts is about 1.4 g/g

Clegg (1981) has classified relationships between water content and metabolic activity according to the scheme shown in Table 7-1. At water contents below 0.3 g/g (grams of water present per gram of initially dehydrated cyst material), Clegg considers the cysts to be in an *ametabolic domain.* No metabolic reactions per se take place, and any chemical transformations that do occur may take place without the contributions of enzyme catalysis. At hydration levels near 0.3 g/g, some metabolic activity is initiated. This metabolic activity is not reflected by an increase in oxygen uptake, however (Table 7-1), and is restricted to but a subset of the full suite of metabolic reactions found at higher water contents. Thus, Clegg speaks of a *domain of localized or restricted metabolism* at water contents between roughly 0.3 to 0.65 g/g.

Once the latter water content is reached, the *domain of conventional metabolism* begins. Oxygen consumption rates become measurable, and rise with increasing water content. The metabolic activities noted in this range of water contents are *qualitatively* the same as found in fully hydrated cysts, even though this conventional metabolism is initiated at water contents slightly less than half of fully hydrated values.

What do these relationships between water content and metabolic activity indicate concerning the nature of cellular water and its relationship to enzymic activity? Clegg argues basically as follows. The initial addition of water to dried *Artemia* cysts appears to represent the water needed to form initial hydration layers around macromolecules, albeit even under highly desiccating conditions, some of this tightly bound hydration water will remain in the cyst. Clegg proposes that this *bound water* addition is completed by the time that approximately 0.15 g/g hydration is achieved. Further hydration of the cysts next contributes to a *vicinal water* phase, defined as water which, while not in the primary hydration layer around macromolecules or other water-organizing solutes (small ions, for example), nonetheless is relatively organized compared to *bulk water* due to longer range interactions with membrane surfaces and macromolecular complexes such as cytoskeletal elements. The vicinal water phase is not completed until hydration levels reach approximately 0.6 g/g. Further addition of water provides a bulk water phase, i.e., a water phase with a relative lack of structure caused by interactions with other cellular components.

Viewing these (hypothetical) states of cellular water in the context of the metabolic activity transitions noted in *Artemia* cysts, Clegg proposes that metabolic activity does not require bulk water. Thus, the initiation of metabolic reactions at approximately 0.3 g/g suggests that local metabolic transformations occurring in vicinal water are possible. However, only with the formation of a bulk water phase does the zone of conventional metabolism occur. A bulk aqueous phase may be necessary to allow effective transport of metabolites, fuel sources, etc. between cellular compartments. Thus, in Clegg's model, the initial bulk water phase can be considered as a channel for communication and transport between different parts of the cell. As more water is added and metabolic rate increases (Table 7-1), the bulk phase may be increasingly able to facilitate the flux of materials throughout the cell, assuming that the viscosity of the bulk phase is reduced as hydration extent increases.

These ideas, while speculative, do provide an attractive basis for appreciating the requirements of the cell for water. Some is needed to hydrate macromolecules and low molecular weight species, while the larger share of the cellular water is required as a channel for exchanging materials throughout the cell. Added to this simple picture, however, is a requirement for maintaining the aqueous phase in a state which is permissive of macromolecular structural stability, and it is in this context that we turn to an examination of another major feature of certain anhydrobiotic systems: the accumulation of polyhydric alcohols (polyols) like glycerol and trehalose.

The Roles of Glycerol and Trehalose

In *Artemia* cysts, glycerol accounts for approximately 4 percent of dry weight, and trehalose constitutes up to 14 percent of dry weight. There exists a strong correlation between the accumulation of these two polyols and survival in the desiccated state. And, as discussed in the chapters on solute adaptations and temperature adaptations, polyols also appear to be important in osmoregulatory and freezing-resistant strategies. What properties do polyols possess that make them useful in situations of water stress?

Polyols may have two distinct functions in anhydrobiotic systems. On the one hand, polyols may serve as water substitutes, forming hydrogen-bonded interactions with polar or charged entities of the cell, and thereby replacing water. In addition, polyols may have the important effect of stabilizing protein structure at low water activities. How this effect is achieved has been shown recently by Gekko and Timasheff (1981a,b). They showed that glycerol is effectively excluded from the highly structured water surrounding proteins. The addition of increasing amounts of glycerol to an aqueous solution containing proteins thus leads to an increase in protein structural stability. The thermodynamic arguments here are straightforward. If a protein unfolds (= denatures), a larger area of protein surface comes into contact with the solvent phase, leading to increased amounts of highly structured water. Since glycerol is excluded from this structured water, the addition of glycerol favors the compact, folded (= native) structures of proteins. Thus, if the loss of cellular water during entry into the desiccated state leads to a destabilization of protein structure due, perhaps, to increased concentrations of low molecular weight species in the proteins' microenvironment, the addition of high concentrations of glycerol can be viewed as a strategy for keeping the proteins in native form during periods of desiccation. When desiccation is over and the need for a rapid reactivation of metabolism occurs, the battery of enzymes needed (for conventional metabolism, in Clegg's terminology) will be present in a functional state.

Anhydrobiosis in Soil Nematodes

Soil nematodes are roundworms, about 0.5–3 mm in length and lacking circulatory and respiratory systems; water and gas movements are by direct diffusion through the cuticle. Although they live in the ground, soil nematodes are aquatic animals requiring a water film around soil particles for their activity. Because of this dependence upon water for movement, it was first speculated that dehydration and survival would be limited to only a few nematode species (an impression also in part due to limited numbers of studies). With the development of quantitative techniques for extraction of dried nematodes from desert soils, it has become apparent that the capacity for anhydrobiosis among nematodes is much more widespread than previously thought. What is more, these studies show that anhydrobiosis is not confined to any particular life stage or trophic group of nematodes, adults and larvae of many genera being found when dried soil samples are rehydrated.

Parenthetically, we should mention that although the phenomenon of anhydrobiosis has been known since the days of Leeuwenhoek, until recently remarkably little progress was made about the adaptations permitting soil-dwelling nematodes, rotifers, or tardigrades to dry up when the soil dries up. A redressing of this situation became possible in the last decade with the development of mass culture methods of nematodes. Crowe and his students, taking advantage of these developments, have supplied the most comprehensive experimental data to date on the entrance into, and arousal from, anhydrobiosis by a mycophagous nematode of the genus *Aphelenchus* (see Crowe and Clegg, 1978).

Morphological and Ultrastructural Changes during Dehydration in Nematodes

The stimulus for entrance into an anhydrobiotic state in nematodes, as in brine shrimp cysts, seems to be reduced water availability, which leads to a series of ordered morphological changes, initiated within the first twenty-four hours and virtually completed within seventy-two hours. The whole animal undergoes a longitudinal contraction and coiling, intracellular organelles such as muscle filaments undergo ordered packing, and membrane systems exhibit ordered change. Although a reduced water availability is a necessary condition for initiating these changes it is not a sufficient one since, on balance, Crowe and his colleagues conclude, all the above changes are regulated by the animal itself. Such endogenous control over morphology is now viewed as a key to survival: *so long as the structural integrity of the organism is preserved and remains intact, the organism is alive.* It is dead when and if that integrity is destroyed. Although this principle, which can be traced back to Keilin (1959), applies as stated primarily to higher levels of organization, structural integrity must be maintained by assuring each cell and tissue an appropriate microenvironment; indeed, the establishment of a proper intracellular (and extracellular?) milieu may be viewed as the main function of metabolism during entrance into anhydrobiosis. What kind of metabolism, then, is observed during dehydration, and what milieu changes are effected?

Metabolic Changes during Entrance into Anhydrobiosis in Nematodes

Nematodes about to enter into anhydrobiosis share with *Artemia* cysts the important characteristic of being essentially closed systems: no nutrients as such enter the system and all metabolic adjustments that accompany dehydration must therefore be generated endogenously. Experimentally, this simplifies the problem of identifying the main metabolic adjustment occurring during dehydration; namely, *a redistribution of carbon arising from storage depots of glycogen and lipid into large intracellular pools of glycerol and trehalose.* The latter two metabolites increase dramatically in concentration during desiccation, reaching levels of about 6 percent and 10 percent of dry weight, respectively, by the time the entrance into anhydrobiosis is completed (seventy-two hours). The rise in glycerol and trehalose is almost quantitatively accounted for by depletion of glycogen and lipid, as may well be anticipated.

Correlation between Polyhydroxy Alcohols and Survival

Interestingly, there is a striking coincidence between the onset of synthesis of glycerol and trehalose and increased ability to survive exposure to dry air. When the glycerol or trehalose contents are plotted against percent survival, a linear relationship is obtained in both cases, with regression coefficients of 0.98 and 0.93, respectively. As in *Artemia* cysts this strongly suggests that survival in the desiccated state is dependent upon glycerol and trehalose contents. In view of the stabilizing effects of polyhydroxy alcohols on the structural integrity of enzymes and nucleic acids, this result is not entirely surprising and is in full agreement with the data on *Artemia* cysts. In addition, Crowe and his colleagues have proposed that these metabolites (glycerol in particular) may play an important role in membrane stabilization during anhydrobiosis. Similar changes in polyhydroxy alcohols are known for other anhydrobiotic organisms as well.

Metabolism at a Standstill during Anhydrobiosis

Although data on metabolic events at different states of hydration, comparable to those for *Artemia* cysts, are not available for nematodes, it is reported that desiccated nematodes do not consume O_2. Three decades ago, Becquerel was able to revive certain nematodes, rotifers, and tardigrades after they had been exposed to temperatures as low as 0.05°K. He estimated that at this low temperature metabolic processes, if they occur at all in the anhydrobiotic state, must proceed at only about 10^{-7} the rate of normally metabolizing specimens. Hinton (1968) argues that this is not metabolism in the normal sense in which that term is used. Thus it can be assumed, at least tentatively, that in nematodes, as in *Artemia* cysts, metabolism is at a standstill during anhydrobiosis.

Arousal from Anhydrobiosis by Nematodes

Metabolic events during rehydration of desiccated nematodes, as may be expected, appear to be the reverse of those occurring during entrance into anhydrobiosis; there is a fundamental difference in terms of rate, however. As we indicated, metabolic adjustments during dehydration are slow, taking up to seventy-two hours to complete. Metabolic homeostasis, once water is made available, is regained, in contrast, within a matter of a few hours, with most of the noted changes occurring within the first hour. The time course of trehalose depletion is essentially identical to that for glycerol and both are mirror images of the repletion of glycogen (Fig. 7-1). These adjustments presumably follow standard pathways of trehalose, glycerol, and glycogen catabolism and anabolism, supported by the sharp elevation in overall metabolic rate that occurs upon rehydration. The difference in time course of metabolic adjustments to dehydration versus rehydration is in fact biologically adaptive, since the first occurs slowly in nature (as the soil dries up,) while the latter occurs much more rapidly (for example, following a rain storm).

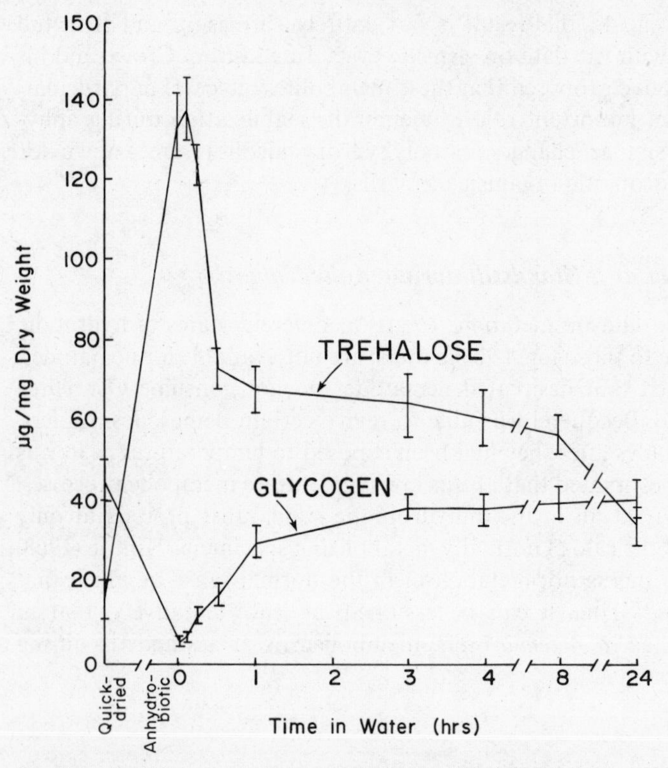

Fig. 7-1. Trehalose and glycogen contents of quick-dried *A. avenae* (assumed to be the same as in freshly harvested worms) and of anhydrobiotic *A. avenae* in air and at intervals following transfer of the anhydrobiotic worms to water. Modified after Crowe and Clegg (1978)

Biological Significance of Anhydrobiosis

Anhydrobiosis is usually assumed to be an adaptive strategy 1) for enhancing dispersal or dissemination of the species, 2) for improving survival chances in severe drought conditions, and/or 3) for timing biological process (feeding, reproduction, and so forth) with favorable environmental periods or conditions. In addition, while in anhydrobiotic state, physiological time in effect stops for the organism (because metabolic processes stop); immensely extended longevity in real (or geological) time, then, may in some cases be an extremely important spinoff of the anhydrobiotic capability.

Finally, although anhydrobiosis may be viewed as an adaptational solution primarily to drought conditions, another important corollary of this state is a *tremendously increased tolerance to numerous other harsh environmental factors.* Anhydrobiotic organisms are famous for their tolerance to extremely low, or extremely high, temperatures and pressures; they are tolerant also to organic chemicals, toxins, and anoxia. Given these kinds of advantages one can understand why anhydrobiosis developed in various phylogenetic lines, apparently arising independently each time it appeared in nature.

Insect Diapause

In order to withstand adverse (usually winter or other harsh seasonal) conditions many insects enter a dormancy state of depressed metabolism termed diapause. Although the overall strategy is similar to that of anhydrobiosis, diapause does not bring metabolism to a complete standstill. Entry into diapause can occur at any of the life stages of insects and is mediated by a variety of endocrine mechanisms. Embryonic and larval diapauses are each induced by the *occurrence* of a specific hormone: a proteinaceous diapause hormone controls diapause in embryo stages and the juvenile hormone does in larval ones. In contrast, pupal and adult diapauses result from the *absence* of particular hormones: the prothoracicotropic hormone in the pupal stage, and the juvenile hormone in adult stages (Riddiford and Truman, 1978).

All forms of diapause, and in fact most forms of insect dormancy (Mansingh, 1971) display a number of common features such as a) extremely low metabolic rates, b) relative or complete inactivity, and c) programmed sets of biochemical adjustments for protecting the organism against low or even freezing temperatures, against lack of water, or against lack of food. The magnitude of these adjustments, by mammalian metabolic standards, is sometimes awesome: the dormant pupa of the *Cercropia* moth, for example, sustains one of the lowest metabolic rates of all animals, while the adult, a finely honed flying machine, displays one of the highest, there being an *absolute* two-thousandfold (!) difference between these two extremes of respiration (Schneiderman and Williams, 1953). Perhaps because of the agricultural importance of insects, the biochemical adjustments associated with diapause, particularly with overwintering forms of it, are now fairly well described. These adjustments are discussed in a variety of reviews (see, e.g., Salt, 1961) and since they share numerous metabolic characteristics in common, it will suffice for our purpose to consider only one such example in detail—the larval dormancy of the gall fly.

Metabolic Organization during Overwintering Dormancy

The third instar larvae of the goldenrod gall fly, *Eurosta*, overwinter inside stem galls on goldenrod plants; exposed above the snowline, the larvae face and survive the rigors of harsh northern winters, with temperatures falling as low as −40°C or even −50°C. Several physiological and biochemical factors have been identified which contribute to overwintering survival. Firstly, although metabolic rates are low, in keeping with low energy demands of the dormant larvae, metabolism clearly remains closely regulated. Some carbon and energy sources (glycogen in particular but also triglycerides) are abundant, having been laid down in preparatory phases, as is typical of all insects showing diapause capacity (Mansingh, 1971). Other carbon sources (small molecular weight carbohydrates, free amino acids, free fatty acids) and immediately utilizable energy sources (phosphagen and ATP) are buffered against change and remain at essentially normal concentrations to well below freezing; only at −30°C is there observed a small drop in phosphagen and ATP levels and in the energy status of the system (see Storey et al., 1981).

The main function of metabolism during this period of gradually falling temperatures appears to be the biosynthesis of at least two polyhydric alcohols, glycerol and sorbitol, and possibly a cryoprotectant protein. In field studies, both glycerol and sorbitol are found in the hemolymph at unusually high concentrations, while in laboratory experiments already at 15°C glycerol concentrations are at 65 percent of their maximum, and reach a plateau at about 235 μmol/gm wet weight as the temperature approaches 0°C. Sorbitol first appears in larvae at temperatures near 0°C; when temperature drops even further, sorbitol levels increase accordingly, reaching a plateau of 145 μmol/gm wet weight by −10°C (Fig. 7-2). As in nematodes, the function of these polyhydric alcohols is widely assumed to be in cryoprotection, either through increasing the amount of "bound" water or through direct interactions with enzymes and other proteins, protecting them in consequence from denaturation. (These roles are to be clearly separated from antifreeze functions discussed in Chapter 10; glycerol and sorbitol do not cause a thermal hysteresis between freezing and melting.) As a result of these metabolically mediated preparations, the hemolymph of these larvae has a supercooling point of −10°C; hemolymph (i.e., extracellular) freezing occurs only below this temperature and is not lethal to the organism. Even after freezing, O_2 uptake continues, although the

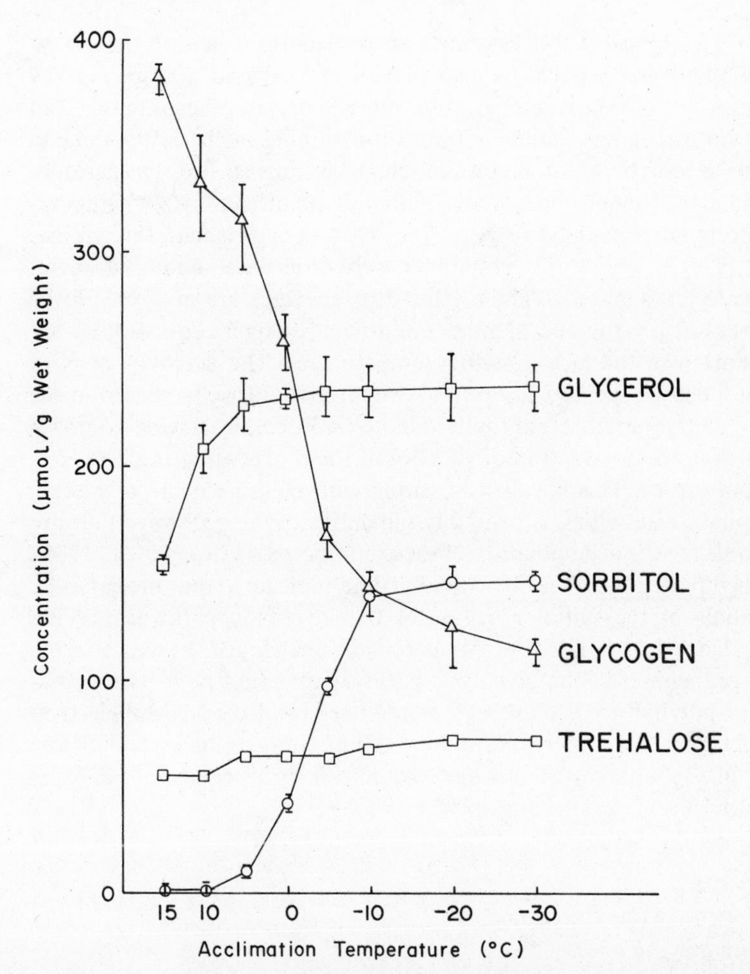

Fig. 7-2. Levels of glycogen, polyols, and sugars in *Eurosta* larvae acclimated to subzero temperatures. Modified after Storey et al. (1981)

activity of the electron transfer system typically is extremely low during dormancy. In contrast, the catalytic potentials of Krebs cycle enzymes such as citrate synthase are retained at normal levels (Storey et al., 1981).

Such differential adjustments in enzymic potentials in fact supply a third indication that *metabolism remains closely regulated during over-*

wintering dormancy and establish an enzymatic basis for the above metabolic responses. This is also evident in increased activities of key enzymes in glycogen metabolism: phosphorylase, hexokinase, and phosphofructokinase, while enzymes functioning in the lower portion of the glycolytic path remain unchanged during low temperature dormancy. Other enzymes whose activities are altered by low temperature acclimation include 3-hydroxyacylCoA dehydrogenase, an enzyme of fatty acid oxidation, glutamate dehydrogenase, and glutamate-pyruvate transaminase. These latter two enzymes are involved in the synthesis of proline and alanine, amino acids which show modest accumulation in the larvae at low temperatures. The activities of both sorbitol dehydrogenase and polyol dehydrogenase are increased in the larvae as temperature gradually falls to −30°C, an enzyme response paralleling the production of sorbitol in these overwintering, freezing-tolerant larvae. This parallelism, along with measurements of specific phosphatase activities, is probably indicative of the pathways that are available for the accumulation of these compounds (Storey et al., 1981).

When pieced together the available data indicate a) that the catalytic potentials of the initial portions of the glycolytic pathway are accentuated, while terminal portions remain unchanged, b) that enzyme steps branching off the glycolytic pathway and leading to the formation of polyhydric alcohols are emphasized, and c) that the electron transfer system is greatly reduced in activity (judging by extremely low rates of O_2 uptake) even while the Krebs cycle catalytic potential appears to remain unchanged.

Glycogen as the Precursor of Polyhydric Alcohols

The picture emerging of extensive metabolic and enzymatic reorganization during low temperature dormancy has been quantified in careful laboratory studies. These show that, as in nematodes entering anhydrobiotic states, there is a fairly close inverse relationship between glycogen depletion and the accumulation of sorbitol plus glycerol. If the more modest changes in glucose and trehalose pool sizes are taken into consideration, the depletion of glycogen exceeds the accumulation of its end products by only about 20 μmol/gm (Table 7-2). According to workers in the field, these metabolic processes are sustained by aerobic glycogen metabolism (see Storey et al., 1981); but what is the source of reducing power for sorbitol or glycerol formation?

Potential Sources of Reducing Equivalents during Dormancy

Normally, there are two main sources to consider: NADH formation at the GAPDH-catalyzed step of glycolysis, or NADH formation at later stages in metabolism. The first could be anaerobic; the second, usually is aerobic, being linked to NADH-oxidation by the electron transfer system (ETS).

In dormant larvae, the first can be confidently ruled out as the sole source of reducing power: if it were, for each mole of sorbitol formed, a mole of triose phosphate would be directed into the lower portion of glycolysis, *which is incompatible with the glycogen-alcohol stoichiometry* (Table 7-2). In effect the stoichiometry dictates that most glucose carbon cannot be metabolized past the triose (glycerol) level and a lot must go to sorbitol. Because the absolute amounts of glycerol and sorbitol can be so large (in the 0.4 *M*! range in gall fly larvae), the incompatibility is not readily explainable by experimental error (even small percentage deviation from theoretically required reducing power would be picked up as fairly large absolute concentrations). So it is important to consider other possibilities. One such involves the Krebs cycle.

The Krebs Cycle as a Source of Reducing Power

If for a moment we imagine the Krebs cycle as not being spatially isolated from the pathways for sorbitol and glycerol formation, it is an

Table 7-2. Changes in Sugar, Polyol, and Glycogen Levels with Acclimation to Low Temperature in Comparison to Metabolite Concentrations in Larvae at 15°C

Metabolite	Change from 15°C (μmol C_6/gm wet weight) 0°C	−30°C
Glycerol/2	+39.3	+42.1
Sorbitol	+41.5	+145.8
Glucose	+16.1	+28.6
Trehalose × 2	+18.0	+32.0
Total	+114.9	+248.5
Glycogen	−123.3	−268.1

SOURCE: From Storey et al. (1981)

easy matter to show that enough reducing power can be generated for the known rates and amounts of alcohol formed. This is best illustrated for sorbitol formation. In order to stay in redox balance (Fig. 7-3), for each 11 moles of hexose moieties derived from glycogen, only 1 need be completely metabolized; yet *in the process enough NADH is formed* (at the GAPDH and pyruvate dehydrogenase reactions plus at all NAD^+-linked steps in the Krebs cycle) *to drive the reductive synthesis of 10 moles of sorbitol.*

Although theoretically possible, the question arises as to whether or not this mechanism is plausible. At least tentatively we suggest that it is for two reasons. Firstly, the enzyme data show that activities of Krebs cycle enzymes during cold dormancy are maintained at essentially normal (prediapause) levels, which would not seem necessary in view of the greatly depressed rates of oxidative metabolism. Our second and related reason for thinking the model is plausible is the reduced ETS capacity observed in diapause. If the capacity of the Krebs cycle surpasses that of the ETS, it is not difficult to envisage redox conditions that would favor the required hydrogen flow. At the limit in this system, sorbitol- (and glycerol-) forming reactions would in effect outcompete the low-capacity ETS for NADH, leaving only $FADH_2$ (formed at the succinate dehydrogenase locus) as the main donor of protons and electrons to the ETS; this could supply the organism with a low rate of oxidative phosphorylation and a small amount of oxidatively formed ATP.

The data on gall fly larvae allow a test of at least one prediction of this interpretation of how metabolism works during cold dormancy. From Figure 7-3 it is evident that the stoichiometry between sorbitol accumulation and glucose catabolism should be 10:1 (i.e., out of 11 glucosyl units, 10 should end up as sorbitol while 1 should be fully metabolized). In gall fly larvae, most sorbitol formation occurs between 0° and −30°C (when glycerol levels are stabilized) and the rise in sorbitol concentration is about 105 μmol/gm over the time interval of the acclimation. Over the same time interval, the drop in concentration of glucosyl units (sum of concentration changes in glycogen, glucose, and trehalose) are equivalent to 117 μmoles/gm. Taken at face value, that means 10.2 percent of the glucose utilized is unaccounted for by sorbitol formation and is presumably further metabolized; the theoretical amount (Fig. 7-3) is predicted to be 9.1 percent, which we consider close enough for support of the model. Exactly the same problem of where the NADH comes from during glycerol formation arises; it too has been entirely overlooked by workers in this field but can be readily solved by the model in Figure 7-3.

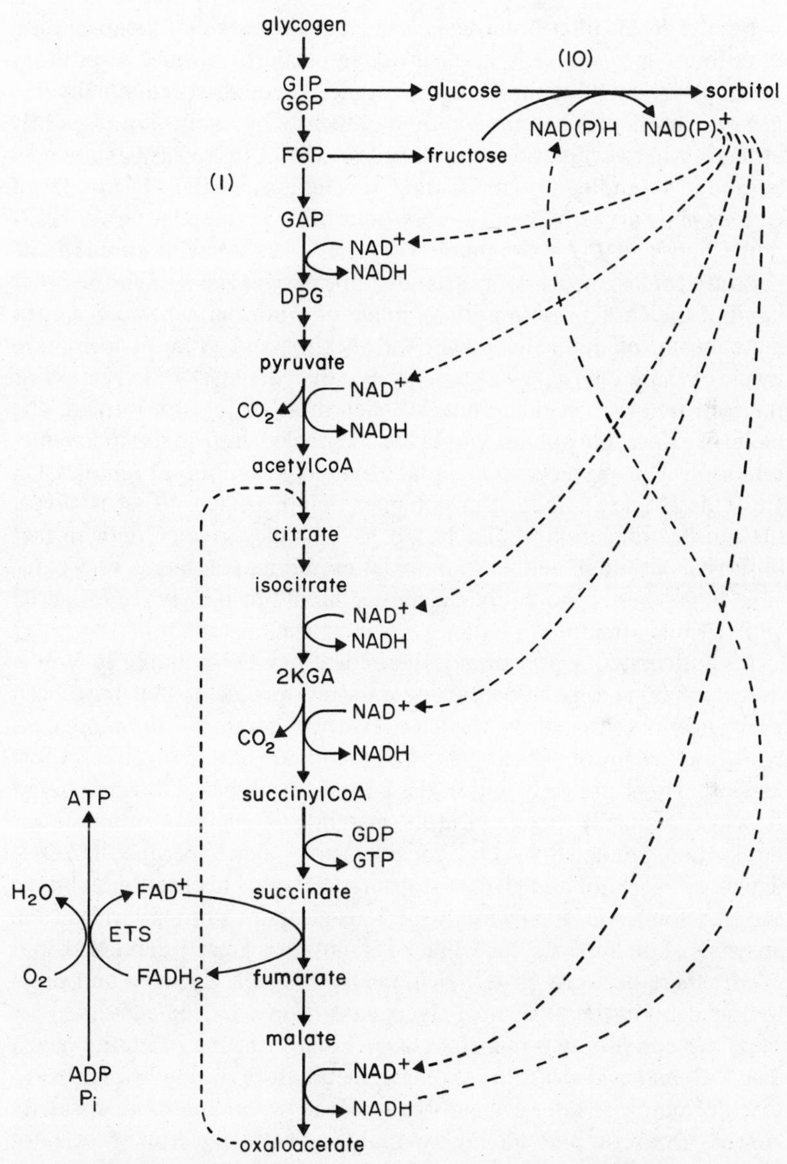

Fig. 7-3. The Krebs cycle shown as a possible source of reducing power for sorbitol formation in *Eurosta* larval diapause. The stoichiometry shown in parentheses (1 mole glucosyl unit oxidized per 10 moles reduced to sorbitol) is consistent with the measured metabolite changes reported by Storey et al. (1981)

Finally, if this interpretation is correct, it suggests that the formation of sorbitol or glycerol is in effect antagonistic to normal respiratory metabolism but that they remain coupled through succinic dehydrogenase. On the basis of that coupling, the rate of respiration of gall fly larvae during acclimation from 0°C to −30°C can be easily shown to be about 4 μmoles O_2 gm^{-1} day^{-1}; estimates of the O_2 uptake of *Chironomus* larvae between −15°C and 0°C (Scholander et al., 1953) range between 0.12 − 9.6 μmoles O_2 gm^{-1} day^{-1} with simple temperature coefficients; so our estimate appears reasonable. On the other hand, if the earlier assumption is made of a normal aerobic glycogen metabolism, *all* reducing power for sorbitol and glycerol formation would presumably have to come from either the GAPDH reaction or the pentose cycle; in this event, for each mole of sorbitol formed, one mole of triose phosphate would be channeled into mainstream metabolism. The predicted O_2 uptake would be about 21 μmoles O_2 gm^{-1} day^{-1}; i.e., about 5 times higher than expected. What is worse, the glucose depletion would be up to 1.8 times greater than in fact observed. Although neither of these latter alternatives seems very plausible to us, there is no doubt the matter should be closely reconsidered with serious attempts to balance redox as well as carbon.

An anticipated aspect of any theoretical model (such as Fig. 7-3) is that it explains a lot of perplexing data and problems that have been previously overlooked. In this case, first and foremost, the model explains the origin of reducing equivalents for sorbitol and glycerol formation almost precisely within the bounds of observed stoichiometry. It also explains a) why respiratory metabolism occurs at much lower levels than predicted by Q_{10} considerations alone (because NADH-dependent sorbitol and glycerol-forming reactions in effect outcompete the ETS for NADH), b) why the organism retains normal levels of the enzymes of the Krebs cycle, while ETS activity is greatly depressed, and c) why there occurs a close stoichiometry between glycogen and polyhydric alcohols (because most glycogen carbon is indeed conserved for these two end products of metabolism). Finally, the interpretation given (Fig. 7-3) makes it clear that glycogen metabolism during low temperature dormancy is not very aerobic at all; it is perhaps more correctly viewed as a complex anaerobic pathway with glycerol or sorbitol serving as carbon and hydrogen sinks entirely analogous to lactate or any other anaerobic end products formed in more conventional ways. That may well be why both these compounds may also accumulate under anoxia: to a modest extent in nondiapause stages, but to a much greater extent in diapause stages when the catalytic potentials of enzymes involved in their formation are elevated. This implies that redox balance at the succinic dehydrogenase step can be maintained independently of ETS (O_2-consuming) function, which indeed is known for other systems. What the above model does not explain, however, is how the NADH-requiring reactions in the cytosol are coupled to the mitochondrial NADH-generating ones of the Krebs cycle. In other systems, specific metabolite shuttling processes achieve a transfer of reducing equivalents across the mitochondrial membrane barriers, but further research is required to identify which shuttles are present and how they are utilized during diapause in insects.

Estivation in Lungfishes

Because among air-breathing fishes they were the first to be closely examined by experimental biologists, the lungfishes (particularly the African *Protopterus* species) are probably the best-known examples of fishes that routinely are capable of prolonged estivation as a means for surviving the drying up of the lakes, ponds, or streams in which they live. To be sure, there are many examples of such estivation, not only in air-breathing fishes, but also among various amphibians. However, we shall concentrate our discussion upon the lungfishes, in part because their estivation is rather well described and because some biochemical data are available.

As in anhydrobiosis, the environmental stimulus for entrance into estivation in lungfishes, as in other aquatic vertebrates, is a lack of water. Although lungfishes often live in large bodies of water that are relatively stable on a year-by-year basis, they also are often found in smaller lakes, ponds, and streams that sometimes dry up during the dry season of the year. Even the larger lakes in which they live vary in depth seasonally. Thus, in various regions of their normal habitat, lungfish may become trapped in pools and swamps that during the dry season ultimately dry up completely. Long before that happens, lungfish burrow into the mud, form a kind of cocoon with a breathing channel to the surface, and there enter an estivation period that can last for variable time periods (from one dry season to several years). In captivity, an African lungfish has been maintained in the estivating state for nine years (K. Johansen, pers. comm.)!

Overall Metabolic Organization of Lungfishes

Lungfishes are rather sluggish beasts, with a low aerobic metabolic rate. They are not very vigorous swimmers and are clearly specialized

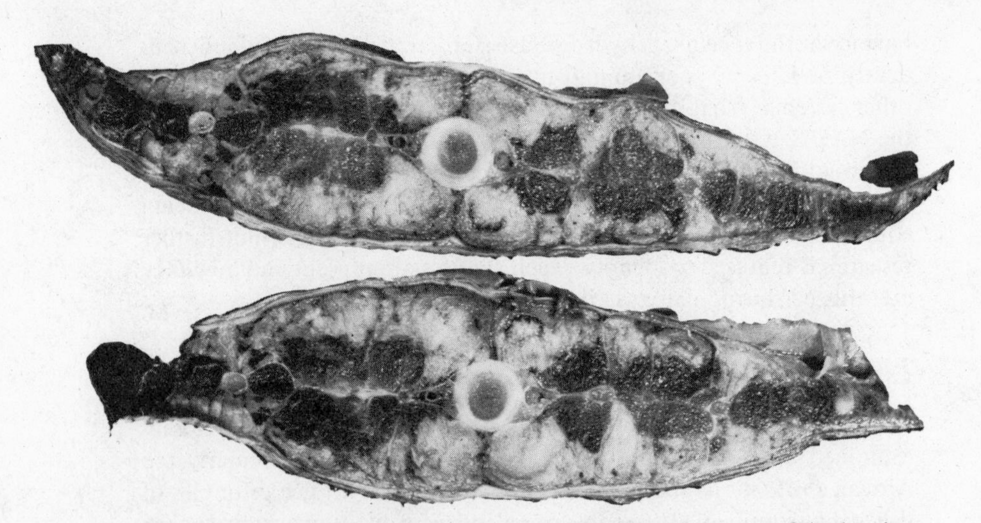

Fig. 7-4. Cross-sectional view of the tail region of the African lungfish, showing extensive depots of fat (dark staining regions). Microscopic and histological analysis shows these depots to represent true adipose tissue (J. F. Dunn, pers. comm.), which is also found in the South American lungfish (W. C. Hulbert, pers. comm.). Photo by J. F. Dunn

in their active phases of life as ambush or quick-strike predators. In the non-estivating phases of life, they fuel metabolism either with glycogen (glucose) or with fat; the first presumably largely for anaerobic metabolism, the second for standard oxidative metabolism. Glycogen is stored in all tissues, but as in all vertebrates, the depots are highest in the liver. Fat, on the other hand, appears to be stored in large masses running in anterior-posterior direction and located in the tail region between the myotomal muscles (Fig. 7-4). In addition to fat and glycogen, normally active lungfish routinely utilize proteins and amino acids as fuels for metabolism.

Enzyme profiles of various tissues and organs of both the South American and African lungfishes are consistent with the above impressions of multiple-substrate-based metabolism (see Hochachka, 1979). Thus the myotomal muscles have relatively high activity ratios of anaerobic/aerobic enzymes, but absolute activities are actually substantially lower for both groups of enzymes than for other, more robust fishes. Even the heart and brain of the lungfish seem to display high glycolytic potential, but lower oxidative potentials than found in other fishes. Liver and kidney, on the other hand, maintain a significant capacity for gluconeogenesis as well as for amino acid metabolism and the urea cycle.

The overall organization of metabolism, even in the non-estivating animal, therefore, is indicative of a low energy turnover, which is indeed observed by direct measurements. During estivation, energy metabolism is even further reduced. According to Homer Smith's original (1930) studies, the metabolic rate of estivating lungfish drops to only about one-third of normal. However, more recent measurements indicate that the metabolic rate during estivation continues to be depressed further and further with time of estivation and may drop by two orders of magnitude (K. Johansen, pers. comm.).

Glycogen Sparing during Estivation in the Lungfish

An analysis of the available fuels in the lungfish makes it clear that even at low metabolic rates, the amount of glycogen is inadequate to support long-term estivation. Moreover, as is observed in other stress situations as well (for example, salmon migration), glycogen and glucose reserves are maintained throughout estivation presumably for those cells and tissues (brain, red blood cells, kidney tubules) which may have an absolute requirement for glucose; muscle glycogen may also be spared for use during emergency situations in arousal from estivation (for burst swimming). As in other vertebrates under these conditions, glycogen reserves must be maintained by gluconeogenesis from protein-derived amino acids; the regulatory properties of at least one control site enzyme in the process (FBPase) are in fact similar to those found for the enzyme in other fishes and mammals. Supporting this activity also are ample levels of the enzymes GPT, GOT, and GDH, required for the mobilization of amino acid carbon toward glucose (Hochachka and Somero, 1973; Dunn et al., 1981, for literature in this area).

Fuel Utilization during Estivation in the Lungfish

During the first part of prolonged estivation, as Homer Smith first showed some half century ago, lungfishes fuel their metabolic machinery with a mixture of substrates; respiratory quotients (RQ values) are predictably near 0.8 during this time. To date, no one has estimated how long lungfishes can estivate using fat as a sole carbon and energy source. Our impression is that an African lungfish could easily estivate through a dry season mainly on fat stored in its tail depots. There are advantages to using fat under these conditions: 1) fats are highly efficient in terms of ATP yield/mole of starting substrate, and hence at the low metabolic rates required during estivation fats should be able to fire the animal's metabolism longer, on a molar basis, than any other substrate; and 2) fats yield no more noxious an end product than CO_2, which of course can be exhaled. Unlike the situation during starvation in some vertebrates (including man), in the fat-primed estivation of the lungfish there is no accumulation of ketone bodies.

Our impression, therefore, is that the preferred carbon and energy sources during most bouts of natural estivation are fats. However, if the estivation is experimentally prolonged or perhaps if the animal's fat supplies are limited (which may have been the case with Homer Smith's experimental animals, since he overlooked the existence of the tail depots of fat in the African lungfish), the animal must ultimately turn to endogenous protein as its main fuel source.

Protein and Amino Acid Metabolism in the Estivating Lungfish

Interestingly enough, despite the fact that protein and amino acids have been recognized as potential sources of energy in the lungfish for fifty years, there are no detailed studies available as to how these metabolites are mobilized. What is the major protein source that is mobilized? What tissues initiate the mobilization? What amino acids are utilized and where? A good working model is available for the spawning migration of salmon, which supplies answers to most of the above questions. And since key metabolic enzymes required for that metabolic organization (GPT, GOT, and GDH) are present in reasonably high activities in all the appropriate tissues and in fact are elevated during estivation, it can at least be tentatively assumed

1. that the main storage house of protein is white muscle,
2. that the mixture of amino acids released into white muscle by proteolysis does not merely spill out into the blood nor is it metabolized completely *in situ*; instead, a partial metabolism of amino acids leads to an enrichment of white muscle supplies of alanine (Fig. 7-5) which is the primary substrate for catabolism at other tissues and organs in the body; and,
3. that, although the main fate of alanine in some tissues may be oxidation, in the liver it is gluconeogenesis which is used to maintain bulk body needs for glucose (Mommsen et al., 1980).

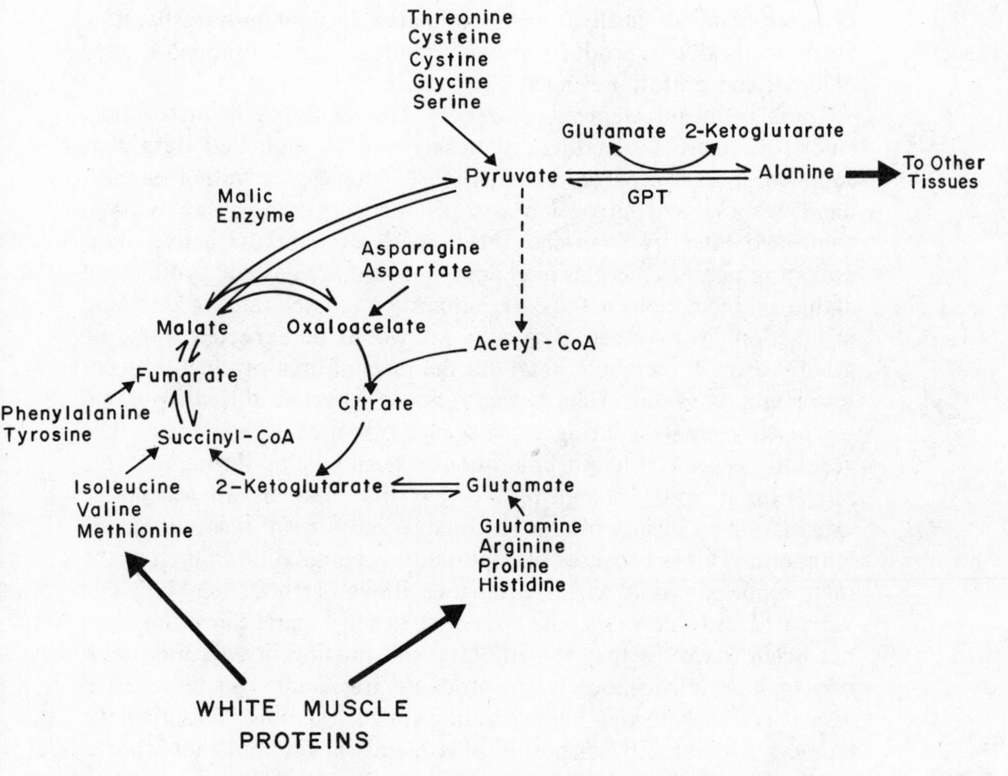

Fig. 7-5. Metabolic map of probable metabolic fate of most of the amino acid pool in fish muscle during active mobilization of protein as an energy source. Modified from Mommsen et al. (1980)

End Products of Amino Acid Metabolism in the Estivating Lungfish

Although the above model of amino acid and protein metabolism in the lungfish remains to be tested, on one score the data are unequivocal: the end products of protein and amino acid metabolism are CO_2, HCO_3^-, H_2O, and glucose, on the one hand; ammonia and urea, on the other. The first group needs no further consideration in this context. However, the matter of waste nitrogen products is of crucial importance for two reasons. Firstly, like the anhydrobiotic organism, the estivating lungfish is an isolated and closed system with O_2, CO_2 and H_2O being the only known metabolites being exchanged with the environment. Hence, end products such as ammonia or urea must be disposed of metabolically or be accumulated throughout the estivation. Secondly, both end products are potentially noxious, although at very different concentration ranges.

These problems were recognized by Homer Smith in his original studies of these fascinating organisms and he obtained data that supplied at least a partial explanation of how the estivating lungfish handles these two nitrogen waste products. According to his data, confirmed later by Jannsens, the lungfish even in the active, non-estivating phase, generates *both* ammonia and urea as end products of amino acid metabolism. On entry into estivation, the rates of ammonia production drop rather drastically (as would be expected from the general drop in metabolic rate) but the rates of urea production seem to remain constant. Hence, the ratio of urea produced/ammonia produced increases during estivation. Since urea is less toxic, it is generally assumed that it continuously accumulates throughout the estivation period in lungfish. Whereas that may be an acceptable explanation of events over a reasonably short term, it is not in the long term. This is because the continuous accumulation of urea at the rates reported would yield blood urea levels of about 0.3 M in one year, 1 M in three years, and over 3 M in nine years! Since this does not occur, it seems that something is still missing in our interpretation of how nitrogenous waste products are handled in these organisms. We suspect that the estivating lungfish avoids "polluting the factory" with urea by disposing of it metabolically and that what is missing in all previous discussions of the problem is the concept of urea recycling. Although urea recycling may or may not turn out to be utilized by the estivating lungfish, it most assuredly is during estivation and hibernation in small rodents and large mammals as well.

Hibernation in Rodents

Necessarily linked to the size of rodents and small mammals in general are 1) high metabolic rates, and 2) limited capacities for energy storage, the latter potentially limiting the former whenever food intake or supplies are restricted. These metabolic problems (of small body size) are amplified most when reduced food and reduced water availability coincide with cold (because a large surface area: volume ratio results in high rates of heat loss): i.e., being small is most disadvantageous in winter.

Hibernation represents an elegant biochemical adaptation that through metabolic depression readily circumvents such overwintering energy problems of small mammals. In the case of the 13-lined ground squirrel, a not atypical example, metabolic rates during hibernation are 1/52 normothermic rates, concomitant with rates of pulmocutaneous water loss down to 1/77 of those in the normothermic condition (Deavers and Musacchia, 1980). Wang (1978) has estimated (for the Richardson's ground squirrel under natural winter conditions) that the hibernating habit saves over 90 percent of the energy that would otherwise be utilized to maintain normal metabolic rates. Interestingly, this saving would even be greater were it not for periodic arousals during which normothermic conditions are temporarily re-established. Wang's calculations show that during the hibernation season of this species (when body weight may decrease by half) fully 90 percent of the energy utilized is associated with periodic arousals and subsequent temporary phases of normothermia. Although these periodic bouts of arousal may seem to be energy-expensive, they may serve important metabolic functions (for example, allowing metabolic or physiological clearance of noxious products such as ketone bodies, urea, etc.); moreover hibernating rodents seem to have a large built-in safety factor and energy reserves are rarely if ever found to be limiting factors in their hibernation. (Many rodents, for example, have been observed not to eat for several weeks following the terminal arousal from hibernation in the spring!) How then are these impressive abilities achieved and how is metabolism in hibernating small mammals organized?

Prehibernation Dietary Adaptations

As we indicated in our introduction, small mammalian hibernators fall into two categories—the feeders and the nonfeeders. The nonfeeders typically gain weight in the fall due to extensive fat deposition. Feeders may or may not fatten prior to hibernation. Even though chipmunks cache food prior to hibernation, they undergo prehibernation fattening that is as extensive as the fattening in the ground squirrel. The caching of food reserves may result primarily from the demand for protein rather than the need for energy. Although the storage of fat as an energy source for utilization during hibernation is well established in ground squirrels, there are no studies that examine the minimum body protein required prior to hibernation. Field observations, however, show a distinct preference in the golden-mantled ground squirrel and several chipmunk species for a diet high in protein, a preference that is most apparent during the fall prior to hibernation (Riedesel and Steffen, 1980). In fact, the deposition of lipid in the fall coincides with a diet of fungi high in protein. The demonstration that protein deprivation in the garden dormouse induces torpor in any season emphasizes the fact that protein may be a limiting factor in the metabolism of hibernators.

Hibernation versus Starvation: Comparative Aspects

From the above, it will be evident that both lipid and protein reserves may influence the onset, duration, and success of hibernation. Of these two fuel sources, however, fat quantitatively makes by far the larger contribution to energy metabolism during hibernation. In the Arctic ground squirrel, for example, changes in lean body mass establish that protein may account for about 10 percent of the energy expenditure during the hibernating season; most of the rest is fueled by fat while tissue stores of glycogen are maintained (by gluconeogenesis from amino acids). The metabolic organization, therefore, appears rather similar to that in man during starvation.

For a frame of reference it is useful to recall that in a typical 70 kg man the caloric equivalents of glycogen, mobilizable protein, and fat are about 1,600, 24,000 and 135,000 kcal, respectively. Energy needs vary from 1,600 to 6,000 kcal/day, depending upon the activity level. Stored fats in man, therefore, are sufficient to meet caloric needs in starvation for one to three months. However, the carbohydrate reserves are exhaustable in a day, so gluconeogenesis is critically important for sustaining tissues and organs (brain, red blood cells, some kidney cells) displaying an absolute glucose requirement. This is indeed considered a first priority of metabolism in starvation in man. The problem is that precursors of glucose are scarce, since most available carbon is stored as fat, which cannot be converted into glucose because acetylCoA cannot be transformed into pyruvate. The glycerol moiety of triacylglycerols can be converted into glucose but only a limited amount is available. The only other potential source of glucose is amino acids derived from the breakdown of proteins. Muscle is the largest potential source of amino acids during starvation. However, in nature survival may well depend upon being able to move about; thus, a compromise must be struck between the demands for glucose and the preservation of muscle protein. This is accomplished by shifting from glucose as a primary fuel source for specific tissues to ketone bodies. Such a change in the metabolic picture occurs later in starvation (after about three days in man) when acetoacetate and 3-hydroxybutyrate (ketone bodies) are formed by the liver. Their synthesis from acetylCoA increases markedly because the Krebs cycle cannot oxidize acetyl CoA at the rate it is generated by β-oxidation. The Krebs cycle and β-oxidation are out of balance because gluconeogenesis depletes the supply of oxaloacetate, which is essential for the entry of acetyl CoA into the Krebs cycle. As a result, the liver produces large quantities of ketone bodies, which are released into the blood, and now become available as fuel for the brain. In man, after about three days of starvation, about one-third of the energy needs of the brain are met by ketone bodies. Under these conditions, the heart also uses ketone bodies as fuel. This condition of ketone usage as a major fuel is referred to as ketosis.

Man's dependence upon ketones increases with length of starvation, so that after several weeks, they become the major fuel of the brain. *This effective conversion of fatty acids into ketone bodies by the liver for use by the brain markedly reduces the need for glucose* (in man, from about 120 gm/day in early starvation to only about 40 gm/day). In addition, a fourfold saving in protein (20 gm/day versus 5 gm/day) is achieved, with *less being degraded for gluconeogenic precursors than would otherwise be necessary*, which is indeed the nature of the compromise struck between the need for glucose and the need for preserving muscle protein.

The same basic organization of metabolism is apparently utilized by small hibernators. However, the degree of ketosis that is allowed to develop seems to be carefully controlled. In the marmot, for example, ketosis gradually develops during deep hibernation but only until the blood concentrations of ketone bodies reach a critical level; at that point, the animal arouses from hibernation. In this species (see Nelson, 1980 for literature), ketosis in fact is considered a metabolic "trigger" that induces periodic arousals during which time circulating ketone bodies are cleared (by re-establishment of normal circulation and by the elevated oxidative metabolism initiated in normothermic conditions). These periodic arousals also favor the clearance of end products of amino acid metabolism which may have accumulated because of their use as glucose precursors.

Role of Amino Acids in Small Hibernators

Although proteins and amino acids are recognized as metabolically critical carbon and energy sources in small hibernators, direct estimates of rates of turnover during hibernation are not available. Plasma concentrations of free amino acids, where monitored, appear to change only modestly; in the 13-lined ground squirrel, for example, leucine, arginine, and alanine are the only amino acids whose plasma concentration rises during hibernation; the rest of the amino acid pool remains relatively constant.

During arousal from hibernation, however, both the levels of plasma free amino acids and their metabolic rates are markedly elevated. It is particularly noteworthy that the levels of plasma alanine rise dramatically; moreover, both alanine oxidation rate as well as its rate of conversion to glycogen (glucose) markedly increase. These processes

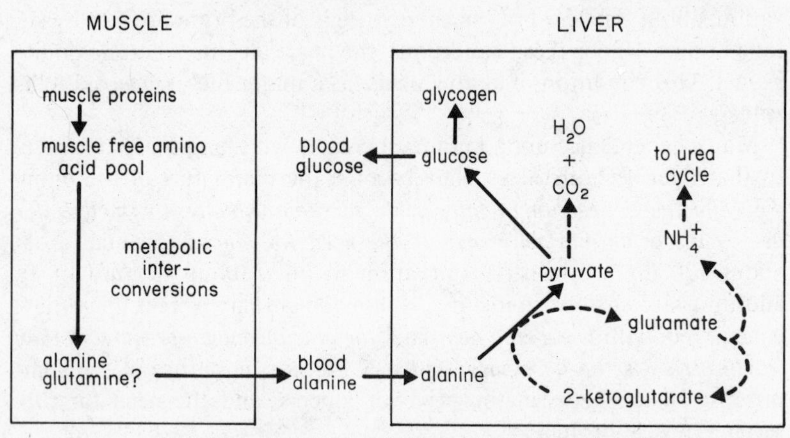

Fig. 7-6. Apparent metabolic functions during bouts of arousal in small hibernators. After Riedesel and Steffen (1980)

contribute to the energy costs of arousal and periodic normothermia, during which time 90 percent of the energy turnover of the winter hibernation season occurs. In metabolic terms, it appears as if the small hibernator utilizes these bouts of normothermia for re-establishing metabolic homeostasis by activating several metabolic processes (Fig. 7-6). Whereas it is well known that the fate of alanine carbon may be either oxidation or incorporation into glucose, the fate of the amino nitrogen is not so clear because several complex events (altered kidney function, urea cycle function, and urease degradation of urea) may all be occurring at once.

Renal Function during Hibernation

Because the hibernating animal is a closed system, internal mechanisms must exist for it to remain not only in carbon and energy balance during hibernation, but also in water balance; interest in this problem (to which we will return below) had led to detailed studies of kidney function in several species of small hibernators. These studies (Deavers and Musacchia, 1980) found that in ground squirrels kidney function is essentially fully blocked during hibernation (by means of reduced arterial pressures and reduced percentage of cardiac output perfusing the kidney); in the marmot the same mechanisms are operative, but to a lesser degree, so renal function continues but at only about 1 percent of normal rates. If kidney function is turned down, or off, but the urea cycle continues to operate during hibernation, urea concentrations would be expected to increase both in tissues and in blood. This expectation is so reasonable that 1) the accumulation of nitrogenous wastes (mainly urea), 2) potential toxicity, and 3) potential osmotic imbalance have been frequently examined as possible factors in the periodic arousals during hibernation. Although there is some controversy in the available literature, it must be concluded, on balance, that *plasma urea and tissue concentrations in fact do not necessarily increase* (*sometimes they actually decrease*!) *during hibernation.* There are two possible reasons why not: either the urea cycle operation is regulated in such a way as to avoid urea accumulation, or the urea formed is disposed of elsewhere than the kidney and by different mechanisms. We must seriously consider both possibilities.

Urea Cycle Regulatory Mechanisms: Role of Acetylglutamate

As indicated in Chapter 2, the entrance of NH_4^+ into the urea cycle proceeds through the formation of carbamyl phosphate, in a reaction catalyzed by carbamyl phosphate synthetase I (CPS-I). CPS-I is abundant in the liver, constituting 15–20 percent of the total protein of rat liver mitochondria. It is well established that CPS-I has an absolute requirement for the activator, N-acetylglutamate. The latter is formed by acetylation of glutamate, and N-acetylglutamate levels depend strictly upon the concentration of glutamate, providing there is a source of acetyl CoA. The only other known ways of controlling N-acetylglutamate production and content involve 1) arginine positive feedback activation of acetylglutamate synthetase, 2) alterations in dietary protein intake (see below), and 3) hydrolysis by acetylglutamate acylase (Sonoda et al., 1982). Thus it is believed that a rapid short-term control mechanism may operate through the regulation of acetylglutamate levels:

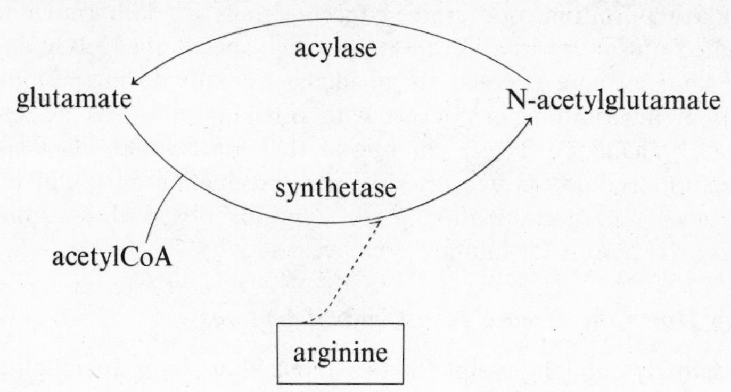

(It should be noted that another form of CPS, termed CPS-II, occurs in tissues such as mammalian spleen. It functions only in pyrimidine biosynthesis, since these tissues are unable to synthesize urea. Interestingly, CPS-II is not activated by N-acetylglutamate. Such an activation appears to have been specifically selected for the controlled channeling of glutamate nitrogen into the urea cycle.)

Ornithine Availability in Urea Cycle Regulation

Another mechanism of urea cycle control may involve ornithine, since an ornithine "sparking" of the urea cycle can be readily demonstrated. It is a well established observation that an ornithine load increases urea synthesis and thereby reduces sensitivities to elevated levels of NH_4^+. Similar "protection" against ammonium toxicity is achieved by introducing other urea cycle intermediates such as arginine; the arginine, cleaved by arginase to urea and ornithine, in this context becomes metabolically equivalent to an ornithine load.

These observations correlate well with studies of perfused rat liver and isolated hepatocytes showing that, through physiological concentration ranges of ornithine, ornithine carbamyltransferase activity limits the overall rate of the urea cycle. This limitation probably arises from a simple K_m control of the enzyme, since the K_m for ornithine (about 0.4 mM for the rat liver enzyme) is above the tissue concentrations of ornithine (about 0.2 μmol/gm). The K_m for the cosubstrate carbamyl phosphate is 0.03 mM, but since the physiological concentrations of this metabolite are controversial, it is not known whether or not this enzyme-substrate interaction is of regulatory significance. Since, at least in the beef liver enzyme, binding of ornithine depends upon prior binding of carbamyl phosphate, it is reasonable to assume that ornithine limitation is a critical and perhaps the principal mainstream metabolite involved in controlling urea cycle activity (Evered, 1981). Plasma ornithine or arginine concentrations are both known to be elevated during hibernation in at least two species (hedgehog and ground squirrel), and, as noted above, increase greatly during periodic arousals in hibernation, in concert with other amino acids as well (particularly alanine). These conditions therefore would all favor continued (indeed accelerated rates of) urea production during hibernation; since urea accumulation is not typically observed, we must consider mechanisms for limiting urea cycle activity.

Turning Down the Urea Cycle: Control Options

Interestingly enough, except for the above short-term, metabolite-mediated mechanisms, none of which are directly inhibitory, the only other way of adjusting the catalytic potential of the urea cycle is to adjust the activity level of its component enzymes. Control of such adjustments is best understood in amphibians and lungfishes where larval metamorphoses and transitions from active to estivating phases are associated with changes from ammonotelism to ureotelism. Enough information is available to suggest that two events are involved in regulating the level of urea cycle enzymes, and these fall into an hierarchical order (see Hochachka and Somero, 1973):

1. The initial event appears to be an activation of transcriptional machinery. Although specific signals at the moment are not clear, hormonal control, particularly by the thyroid, is definitely involved, at least in *Rana catesbeiana*. In *Xenopus*, reduced water availability is an important environmental signal, since the same changes in nitrogen metabolism are induced by dehydration and by hypertonicity of the medium. Studies by Janssens indicate that increased levels of NH_4^+ in tissues and blood constitute an important metabolic signal for activating enzyme synthesis. During early metamorphosis, RNA polymerase activities increase dramatically, followed by the production of various m-RNA species. These presumably are involved in specifying the urea cycle enzymes which are being synthesized at this time. The increase in activities of the urea cycle enzymes can be impressively large. A ten- to twentyfold increase, for example, is found in the case of a species of *Rana*. A sixfold increase in CPS-I activity is observed in *Xenopus* during estivation. A five- to tenfold increase in aspartate transaminase activities occurs in estivating lungfish. Estimates of *de novo* synthesis rates of CPS-I indicate comparable increases during metamorphosis in *Rana*. A number of observations suggest that the production of the urea cycle enzymes is a single, coordinate event. Thus, a) the time of RNA polymerase activation, b) the time of initiation of enzyme synthesis, c) the time course of synthesis, d) the approximate percentage increase in the activities of the various enzymes, and e) the derepression signals for these events (thyroid, NH_4^+ levels, and water availability) are all identical or remarkably similar. Such behavior would be predicted if all the component enzymes were specified by a closely integrated section of the genome.

2. Despite the similarity in response pattern of the enzymes required for urea biosynthesis, there are a number of findings which suggest that a finer level of control is involved than mere synthesis of the enzymes. The best available data concern CPS-I. The biosynthesis of carbamyl phosphate synthetase appears to be the usual m-RNA dependent synthesis at the ribosomal site. This process is sensitive to the usual inhibitors of both transcription and translation. A second step (b) is an epigenetic one and involves transfer of inactive precursor

"subunits" into the mitochondria, followed by the assembly of these subunits into the active oligomeric enzyme. The second step of this process also may be activated by thyroid hormone. Comparable assembly processes have been described in somewhat lesser detail for glutamate dehydrogenase and may also account for arginase differences in premetamorphic versus postmetamorphic amphibians. It is indeed probable that some epigenetic control is put upon the assembly of other oligomeric enzyme components of the urea cycle (Mori et al., 1982) as well as upon their translocation into proper intracellular location (two of the enzymes are mitochondrial, the rest are cytosolic). Since epigenetic control of each of the urea cycle enzymes may differ, one would anticipate the observed differences in the percentage increase in activities of different urea cycle enzymes during enzyme induction.

Turning Down The Urea Cycle in Mammals

The same kind of control of the catalytic potential of the urea cycle seen in lower vertebrates also appears to be operative in mammals. Although the percentage change is not as large as in lungfish and amphibians, it is well known for rats that *the total liver activities of all the urea cycle enzymes as well as N-acetylglutamate synthetase are directly proportional to the protein intake of the organism.* In rats subjected to acute changes in dietary protein, the correlated adjustments in enzyme levels of the urea cycle are completed in four to eight days; so unlike the metabolite-mediated controls on the cycle, this is a long-term regulatory mechanism that seems to involve adjustments in the actual content of enzyme protein (Evered, 1981). *Reducing dietary intake of protein, therefore, is in mammals a plausible way to turn down the urea cycle.*

The question arises as to whether or not such long-term adjustments may account for the lack of urea accumulation during hibernation in mammals. In small hibernators that do not feed during arousal periods, this may be possible (although the dietary need for protein may be made up by accelerated use of endogenous protein reserves); this possibility is in need of being tested in future research. In small hibernators that feed during their periodic arousals through the winter sleep, this seems less likely, in part because their metabolic rates are depressed during hibernation and in part because they retain a dietary input of protein on approximately a weekly basis. If the same control mechanisms are operative as in the rat, one may very well anticipate oscillations in the catalytic potential of the urea cycle lagging behind the arousal (feeding) periods. Even if these occurred, however, they would probably be dampened in amplitude and period by the hypothermia of the hibernating animal. On balance, therefore, *one would not anticipate any substantial and sustained drop in the content of urea cycle enzymes during hibernation.*

Thus we are faced with an interesting situation: during hibernation of small mammals, renal function is either completely turned off or is at least greatly inhibited. At the same time, there is no reason to anticipate any serious reduction in the content of enzymes of the urea cycle, nor therefore in their catalytic potentials. Metabolite conditions, on the contrary, are if anything more suitable for urea cycle activity during hibernation than during the normal state (ornithine, arginine, and alanine levels, for example, are often elevated, particularly during arousal bouts). One would therefore expect urea formation to continue during hibernation, abated only by a lowering of cellular temperature.

Two problems arise. Firstly, what function, if any, does the urea cycle play during hibernation; why, in other words, does it continue to operate? And secondly, what mechanisms prevent gradual accumulation of urea? The answer to the first question although not fully clarified, appears to relate to the control of HCO_3^- and pH regulation. As to why urea does not accumulate, the answer is that it is recycled. Interestingly, although these processes appear distinct, it will become evident that they are in fact functionally related.

Role of Urea Cycle in Removal of Metabolic Bicarbonate

Although urea as a sink for waste nitrogen has long been appreciated, the suggestion by Atkinson and Camien (1982) that another function of the urea cycle—removal of HCO_3^- by proton production—is also biologically important has not been widely recognized. This may have been overlooked 1) because the urea cycle as a mechanism for transferring protons to HCO_3^- is not very obvious from the equation for urea synthesis, and 2) because the metabolic generation of HCO_3^- has not been generally appreciated. Nevertheless, the oxidation of carboxylate anions at physiological pH must yield HCO_3^- (Atkinson and Camien, 1982). In most animals, proteins constitute the main source of metabolic bicarbonate, for the oxidation of simple amino acids containing $—COO^-$ and $—NH_3^+$ groups yields HCO_3^- and NH_4^+ in approximately equimolar amounts. Additional

HCO_3^- may arise from carboxyl or carboxamide groups in the side chains of aspartate, asparagine, glutamate, and glutamine, as well as from the salts of metabolic acids (in nonruminant herbivores) or, in ruminants, from the two main carboxylate anion energy sources (acetate and propionate). In a 70 kg man, about 1 mole of HCO_3^- is produced and must be eliminated, daily. If it were not eliminated, assuming 3.5 liters of plasma volume and an interstitial fluid volume of 10.5 liters, the accumulation of 1 mole of this metabolic end product would raise the pH to about pH 8.0. Although the impact of this amount of HCO_3^- could be minimized by buffers, it is important to remember that buffers protect only against fluctuations and cannot provide a sink for continuous disposal of base. That is why animals must eliminate this metabolic end product at rates essentially similar to rates of production. Of the various mechanisms possible (excretion at the lungs, kidneys, or digestive tract), *protonation to form CO_2*, which can then be blown off, is by far the most important. What this means is that in man about 1 mole of protons must be obtained for this process per day. In hibernating ground squirrels, with oxidative metabolism depressed to 1/50 of normoxic (Deavers and Musacchia, 1980), about 0.6 mmoles HCO_3^- would be produced/day, but about 0.03 moles HCO_3^-/day would be produced during arousal bouts. Thus the hibernator must obtain at least these amounts of protons/day during routine hibernation. From where do they arise?

NH_4^+ as a Source of H^+ for Bicarbonate Elimination

Since HCO_3^- is produced in essentially equimolar amounts with NH_4^+, the astute student might well wonder why the latter does not directly donate the required protons for the conversion of HCO_3^- to CO_2. The reason this cannot occur is because the system of CO_2 + H_2O (pK_a = 6.1) is more than 1,000 times stronger as an acid (a proton donor) than is NH_4^+ (pK_a = 9.25). Unfortunately, aerobic metabolism of animals produces no other convenient proton donor, so the regulatory problem is how to transfer protons from NH_4^+ to HCO_3^- against a substantial thermodynamic barrier. We have seen such problems arise frequently in metabolism before and the mechanisms utilized to circumvene them—coupling uphill processes to ATP hydrolysis—in principle are often the same. In this instance, the mechanism is the urea cycle, in which 2 moles HCO_3^- are eliminated/mole of urea synthesized. One HCO_3^- is in effect incorporated into the product, while the other is simply protonated to form CO_2:

$$HCO_3^- + 2NH_4^+ \rightleftharpoons H_2NCONH_2 + 2H_2O + H^+$$
$$HCO_3^- + H^+ \rightleftharpoons H_2O + CO_2$$
$$\text{Sum: } 2HCO_3^- + 2NH_4^+ \rightleftharpoons H_2NCONH_2 + CO_2 + 3H_2O$$

As we have seen in Chapter 2, the synthesis of urea by the urea cycle consumes four ATP equivalents, but in terms of acid-base regulation, *the fundamental stoichiometry is 2 moles of* H^+ *transferred from* NH_4^+ *to* HCO_3^- *per mole of urea formed.* The overall process is thermodynamically very favorable, which is why, in this context, the urea cycle can be viewed as an ATP-driven proton pump that can transfer protons from NH_4^+ to HCO_3^- against a large energy barrier; at the same time, it disposes of both of these end products of protein catabolism. Since protein and amino acids remain an important and utilized substrate during hibernation, these considerations nicely explain why the urea cycle continues to operate at this time.

We now turn to the question of why urea does not accumulate during hibernation, when kidney function is essentially fully blocked. To extend our ground squirrel example (Deavers and Musacchia, 1980), it is easy to show that at the expected rates of HCO_3^- production, urea cycle rates should accumulate urea up to about 1–2 μmol gm^{-1} day^{-1} or about 7–14 μmoles/gm during a week of hibernation. Changes in blood and tissues urea concentrations should therefore be fairly marked. Yet they often are not observed and the question of why not must be considered.

Small Hibernators Recycle Urea

At least one important reason why urea does not accumulate during hibernation despite continued protein and amino acid catabolism is because it is disposed of metabolically. This capacity can be nicely demonstrated in ground squirrels by urea-loading (by intraperitoneal injection of 0.3 gm urea/100 gm body weight), which leads to essentially immeasurable plasma accumulation of urea; the same urea loading in the rat leads to a threefold increase in plasma urea levels, and requires up to twenty-four hours for re-establishing urea homeostasis (Fig. 7-7). The lack of urea accumulation in the ground squirrel is not due to more efficient renal excretion of urea, but rather to the catabolism of urea. In the 13-lined ground squirrel, fasting for four days combined with twenty-four hours of water deprivation results in about a tenfold increase in urea catabolism and fully a thirtyfold *reduction* in the amount of ^{14}C-urea excreted by the kidney (Riedesel and Steffen, 1980).

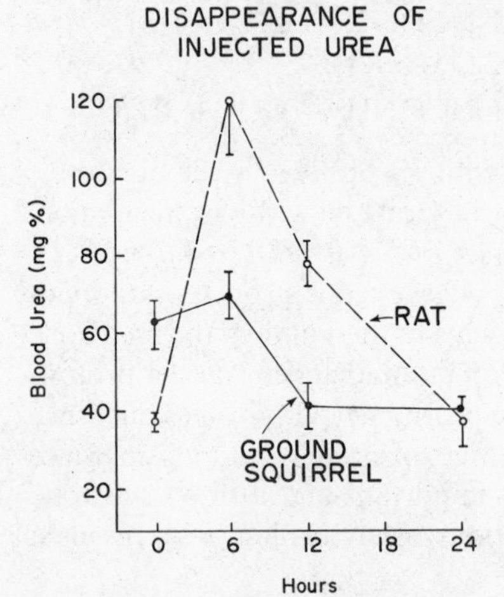

Fig. 7-7. Time course of blood urea changes following urea loading in the ground squirrel (a hibernator) and in the laboratory rat. From Riedesel and Steffen (1980)

These experiments therefore unequivocally demonstrate the capacity for urea catabolism in ground squirrels and explain a) why urea is not accumulated during hibernation even if the urea cycle remains functional, and b) why the amount of protein utilized during hibernation cannot be accounted for by accumulated or excreted urea. To appreciate why urea is recycled, we must first inquire as to how the process works.

Urea Recycling Depends upon Symbiotic Microorganisms

In considering the path of nitrogen during urea recycling we should recall that animal tissues typically do not contain urease; for this reason, recycling of urea involves critical metabolic steps being carried out by microorganisms of the digestive track. Antibiotics directed against these in ground squirrels, for example, greatly reduce the rate at which ^{14}C-urea can be hydrolyzed. The process of urea recycling, however, is best worked out in ruminants where the path of urea in the overall process can be described as follows: liver urea → blood urea → equilibration with extracellular water → salivary glands → urea in saliva → rumen urea. Once in the rumen, urea is hydrolyzed to HCO_3^- and NH_4^+ by the action of urease. The fate of NH_4^+ is reincorporation into the microbial amino acid pool and subsequently into microbial protein, which is ultimately digested and reabsorbed during transit through the small intestine (Schmidt-Nielsen, 1979).

The quantitative significance of urea recycling will be appreciated if it is recalled that about a half the total body water passes through the salivary glands and therefore the rumen each day. But why is this potential retained in hibernators? One simple answer is that since the urea cycle must operate to control HCO_3^- mediated akalinization, urea recycling serves simply and elegantly to minimize the debilitating effects of urea accumulation during hibernation. This function may become particularly important in prolonged, nonfeeding hibernation periods. Another possible function is to integrate acid-base regulation of the host and the symbiotic microorganisms.

Role of Urea Recycling in Acid-Base Regulation of Host and Symbiont

This concept can be illustrated by considering the problem of HCO_3^- production in animals, such as many hibernators, whose diet is composed primarily of plant material. In such nonruminant herbivores, the difficulty stems from large amounts of salts of metabolic acids being consumed in the diet. Since an equivalent of HCO_3^- is produced for each carboxylate group, the amount of HCO_3^- to be eliminated may exceed the amount of waste NH_4^+ being formed. In ruminants, the same problem arises even in more extreme form because two carboxylate anions, acetate and propionate, constitute the main carbon and energy source. Subsisting largely on acetate and propionate, these animals also would be expected to produce very large amounts of HCO_3^- and thus to produce large amounts of urea. They do, but because they recycle the urea formed, they excrete relatively little in the urine. Thus urea cycle function in acid-base regulation of the host is achieved without the attendant problems of high amounts of urea accumulation in tissues and of large amounts excreted in the urine.

In addition to being advantageous to the host, urea recycling is useful to the rumen microorganisms. This is because the hydrolysis of urea leads to the release of HCO_3^- to neutralize fermentation-based acidification, which, if unchecked, would soon fully inhibit further rumen metabolism. Since the production of base in the tissues is approximately equimolar with that of acid in the rumen, the recycling process contributes to acid-base homeostasis in both loci (Atkinson and Camien, 1982). In hibernators, the stoichiometry cannot be as closely specified; nevertheless, exactly the same considerations apply.

Role of Urea Recycling in Water Retention

Another advantage of urea recycling may relate to water balance during hibernation. This idea stems from the observation that microbially assisted recycling of urea, although now known for several mammals (sheep, goats, several monogastric species including rabbits and man), appears to be particularly important in desert species like the camel, which may be routinely exposed to either a protein-poor diet or to reduced water availability, or both. Under such conditions, the camel excretes little or no urea into its urine because of a vigorous and efficient recycling of this metabolite (Schmidt-Nielsen, 1979). The above studies of urea catabolism in ground squirrels simply imply that exactly comparable problems (of reduced protein and water intake) arising during hibernation may be solved in an analogous manner. In both groups of organisms, *the recycling process saves the organism that water which would otherwise be lost with urea* in the renal formation of urine. It turns out that is only one way in which the catabolism of proteins and amino acids contribute to water homeostasis during hibernation.

Sources of Water during Hibernation

Since a hibernator is in effect a closed system with no provision for water intake, internal provisions must be made available to balance with continuous water loss. The seriousness of this problem is indicated by the observation that water loss amounts to one-quarter of the total drop in body weight during the hibernating season. This loss would be even greater were it not for several mechanisms harnessed to minimize its extent. Quantitatively, the most important of these is simply the drastic reduction in metabolic rate, and thus in evaporative water loss, during hibernation. The hibernating posture also serves to minimize the problem by promoting rebreathing of humidified air and thus reducing the net water loss. Urea recycling, as already mentioned, saves the water which would be otherwise lost at the kidney in excreting it. Moreover, a significant advantage arises from the mobilization of muscle proteins in addition to fat as fuels for metabolism during hibernation. The latter situation is complex (Riedesel and Steffen, 1980) and deserves a careful explanation.

Because fat is the most reduced of utilizable fuels for metabolism, in its complete oxidation it yields the most metabolic water. However, very little free water is released from adipose tissues during the mobilization of triacylglycerols; stored lipids are relatively free of water. In contrast, free water represents about 75 percent of the mass of muscle tissue, which is why catabolism of muscle protein yields more water per calorie of energy obtained than does fat. If only fat were to be catabolized during hibernation, the metabolic water formed would be almost equivalent to the evaporative water loss; the organism could be in water balance only if it had no other needs or avenues of water loss (such as feces or urine). Catabolism only of muscle protein, on the other hand, would put the organism into positive water balance. What in fact is observed, a mixed metabolism (about 90 percent fat-based; 10 percent protein-based), leaves the small rodent hibernator in a positive water balance (equivalent to 0.85 ml water/10 kcal of metabolism), allowing a nice margin of safety, which may become particularly important during early arousal when metabolic rates and rates of evaporative water loss both must rise one in step with the other.

Hibernation in Large Mammals (Bears)

In many of their metabolic and physiological processes during hibernation (decreased heart rate and metabolic rate, fat dependence, and so forth), bears are similar to smaller mammalian hibernators. Nevertheless, there are several reasons why the hibernation of bears is unique and why it represents as highly a refined response to starvation as is found in any hibernating (or nonhibernating) mammal. The bear hibernates for three to seven months at a near normal body temperature, 31 to 35°C. Although its metabolic rate is reduced during hibernation, it expends 4,000 kilocalories per day (calculated on the basis of body fat utilization), yet apparently neither eats, drinks, urinates, nor defecates. It is fairly easily aroused into an active state, aware of its surroundings, and able to defend itself. Female bears give birth to cubs and nurse them under these stressful conditions (see Nelson, 1980).

Although the bear starves throughout hibernation there is no net accumulation of common metabolic end products of protein metabolism in the blood. The concentrations of amino acids, total protein, urea, and uric acid remain unchanged throughout winter as do blood volume, hydration, and hematocrit. Because nitrogeneous end products do not accumulate in blood and are not excreted in urine or stored in feces, lean body mass remains relatively constant, unlike the situation in small hibernators.

As in small hibernators, however, there is not enough body carbohydrate to support sustained hibernation, and this leaves fat as the energy source for metabolism. That is why a starting RQ of about 0.8 drops during hibernation to between 0.6 and 0.7. A vigorous fat mobilization is also evident in the high blood concentrations of cholesterol, triglycerides, fatty acids, and phospholipids during hibernation. The finding of a decrease in RQ below that of 0.71 (representing pure fat combustion) may suggest that fixation or retention of CO_2 is occurring, possibly implying some acidification.

Finally, since fat appears to be the only substrate utilized for energy requirements, metabolic water produced from it must be the main source of water for replenishment of respiratory losses. Since total body water, red cell and plasma water, hematocrit, and blood volume remain within expected normal limits, metabolic water must be sufficient for these needs.

None of these characteristics is as well developed in small hibernators, and it is important to inquire why. One secret of the bear's success may be its large size. Because of large size, it displays a reduced mass specific metabolic rate, and hence a relatively less serious problem of evaporative water loss. But as we have seen, even ground squirrels are nearly in water balance during hibernation on metabolic water formed from fat metabolism, so this advantage would not seem great enough to account for the hibernating capabilities of the bear. Nor can fat depots per se be the full answer, since bears and small hibernators both store more than enough fat to carry them through the hibernating season. The most important reason for the bear's outstanding success seems to be a more closely regulated protein metabolism, allowing maintenance of all necessary protein-dependent metabolic functions without going into a net protein deficit as occurs in small mammals.

Organization and Function of Protein Metabolism in Hibernating Bears

As we indicated in our discussion of small hibernators, protein-derived amino acids sustain a number of key metabolic functions (such as serving as precursors for glucose). As in other species, these functions in the bear must also be sustained. Therefore, it is not surprising to realize that protein metabolism continues (in fact is accentuated) during hibernation. The paradox is that this occurs with hardly a measurable loss of lean body mass (i.e., of body protein pools). Nelson and his colleagues have examined this problem and have identified an interesting metabolic organization in hibernating bears that resolves the paradox. Its centerpiece seems to be a close interaction between protein and fat catabolism during hibernation. The catabolism of fat obviously supplies the bulk of the animal's energy needs as well as glycerol, whose turnover rate increases about five- to sixfold during hibernation (even if whole-organism metabolic rate is drastically reduced). An important fate of this glycerol—perhaps the most important—is conversion to glucose, although, as emphasized by Nelson (1980), glycerol carbon also appears in alanine.

The accelerated catabolism of protein, on the other hand, is thought to supply amino acids for two main processes: for any remaining gluconeogenesis that is needed and for biosynthesis of other proteins, the energy for both processes coming from fat oxidation. In this organization glycerol and alanine (or other amino acids) are both potential precursors of glucose. The observed incorporation of glycerol into glucose during hibernation is therefore considered functionally highly significant, since this in effect spares alanine (and other amino acids), making them available for catabolism or protein biosynthesis. Of these two pathways for disposing of amino acids, the biosynthetic route is accelerated in hibernation, while the oxidative one is markedly inhibited. Thus rates of incorporation of amino acids into proteins rise about threefold during hibernation, while their rates of oxidation and incorporation into urea are reduced.

If this metabolic organization were in fact pushed to the limit, the flow of amino acid carbon to glucose and CO_2 and the flow of amino nitrogen to urea would all be fully blocked. In hibernating bears this does not occur (perhaps because there is simply too critical a need for alanine-primed formation of glucose; i.e., not enough glycerol is available to sustain the organism's glucose and glycogen needs). However, the price that is paid for the spillover of amino nitrogen into urea is modest, for the hibernating bear also is able to recycle the urea nitrogen (via microbially assisted hydrolysis of urea and reincorporation of the released NH_4^+ into amino acids and proteins). When these are digested, they make their way back into the same metabolic routes as do endogenous amino acids, of which the most important is reincorporation into endogeneous proteins (Fig. 7-8).

This organization of fat and protein catabolism allows the hibernating bear to function as a relatively closed system, exchanging only O_2, CO_2, water, and heat with its environment, and with minimal perturbation of internal milieu (because noxious end products such as urea and ketone bodies are not accumulated). The final question we wish to consider, then, is how well it might work in the extreme case, where essentially all glucose supplies for the brain would be derived from glycerol. If we assume that all the fat utilized is triglyceride containing 16-C saturated fatty acids, then its complete hydrolysis would

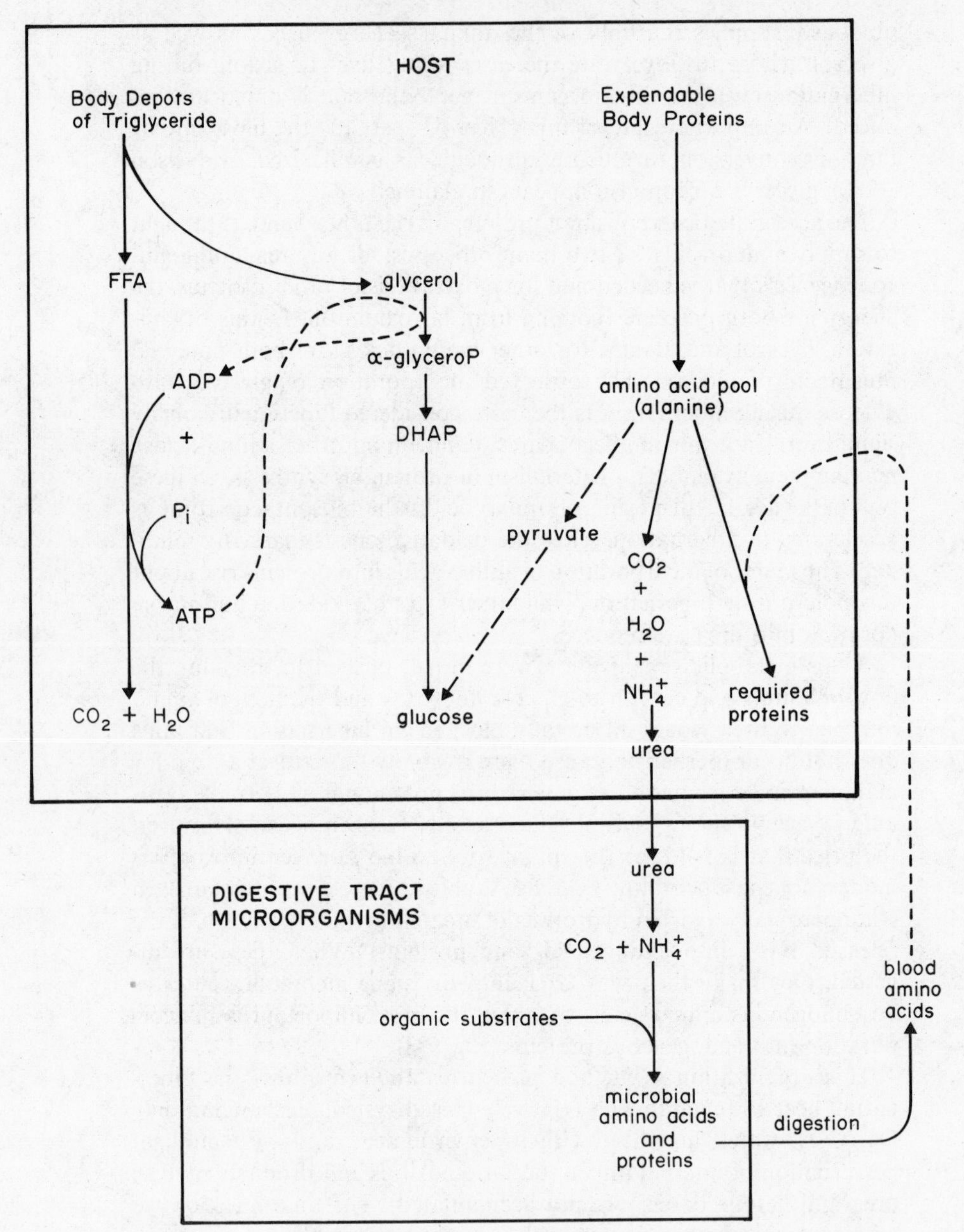

Fig. 7-8. Metabolic organization during hibernation in bears. Summarized from Nelson (1980)

lead to about 387 moles ATP/mole of triglyceride. If all the glycerol released from this triglyceride were converted to glucose (a process requiring only 2 moles ATP/mole of glucose formed, both of which would be required by glycerol kinase and which could of course be readily supplied by fatty acid oxidation), this glucose could generate in the brain 36 moles ATP/mole glucose. That is, if the brain had to operate only on glucose derived from body triglyceride stores, it could maintain a metabolic rate that was about 4.5 percent of the total fat-fueled, whole-organism metabolic rate. Is this enough? Unfortunately, we do not know for sure, but judging from man, it probably is. In man, the brain consumes about 20 percent of the whole-organism basal O_2 uptake, but the relative mass and metabolic rate of the human brain have been scaled upwards (Mink et al., 1981) and there is now known to be significant heterogeneity in the rates of metabolism of different parts of the human brain depending upon their neural activity (Phelps et al., 1981). The hibernating bear, by comparison, has a relatively smaller brain and is essentially comatose; it might be reasonably expected, therefore, that its brain energy demands be a smaller fraction of whole-organism metabolism than in awake and conscious man.

Control of Metabolic Depression in Hibernators

The mechanisms by which the large decreases in metabolism during hibernation are effected remain a topic of some controversy. Two basic classes of theories have been advanced to account for metabolic depression; one proposed mechanism involves chemical depressants of metabolism (pH or hormonal effects), and the other mechanism is a simple Q_{10} effect involving nothing more than the influences of decreases in body temperature on metabolic rates.

Both types of mechanism appear to be important, although the recent work of Snapp and Heller (1981) with the ground squirrel, *Citellus lateralis*, suggests that the low metabolic rates of this species during hibernation are essentially fully explained by simple Q_{10} effects. By comparing the oxygen consumption rates of hibernating specimens (near 8°C) with rates found for sleeping, normothermic specimens, Snapp and Heller determined that a Q_{10} near 2.5 characterized the drop-off in metabolic rate during hibernation. This is a typical Q_{10} value for metabolic processes, and it seems fair to conclude that additional depressing influences need not be invoked, at least in this species.

Despite the apparent sufficiency of Q_{10} effects to reduce the metabolism of a hibernator, additional regulatory mechanisms may be needed to alter the *relative* contributions of different metabolic pathways during entry into hibernation. For example, thermogenic activity due to shivering and brown adipose tissue metabolism must be curtailed. Also, in view of the fact that lipid serves as the major fuel during hibernation, a reduction in carbohydrate catabolism during hibernation seems essential. A role for pH-temperature effects in these regulatory tasks may be critical. Malan (1978) has suggested that the relative acidosis found in the body fluids of hibernators (blood plus muscle and brain, but not liver and heart) due to CO_2 retention may be an important depressant of metabolism. Indeed, the pH values of body fluids tend to be about 0.3 pH units lower at hibernating temperatures than at these same body temperatures in active ectothermic animals, and this degree of acidification is known to be adequate to depress metabolism strongly in a variety of biological systems, including inactive eggs, bacterial spores, and cysts (see Nuccitelli and Heiple, 1982, for review). How does Malan's hypothesis fit into an overall scheme of metabolic regulation in hibernators?

The acidotic state of the hibernating mammal may lead to several specific depressions of metabolic activity. The metabolism of brown adipose tissue is reduced at low pH (Malan, 1978). The effects of low pH and low temperature on PFK are such as to cause the active, tetrameric form of the enzyme to dissociate into inactive dimers (Hand and Somero, 1982, 1983a). Inhibition of PFK would block shivering thermogenesis and minimize depletion of muscle glycogen stores. The latter must be conserved during hibernation to provide energy for shivering thermogenesis during arousal and for bouts of locomotory activity following regain of the normothermic state. Thus, whereas *overall* metabolic rate in a hibernator may be set by simple Q_{10} effects on enzymes involved in oxygen consumption, the specific reduction in metabolic systems (brown adipose tissue O_2 uptake, shivering, muscle glycogenolysis) that must be curtailed during hibernation may be largely due to pH effects.

One facet of pH regulation which merits emphasis is the rapidity of body fluid pH changes during entry into, and arousal from, hibernation. In hibernating mammals like ground squirrels, a change in CO_2 concentrations in the blood can be rapidly achieved by altering rates of ventilation. As shown clearly by the work of Snapp and Heller, RQ values (CO_2 produced:oxygen consumed) during entry into hibernation are extremely low, while RQ's during the initial phases of arousal are exceptionally high. These RQ changes do not result from shifts in the type of foodstuff used for fuel, but instead reflect either the retention (during entry), or removal (during arousal), of large amounts of CO_2 from the blood. The changes in body fluid pH that occur over periods of several minutes to one to two hours may lead, therefore, to comparably rapid adjustments in pH-sensitive metabolic processes such as glycolysis in muscle and oxidative metabolism in brown adipose tissue. In this regard it is interesting that the reassembly of PFK dimers into functional tetramers *in vitro* also occurs rapidly, requiring approximately one hour to complete enzyme reactivation (Hand and Somero, 1983a, see Chapter 11).

Chemical signals other than pH may also contribute to the regulation of metabolism during hibernation. For example, a hypometabolic peptide is extractable from brain tissue of hibernators (Swan and Schatte, 1977). Also, small peptides like the endorphins may help govern metabolic activity (see Margules, 1979). Margules has proposed an antagonistic effect between β-endorphin and a hypothesized endoloxone substance as a mechanism of regulating the disposal of food reserves. The former hormone favors conservation of fuel reserves, the latter elicits their expenditure. The balance between these two antagonists could determine the set-points for a suite for physiological functions, including body weight, body temperature, and respiratory rate. However, it is not yet known how much of a role such mechanisms play in the control of metabolism in various stages of the life of hibernators.

CHAPTER EIGHT

Mammalian Developmental Adaptations

Nearly all of the basic strategies of biochemical adaptation outlined in Chapter 1 are utilized during development of organisms. Anyone who has witnessed the growth and development of animals, or even of his or her own children, from newborn to adult stages will appreciate the marvels of such biochemical adaptations to time. These adjustments begin with egg fertilization and zygote division and usually proceed at modest pace. Periodically, however, abrupt and drastic developmental transitions are imposed upon the species. For mammals, it is difficult to imagine a more traumatic and abrupt environmental change than that occurring at birth. With the loss of the umbilical circulation at delivery, the fetus suddenly and forever is cut off from a continuous source of food, O_2, shelter, and warmth and is forced to adapt rapidly to an entirely new life style. Such abrupt transitions, of course, are not restricted to mammalian species; insect development often involves equally drastic changes in mode of life and environment (larval versus pupal versus adult stages, for example, can entail food sources, environments, and survival strategies that are as different from each other as from entirely different species). However, for obvious reasons, the biochemical problems faced in developmental transitions, and the mechanisms used to resolve them, are best understood for mammals (including man), so we shall focus our attention on them, restricting ourselves mainly to metabolic aspects.

Since development and maturation are continuous processes, conditions and needs of existence are continuously changing; yet for mammals, several developmental periods can be readily distinguished according to the pathways along which nutrients are delivered (Hahn, 1982):

In Phase 1, the zygote is fed directly from the surrounding tissues, placental blood supply not yet being adequately developed.

In Phase 2, once the placenta and the circulation are developed, fetal nutrition sets in and all nutrients are supplied via the blood; most fetal waste products are removed by the same route.

In Phase 3, the early postnatal or suckling periods, the neonate is fed on breast milk.

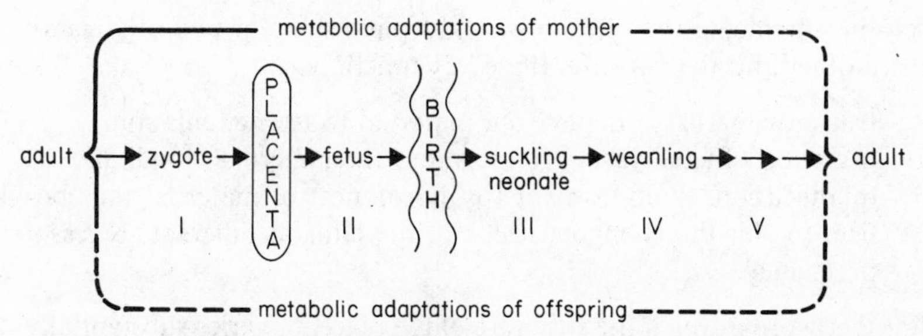

Fig. 8-1. Diagrammatic representation of major mammalian developmental periods

In Phase 4, the animal enters a transition period when milk is still consumed but other food is eaten more and more frequently.

Finally, in Phase 5, the animal enters the postweaning stages of life when all nutrients are obtained by eating.

These processes and stages can be illustrated diagrammatically in a simple time-series (Fig. 8-1). The diagram is useful because it emphasizes two fundamental questions. Firstly, what is the impact of the fetus and neonate on mother and what mechanisms does she harness to deal with that impact? And secondly, what are the problems faced by the offspring in prenatal and early postnatal stages? Our primary concern in this chapter is with the latter, but from time to time we will also consider the former.

Although maternal adjustments to pregnancy are known to occur very soon after development begins, neither the true metabolic requirements of early embryos (in Phase 1 above) nor the exact pathways by which they are fed are well understood. For this reason, even if Phase 1 represents a currently active area of research, it will not be further discussed here. In contrast, once the placenta and its maternal and fetal circulations are established (in Phase 2 above), the mother-offspring system is more amenable to study, and much more precise metabolic pictures have emerged as a result. As we shall see, their centerpiece (for Phase 2 above) is always the placenta, which is a good place for us to begin our analysis.

The Strategic Position and Role of the Placenta

At the outset it is important to emphasize that the placenta per se is a mammalian adaptation for facilitating *in utero* development. It is not simply a passive organ for fetal attachment, but plays a dynamic role in interfacing the metabolism and physiology of two organisms: the mother and the fetus. Its three key functions are

1. transmission of nutrients from maternal to fetal circulation,
2. excretion of fetal waste products into the maternal blood, and
3. modification or adjustment (by hormones) of maternal metabolism to suit the changing needs of the fetus at different stages of pregnancy.

In other words, even if the structure of the placenta varies substantially between species, in all cases it serves to mediate interactions between fetus and mother. In this book, we cannot comprehensively cover all aspects of these mediating functions; neverthless, we can illustrate such interactions, gaining some idea of their importance and nature, by considering the action of placental lactogenic hormone (PLH). All species so far examined secrete PLH into the maternal circulation during the second half of pregnancy. The secretory patterns vary in timing from species to species. Yet again in all of them, and certainly in man, PLH appears to serve three functions during the latter part of pregnancy (Fig. 8-2):

1. it stimulates the corpus luteum of the ovary to secrete progesterone during the second half of pregnancy (a kind of positive feedback loop that assures the maintenance of the placenta for as long as it is needed);
2. it promotes the growth of the mammary gland in preparation for lactation (a kind of preadaptation of maternal metabolism), and
3. it increases the availability of glucose, amino acids, and minerals to the fetus by diminishing maternal responsiveness to her own insulin and thus allowing these metabolites to be released from her tissues. In biochemical terms, the target tissues for PLH of course are maternal, but in biological terms, the target tissues are also those of the offspring, for the net effects of the hormone actions primarily subserve the needs and welfare of the offspring. Thus, in such a fundamental way the impact of the offspring on the mother is mediated through the placenta. Similarly, the placenta mediates the maternal impact on the fetus; by far the most important way this is done is by mediating the transfer of materials (nutrients) from maternal to fetal circulations.

Carbon and Energy Sources for the Fetus

It is widely accepted that the main source of energy for the mammalian fetus is glucose derived from maternal blood. Glucose concentrations in maternal plasma are always substantially higher than in

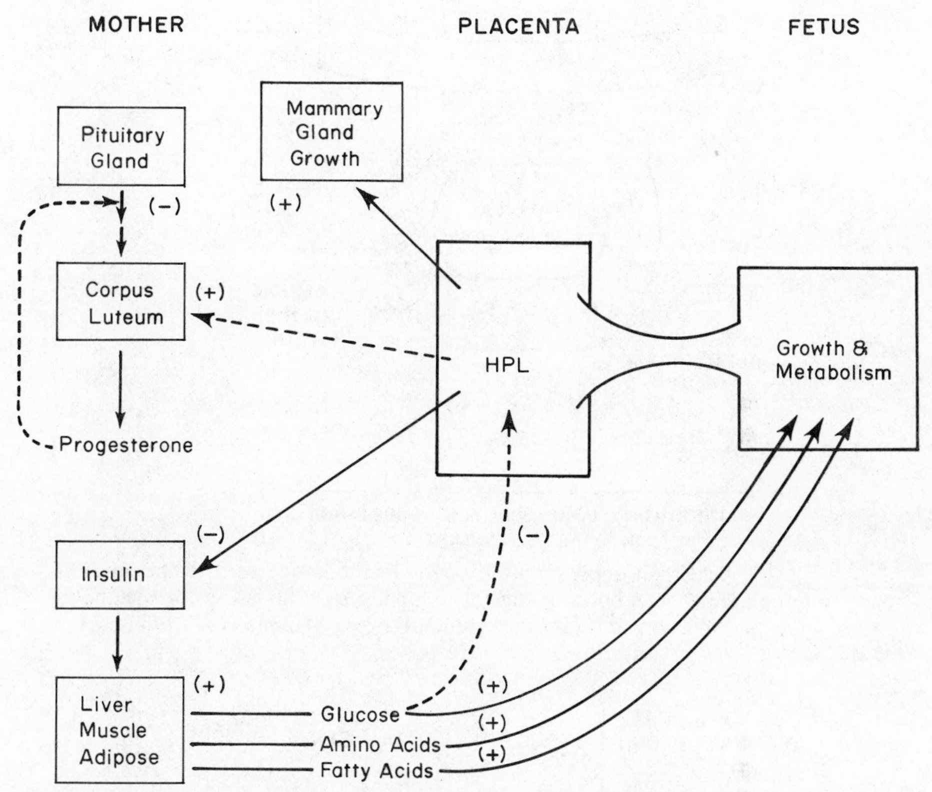

Fig. 8-2. Strategic role of the placenta illustrated through the metabolic and physiological roles of human placental lactogen (HPL). The symbol (−) denotes inhibitory action, while (+) indicates stimulatory effects. From Munro (1980) with modification

fetal arterial blood arriving at the placenta, thus a downhill concentration gradient is always preserved and favors the facilitated transfer of glucose from mother to fetus (Figs. 8-3, 8-4). The glucose demands on the mother during gestation are large and appropriate adjustments must be made to accommodate these. To choose a specific example, in the last two weeks of gestation the pregnant sheep sustains a placental-mediated drain of about 0.3 mmoles of glucose/min. This high rate of glucose removal from maternal pools *accounts for as much as half of the total glucose turnover* of the mother (Fig. 8-4). It is therefore not surprising that pregnant ewes, despite adjustments for this added physiological load, are much more susceptible to hypoglycemia and ketosis than nonpregnant animals.

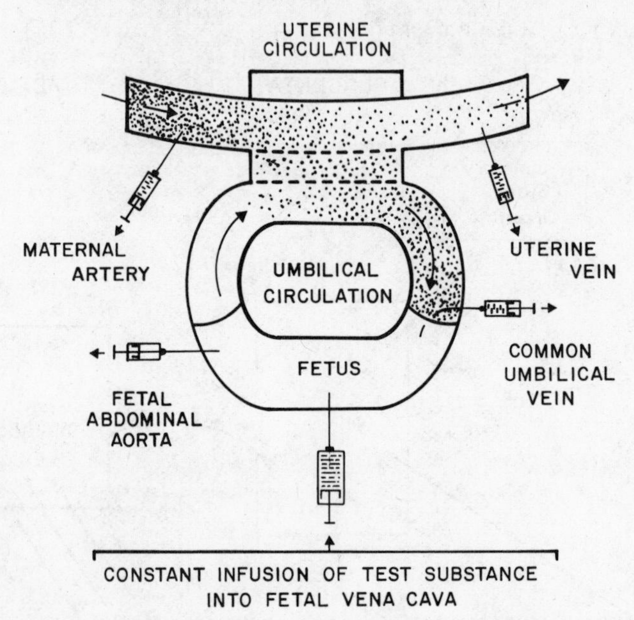

Fig. 8-3. Schematic representation of the methodology used in the study of uteroplacental metabolism, showing sampling sites for assay for metabolite levels and infusion sites for kinetic studies. From Meschia et al. (1980)

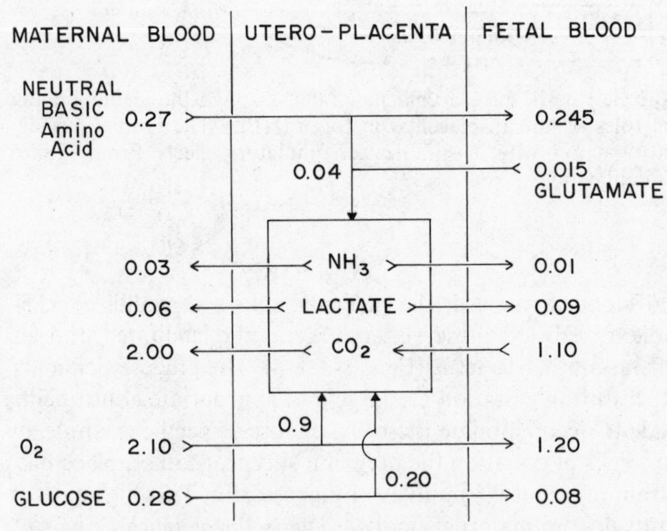

Fig. 8-4. Summary of net substrate fluxes in and through the ovine utero-placenta in the last two weeks of gestation. Amino acid fluxes are in nitrogen millieq/min. All other fluxes are in millimoles/min. for a 4-kg fetus. Modified from Meschia et al. (1980)

What is surprising, on the other hand, is the way in which the glucose removed from maternal blood is partitioned between the placenta and the fetus. In the same case of a 1 kg placenta containing a 4.5 kg sheep fetus, recent data unexpectedly demonstrate that *the fetal lamb receives only one-third of the glucose that the uterus takes up from the maternal circulation.* The rest goes to the placenta. Thus the placenta is an important site of glucose utilization, but what is the subsequent metabolic fate of glucose?

At least in the sheep, but probably in other species as well (Meschia et al., 1980), the two most important end products of glucose metabolism by the placenta are lactate and CO_2. That is, the two major fates of glucose in the placenta are fermentation to lactate (via anaerobic glycolysis) or oxidation (via the Krebs cycle). Estimates for the fetal lamb indicate that the lactate produced by placental glycolysis accounts for about 40 percent of the glucose extracted from maternal blood, leaving enough glucose for oxidation to account for about 90 percent of the O_2 consumed by the placenta. The lactate formed here in turn has two fates: either transfer back into the maternal circulation or transfer into the fetal circulation. Taken together, these data mean that as a result of placental involvement, two major substrates (*glucose plus lactate*), not just one, are being supplied to the fetus as potential energy sources. Interestingly, the two enzymes initiating fetal metabolism of both these substrates themselves undergo fundamental developmental changes.

Developmental Patterns of Hexokinase Isozymes

The first step in the metabolism of glucose in all tissues involves conversion to glucose-6-phosphate (G6P), a reaction catalyzed by hexokinases. It has been known for some time that several isozymic forms of hexokinase are present in most mammals and indeed in most vertebrates (Katzen and Soderman, 1975). Isozymes A, B, and C are the so-called low-K_m forms, while hexokinase D is a high-K_m form of the enzyme, often termed glucokinase in the literature. Although semiquantitative ontogenetic studies of these isozymes were first performed some fifteen years ago, it was only more recently that these were quantitatively specified. Thus in liver of the fetal rat, the predominant hexokinase activity is due to isozyme A which peaks out in its activity just prior to birth. Isozyme B increases in activity, reaching maximum levels several days after birth. Isozyme C, which is all but absent before birth, shows a peak of activity during the second week of life and then declines rather abruptly after about ten days. Isozyme D, the high-K_m

glucokinase, is in effect absent prior to birth and is not induced for at least two weeks following birth.

It is not difficult to explain why the fetus retains only low-K_m isozymes of hexokinase, for at this stage in development the concentrations of blood glucose rarely fluctuate much and never reach the high levels typically found in portal blood of adult mammals (because the fetal gut is of course not functional and the fetus does not consume glucose orally). What is not clear, however, is why there occur specific developmental patterns for isozymes A, B, and C, which kinetically are much more similar to each other than any is to glucokinase. One possibility relates to localization within cells. Much provocative evidence is available suggesting that hexokinase activity *in vivo* is partially regulated by control of an equilibrium between free and bound forms of hexokinase (see Wilson, 1980); just as the kinetic properties of hexokinases differ somewhat, so do their binding properties, and it is this kind of characteristic that may determine when in ontogeny a given form of the enzyme is most advantageous.

Lactate Dehydrogenase Isozymes during Development

All mammalian embryos go through a stage when mitochondria are immature and few in number. As a consequence of this, and perhaps because of a relatively hypoxic environment, dependence upon anaerobic glycolysis is high. In the mouse, for which comprehensive data are available, all embryonic tissues first exhibit a predominance of the muscle isozyme or LDH-5. This situation is altered to some extent during fetal development, yet it is notable how long it is retained, for final LDH isozyme patterns are not realized until well after birth (see below). Since LDH-5 is kinetically best suited for function as pyruvate reductase in anaerobic glycolysis, its widespread occurrence in embryonic and fetal tissues correlates nicely with the well-known anoxia tolerance of the mammalian fetus and neonate. However, this is only part of the story; the most important basis for the improved anoxia tolerance of the fetus is to be found in the availability of endogenous substrate. Many, perhaps all, tissues of the fetus gradually build up glycogen reserves during development, usually reaching maximum levels in the late gestation or at birth. The length of time anoxic conditions can be sustained appears to be determined by glycogen stores of the heart and by the continuous availability of blood glucose. As we have seen, this combination (of *ample substrate supply and a functional circulation coupled with maintenance of kinetically appropriate enzyme machinery*) is utilized by organisms in other hypoxic environments and

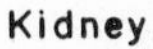

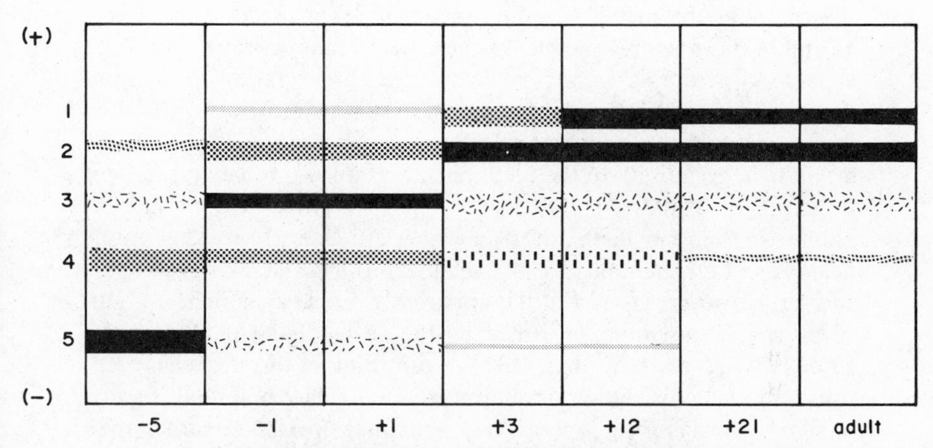

Fig. 8-5. Diagrammatic representation of changing LDH patterns during the development of the mouse kidney. Modified from Markert (1963)

is impressively effective. In the anoxic fetal heart of the guinea pig, for example, the combination is able to sustain energy metabolism at nearly 50 percent of the normoxic rate, while in the adult heart, maximal anaerobic glycolytic rates can sustain ATP turnover at only about 15 percent the normoxic rates (Rolph and Jones, 1981).

The gradual release of the fetus from an exaggerated dependence on anaerobic glycolysis is elegantly illustrated by the complete maturation of the LDH isozyme patterns. As development proceeds a gradual shift in pattern occurs so that enzyme activity is progressively transferred toward the LDH-1 end of the spectrum. The extent of this shift varies enormously in different tissues. In the heart, for example, enzyme activity is progressively shifted from LDH-5 toward LDH-1 through development. In the adult heart nearly all LDH activity is found in isozymes LDH-1 and LDH-2 with almost none remaining in the dominant LDH-5. Other tissues show intermediate degrees of change from the embryonic emphasis on LDH-5 to the increased abundance of LDH-1 generally characteristic of adult tissues (Fig. 8-5).

Metabolic Significance of LDH Isozyme Changes in Ontogeny: Two-Substrate Option

In considering the meaning of the above ontogenetic changes in LDH isozyme patterns, it is important to recall that the kinetic characteristics of heart-type LDHs for both the forward and back reactions are more suitable for function in an aerobic metabolic field. When the isozyme and the metabolite data are considered together, therefore, an interesting picture emerges of the fetus becoming more and more capable of accepting lactate as a substrate for oxidation (because of the suitability of LDH-1 for lactate oxidase function) at a time when lactate released into the fetal circulation is reaching its highest levels (i.e., late in gestation). In the adult, lactate is known to be an excellent substrate for oxidative metabolism by many tissues (heart, red muscle, kidney, to mention three), most of which are characterized by ample heart-type LDH subunits; so this arrangement in the fetus is not surprising. However, in view of the apparently ample availability of glucose, what, if anything, is gained by the development of the specific LDH isozyme patterns observed? We can think of three excellent reasons why the fetus should develop a lactate oxidase potential. Firstly, it allows salvaging the energy still remaining in lactate rather than passing it back to the mother. Secondly, as in some adult tissues (such as the heart) it may be possible to maintain higher metabolic rates on lactate oxidation than on glucose oxidation, an advantage that may take on importance in certain fetal tissues in late gestation. But aside from these kinds of obviously useful functions for ample lactate oxidase potential, by far the most critical one is that *it allows effective metabolic clearance of lactate loads* (*internally or externally generated*) *by the fetus itself*, without requiring offloading lactate into the maternal circulation and without having to sustain any unusual perturbations of intracellular and blood pH. Perhaps for a combination of such reasons, the fetus in late gestation does not rely solely upon maternal blood glucose for its energy needs, but also avidly utilizes placenta-derived lactate as a secondary source of energy. This two-substrate diet seems to be adequate to meet nearly all the energy demands of the fetus, so the question arising next concerns other potential substrates. What is the role of free fatty acids and free amino acids, all of which are of known importance in the adult? Let us consider these in turn.

Fatty Acid Oxidation during Development

Because the fetus is in effect on a high carbohydrate diet, it is not surprising that fatty acid oxidation rates are low, approaching zero in some cases. In the rat, for example, there are a number of factors that limit fetal utilization of fatty acids (Hahn, 1982): 1) the number of mitochondria in fetal tissues is relatively low, which in itself implies low oxidative capacities; 2) the carnitine content of tissues is also low; 3) the catalytic potentials of carnitine transferases are low; and 4) the transfer rates of fatty acids from maternal to fetal circulation are low. This situation in the rat fetus, however, may represent an extreme one. Fatty acid transport rates across the placenta, at least in man, monkey, and the guinea pig, are substantially higher than in the rat. In such cases, where the fetus is born in a relatively mature state, development toward an adult-type of metabolic organization may have progressed further by the time of birth than in the rat fetus.

In peripheral tissues, the main end products of fatty acid oxidation are CO_2 and water; in liver, ketone bodies may be the major end products of fatty acid metabolism. Interestingly, endogenous rates of ketone production by mammalian fetuses are typically low. Fetal rat liver homogenates, for example, can produce about 2 μmoles ketone bodies $gm^{-1}hr^{-1}$ compared to over 20 by similar preparations of adult liver.

Whereas ketone accumulation may be low, the rate of ketone metabolism may become quite high. Unlike fatty acids, which are transported only relatively slowly across the placenta in most, if not all, mammals, ketone bodies seem to pass from mother to fetus much more easily. (In sheep where this cannot occur, acetate is thought to take the place of ketone bodies). During maternal starvation or maternal diabetes, blood ketone levels rise much more than in nonpregnant individuals; as a result, more ketones enter the fetal blood and appear to be utilized at high rates, either as fuels for oxidative metabolism or as precursors for phospholipid synthesis and growth. Under these conditions, when glucose supplies are probably limited, the functional advantages of these metabolic capacities are clear: ketone bodies from the maternal circulation either partially replace glucose as a substrate source (and thus spare glucose for particularly glucose-needy times and tissues) or wholly replace glucose as a substrate source in the extreme situation of maternal hypoglycemia. It may very well be that capacities for ketone metabolism correlate with the life style of the mother, in particular with the likelihood that the species sustains maternal starvation periods during gestation (see, for example, Chapter 8, section on the hibernating bear).

Amino Acid Metabolism during Fetal Development

Although the major energy source for development of the mammalian fetus may well be carbohydrate, because of its growth requirements, no substrates are taken up from maternal blood more avidly by the fetus than are amino acids. The uptake of amino acids proceeds against an overall concentration gradient, the concentration of free amino acids in fetal blood typically being substantially higher than in maternal blood. The details of the active transport of amino acids across the placenta are not yet worked out; however, the movement patterns of different groups of amino acids, and the role of the placenta in the transfer, are now being elucidated. These studies (Fig. 8-4) indicate that in the last month of gestation in the sheep *neutral and basic amino acids enter the placenta from the maternal circulation; small amounts of glutamate and urea as waste-nitrogen carriers, on the other hand, enter the placenta from the fetal circulation.* Estimates of the partitioning of amino acids between fetal and placental metabolism indicate that the fetus is the predominant site of amino acid utilization, unlike the situation with glucose (see above).

Once taken up by the fetus, amino acids could have three main metabolic fates: 1) incorporation into new protein during fetal growth, 2) deamination followed by gluconeogenesis, and 3) deamination followed by complete oxidation. Of the three, the first, of course, is by far the most important quantitatively, which is not surprising given that growth is the most important homeostatic mechanism of the fetus. This is strikingly indicated by ammonia measurements: of the amino nitrogen entering the fetus, only somewhat more than 10 percent is returned to the fetal and maternal circulations as ammonia (Meschia et al., 1980); urea production and excretion rates are similarly low. Consistent with this picture is the observation that urea cycle enzymes, although present, occur in low activities in fetal liver. Carbamyl phosphate synthetase (CPS) activity is of particular interest since it occurs in two isozymic forms: CPS-I forms carbamyl phosphate for the urea cycle, CPS-II, for pyrimidine synthesis. The former occurs in low activity, while the latter is very active in fetal liver, consistent with high biosynthetic demands at this time.

Gluconeogenesis during Development

In view of the ample abundance of gluconeogenic amino acid precursors, it is interesting to note that gluconeogenesis is a sluggish process in the fetal liver. The reason for this lies in the enzyme pathway per se, for except in ruminants, fetal liver contains only low levels of enzymes involved in gluconeogenesis (pyruvate carboxylase, PEP carboxykinase [PEPCK], fructose bisphosphatase [FPBase], and glucose-6-phosphatase [G6Pase] have all been assayed). From a nutritional standpoint, this is quite understandable since normally glucose is being supplied exogenously as the major energy source for the organism; but under unusual stress conditions, the fetus may be called upon to make its own glucose, and it is therefore instructive to inquire how far the fetus is capable of responding to a glucose lack. The answer is: not very far!

After four days of maternal starvation, fetal blood glucose levels decrease by nearly a half (as a result of the fall in maternal blood glucose); yet the fetus does not mobilize its liver glycogen stores! Although it can increase gluconeogenic rates, it does so nowhere nearly as effectively as even the neonate. At least one basis for this is enzymic: with fetal starvation, PEPCK activity increases only fivefold, whereas the rise in the neonate can be one hundredfold! Similarly, starvation induces modest drops in fetal insulin levels in the blood, concomitant with a rise in the level of glucagon; these changes, which in the newborn and adult serve to activate glucogenesis, are obviously in the right direction, but in the fetus they do not go far enough and the changes are small when compared to those occurring in response to similar stress after birth (see Hahn, 1982). These studies indicates how helpless the fetus is (compared to the neonatal or adult capacities to adjust gluconeogenic rates in the face of starvation). The study also serves to emphasize how it uses an alternate adaptational strategy: instead of compensating (through glucogenesis) for lowered glucose supplies, *the fetus conserves glucose by turning to other substrates (ketones) whose availability in maternal blood is concomitantly rising.* By using ketones under these stressful conditions, the fetus gains two advantages; firstly, it minimizes the need for glucose in oxidative metabolism; and secondly, since ketones, like glucose, can also be used in lipid synthesis, it retains the capacity to lay down fat at a time when the mother is starving. Lipogenesis in the fetus is in fact an important job, for enough fat must be laid down to sustain the organism through early postnatal periods when it may be deprived of nutrients and additionally may have serious energy drains into thermogenesis; so it is important to consider the process by which fetus normally lays down fat.

Lipogenesis during Development

For making lipid, there are two options theoretically available to the fetus: either to use fatty acids directly obtained from maternal circulation, or to make the lipid endogenously utilizing metabolite precursors. Of these two, the former is quantitatively less significant. Even if free fatty acids can be and are transferred from maternal to fetal circulations, this process does not occur at a high enough rate to satisfy fully the needs for lipid deposition (although fatty acids obtained by this process may be utilized for this process as well as for phospholipid and membrane synthesis). The transport limitation means that the exogenous fat consumed by the fetus is rather small and that, for practical purposes, *the fetus is on a high-carbohydrate, high-protein* (*amino acid*), *low-fat diet*. In adults, the carbohydrate in such diets raises the rate of lipogenesis, and this is also true for the fetus. Hence, the activities of the main enzymes of fatty acid synthesis—citrate cleavage enzyme, acetylCoA carboxylase, and fatty acid synthetase—are all higher in fetal than in neonatal rat liver and brown adipose tissue, two tissues that have been closely examined. Data are not availabile on the partitioning of glucose and lactate (the two carbohydrate substrates delivered to the fetus) between oxidation and lipogenesis. However, the latter process must consume a reasonable fraction of the two precursors, for enough fat is laid down to sustain the organisms through the first, most stressful hours (and sometimes even days) following birth.

Maternal Adaptations to Pregnancy

Since fetal growth and development depend upon carbon and energy sources from the maternal circulation, it is not surprising that maternal adaptive mechanisms during pregnancy are for the most part directed toward assuring sufficient supplies of both O_2 and nutrients to the fetus. This is achieved by a carefully orchestrated set of metabolic adjustments in the mother, timed to the stages of gestation and hence to fetal needs (see Rosso and Cramoy, 1982). During early stages of pregnancy (the so-called maternal phase, when demands of the fetus are still modest), the maternal organism prepares for future fetal needs; this can be described as an anabolic phase of pregnancy because it is characterized by weight increases, by a massive expansion of the blood volume, and by the deposition of protein and fat reserves. Emphasis on fat storage is indicated by 1) increased maternal feeding, and 2) by increased lipogenesis from glucose in peripheral adipose tissue, a process promoted by high insulin levels at this time. The growth of mammary tissue is also well underway during this phase of pregnancy even if demands upon it are yet to be made.

The mother enters the catabolic phase of pregnancy toward the middle or last third of pregnancy, when fetal demands for nutrients and O_2 reach their highest level, and maternal blood glucose levels consequently are continuously diminishing. A number of adaptational options are available to the mother. With maternal supplies of carbohydrate becoming more and more drained by the fetus, the partitioning of glucose between different tissues and organs must be modified, leaving an ever-increasing fraction for the brain. Other tissues and organs (heart, muscles, liver) may need to rely more upon alternate substrate sources, in particular upon fat (either endogenous from adipose tissue or from dietary sources). Some organs, such as the heart, may be able to take advantage of periodic availability of lactate (from fetal input into maternal circulation).

Although many of these kinds of details of maternal metabolism during pregnancy are yet to be worked out, what is now well known, particularly for the human species, is that blood glucose levels decrease during pregnancy, particularly during the catabolic phase of it (see Rosso and Cramoy, 1982). Since glucose feedback control of gluconeogenesis is relieved, and glucocorticoid levels are high, conditions are conducive for elevated rates of gluconeogenesis (from exogenous and endogenous precursors) to assist in maternal maintenance of glucose homeostasis. However, rates are not high enough and regulation is gradually set to lower and lower levels of blood glucose. Maternal dependence upon glucose, further reduced by decreased responsiveness to insulin, therefore must be minimized during this stage of gestation. That is one important reason why fat is mobilized by triglyceride hydrolysis in adipose tissue, leading to elevated free fatty acid levels in blood (nearly a doubling in pregnant women). Triglyceride levels in blood are also elevated at this time, possibly due to a reduced uptake by adipose tissue (i.e., lower levels of lipoprotein lipase, a membrane-bound enzyme associated with triglyceride uptake by adipose tissue). This combination of events improves the supply of fatty acids for both maternal and fetal metabolism. The elevated fat catabolism in the mother, particularly during fasting, also leads to increased availability of ketone bodies, which, as we have seen, can be utilized for her own metabolism and for the fetus. Finally, at this stage, nutritional input of proteins and amino acids is augmented by maternal mobilization of endogenous protein to favor an ample supply and delivery of amino acids to the fetus right to its birth.

Scaling Constraints on Maternal Metabolic Adaptations

In spite of a lot of plasticity in developmental patterns (in gestation time, in differentiation stage at birth, in litter number, and so forth), there are a number of curious size-related constraints that appear to be imposed upon the maternal-fetal unit considered together (see essay by Rahn, 1982). One of these—perhaps the most obvious—is that the weight of the *term* fetus or of the litter is proportional to the maternal metabolic rate, *which means that the relative cost or relative caloric investment of the mother is the same (about 20 percent of basal rates of metabolism) regardless of adult size*. Because of its high mass-specific metabolic rate, a small (10 gm) mammal is able to sustain offspring that at birth weigh about 30 percent of the adult weight, while in contrast, because of scaling adjustments of metabolism, a 100-ton whale gives birth to offspring that are only 1.5 percent of the adult weight (Rahn, 1982).

A second biochemically more intriguing constraint relates to the scaling of fetal metabolism to size. Because of large differences between the size of the fetus and its mother, one would expect the mass-specific metabolic rate of the former to be substantially higher than that of the latter. In fact, *their metabolic rates/unit mass are the same*!—which means in effect that "something" prevents full realization of the catalytic potential of oxidative enzymes in the fetus. This unexpected scaling effect, observed also in birds, implies that in both, the offspring have compromised to develop at a relatively low metabolic rate: in birds these rates do not exceed the capacities of the shell conductance to supply O_2, while in mammals these rates do not compromise the mother by demanding too large an O_2 delivery to the feto-placental unit.

The intriguing question is: what keeps the fetal metabolic rate down? One suggestion, made by Rahn (1982) and Kleiber (1965), is that the fetus is simply O_2-limited. This may be a contributing factor, but because tissue oxidative potentials of the near-term mammalian fetus are often nearly fully developed, there would be a pronounced Pasteur effect if O_2-limitation were the sole mechanism for holding fetal metabolic rates down. Yet a pronounced Pasteur effect is not usually observed. Although normally O_2-dependent tissues such as the heart and brain have higher anaerobic capacities in the fetus than in the mother, these are utilized only in emergency situations; normally, these organs at term no doubt operate largely (and perhaps solely) upon oxidative metabolism. If, in the case of the pregnant rat, eight fetuses at term consumed O_2 at the rate equal to that of a one-day-old infant, the metabolic rate of the pregnant rat would be forty-seven times her basal rate (Kleiber, 1965), which of course does not occur. So it is probable that some internal regulatory mechanism maintains fetal metabolism at a relatively low rate; if this did not occur, the fetal drain of O_2 across the placenta would be far more drastic than observed. Whatever the mechanistic basis for this regulation, it evidently is released very soon after birth, when the true catalytic potential of its oxidative machinery is quickly realized and the neonate rapidly attains a metabolic rate commensurate with its size.

Metabolic Adaptations to Birth

Delivery entails three outstanding problems with which the newborn must immediately cope: 1) deprivation of constant and very specific kinds of carbon and energy sources, 2) physical and thermal stresses, causing increases in the circulating levels of catecholamines and thyroid hormones, and 3) variable degrees of hypoxia. Because of the ample availability of glycogen in many tissue depots of the fetus late in gestation, the solution to the latter problem is readily understood. Glycogen is utilized in an aerobic metabolism if all proceeds normally during delivery; however, almost always a modest-to-severe hypoxia occurs during this time (blood lactate levels rising seven to eight times maternal ones). As has been well documented for many species, the high glycogenolytic capacity of the late fetus and early newborn, particularly of the heart, renders the term fetus notably resistent to hypoxia or anoxia. The same options and mechanisms are available to the newborn as to adult vertebrates facing problems of limited O_2.

Similarly, the fetus is well prepared to handle the thermal stress imposed upon it with delivery. Generally, in late fetal and early neonatal stages, the amount of brown (thermogenic) adipose tissue that is retained is proportionately much higher than in adults of the species. Elevated levels of catecholamines, typically observed at this time, trigger calorigenesis, which constitutes one line of defense against cold stress.

Solutions to problems (2) and (3) above, therefore, are relatively simple and are now well understood; in metabolic terms, problem (1) above (deprivation of maternal nutrient source) is by far the most complex. It involves at least two components: an initial starvation period of variable length, followed by the adjustments required on switching from a high-carbohydrate, low-fat food source to a high-fat, low-carbohydrate one (milk). Interestingly enough, the fetus seems well prepared even for these more complex ramifications of delivery.

Energy Sources in the First Hours of Life

The first and most obvious change that is expected (and indeed is observed) as soon as the supplies of nutrients from the mother are curtailed is a decrease in the level of blood glucose in the term fetus. It is the first energy source tapped as the mother's supply is lost. Given that the blood volume is only 7 percent of the animal's weight, the amount of energy available from this source is negligible, and can only be useful during the first few minutes after birth. Fifteen years ago, Hahn and Koldovsky (1966) correctly concluded that very soon after birth endogenous fat becomes the main supplier of energy for the newborn. A simple series of calculations shows why this is so.

The metabolic rate of a newborn rat is about 0.1 cal 5.5 gm body weight^{-1}hr^{-1} (0.45 cal gm^{-1}day^{-1}). If carbohydrate were its sole source of energy, one hour after birth the newborn would use up two-thirds of its *total* body glycogen and within two hours with no feeding no carbohydrate would remain in the body!

Another possible source of energy that could become important prior to suckling is glucose formed *de novo* from lactate or free amino acids, particularly from alanine. To some extent this does occur, but neither the timing nor the amounts are quite right to account for early energy needs. Thus within the first hour following birth, gluconeogenesis from lactate commences (blood lactate levels, typically three to four times higher in the late fetus than in the adult, rise to even higher levels during delivery) but the process does not reach peak rates until six hours after birth. So the timing is a bit off to be helpful during the first hour or two, which is part of the problem. The other part stems from the limited amount of glucose that can be formed, and this seems to be largely determined by the limited catalytic potentials of PEPCK, a known regulatory site in the pathway. Thus during the first postnatal hour there is hardly any change in PEPCK in the liver; it begins to rise about two hours after birth, reaching its maximum capacity about six to ten hours later. This development pattern means that not only lactate-primed glucose synthesis, but also alanine-based gluconeogenesis is limited at this time, since their conversion to glucose proceeds through the PEPCK step. An alanine contribution to glucose-based energy metabolism is made all the less significant by the low availability of alanine in early postnatal hours (blood levels of alanine and other amino acids drop precipitously in the first hour after birth).

These considerations, although based on indirect data, imply that gluconeogenesis in the first few hours following birth would be a variable and inadequate source of glucose for energy metabolism, and as it turns out this conclusion is supported by direct measurements (either in terms of lactate conversion to glucose or in terms of glucose turnover rates). These show that the low gluconeogenic rates typifying the term fetus do not reach their peak capacities of about 0.2 μmol gm^{-1}hr^{-1} (equivalent to about 0.04 cal/newborn rat of 5.5 gm weight/hr.) until at least six hours after birth. Although these maximal rates of glucose generation represent a potentially significant contribution to energy metabolism at this time, the amount is not enough (maximally only about one-third that required); of course the amount of glucose that can be formed by gluconeogenesis is even less during the first hour or two after birth.

Hence, we are left with fat as the only alternative source of energy following birth but prior to suckling. Lipolysis, initiated by catecholamine activation of hormone-sensitive lipases, gives rise to both glycerol and free fatty acids. Although both substrates could contribute to energy metabolism, it has proven difficult to quantitate their relative contributions immediately after birth because measurements on rates of triglyceride breakdown in adipose tissue are not easily obtained. Nevertheless, indirect evidence is fairly convincing. Thus soon after birth both glycerol and fatty acid levels in the blood rise quite dramatically. The rise would be stoichiometric (1:3) if glycerol and fatty acids were subsequently metabolized at similar rates. In fact, the rise is not stoichiometric, levels of glycerol increasing five fold, free fatty acid levels increasing only 1.5-fold. Since the oxidation of fatty acids yields much more energy than that of glycerol, these data imply their much more vigorous catabolism at this time.

Finally, as we mentioned above, ketone bodies are an end product of fatty acid metabolism in the liver; although their level drops in the first hour after birth, it rises again quite dramatically afterwards. As in the glucose-limited fetus, ketones at this time may be used in place of glucose by various tissues in the body and this may be a way of conserving limited supplies for specifically glucose-needy tissues.

In qualitative terms, most species thus far studied show metabolic patterns comparable to those in the rat during the first few hours after birth. So it is fair to say in summary that currently known major metabolic adjustments in response to delivery include 1) *a fall in blood glucose and amino acid levels*, and 2) *activation of lipolysis and gluconeogenesis*. These adjustments, which are mediated hormonally through the actions of insulin, glucagon, catecholamines, glucocorticoids, and thyroid hormones, manifest themselves as *rapid elevations in the blood levels of free fatty acids, glycerol, and ketone bodies. Endogenous fat thus constitutes the main source of energy in the first few hours after birth* and these metabolic adaptations are in fact continued when suckling begins.

Metabolic Adaptations in the Suckling Period

After the initial shock of delivery and facing a new environment unfed, the newborn commences feeding and now must adapt to a high-fat milk diet. Essentially, this nutrient source and the metabolic machinery needed for handling it represent a continuation of the perinatal starvation period; the difference is that fat, which is predominantly utilized, is now derived from the mother's milk instead of from endogenous body reserves.

Interesting differences occur between species that are born relatively mature with a large amount of storage fat and those that are born in very immature states with small amounts of fat in their body. The human fetus, for example, with 16 percent fat at birth, constrasts sharply with the relatively immature rat fetus, containing less than 2 percent fat. Although both live off fat as soon as they are born, neonatal man can use his own fat for a long period of time, whereas the newborn rat very soon has to utilize the fat contained in its mother's milk.

Similarly, the amount of fat in the milk of a species correlates with the degree of maturity at birth; the more mature, the less the percentage fat content of the milk. This generalization breaks down, however, if species-specific needs arise for the newborn. Species that are born mature but have to expend an unusual amount of energy on staying warm (seals and other marine mammals, moose, elk, and arctic mammals in general) have a high percentage of fat in their milk, again implicating fat as the main energy substrate for newborn mammals.

Although as originally estimated by Hahn and Koldovsky (1966) fat is the major source of energy during suckling (most milk proteins being used for growth), the problem remains of the high energy demands of critical tissues such as the brain, which typically cannot utilize fatty acids. This problem is resolved for the suckling neonate by the availability of three potential sources of energy for the brain: milk-derived (i.e., lactose-derived) glucose, glucose derived from gluconeogenesis, and ketone bodies. As in the perinatal period immediately following delivery, so in the suckling stage, the high availability of fat leads to high levels of blood ketone bodies, which can be used as substrate sources for various tissues including the brain, particularly should glucose reserves become limiting. The suckling rat is buffered from glucose limitation problems by lactose in the milk (mobilized in the liver by β-galactosidase-catalyzed hydrolysis to glucose and galactose) and by gluconeogenesis. Although alanine and lactate can be used for the *de novo* formation of glucose, the potential is not fully developed for utilizing alanine, probably because of low activities of liver alanine aminotransferase: three-day-old infant rats, for example, have only one-sixth the alanine-based gluconeogenic capacity of nineteen-day-old rats. Thus it is accepted that during suckling stages, lactate is probably the main gluconeogenic precursor, at least in species such

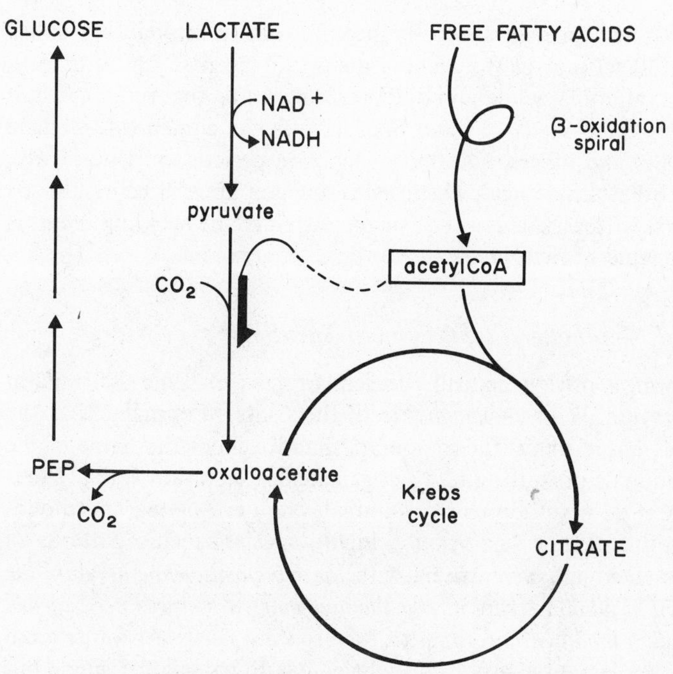

Fig. 8-6. Summary of metabolic events in liver of neonatal rats at suckling stage

as the rat. Lactate presumably derived from glycogen metabolism of peripheral tissues is available for liver glucose formation because mitochondrial decarboxylation of pyruvate (catalyzed by pyruvate dehydrogenase) occurs at a low rate, while pyruvate carboxylase activity, rising rapidly after birth, is quite high by the suckling stage. Under these conditions, fatty acids are thought to supply the acetylCoA for the Krebs cycle while lactate (via LDH and pyruvate carboxylase) supplies oxaloacetate for both the Krebs cycle and gluconeogenesis (Fig. 8-6).

Although some alanine (and presumably other glucogenic amino acids) can be utilized for *de novo* glucose formation, it is clear that alanine plus other free amino acids made available from milk (directly or as milk protein) are used mainly for protein synthesis (i.e., growth). This explains why urea cycle enzymes and alanine aminotransferase occur at low activities during this time and rise toward adult levels only around the time of weaning. Attempts to assess directly the partitioning of alanine between deamination pathways (leading to glucose or to CO_2) and protein synthesis indicate that in ten-day-old infant rats only 30 percent of the alanine amino nitrogen is recoverable as urea plus ammonia, while about 75 percent can be recovered in adult rats. From these kinds of data, Snell (1980) has concluded, as have Hahn (1982) and others, that during suckling periods most of the alanine and other amino acids absorbed by the gut are still being utilized for protein synthesis. This is not surprising since the suckling infant is in a very rapid growth phase.

Maternal Metabolic Adaptations to Suckling

Lactation, a phylogenetically ancient process of some 200 million years duration, is so characteristic of the Class Mammalia that the organ it is based upon (the mammary gland) echoes the name of the class of animals possessing it. Phylogenetically, lactation actually predates placental gestation (early mammals were egg-laying and monotremes still are); not surprisingly, highly species-specific patterns of "adaptive suckling" have arisen, with *the composition of the milk, the mechanism of administering it, and the mammary structures per se, each and all being modified according to each species' needs.* As an extreme example, the lactation process in whales has to cater for a single but huge offspring, with high caloric requirements, nursing rapidly while submerged; at the same time, the mother must conserve water because of the high salt content of the sea. A balanced compromise solution to these conflicting requirements (see Rosso and Cramoy, 1982) is achieved by the formulation of a highly concentrated milk, with a fat content of 50 percent or more, and a very powerful let-down reflex on the part of the mother; thus, by literally and quickly pumping up the offspring with cream, water is conserved by the mother, the high caloric requirements of the offspring are stabilized, and the entire suckling process can occur easily within the time limits of submergence!

These elegant adaptations of whales and other marine mammals are only unusual in the degree of specialization, for all mammals display a similiar close coordination of maternal lactation with offspring needs. In all mammals as far as is known the prolactin-based reflex is also a part of the "standard machinery," with prolactin secretion being proportional to the suckling stimulus. In other words, the greater the suckling stimulus (the number of feeds times the length of feeding times the vigor of the offspring), the greater the amount of milk secreted. And finally, in all mammals, the two major nutrient sources in milk are fat and carbohydrate (although many other substances such as casein, amino acids, vitamins, ions, and even antibodies for protection against infection are also secreted by the mammary gland). Important metabolic adjustments occur in the mammary gland for the production of these two major carbon and energy sources, each of which we shall discuss in turn.

Adjustments in Mammary Gland Lipogenesis during Lactation

In the liver and adipose tissue of nonruminant mammals, the usual pathway of lipogenesis begins with glucose and involves its catabolism through pyruvate to acetylCoA, which condenses with oxaloacetate to form citrate. Citrate is then transported out of the mitochondria where it is cleaved to acetylCoA and oxaloacetate; the latter is converted to malate which can serve as substrate for malic enzyme, generating reducing power (as NADPH) for driving the incorporation of acetylCoA into fat. This is the so-called citrate cleavage pathway and its key characteristic is the use of citrate as a mechanism for transporting 2-carbon units out of the mitochondria to the cytosol where they can serve as substrates for acetylCoA carboxylase, the first committed step in fatty acid synthesis. This pathway of lipogenesis appears to be highly adaptable (see Hochachka, 1973). For example, in ruminants, citrate cleavage and malic enzyme occur in unusually low activities in liver, adipose, and mammary tissues, and the pathway here is not considered functional; ruminants instead utilize dietary acetate directly as a source of extramitochondrial acetylCoA for fatty acid formation, and NADPH-linked isocitrate dehydrogenase as a source of hydrogen for the process. In striking contrast, the fetal ruminant, which of course has a continuous source of glucose from its mother, contains the usual mammalian complement of enzymes needed for the citrate cleavage pathway.

In the mammary gland of the rat and other mammals, large increases in citrate cleavage enzyme activity occur quickly after the onset of lactation, while weaning results in a rapid decline of the activity (usually the decline takes about a day). Since citrate cleavage enzyme is under allosteric regulation and is known to represent an important control site in lipogenesis in the liver, these kinds of data suggest that in mammary gland tissue lipogenesis is also controlled in part at least by the activity level of citrate cleavage enzyme. In contrast to liver and adipose tissues, however, in mammary tissue the reducing power for fatty acid synthetase appears to be largely derived from the G6P dehydrogenase reaction catalyzed by a specific isozyme which is induced at lactation. The activity of this particular form of G6P dehydrogenase can increase up to twentyfold at this time, itself an indication of how vigorous a metabolism the mammary gland sustains during the suckling period.

Mammary Gland Adjustments for Lactose Synthesis

Enzymatic adaptations of the mammary gland for turning on lactose synthesis at the right time are equally interesting. The disaccharide lactose, like casein, is made only by the mammary gland in a reaction catalyzed by lactose synthetase:

$$\text{UDPgalactose} + \text{glucose} \longrightarrow \text{lactose} + \text{UDP}$$

The transfer of galactose to glucose is mediated by two proteins, referred to as A and B. The B-protein is identical with one of the common milk proteins, α-lactalbumin, and by itself has no known enzymatic activity. The A-enzyme apparently is associated with the microsomes and is a galactosyl transferase which catalyzes the reaction:

$$\text{UDPgalactose} + \text{N-acetylglucosamine} \longrightarrow \text{N-acetyl-lactosamine} + \text{UDP}$$

In contrast, in the presence of the B-protein, the A-enzyme is able to utilize glucose as an acceptor of the galactosyl portion of the UDP-galactose. In essence, *the B-protein directs the A-enzyme to catalyze lactose rather than N-acetyl-lactosamine formation, by changing substrate specificity (to a glucose preference) and by increasing substrate affinity (decreasing the* K_m *for glucose).*

Although in explants of mammary gland a maximum and simultaneous synthesis of both A and B components of lactose synthetase can be elicited by the proper administration of three hormones—insulin, a glucocorticoid, and prolactin—this pattern is not observed *in vivo*. Instead, because of the sequential and controlled production of these hormones, the A-enzyme activity increases rapidly at about the middle of pregnancy, reaching a maximum shortly before parturition. The B-protein activity increases markedly shortly after parturition, in this way ensuring that maximum lactose synthesis rates cannot be achieved until they are required; i.e., when suckling begins in earnest (see Hochachka, 1973, for further discussion).

Finally, we should mention that these biosynthetic and metabolic events in the mammary gland are closely associated with ultrastructural changes in the alveolar cells of the mammary gland, and with other specific biosynthetic events (such as casein biosynthesis) occurring at this time. This coordinated control, moreover, is retained until weaning of the young.

Metabolic Adaptations to Weaning

The weaning period in mammalian development can be defined as the period between the time of first consumption of food other than breast milk and the time of complete cessation of suckling. In the rat, the weaning period stretches between about day 16 and day 30. In species that are born relatively mature, the weaning period is rather more difficult to define. In man, for example, the weaning period may stretch from very early after birth to three years or more of age, depending upon cultural and socioeconomic conditions. For this reason, metabolic adjustments occurring during this time period may occur gradually and are more difficult to catalogue. They are nevertheless fundamental for it is only in this stage that adult metabolic patterns (if not their full potentials) are realized. Probably our most instructive insights arise from studies of the weaning period in the common laboratory rat.

The standard diet of the laboratory rat is usually Purina Chow or a diet similar in composition. Compared to pure breast milk, this is a high-carbohydrate, low-fat diet, so as the rat is weaned it gradually consumes fractionally more and more carbohydrate and less and less fat. This indeed is a very general trend among many mammals; the metabolic reflections of this trend include *a decrease in gluconeogenic rates and capacities* (*due to dropping levels of key enzymes in the pathway*) *and an increase in lipogenic rates* (*due to increasing levels of key enzymes in lipid synthesis*). In some respects, therefore, the patterns of lipid and glucose metabolism return to prenatal patterns (Figure 8-7). Such, however, is not the case for protein and amino acid metabolism.

Up to the weaning stage, as we have seen, the main fate of the free amino acid pool seems to be utilization for protein synthesis, with only a small fraction of the amino nitrogen appearing as urea or ammonia. At weaning this situation changes dramatically, for the potentials of alanine aminotransferase, malic enzyme, and the urea cycle enzymes in the liver are gradually being expanded (Table 8-1). With weaning, in other words, the metabolism of amino acids becomes more complex. Since the organism is still growing, protein synthesis remains an important pathway for amino acid metabolism. However, amino acid deaminations followed by further metabolism (in oxidative pathways,

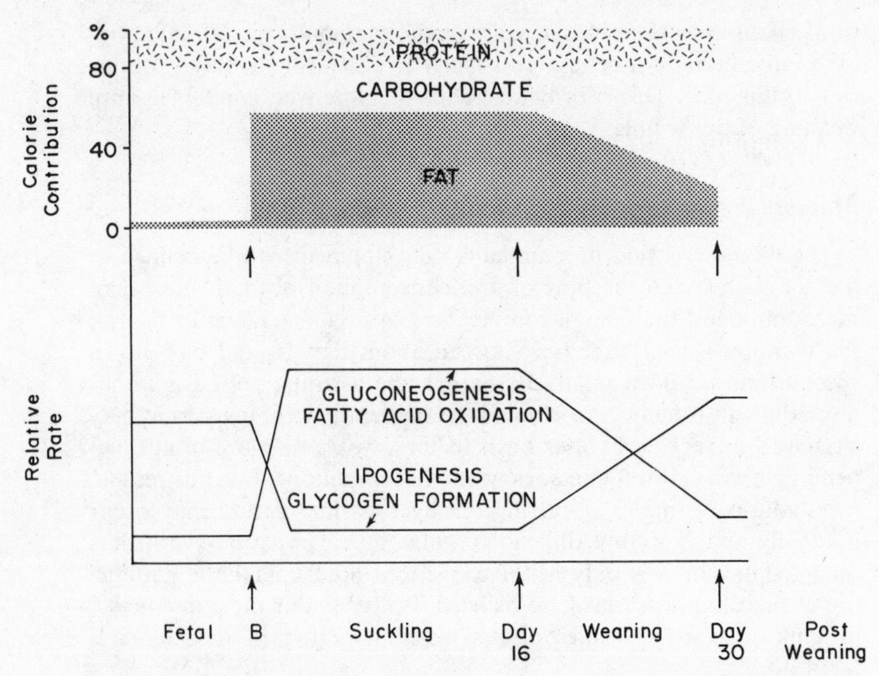

Fig. 8-7. Changes in nutrient composition of food consumed by fetal, suckling, weanling, and weaned rats, showing the relationship to changes in metabolic organization. Metabolic rate is expressed as % caloric contribution by protein, carbohydrate, and fat during development. Developmental periods: B, Birth; day 16, start of weaning period; day 30, end of weaning period. From Hahn (1982) with modification

gluconeogenesis, lipogenesis, and other series such as the purine nucleotide cycle, the urea cycle, etc.) now form effective alternative routes competing for the incoming (dietary) pool of free amino acids. In metabolic terms, the adult patterns of metabolism are set in this stage, even if the adult levels of catalytic activities of many of the component metabolic enzymes are yet to be realized later in life.

Summary of Overall Metabolic Patterns during Development

At this point it may be useful to summarize briefly the overall nutrient and metabolic changes that typically occur during development from fetus to adult. Basically, the main energetic parameters the developing organism has to deal with are varying supplies of fat and

Table 8-1. Enzyme Activities of the Liver of Fetal, Suckling, and Weaned Rats

	Development Period		
	Fetal	Suckling	Weaned
Phosphoenolpyruvate Carboxykinase	Low or absent	High	Decreasing
Pyruvate carboxylase			
Fructose-bisphosphatase			
Glucose-6-phosphatase			
Carnitine transferases			
Enzymes of ketone metabolism			
Pyruvate kinase	High	Low	High
AcetylCoA carboxylase			
Fatty acid synthetase			
Citrate cleavage enzyme			
AcetylCoA synthetase			
β-Methyl, β-hydroxyglutaryl-CoA reductase			
Alanine transaminase (cytoplasmic)	Absent	Low and rising	High
Urea cycle enzymes			
Malic enzyme	Absent	Absent, then rising	

SOURCE: Modified from Hahn (1982)

carbohydrate. During fetal development, the main sources of energy are carbohydrate (glucose and lactate) and the organism is essentially on a high-carbohydrate, low-fat diet. For a variable, but usually fairly short, time period after birth, the neonate enters a period of starvation; then with the initiation of suckling, it must contend with a strikingly altered nutrient source, rich in fat but low in carbohydrate. This situation continues into the weaning stage, where the ratio between dietary carbohydrate and fat again is reduced.

A point of considerable importance is that through most of the above developmental stages (at least up to weaning and often even in this stage) the developing organism really cannot choose its food; it is quite literally stuck with what it is given. This greatly limits the number of adaptational options available to the developing mammal, which is why many, and perhaps all, of the adjustments in carbohydrate and fat metabolism occurring through development are immediate and direct consequences of the change in available nutrients (Figure 8-7).

The situation with respect to amino acid metabolism differs somewhat from the above. One reason for this is that the availability of amino acids (either directly or as proteins in the diet) is basically fairly constant at least through weaning. Although input of amino acids can be viewed as being relatively stable, their subsequent metabolism as a function of developmental time is not. At least through the suckling period the major fate of nutrient amino acids is incorporation into proteins (for growth and differentiation). Weaning has a large impact upon this picture, for at this time the varied needs and activities of the organism now call for a much more complicated amino acid metabolism, with a much smaller fraction of the total amino acid pool being directed toward proteins and growth, a much larger fraction being directed toward catabolic, anaplerotic, and other biosynthetic functions.

Enzyme Basis for Developmental Adaptations

In the above sections we have briefly summarized numerous metabolic adjustments that correlate with, and often clearly are an integral part of, survival through particular stages in development. How are these changes achieved and how are they regulated? As in other aspects of this large, and still actively growing area of research, we can at this time only point to broad categories of mechanisms. These fall into two major classes: genetically programmed mechanisms and environmentally signaled ones. The balance between these, however, is not always very clear.

Consider the rat, for instance. Although born in a very immature state (no fur, sealed eyes, closed ears), it nevertheless readies itself for delivery in numerous ways. Enzyme activities destined to rise to high levels in liver, brain, heart, and probably many other tissues commence to increase usually during the last two or three days of gestation. Are these changes simply the result of reading a genetic timetable or are they environmentally induced? It turns out that both factors come into play. Experiments inducing delivery either prematurely or postmaturely, for example, modify the normal development pattern; thus at least to some extent it is the "environmental circumstances" that determine the maturation of at least some developmental steps. Others presumably are largely programmed genetically.

The conflict between the two opposing (genetic versus environtal) views must not be exaggerated out of realistic proportions, for whatever the starting signals, there is widespread agreement as to the critical roles of hormones and, to a lesser extent, metabolites in mediating developmental adjustments. This can be nicely illustrated by considering birth. As mentioned above, delivery is a process involving substantial stress which causes the release of catecholamines in the fetus. These in turn activate adenyl cyclases in liver and adipose tissue, leading to glycogen breakdown in the liver (supplying glucose for energy at a time when maternal sources have been cut off) and fatty acids as fuel for thermogenesis. The catecholamines also increase pancreatic glucagon release and decrease insulin release from the islet cells; at the same time, the decrease in blood glucose following delivery itself may trigger hormonal changes, such as elevated glucagon and glucocorticoids, thus facilitating glucose homeostasis. If these hormonal changes are to be effective, of course there is also the requirement for the (earlier or simultaneous?) development of hormone receptor sites (usually on the cell surface), a problem that has been receiving increasing attention in recent years.

Whatever the signals are, and however they are mediated, their main pathways for mediating metabolic change are via adjustments in the kind or the amount of enzymes present in any given tissue at any given stage in development. This can be illustrated with the rat brain, which, as we have seen, sustains an unusually high capacity for ketone and for glucose metabolism during early stages in development. This is achieved by a carefully programmed pattern of enzyme induction involving several critical steps in metabolism, with pyruvate dehydrogenase maturation lagging (by about a week) behind the induction of Krebs cycle enzymes such as citrate synthase as well as behind the induction of pyruvate carboxylase and β-hydroxybutyrate dehydrogenase. As a result, glucose metabolism in the immature rat brain is preferentially guided into performing an anaplerotic role, supplementing the loss of Krebs cycle intermediates used for biosynthetic purposes; ketone bodies on the other hand feed 2-carbon fragments (as acetylCoA) into the Krebs cycle (Fig. 8-8). In the extreme situation, the only function of glucose may be such augmentation of Krebs cycle intermediates, with ketone bodies supporting essentially all of the brain's energy needs. Although complex, this represents an elegant way of minimizing the organism's needs for glucose at times when glucose availability may in fact be frequently limited.

In a comparable way, each tissue and organ has its own specific enzyme and metabolic adaptation story to tell. In some, simply regulating the levels of standard enzyme machinery is adequate to resolve the developmental problems being faced. In other tissues, not only are

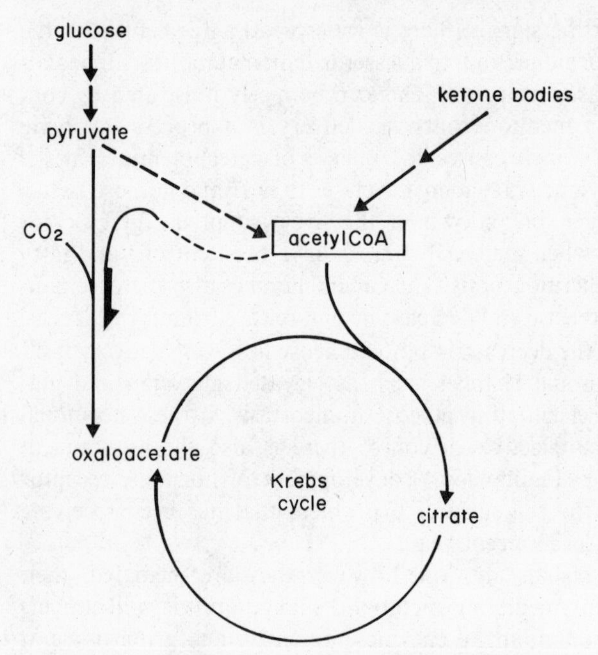

Fig. 8-8. Anapleurotic contribution of glucose to brain metabolism in the immature rat

actual catalytic potentials regulated at specific loci in metabolism, so are the kinds of isozymes retained. In all cases, however, timing is at the heart of developmental adaptations: timing of hormone receptor maturation, timing of hormone release, timing of enzyme induction, and ultimately even timing of enzyme action and substrate mobilization. As a result of closely controlled tuning and timing of its own metabolism, the developing offspring interacts smoothly and most effectively with its mother. Considered together, the mother-offspring pair thus constitute an impressively balanced, homeostatic system whose purpose is successful continuation of life.

CHAPTER NINE

Respiratory Proteins

Introduction

As discussions in several of the preceding chapters have shown, the transport of oxygen from the ambient environment to the respiring tissues is a critical feature in the metabolic architecture of organisms. Adequate oxygen supplies to the cells permit efficient aerobic processes to occur, while periods of oxygen deprivation necessitate reliance on less efficient pathways of ATP generation. To ensure a continuous supply of oxygen for the cells, selection has favored adaptations at all levels of biological organization. Gas exchange surface area may be adapted to the mode of oxygen uptake, blood volumes may be adjusted to maintain a satisfactory gas transport capacity, and certain fishes may undergo transitions from water-breathing to air-breathing as the oxygen content in the water changes. At the molecular level one also observes pervasive modifications in the oxygen transport system, and these adaptations involve a class of proteins, termed "respiratory proteins," which are typically the major vehicle for oxygen transport in animals. Although some oxygen may be transported as a dissolved species, free in solution, most animals utilize one or more types of respiratory proteins to carry the bulk of their oxygen. The adaptations of these respiratory proteins for function in different environments, in organisms having different metabolic rates, and in species that undergo major life-style transitions during their development serve as the focus of this chapter.

There are several reasons why a detailed study of respiratory proteins is an instructive exercise in our attempt to outline the major strategies of molecular adaptation. First, more is known about structure-function relationships for these proteins than for any other type of protein. As we shall discuss later, many of the functional differences found, for example, among vertebrate hemoglobin variants, can be interpreted precisely in terms of amino acid substitutions. A second reason why the study of respiratory proteins is an especially relevant function for this volume is that the classes of adaptations noted in these proteins clearly reflect what we have proposed to be the major types of biochemical adaptations. For instance, hemoglobin variants

Table 9-1. The Major Structural and Functional Characteristics of the Oxygen Transport Proteins

	Hemoglobins			
	Vertebrate	Invertebrate	Hemocyanins	Hemerythrins
Occurrence	Most species	Several phyla	Molluscs, arthropods	Sipunculids, brachiopods, priapulids, annelids
Structural properties				
O_2 binding site	1 heme: O_2	1 heme: O_2	2 Cu:O_2	2 Fe:O_2
Color-oxygenated state	Red	Red	Blue	Pink-purple
Color-deoxygenated state	Red-purple	Red-purple	Colorless	Colorless-yellow
Subunit molecular weight	17,500	12,000–400,000	70,000–75,000 (arthropods); 350,000 (molluscs)	12,800–14,400
Number of subunits	1, myoglobin 4, blood Hb	highly variable, up to 180	6–48, arthropods; 20, molluscs	3–8
Cellular (C)/Extracellular (EX)	C	C and EX	EX	C
Functional properties				
Cooperativity (Hill's n value)	1, myoglobin up to 3, Hb	up to 5	up to 6 (arthropods) up to 5 (molluscs)	1, myohemerythrins up to 1.5, vascular
Bohr effects (positive: + or negative: −)	O:myoglobin + or −:Hb	+	+ or −	+ or −
Modulators	Anions: organic and inorganic, e.g., 2,3-DPG	Na^+, Cl^- Mg^{2+}, Ca^{2+} ATP (in *Glycera*)	Inorganic ions, e.g., Ca^{++}, Cl^-; lactate	?

with different oxygen binding affinities and regulatory properties illustrate how qualitatively distinct protein variants function to adapt organisms to different environments. Changes in the concentrations of respiratory proteins show how important this type of macromolecular adaptational strategy is in many situations. Lastly, many of the adaptations in vertebrate hemoglobin systems in particular involve adjustments in the milieu in which the proteins function. A third reason for examining respiratory proteins in detail is that one gains an especially clear image of convergent evolutionary processes at the level of molecular function. Despite the variety of respiratory proteins in different animals (Table 9-1), certain fundamental aspects of function are seen, again and again, as one surveys the adaptations that are critical for oxygen transport.

The Basic Tasks of Respiratory Proteins: Analogies with Enzyme Function

To a very large degree, the adaptations noted in respiratory protein systems mirror those characteristics of enzymes adapted for function in different physical environments and for varied metabolic intensities. Both enzymic and respiratory proteins must satisfy a certain requirement for *capacity of function.* For enzymes this translates into an adequate rate of substrate conversion to product; for respiratory proteins this requirement means a satisfactory carrying capacity for oxygen. Both enzymic and respiratory proteins must have appropriate ligand binding affinities. Enzymes must be capable of binding substrates and cofactors at appropriate levels and respiratory proteins must be capable of efficiently extracting oxygen from the environment and, then, of releasing this oxygen where it is needed. For enzyme-ligand interactions, the apparent Michaelis constant (K_m) is used to describe binding relationships (K_m equals the concentration of ligand at which half of the maximal velocity of the reaction occurs); for respiratory proteins, an analogous parameter, the P_{50} value is used. The P_{50} value is the concentration of oxygen at which the respiratory protein is half-saturated (Fig. 9-1). Both enzymes and respiratory proteins must be sensitive to the needs for their function; that is, both classes of proteins must be sensitive to regulatory signals. In the case of enzymic function, regulatory signals lead to changes in catalytic rate; for respiratory proteins, regulatory signals trigger changes in oxygen affinity and, in some cases, oxygen carrying capacity. Finally, in common with many multisubunit regulatory enzymes, respiratory proteins exhibit cooperativity in ligand

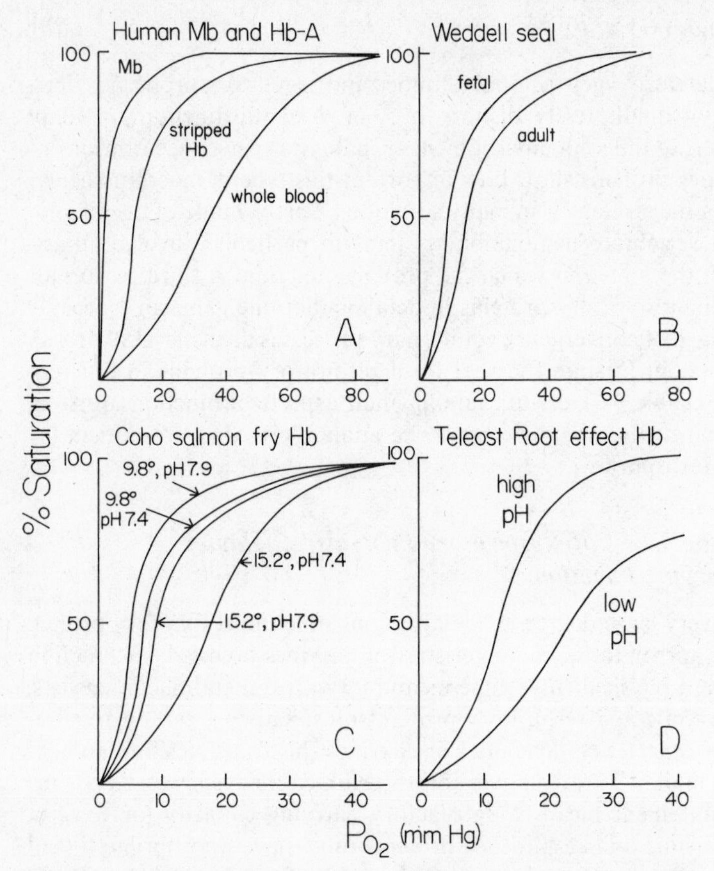

Fig. 9-1. Oxygen binding curves for different vertebrate hemoglobins (Hbs) as functions of Hb variant, modulator concentration, and temperature.

A. Human myoglobin (Mb) and Hb-A (whole blood and stripped [modulator-free]). Note the low P_{50} (oxygen concentration at 50 percent saturation) for Mb and the decrease in P_{50} which occurs when Hb-A is stripped of modulators (2,3-DPG is the principal modulator removed during dialysis).

B. Fetal-Hb and adult-Hb components of the Weddell seal (after Lenfant et al., 1969). Note the lower P_{50} value of the fetal-Hb.

C. Coho salmon fry Hb at different temperatures and pH values (after Giles and Randall, 1980). Note the increase in P_{50} as pH is decreased at constant temperature (Bohr effect) and as temperature is increased at constant pH (indicative of a positive enthalpy change during oxygenation, i.e., binding of O_2 is exothermic). Reprinted with permission from Giles and Randall (1980). Copyright 1980 Pergamon Press, Ltd.

D. The Root effect exhibited by a Hb from a teleost fish having a swimbladder which is filled via oxygen secretion. Note the decrease in oxygen carrying capacity as pH decreases (see text for details).

Note the differences in inherent oxygen affinities (as measured by P_{50} values) between the Hbs of air-breathing and water-breathing species, differences which reflect the relative concentrations of oxygen in these two media.

binding (Fig. 9-1). This cooperative response permits large-scale loading and unloading of oxygen over a small range of oxygen concentrations. A noncooperative respiratory protein lacks this ability to alter the fractional saturation state so markedly over a small oxygen concentration range (Fig. 9-1). The measure of cooperativity in oxygen binding is the "n" value, the Hill constant.* For vertebrate hemoglobins, n values range from 1, in the case of muscle hemoglobin (myoglobin, Mb), which is monomeric, to approximately 2–3, in the case of typical adult hemoglobin (Hb-A), which is tetrameric.

Correct capacity, proper ligand affinities and appropriate regulatory responses thus are common features of enzymic and respiratory protein systems. We turn now to an analysis of how these key properties are established and adaptively modified in diverse respiratory proteins to ensure adequate oxygen delivery under an extremely wide range of environmental and physiological conditions.

The Function and Regulation of Vertebrate Hemoglobins

The best understood respiratory proteins are the vertebrate hemoglobins. These diverse proteins (Table 9-1) have been examined functionally and structurally to such a extent that many of the most important binding and regulatory properties can now be given detailed interpretations in terms of the primary, secondary, tertiary, and quaternary levels of Hb structure. For the reader interested in the historical development of our knowledge about hemoglobin function and structure, the excellent historical essay by Edsall (1980) is highly recommended.

Vertebrate hemoglobins have subunit molecular weights near 17,000 M_r (Table 9-1). Myoglobin is monomeric while circulating (erythrocyte) hemoglobins are almost always tetrameric. Tetrameric structure and, probably, the presence of two types of subunits (termed α and β for Hb-A) are necessary structural features for cooperativity in oxygen binding. Like regulatory enzymes, tetrameric hemoglobins typically

* As discussed in Chapter 3, the index of cooperativity in ligand binding, the Hill constant, is now commonly denoted by "h." The number of binding sites of the system (enzyme or respiratory protein) is commonly designated by "n." In most of the literature dealing with oxygen binding to respiratory proteins, however, "n" is used to denote the Hill constant, and in this chapter we will retain this convention. We again stress, however, that the Hill constant, whether denoted as "h" or "n" is an empirical measure of binding cooperativity, and must not be assumed to equal the number of binding sites on the molecule.

display both homotropic and heterotropic interactions with ligands (see Chapter 3). Cooperative oxygen binding, a positive homotropic interaction, is characteristic of most tetrameric vertebrate hemoglobins. Binding of the first oxygen molecule promotes a change in the higher orders of Hb structure (Perutz, 1970) which leads to energetically more favorable binding of subsequent oxygens. This change in Hb structure is frequently termed a shift from a "tight" (T) to a "relaxed" (R) state.

A varied suite of heterotropic interactions is also characteristic of vertebrate Hb function (Table 9-1, Fig. 9-1). For most vertebrate Hb's, protons are an extremely important regulatory signal. Via the Bohr (Fig. 9-1C) and Root (Fig. 9-1D) effects, increases in proton activity (decreases in pH) lead to a drop in oxygen affinity and carrying capacity, respectively. Organophosphates such as 2,3-diphosphoglycerate (2,3-DPG) are also very important modulators of Hb function. Organophosphate modulators bind to tetrameric Hb in a one: one molar ratio, and greatly reduce oxygen affinity. For hemoglobins that are sensitive to modulation by protons and organophosphates, a synergism between the effects of these two modulators is typically observed. Since 2,3-DPG (or, in fishes, ATP and GTP and in birds and reptiles, inositol hexaphosphate—IHP) binding is stronger to the deoxygenated form of Hb, decreases in pH will favor organophosphate binding. Consequently, as arterial blood enters an actively respiring tissue, the decrease in blood pH due to CO_2 or metabolic acid entry from the tissue will facilitate oxygen unloading (Bohr effect), and the binding of organophosphate modulators. The latter interaction will act to keep the Hb in a deoxygenated state.

Proton Binding and Hb Modulation

The structural bases of the Bohr effect and 2,3-DPG interactions are now quite well understood. The binding of 2,3-DPG involves several amino acid residues (Arnone, 1972), including a cluster of positively charged residues near the center of the Hb tetramer, and the terminal amino groups of the β chain. Protonation of these terminal amino groups favors 2,3-DPG binding; hence the pH sensitivity of this regulatory interaction. The larger share of the stabilization free energy involved in 2,3-DPG binding involves the cluster of positively charged residues, however. Salt bridges between these residues and the negatively charged groups on 2,3-DPG stabilize this complex. The complementarity in geometry between 2,3-DPG and the positively charged groups is high in deoxygenated Hb, but low in oxygenated Hb due to shifts in Hb conformation during oxygen binding and release.

The Bohr effect (Fig. 9-1C) is also mediated by reversible changes in hemoglobin conformation and protonation state. The essential amino acid residues involved in the Bohr effect are the C-terminal histidines of the β chains and the N-terminal valines of the α chains, albeit additional histidyl residues in both the α and β chains also have been shown to contribute to the Bohr effect (Ohe and Kajita, 1980). The C- and N-terminal residues have been shown by X-ray crystallography to change their orientation as a result of oxygen binding, and this change in steric configuration is accompanied by a fall in the residues' pK values. Thus, *oxygen binding is associated with the release of protons from the Bohr groups (oxygenated Hb is a stronger acid than deoxygenated Hb) and, conversely, oxygen unloading is accompanied by the uptake of protons by the Bohr groups.* The precise mechanistic basis for these shifts in histidyl pK values has also been shown by X-ray crystallographic studies (Perutz, 1970; Ohe and Kajita, 1980). In the deoxygenated state, the Bohr effect histidyls are not freely accessible to solvent (water), but instead form salt bridges with other amino acid residues. These interactions stabilize the protonated form of the imidazole ring, i.e., they facilitate a high pK value. Upon oxygenation, the Bohr effect histidyls become exposed to water as the salt bridges rupture. The pK values of the imidazole groups decrease substantially. For example, the C-terminal histidyl of the β chain (residue number 146) undergoes a 1.1 pH unit shift of pK during oxygenation/deoxygenation events (Ohe and Kajita, 1980).

With these basic functional and structural features of vertebrate hemoglobins as a background, we now examine a series of respiratory protein systems in which the carrying capacity, oxygen affinity, and regulatory characteristics of the systems are adapted for oxygen transport under a wide spectrum of environmental and physiological conditions. We subdivide our treatment in such a way as to emphasize how adaptations in respiratory protein systems may include 1) the development of new protein variants, either on evolutionary or ontogenetic time-scales, 2) the alteration in protein concentration, and 3) the adjustment of modulator concentrations in concert with alterations in oxygen demands.

Adaptations of Vertebrate Hemoglobins: Hb Variants

The Multiple Hemoglobins of Salmonid and Catastomid Fishes. In many vertebrates, especially in ectothermic species, more than one hemoglobin variant is found in the blood. Analysis of the functional and structural differences among these multiple Hb components offers a particular clear perspective of the role of protein polymorphism in adaptation to different environments and physiological states. For at least some of the multiple hemoglobins of fishes, notably salmonid species like trout and catastomid fishes (suckers), there is compelling evidence for an important "division of labor" among the variant hemoglobins. This multiplicity of functional roles is especially well defined in the case of the four-membered Hb family of the trout, *Salmo irideus* (Brunori, 1975).

The Hb variants of *S. irideus* are termed "trout Hbs I-IV," based on their relative electrophoretic mobilities. We will examine trout Hb-I and trout Hb-IV which differ greatly in their functional properties and exemplify the "division of labor" likely to exist in this, and other, multiple Hb systems of fishes. The major differences between trout Hb-I and Hb-IV are given in Table 9-2. The basic functional distinction

Table 9-2. Properties of "Typical" and "Backup" Hemoglobins, including Trout Hb-I and Trout Hb-IV

Functional property	Trout Hb-I	Trout Hb-IV
Cooperativity in oxygen binding	Present	Present
Bohr effect	Absent	Present
ATP sensitivity	Absent	Present
Heat of oxygenation	Very low	Normal
Root effect	Absent	Present

Sequences of the C- and N-terminal Regions

Species	N-terminal region	C-terminal region
α-chain		
Human	Val-Leu-Ser-Pro-Ala-	-Ser-Lys-Tyr-Arg
Trout IV	Val (?)	-Lys-Tyr-Arg
Trout I	Acetyl-Ser-Leu-Ser-Ala-Lys	-Lys-Tyr-Arg
Catastomus	Acetyl-Ser-Leu-Ser-Asp-Lys	-Gln-Lys-Tyr-Arg
β-chain		
Human	Val-His-Leu-Thr-Pro-	-His-Lys-Tyr-His (146)
Trout I	Val (?)	-Ser-Arg-Tyr-Phe
Trout IV	Val-Asp-X-Thr-Asp-	-Arg-Gly-Tyr-His
Catastomus (anod)	Val-Glu-Try-Thr-Asp-	-Arg-Gly-Try-His
(cath)	Val-Glu-Try-Ser-Ser	-Phe

SOURCE: Brunori, 1975

between these two Hb variants is the absence of heterotropic ligand interactions for Hb-I and the occurrence of a broad family of such interactions for Hb-IV. Thus, Hb-I displays no Bohr effect, is insensitive to organophosphate compounds, has no Root effect and, in addition, oxygen binding is temperature-insensitive. Hb-IV, in contrast, is highly responsive to changes in pH and organophosphates and has a marked temperature dependence of oxygen association.

What are the likely functional roles of these two Hb variants, which co-occur in the same erythrocyte? Can we rationalize "why" this species would benefit from possessing a hemoglobin like Hb-I which appears to lack all of the important regulatory properties we earlier said were important for the control of oxygen loading and unloading? In the case of trout Hb-IV, the significance of its regulatory properties is readily apparent. The pronounced Bohr effect will permit effective oxygen unloading at the respiring tissues, where local increases in acidity will promote oxygen dissociation. The very strong Root effect displayed by Hb-IV (the hemoglobin is only partially saturated at pH values near 6.5 even in the presence of 1 atm oxygen; Brunori, 1975) may be particularly important in the transfer of oxygen into the swimbladder. The combination of reduced oxygen affinity and oxygen carrying capacity resulting from a fall in blood pH, as occurs in the swimbladder circulation, would facilitate a quantitative unloading of O_2 from Hb-IV and a concomitant flow of oxygen into the swimbladder. Thus, trout Hb-IV appears well adapted for function in a fast-swimming fish that is dependent on a gas-filled swimbladder for buoyancy regulation.

What of the functional role of Hb-I, which may constitute about 35 percent of the hemoglobin of *S. irideus*, along with Hb-II, which has similar functional properties (Brunori, 1975)? To a large extent Hb-I types of fish hemoglobins can be regarded as "backup" or "emergency" hemoglobins which allow continued oxygen supply under conditions where hemoglobins like Hb-IV begin to lose their transport capabilities. Such conditions are typified by states of intense locomotory activity. During bouts of vigorous swimming powered by anaerobic glycolysis in white muscle, the generation of large quantities of lactate will lead to sharp reductions in blood pH (see Chapter 4). At these times, loading of oxygen at the gills by a hemoglobin having a strong Bohr effect, like Hb-IV, may be largely precluded. Were it not for the presence of a pH-insensitive Hb like Hb-I, the fish might be in danger of asphyxiation. The presence of Hb-I types of hemoglobins thus allows continued gas transport under low pH conditions which, while likely to be encountered only during short periods of strong swimming, could prove lethal without this backup system. The inherent oxygen affinity of Hb-I is also higher than that of Hb-IV (Table 9-2), so a high degree of saturation of Hb-I is likely to occur in the gills. The strong co-operativity of oxygen binding/unloading noted for Hb-I (and Hb-IV) will ensure quantitative unloading at the respiring tissues.

The lack of an organophosphate (ATP or GTP) response by Hb-I also may have adaptive significance. As proposed by Brunori (1975), an insensitivity to ATP by trout Hb-I may have important consequences for oxygen unloading by the blood. Consider the following scenario which assumes that Hb-I has the ATP sensitivity of Hb-IV. As the blood enters a tissue where pH is low, Hb-IV will rapidly unload its oxygen due to Bohr and Root effects. The deoxygenated Hb-IV will then bind much of the ATP in the erythrocyte, a binding event that is facilitated by low pH and by the removal of oxygen from the Hb molecule. Were trout Hb-I modulated by ATP, the ATP binding by Hb-IV could effectively remove ("strip") ATP from Hb-I, with the result that oxygen affinity of Hb-I would greatly increase. This increase in oxygen affinity is, of course, maladaptive under these metabolic conditions that dictate a need for oxygen unloading. Thus, the insensitivity of trout Hb-I to ATP can be viewed as an important mechanism for ensuring continued oxygen unloading under low pH conditions that favor the tying-up of erythrocyte ATP by the Bohr- and Root-effect Hb IV.

Lastly, the absence of a temperature dependence for oxygen binding by Hb-I also can be viewed as adaptive. For an ectothermic species, the presence of a Hb variant with temperature-independent function could serve to buffer the effects of body temperature changes on the oxygen delivery system. Trout Hb-I is not unique in this respect. The hemoglobins of warm-bodied fishes like tunas and certain sharks appear to share this temperature-insensitivity of oxygen binding (Rossi-Fanelli and Antonini, 1960; Andersen et al., 1973). For the warm-bodied fishes, this temperature-independent oxygen binding may be important for ensuring that unloading of oxygen is kept under control of normal regulatory factors, e.g., protons and organophosphates, and is not adversely governed by temperature changes taking place as the blood flows from the gills (which are at ambient temperature) to the deeper tissues (which may be up to 15°C warmer than ambient seawater). Were the Hbs of warm-bodied fishes to possess the temperature sensitivities noted for most Hbs (e.g., for Hb-A, the P_{50} value exhibits a Q_{10} near 2), massive unloading of oxygen might occur as the cool blood warms upon entry to the deeper tissues, whereas the greatest needs for oxygen may be considerably downstream from this point. Loss of arterial oxygen to venous blood at the countercurrent heat exchanger might be an especially critical problem if the Hb of warm-bodied fishes had "typical" temperature sensitivities.

There also are other examples of fish hemoglobins that lack the pH and organophosphate sensitivities that are found in "typical" hemoglobins. Many fishes that engage in periods of highly active swimming possess Hb variants similar to trout Hb-I. For example, Powers (1972) found that *Catastomus clarkii*, which lives in fast-moving streams, has a hemoglobin component that lacks a Bohr effect. This Hb variant is not present in fishes of another subgenus (*Pantosteus*), which inhabit quiet pools. Thus, the evolution of "backup" or "emergency" hemoglobins in highly active fishes appears to be a good example of convergent evolution at the molecular level. This statement, in fact, is true at both the functional and structural levels.

Based on the earlier discussion of the amino acid residues involved in establishing the pH sensitivities of Hb function, it is possible to predict the types of amino acid substitutions which are likely to be required for the conversion of a pH-sensitive Hb to a variant lacking pH sensitivity. Of probable importance are the terminal histidyl and valyl residues. In trout Hb-I and *C. clarkii* "backup" hemoglobins, changes in these residues have been discovered (Table 9-2). The critical valyl residue is replaced by an acetylated serine residue. This change removes the terminal amino group, a proton binding group known to contribute to the Bohr effect and to organophosphate binding. Furthermore, in trout Hb-I, the C-terminal histidyl residue (146) of the β chain is replaced by a phenylalanyl residue. This substitution eliminates the Bohr effect potential of the C-terminal histidyl. In addition, a major source of the temperature dependence of oxygen binding is eliminated because the dissociation of a proton from an imidazole group occurs with an enthalpy change of approximately 6–7 kcal/mole. Although other amino acid substitutions may contribute to the altered functional properties of hemoglobins like trout Hb-I, these few amino acid substitutions at the termini of the α and β chains provide a marked alteration in Hb function with *less than a 1 percent change in amino acid sequence.*

Developmental Changes in Hemoglobins of the Coho Salmon (*Oncorhynchus kisutch*). Reliance on multiple Hb variants is characteristic not only of adult fishes that experience different constraints on oxygen delivery, but also of many organisms having different oxygen delivery problems during varied stages of their life cycles. Most readers will be familiar with so-called "fetal" hemoglobins, which are found not only in humans, but in other endothermic and ectothermic species for which the embryonic stage has a different access to oxygen from the adult stage. Typical fetal versus adult Hb properties are shown in Figure 9-1B. The fetal Hb generally has a higher oxygen affinity, enabling it to compete effectively for oxygen with the adult circulation. Much as myoglobin is capable of extracting oxygen from circulating Hb, the fetal-types of Hb variants use an enhanced oxygen binding ability to shift oxygen from the maternal blood to the fetal circulation.

In the coho salmon *O. kisutch*, developmental stage-specific Hbs share certain properties of the classical fetal-to-adult differences and reflect adaptations to other habitat differences experienced by the young (fry) and adult fish (Giles and Randall, 1980). Table 9-3 summarizes the major functional differences between the fry and adult Hbs of *O. kisutch*. Coho fry hemoglobins have an exceptionally high oxygen affinity, a property shared with many fetal-type Hbs. The fry Hbs also display a large heat of oxygenation and a strong Bohr effect. The adult Hbs, in contrast, have a low temperature- and pH-dependence and a higher P_{50} value. What are the functional advantages of these variants?

In the case of the adult-type Hbs, Giles and Randall (1980) argue that a low sensitivity to temperature variations and changes in pH would be adaptive for a fish like the adult coho salmon which encounters wide ranges of temperature during migrations and which must cope with large shifts in blood pH during intense bursts of upstream swimming. Thus, somewhat like trout Hb-I, the adult Hbs are capable of functioning over a wide range of physiological conditions. For the fry Hbs, the adaptive significance of a stronger Bohr effect, a higher inherent oxygen affinity and a greater temperature sensitivity of oxygen binding may all be reflections of the demands on the gas transport system arising from a higher weight-specific oxygen consumption rate. Fry consume severalfold more oxygen per gram of mass than adults, necessitating an elevated requirement for oxygen transport per unit of

Table 9-3. Functional Properties of Fry and Adult Hemoglobins of the Coho Salmon (*Oncorhynchus kisutch*)

Functional property	Fry Hb	Adult Hb
Bohr effect	Very strong	Present
P_{50}	Very low	Typical of adult salmonids
ATP modulation	Not examined	Very minor effect
Heat of oxygenation	Very high	Moderate

SOURCE: Giles and Randall, 1980

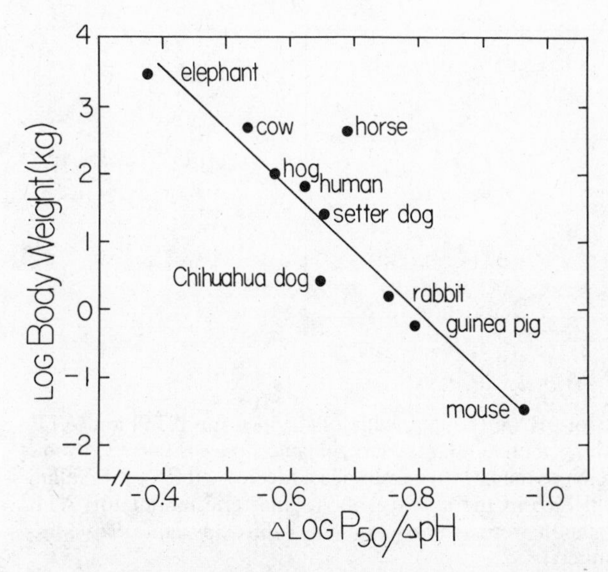

Fig. 9-2. The relationship between body size and degree of Bohr effect for Hbs from different mammals (after Riggs, 1960). (Reproduced, with modification, from *The Journal of General Physiology* (1960) 43: 737–752 by permission of The Rockefeller University Press.)

mass. Since the blood volume as percentage of body mass does not appear to differ in fry and adults, and since an increased rate of heart pumping can be viewed as a suboptimal means for enhancing oxygen supply (due to energy costs, for example), it is appropriate to examine the functional properties of the fry hemoglobins to see if they have assumed the role of elevating the oxygen delivery capacity of the system. Giles and Randall suggest that this is the case. The high oxygen affinity of the fry Hbs ensures complete Hb saturation at the gills. The strong Bohr effect facilitates quantitative unloading of oxygen at the respiring tissues. Comparisons of different-sized mammals have shown a strong negative correlation between the magnitude of the Bohr effect (change in P_{50} per change in pH) and body size (Riggs, 1960). Smaller organisms, with their vastly higher mass-specific metabolic rates typically have a much greater shift in P_{50} per unit change in pH, with a greater than two-fold difference noted among small and large mammals (Fig. 9-2).

Hemoglobins with Reversed Bohr Effects: The Hemoglobin of Amphiuma means. The Bohr effects discussed to this point are "positive" Bohr

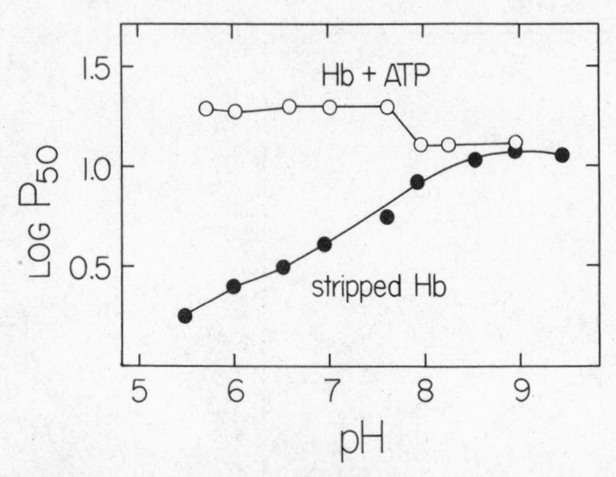

Fig. 9-3. The effects of pH on the P_{50} value of stripped (no ATP) and ATP-associated Hbs of the primitive salamander, *Amphiuma means* (after C. Bonaventura et al., 1977). Note the negative Bohr effect (decrease in P_{50} with falling pH) exhibited by this Hb in the absence of the allosteric modulator, ATP. (Reprinted by permission from *Nature* 265:474–476. Copyright 1977 Macmillan Journals Limited.)

effects: decreases in oxygen affinity as pH falls through the physiological pH range. At extremely low pH values, say below 6, many hemoglobins exhibit a reversed Bohr effect; oxygen affinity begins to increase as acidity rises. While these reversed Bohr effects may be without physiological significance in most cases, there are examples of hemoglobins (and hemocyanins; see below) where an increase in oxygen affinity at pH values near the lower limits of the physiological pH range may be highly adaptive. One such example is found in the primitive amphibian, *Amphiuma means* (C. Bonaventura et al., 1977). In the absence of organophosphate modulators, the single Hb species found in *Amphiuma* displays a marked reversed Bohr effect (Fig. 9-3). The putative significance of this departure from normal Hb relates to the habitat conditions faced by *Amphiuma* when water levels drop. At these times, *Amphiuma* may be forced to burrow into the mud and remain relatively torpid. Bonaventura et al. (1977) propose that, under these conditions, both pH and blood organophosphate concentrations are apt to be low. Anaerobic metabolism will lead to acidification of the animal and a reduced production of the Hb modulator, ATP, which is generated by oxidative phosphorylation in the amphibian RBC. If conditions of low pH and low erythrocyte [ATP] pertain, then the Hb of *Amphiuma* will attain an increased ability to bind oxygen, and this will be adaptive in view of the low oxygen concentrations in the mud burrow.

The amino acid substitutions leading to the reversed Bohr effect in this Hb are not fully understood. One major difference between *Amphiuma* Hb and positive Bohr effect Hbs is the absence of the C-terminal histidine (β-146) of the β chain in *Amphiuma* Hb. As discussed earlier, this residue plays a major role in the positive Bohr effect. Detailed kinetic studies by Bonaventura et al. (1977) showed that the Hb of *Amphiuma* differs markedly from most Hbs in terms of the pH effects of oxygen loading and unloading. Reductions in pH typically lead to a reduced rate of oxygen association and an increased rate of oxygen dissociation in positive Bohr effect Hbs, while in *Amphiuma* Hb, lowered pH enhances association and retards dissociation rates.

These few examples from among many (see Volume 20(1) of *The American Zoologist*, Johansen and Weber, 1976, and J. Bonaventura *et al.*, 1977, for additional cases of Hb adaptation) illustrate the wide spectrum of functional characteristics displayed by vertebrate hemoglobins which must function in environments of different oxygen tensions and under a variety of physiological states, e.g., quiescent and highly active states. There are, however, a great many other adaptations of Hb systems which do not involve novel Hb variants, but instead are effected by adjustments in Hb concentration and/or modification in the microenvironment within the erythrocyte. We now consider certain of these latter types of adaptations in oxygen transport systems, and these adaptations will be seen to be important during both long-term evolutionary adaptation and shorter-term acclimatory adjustments to changed oxygen tensions.

Hemoglobin System Adaptations via Modulator Effects

The Fundulus System: Enzymic Adaptations for Facilitating Hb Function. An especially well-studied case of Hb system adaptation via modulator (ATP) concentration changes is that of the killifish, *Fundulus heteroclitus* (Greaney and Powers, 1978; Powers et al., 1979). Adaptations to short-term hypoxia and genetically based differences in latitudinally separated populations of this species have been studied. In the case of acclimation to hypoxic conditions, the erythrocytes of *F. heteroclitus* show two distinct changes (Fig. 9-4). The total hematocrit rises, giving the fish an enhanced oxygen carrying capacity. In addition, the concentration of ATP in the erythrocytes decreases (Fig. 9-4). The latter change will lead to a reduction in P_{50}, an adaptive modification of the system under hypoxic conditions. These two acclimatory effects, which have been noted in other aquatic species acclimated to

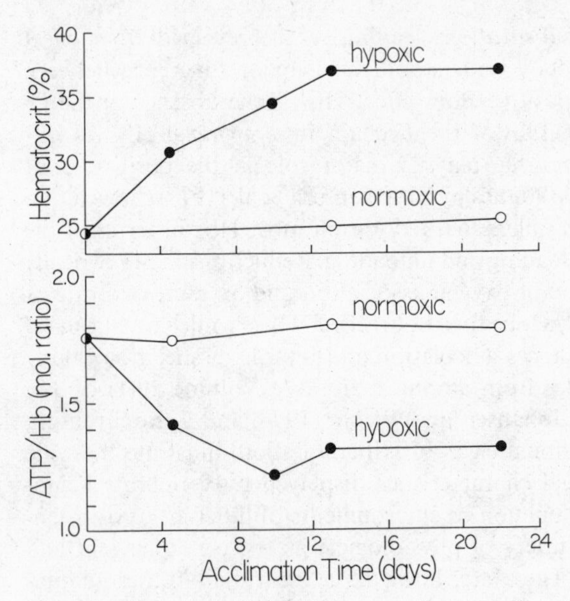

Fig. 9-4. Changes in blood parameters during acclimation of the killifish, *Fundulus heteroclitus,* to hypoxic conditions (after Greaney and Powers, 1978)
Upper panel: Time course of change in hematocrit (% blood volume)
Lower panel: Time course of change in the ratio of the allosteric modulator. (ATP) concentration to Hb concentration (expressed as a molar ratio)
Note that the amount of circulating Hb (as expressed by percent hematocrit) and the relative amounts of ATP per Hb molecule both change adaptively during hypoxic acclimation

low oxygen tensions (see Johansen and Weber, 1976, for review), show that adaptive changes in Hb systems can involve only concentration and modulator changes, and need not entail the synthesis of new Hb variants.

Further studies of *F. heteroclitus* have revealed an important genetic component in adapting populations from different latitudes, i.e., thermal conditions, to ambient oxygen concentrations. The protein variant of significance in this latitudinal adaptation is not Hb, however, but rather the LDH isozyme (LDH-B; H_4-LDH) found in the fish's erythrocytes (Powers et al., 1979). Only the heart or "B" type of LDH is found in the erythrocytes, but two electrophoretically and kinetically distinct variants of the enzyme are found (Place and Powers, 1979). The "a" allozyme occurs in highest frequency in southern (warm-adapted) populations of the species, while the "b" allozyme exhibits the opposite geographic cline. The significance of this cline in LDH gene frequen-

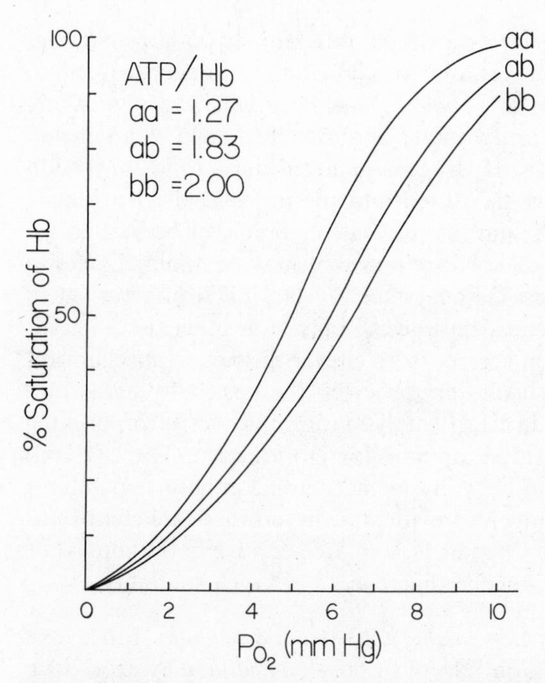

Fig. 9-5. Oxygen equilibrium curves of whole blood of *Fundulus heteroclitus* individuals possessing different lactate dehydrogenase (LDH) genotypes (after Powers et al., 1979). Among the three LDH genotypes, the "aa" genotype (found in highest frequency in lower latitude [= warmer] habitats) has a lower ratio of ATP:Hb and, consequently, a higher oxygen affinity (lower P_{50} value) than the other two genotypes ("ab" and "bb"). (Reprinted by permission from *Nature* 277:240–241. Copyright 1979 Macmillan Journals Limited).

cies appears to involve control of erythrocyte ATP concentrations (Fig. 9-5). Thus "bb" homozygotes contain significantly higher ATP concentrations in their erythrocytes than the "aa" individuals; "ab" heterozygotes are intermediate in [ATP]. Correlated with these differences in erythrocyte [ATP] are significant differences in the blood P_{50} values (Fig. 9-5). It is important to stress that these differences in P_{50} values are for whole blood, not "stripped" (modulator-free) Hb. All populations have the same Hb type, so only modulator effects can account for the differences in P_{50} in the different populations.

The exact causal link between LDH genotype and [ATP] is not known. However, the "bb" LDH has a significantly higher substrate turnover number than the "aa" variant, and this difference could lead to different rates of ATP synthesis in the erythrocytes having the different allozymes (Place and Powers, 1979), at least when the erythrocytes are incubated under identical temperatures. This hypothesis assumes that lactate oxidation makes a major contribution to energy metabolism in the erythrocytes of *Fundulus*. Lactate originating in working muscle could be taken up by the erythrocytes and used as an energy source for ATP synthesis. Hydrogens removed during the conversion of lactate to pyruvate could be fed into the mitochondrial oxidative phosphorylation system, and the pyruvate could either be burned via the tricarboxylic acid cycle (which, however, may be minimally active in *Fundulus* erythrocytes; Greaney and Powers, 1977) or returned to the blood (and subsequently returned to muscle or other tissues needing a source of hydrogen acceptor). Whatever the precise mechanisms of LDH function in *Fundulus* erythrocytes, the possibility exists that adaptive modifications in LDH catalytic rates may serve to poise the erythrocyte [ATP] at values optimal for Hb function. The LDH allozyme differences found between northern and southern populations of *Fundulus* thus are temperature-adaptive in terms of rate compensation to temperature (see Chapter 11) and are adaptive in the context of latitudinal, temperature-related differences in oxygen solubility.

Modulation of Hb Function in the Eel, Anguilla anguilla. Much like *F. heteroclitus*, the common European eel, *A. anguilla*, may experience periods of hypoxia in its habitat. Hypoxia acclimation studies by Wood and Johansen (1972, 1973, 1974) have demonstrated that *Anguilla*, like *Fundulus*, adjusts the oxygen transport functions of its blood via changes in hematocrit and organophosphate concentrations. An approximately 50 percent rise in oxygen carrying capacity was noted in hypoxia-acclimated eels, and the P_{50} value of the blood fell from 16.6 mm Hg in control fish to 10.6 mm Hg in hypoxia-acclimated specimens. The latter change was due in part to a decrease in the organophosphate:Hb ratio, an effect also observed in *Fundulus* (Figs. 9-4 and 9-5). In addition to the effects of changing organophosphate concentrations per se, Wood and Johansen stress the importance of concomitant alterations in erythrocyte hydrogen ion concentrations due to ATP and GTP changes. When the concentrations of ATP and GTP are reduced (recall that these molecules are nonpermeating of the plasma membrane), hydrogen ion concentrations must fall at the same time to maintain charge balance in the cell. This effect of changes in organophosphate concentrations on erythrocyte pH is purported to be of greater significance in raising oxygen affinity than the changes in [ATP] and [GTP] themselves (Johansen and Weber, 1976).

Postnatal Changes in Modulator Concentrations. The differences in oxygen availability to embryonic and adult life stages of an organism are

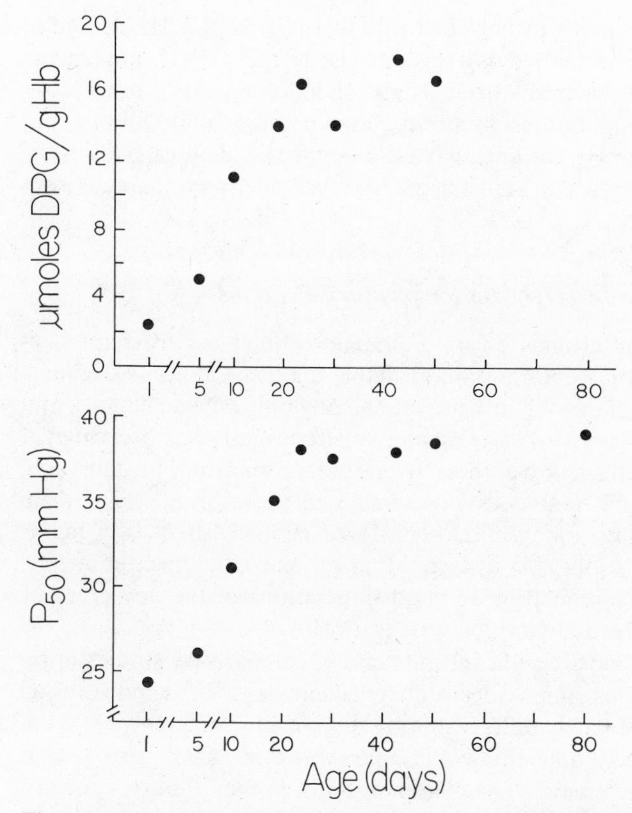

Fig. 9-6. The relationship between blood 2,3-diphosphoglycerate (2,3-DPG) concentration and oxygen binding ability (P_{50}) during postnatal development of the rat (after Dhindsa et al., 1981). Dark vertical lines to the right side of each frame indicate normal adult values for the parameter in question. (Reprinted by permission from Dhindsa et al. (1981). Copyright 1981, Pergamon Press, Ltd.)

often reflected in unique "fetal" and "adult" Hb variants (Fig. 9-1B), with the "fetal" Hb characteristically displaying the higher oxygen affinity. A similar difference in whole blood oxygenation behavior could, of course, be produced by life-stage-specific adjustments in modulator concentrations. This alternative strategy does, in fact, play an important role in certain species.

Figure 9-6 illustrates the changes in 2,3-DPG concentrations and blood P_{50} value during postnatal development of the rat (Dhindsa et al., 1981). The regular increase in [2,3-DPG] is reflected in a parallel rise in the P_{50} value of the blood. The concentration of 2,3-DPG rises from 2.53 μmol/g Hb in newborn rats to 13.93 μmol/g Hb in adults. The rise in P_{50} accompanying this increase in [2,3-DPG] ranges from 24.6 mm Hg in newborns to 38.7 mm Hg in adults. As in other cases of Hb system adaptations discussed above, changes in modulator concentrations or in Hb variants may yield essentially identical end-results in terms of appropriate adjustments to oxygen delivery characteristics of the system.

Adaptations in Invertebrate Respiratory Proteins

Our focus on tetrameric vertebrate hemoglobins in this chapter reflects the wealth of information available for these molecules relative to the data available for invertebrate hemoglobins, hemocyanins, and hemerythrins (Table 9-1). Fortunately, there recently has been a great acceleration in the study of these invertebrate respiratory proteins and, while many important questions remain to be resolved about them (see Bonaventura and Wood, 1980, for a concise list of the "major unresolved questions" about respiratory proteins), the invertebrate respiratory proteins do provide good illustrations of the key types of adaptations already found for vertebrate Hbs.

Perhaps the most significant outcome of comparative studies of respiratory proteins from widely different animals is the demonstration that proteins of vastly different molecular weights, different metal ion constituents, and different evolutionary histories have very similar adaptive characteristics. The conservation of proper affinity, capacity, and regulatory characteristics is found for all classes of respiratory proteins. For example, much as myoglobin and blood Hb differ in their relative oxygen affinities, the hemerythrins of certain invertebrates show body-compartment-specific variants with adaptively different oxygen affinities. Manwell (1960) found that the vascular hemerythrin of the sipunculid, *Themiste zostericolum*, had a lower oxygen affinity than the coelomic hemerythrin variant. Oxygenation of the blood occurs in the tentacles of the animal, and the oxygen so gained by the vascular hemerythrin will be withdrawn effectively from the blood by the coelomic hemerythrin.

Interspecific differences in oxygen affinity reflecting different habitat oxygen tensions are also characteristics of the invertebrate respiratory proteins. For example, the oxygen binding characteristics of polychaete vascular hemoglobins ("erythrocruorins") from species living in hypoxic and well-oxygenated environments differ in P_{50} values. The erythrocruorin of *Marphysa sanguinea* has a significantly lower P_{50} than that of *Diopatra cupres*, reflecting the fact that the former species lies in a relatively stagnant burrow, while the latter species inhabits a tube which is vigorously ventilated (Weber et al., 1978).

Another similarity among the different classes of respiratory proteins is the occurrence of both positive and reversed Bohr effects. Not all respiratory proteins of a given class exhibit a Bohr effect. However, regulation of oxygen affinity via changes in proton activity are very common in all classes of these proteins, and many examples of convergent evolution in function are known. For example, the hemocyanin of the horseshoe crab, *Limulus polyphemus*, displays a reversed Bohr effect similar to that found for the Hb of *Amphiuma* (Fig. 9-3; Bonaventura and Bonaventura, 1980). Both species may spend considerable periods of time burrowed in hypoxic muds, and both species appear to have "discovered" a common solution to the oxygen acquisition problems created by these environments.

Acclimation effects may also be similar in different classes of respiratory proteins. For example, Rutledge (1981) reported that temperature acclimation of the crayfish, *Pacifastacus leniusculus*, led to apparently adaptive changes in hemocyanin function. Warm (25°C)-acclimated specimens' hemocyanins had higher oxygen affinities than those from cold (10°C)-acclimated individuals, a change that would compensate for the reduced oxygen concentrations of the warmer water. The molecular basis of this acclimatory effect on P_{50} is not known; modulator-induced changes or altered hemocyanin variants are potential contributors to this type of adaptation.

One manner in which vertebrate Hbs generally appear to differ from invertebrate respiratory proteins is in the role of low molecular weight organic modulators of oxygen affinity. It remains unclear to what extent organophosphates like ATP, GTP, and 2,3-DPG influence oxygen binding to hemocyanin and hemerythrin. The latter two proteins often are strongly affected by inorganic ions, notably divalent ions like Ca^{2+}, which may alter oxygen affinity via changes in the subunit aggregation states of these proteins (Bonaventura et al., 1977b). Therefore, while tetramer-dimer equilibria may be important in some aspects of vertebrate Hb function, the reversible subunit polymerizations involving hemocyanins may be especially critical features in the regulation of gas transport by these proteins.

Summary: Lessons from the Study of Respiratory Proteins

The different mechanisms used to achieve satisfactory gas transport in highly varied environments provide outstandingly clear examples of the major adaptational strategies which serve as the themes of this volume. Many adaptations in respiratory protein systems are known to involve adjustments in the primary structures of these proteins, and, suffice it to say, there are few proteins for which such clear illustrations of structure-function relationships have been obtained. We noted that relatively minor changes in amino acid sequence are sufficient to alter radically the functional properties of respiratory proteins, e.g., vertebrate hemoglobins. Those amino acid residues that determine such traits as Bohr effects and organophosphate sensitivities represent but a minor fraction of the total residues of a respiratory protein. Thus, while large numbers of amino acid substitutions may be selectively "neutral" in terms of respiratory protein function, other residues, notably those involved in establishing regulatory properties, reflect intense selective pressures. The fact that only a few amino acid residues need to be changed to bring about large adjustments in regulatory function implies that a high degree of plasticity is likely to exist in respiratory protein adaptation and evolution. That is, adaptive change in respiratory protein systems may be a relatively "simple" matter of adjusting one to several amino acid residues. Perutz et al. (1981) stress this point in their discussion of the interesting allosteric properties of crocodilian hemoglobin. This type of Hb lacks sensitivity to organophosphates and carbamino CO_2^-, but is strongly modulated by bicarbonate ion (HCO_3^-). Perutz et al. show that these changes can be accounted for by only five amino acid substitutions which, in turn, require only seven nucleotide base changes. They further point out that most of the approximately 100 amino acid substitutions that distinguish crocodilian from human hemoglobin are conservative and likely to be without effect on oxygen binding and its regulation. Structure-function analyses at this high resolution level thus may contribute importantly to the debate over the relative amounts of "neutral" and "selectively important" amino acid substitutions in proteins. And these types of analyses provide at least a qualitative appreciation for the amount of protein "remodeling" that is needed to alter adaptively the molecule's functional properties.

Studies of vertebrate hemoglobin systems also shed light on the vital roles played by adjustments in the composition of the milieu in which macromolecules function. Adjustments in the concentrations of organophosphate modulators are found to be instrumental in adapting Hb function in many different circumstances where ambient oxygen availability varies. Although the metabolic schemes responsible for controlling the concentrations of the diverse organophosphate modulators differ among species and, in no one case, are well understood in fine detail, there is strong reason to believe that the evolution of respiratory proteins, on the one hand, and organophosphate synthesizing pathways, on the other hand, is a joint process. For example, the relationship discussed earlier between LDH phenotype and [ATP] in the erythrocytes of *Fundulus heteroclitus* suggests that enzyme polymorphisms may play critical roles in the fine-tuning of respiratory protein function. An important heuristic lesson here is that, in attempting to interpret the roles of enzyme polymorphisms, the investigator must maintain a wide field of vision. The enzyme polymorphisms may not be adaptive strictly in terms of overall cellular metabolism, e.g., in terms of temperature compensation, but rather may be part of a hierarchical system in which changes in several metabolic reactions are directed toward conservation of a distinct function such as gas transport. Again, referring to the debate between "neutralists" and "selectionists," the possible roles of polymorphisms must be discussed in a wider context than has usually been the case in the past.

What lessons about the adaptation and evolution of "proteins in general" can be derived from the studies of respiratory proteins? Are the amounts of "neutral" and "selectively important" variations present in one class of protein representative of those levels of variation in other proteins? Are there unique features of respiratory protein function that necessitate high levels of polymorphism? We can suggest some tentative answers to these questions via an analysis of one of the major distinctions between enzymic proteins and respiratory proteins.

At many places in this volume we have emphasized the high degree of conservation in intracellular substrate concentrations among tissues, cell types, and species. For example, concentrations of pyruvate in vertebrate skeletal muscle are conserved within an approximately threefold range, under quiescent conditions, in fishes, amphibians, reptiles, and mammals (see Table 11-5). Paired with this high conservation of substrate concentrations is a marked interspecific similarity in apparent Michaelis constant values. These two conservative trends are closely related, of course, and they reflect the importance of maintaining subsaturating concentrations of substrates for proper metabolic regulation.

In the situations in which adaptive modifications of respiratory proteins are most apparent, a different relationship between "substrate" (oxygen) concentration and protein function is found. Oxygen concentration varies widely among habitats and, in certain cases, body compartments. As in the case of K_m-substrate concentration relationships for enzymes, oxygen transport proteins must possess the correct binding properties to facilitate adequate loading and unloading of oxygen. However, in view of the wide differences in available oxygen concentration, a single affinity value at physiological temperatures is obviously not possible for the respiratory proteins of different species, or, in many cases, for respiratory proteins of different life stages of a single species. Instead, respiratory protein variants with adaptively different P_{50} values must be developed. Fetal and adult Hbs are a good example of this type of adaptation, as are the hemerythrin variants of the sipunculid, *Themiste zostericolum* (Manwell, 1960). Thus, to the extent that the available oxygen concentration is variable, the need for multiple forms of respiratory proteins may be substantial. This type of selective factor leading to multiple protein forms must be distinguished from selection for regulatory variants. For both enzymic and respiratory proteins, the requirements for sensitivities to allosteric modulators are a separate feature of protein evolution, distinct from those adaptations that reflect a proper relationship between an affinity parameter and substrate (or oxygen) concentration.

The discovery that critical residues of hemoglobins are the major loci of adaptive differences in regulatory properties, whereas many other residues appear to be subject to conservative substitutions without impact on oxygen transport, suggests certain conclusions for the evolution of "proteins in general." As is now clear for the many proteins that have been sequenced from two or more different species, certain regions of enzymes are highly conserved, while others vary widely. Active site residues (catalytic and regulatory sites) are usually highly conserved among species, and when variations in active site residues are found, functional changes are typically noted as well. Where additional study is especially needed is of the regions of multimeric proteins where subunit interactions take place. Neither for respiratory proteins nor enzymic proteins are there many comparative data to allow a clear image of subunit contact site adaptations, e.g., in proteins that function at different temperatures and pressures. In the chapters dealing with adaptations to these two physical parameters, we note that muscle actin polymerization reflects the physical conditions encountered by the species (see Chapters 11 and 12). These differences in polymerization energetics could derive from differences in subunit contact sites and/or differences in subunit structural stability due to residues spread throughout much of the protein. It is appropriate to ask whether respiratory proteins adapted to different physical environments have comparable differences in polymerization energetics. If they do, are the amino acid residues responsible for these differences localized strictly at subunit contact sites or, alternatively, are they found in many regions of the proteins? This question may, like other questions raised above, help us answer questions concerning the balance between "neutral" and "selectively important" amino acid substitutions. We believe it is possible, even likely, that certain amino acid substitutions for which no functional significance is known may, in fact, be of importance in establishing subtle, yet adaptively important, differences in subunit-subunit interaction properties.

In conclusion, the further detailing of structure-function relationships in respiratory proteins is very apt to lead to unprecedented insights into molecular design. The amino acid substitutions underlying differences in regulatory properties, oxygen carrying capacities, and subunit interactions provide currently unequaled insights into interplay between protein structure and protein function. These insights, in turn, may allow resolution of many of the major outstanding questions in adaptational biochemistry, notably those dealing with the relationship between protein polymorphism and adaptive change.

CHAPTER TEN

Water-Solute Adaptations: The Evolution and Regulation of Biological Solutions

Introduction

When considering the compositions of the extra- and intracellular fluids of organisms, and the mechanisms used by organisms to regulate these compositions, one fact is of paramount importance: in no instance is the chemical composition of a biological solution identical to the composition of the surrounding medium. This fact has several extremely important implications which serve as the focus of much of the discussion to follow. First, the state of disequilibrium which exists between biological solutions and the surrounding medium tells us that considerable energy expenditure may be necessary to prevent physiologically disadvantageous changes in the compositions of the extra- and intracellular fluids. Many organisms tend to gain or lose water, and all organisms must continuously cope with the necessity of maintaining solute concentration gradients, e.g., in K^+ and Na^+, between themselves and the external medium.

The disparity which exists between the compositions of biological solutions and the compositions of fresh water and seawater, tells us another extremely important point about the nature of biological solutions: organisms do not accumulate solutes in direct proportion to the availability of these solutes in the external environment. Rather, as we illustrate in detail throughout this chapter, solutes differ radically in their suitability—their "adaptiveness" as it were—for use in biological systems. We will show that the qualitative and quantitative aspects of biological solution composition reflect the evolutionary selection of solute species and solute concentrations which are able to provide a solution microenvironment in which the structural and functional properties of macromolecules such as enzymes approach optimal values. Surprisingly little attention has been given to this level of molecular evolution—solute system evolution. Since the publication of L. J. Henderson's classic treatise, "The Fitness of the Environment," in which the "adaptiveness" of compounds like water and certain chemical elements for biological roles is lucidly treated, most emphasis in evolutionary biology has been centered on higher-level phenomena.

The treatment of biological solution evolution which follows is an attempt to redress the lack of attention given to the design of the solutions in which most biological functions take place.

To understand the basic principles involved in the design of biological solutions, especially the intracellular fluids, it is appropriate to build our discussion around three questions:

1. What types of solutes and what concentrations of these solutes are found in the extra- and intracellular fluids?
2. What are the criteria that determine whether a particular type of solute, or combination of solutes, is suitable for use in a biological solution? What determines the potential "fitness" of a solute?
3. What are the principal regulatory mechanisms that are utilized in the selective accumulation and retention of biologically "fit" solutes?

Our approaches to these three questions will differ significantly from treatments customarily found in comparative physiology textbooks, as we will pay most attention to question #2, and give relatively little space to the other questions. Detailed treatments of chemical compositions and regulatory mechanisms can be found in recent books edited by Gilles (1979) and Maloiy (1979). By focusing in particular on question #2, we hope to give the reader a new perspective on biochemical design and a deeper understanding of the critically important interactions that exist among macromolecules, water, and low molecular weight solutes. Through appreciating these interactions, which likely are universal in living systems, the reader may also come to appreciate the possible merits of a conjecture raised at several points in our discussion, namely that through proper adjustments in the qualitative and quantitative compositions of biological solutions, much of the task of adaptation is shifted from macromolecules to low molecular weight solutes. This effect, in turn, will be argued to have potentially major influences on evolutionary rates, e.g., in organisms colonizing an environment with grossly different osmotic concentration.

Basic Osmotic Strategies

At the outset of this discussion of biological solutions, their evolution and their regulation, it is important to realize that the extra- and intracellular fluids may differ from the ambient solution in quantitative and qualitative ways. Organisms having the same total concentration of solute species as the external environment are in

Table 10-1. Intracellular Osmolyte Concentrations of Diverse Organisms, Illustrating the Major Classes of Solute Accumulation Patterns

Organisms	Osmotic concentration (mmole/kg water)		Inorganic ions (mmole/kg water)		Organic osmolytes (mmole/kg water)					
	Cell	Environment	$[Na^+]$	$[K^+]$	Amino acids	Betaine	TMAO	Sarcosine	Urea	Polyols (molal)
Osmotic conformers										
Plants										
Unicellular algae										
Dunaliella viridis		4.25 *M* NaCl								4.4
		2.55 *M* NaCl								2.4
Multicellular species										
Triglochin maritima		0.2 *M* NaCl			350					
		Fresh water			40					
Hordeum vulgari						*				
Animals (muscle tissue)										
Invertebrates										
Balanus nubilus (barnacle)	1005	Seawater	45	169	503	82				
Eriocheir sinensis (crab)	588	Fresh water	55	71	158	18	47			
	1118	Seawater	144	146	341	14	75			

Parastichopus sp.									
(echinoderm)	1246	Seawater	71	217	221	208			
Sepia officinalis									
(mollusc)	1377	Seawater	31	189	483	108	86		
Vertebrates									
Myxine glutinosa									
(hagfish)		Seawater	96	140	291		87		2
Latimeria (coelacanth)		Seawater	30	90			290		422
Squalus acanthias									
(dogfish)		Seawater	18	130		100	180		333
Dasyatis americana (ray)		Seawater	7	201			255		604
		50% seawater	30	171			186		377
Raja erinacea (ray)		Seawater	10	162	214		64	62	398
		50% seawater	4	134	144		36	23	264
Eubacteria									
Klebsiella aerogenes		1 *M* NaCl		625					
Archaebacteria									
Halobacterium salinarium		3.4 *M* NaCl	400–800	4500					
		5.1 *M* NaCl	400–800	7500					
Osmotic regulator									
Animal: teleost fish									
Pleuronectes flesus		Fresh water	10	157	44		20		
		Seawater	15	158	71		30		

SOURCE: Yancey et al. (1982); consult this paper for references to the original literature

* Exact cytosol concentration are not known due to contribution of vacuole to total cell volume

osmotic equilibrium with the environment, i.e., they are *isosmotic*. Organisms which are dilute relative to their environment are termed *hypoosmotic*; organisms which are concentrated relative to their environment are *hyperosmotic*. Table 10-1 lists the osmotic concentration (and qualitative composition) of the intracellular fluids of selected procaryotes, green plants, and animals. Wide variation in total osmotic content is apparent. Even among marine animals, widely different approaches to osmotic regulation appear: marine invertebrates are isosmotic, marine cartilaginous fishes (sharks, skates, and rays) are either isosmotic or slightly hyperosmotic, and marine teleost fishes are hypoosmotic. Clearly, there is no one optimal solution to the osmotic regulation problem. There are, however, important similarities among these marine animals—indeed, among all animals and most plants and bacteria—in the qualitative composition of the cellular fluids, and the variation in total osmotic content must not blind us to this important fact, which we return to below.

Qualitative Compositions of Biological Solutions: Basic Considerations

To develop a broad, synthetic overview of the design principles involved in the evolution of biological solutions, we must first consider the varied array of osmotic agents (osmolytes) which are accumulated within the body fluids of organisms. The variety of osmolytes found in animals, plants, and bacteria may at first seem bewildering to the reader and a reflection of a selection process in which each organism does things in its own particular way, oblivious to the "design principles" deduced by the physiologist and biochemist. There are some relatively obvious unifying threads in the fabrics of biological solutions, however, and we shall consider these first, prior to discussing the more complex designs noted, especially, in the case of organic osmolytes (amino acids, urea, methylamines, and carbohydrates).

The common base to all biological solutions is, of course, water. L. J. Henderson (1913) and, more recently, Wald (1964), have treated in detail the "fitness" of water for use as a biological solvent (high heat capacity, ability to dissolve a wide spectrum of compounds, low density of ice, pK value, and so forth). As we argue later, one of the important criteria involved in biological solution design is the influence of solutes on water and macromolecular-water interactions. Thus, while we will tend to channel water into the background of our discussion for a time, it will remain the unifying entity in our treatment.

A second property of biological solutions which appears to be ubiquitous is a pH value close to the neutrality point of water (pN) and the pK of histidine imidazole groups. The importance of this particular pH value cannot be overestimated. We return to a detailed discussion of the need for such a pH value later in this chapter and again in the chapter on temperature effects. For the moment it suffices to emphasize that a very strong conservation of particular pH values is found among all living forms studied to date.

A third characteristic of biological solutions which is common, but not universal, is the maintenance of a high ratio of $[K^+]:[Na^+]$ in the *intra*cellular fluids (Table 10-1). In multicellular animals, for example, the intracellular fluids and blood differ markedly in K^+ and Na^+ concentrations. Even in the most osmotically concentrated organisms known, the halophilic bacteria (*Halobacterium halobium*, etc.), intracellular $[K^+]$ is kept much higher than intracellular $[Na^+]$ and extracellular (medium) $[K^+]$. We will consider the benefits of this relationship later.

A fourth characteristic of biological solutions which is relatively common is related to the phenomenon just discussed. As inspection of Table 10-1 reveals, among animal species at least, a relatively close conservation in total intracellular inorganic ion concentration is found among species which differ markedly in total osmotic concentration. We can phrase this conclusion in a different way: among organisms differing widely in total osmotic concentration, the greater fraction of the difference in solute concentration is attributable to organic, not inorganic, solutes. This generalization leads us to consider the nature of these organic solutes and, especially, the chemical and physical factors that establish the "fitness" of organic osmolytes for use at high concentrations in the intracellular fluids.

An Overview of Strategies of Organic Osmolyte Accumulation

A priori, one might list a number of factors that could serve as important criteria determining the suitability of different organic molecules for use at high concentrations in osmotic balance. First, ready (and cheap) availability could be important. A solute which is a waste product might be an economical choice for use as an osmolyte. Urea, present at concentrations of up to 400–500 mM in marine cartilaginous fishes and the coelacanth, *Latimeria chalumnae*, would seem to fall into this category. However, free amino acids and glycerol, two commonly used organic osmolytes, definitely are metabolically expensive and otherwise useful (e.g., in protein synthesis and energy metabolism, respectively). Thus, more than a single criterion must be involved in the design of solute systems. In fact, the failure of organisms to adapt osmotically using only the readily available inorganic ions in their environment argues strongly that ease of accumulation in terms of metabolic energy expenditure is not a critical factor influencing the compositions of biological solutions.

A second factor that could contribute to the organic osmolyte composition patterns noted in widely different organisms is the efficiency, e.g., speed, with which adjustments in osmotic content of intracellular fluids can be made. It is reasonable to propose that the synthesis and degradation of certain organic molecules can be turned on and off quickly, thereby allowing an organism to reach osmotic equilibrium rapidly, before large-scale hydration or dehydration of the cellular fluids can occur. Indeed, we will see clear examples of the speed with which osmotic balance can be obtained via adjustments in the activities of enzymes controlling biosynthesis of, for example, glycerol and isofloridoside.

The selection of particular organic molecules for use as osmotic agents is not based primarily on the efficiency with which their concentrations can be regulated, however. We argue that regulatory efficiency and metabolic cheapness are no more than secondary factors influencing the design of biological solutions. At the very basis of solution design is a principle that is of overriding importance and, not surprisingly, is reflected in the solute compositions of bacteria, plants, and animals. *This design principle concerns the establishment of a solute microenvironment for macromolecules in which their structural and functional properties are optimized for catalysis, metabolic regulation, information transfer, and mechanical work, and in which a balance is achieved between stability and instability of structure.* The selection of particular organic osmolytes is made primarily on the basis of this microenvironmental design principle, and the ease and speed of osmolyte accumulation noted in many organisms was elaborated after the initial commitment to employ these particular compounds was made.

We will consider the diverse family of organic osmolytes in terms of how they affect—or fail to affect—macromolecular structure and function, and this anaylsis will reveal that two different approaches are used to establish a hospitable microenvironment for macromolecules. On the one hand, cells may accumulate organic osmolytes having virtually no effect on macromolecular function. These solutes, which contribute significantly to osmotic balance, but which are essentially "invisible" to macromolecules as far as perturbation of their functions is concerned, will be termed compatible solutes, using the terminology suggested by Brown and Simpson (1972). The *compatible solute strategy* of osmotic regulation thus consists of accumulating, at appropriate concentrations to reach osmotic balance with external fluids (blood or ambient water), compounds such as certain free amino acids and glycerol, which do not alter the abilities of macromolecules such as enzymes to conduct their functions. The chemical bases of solute compatibility will be examined, and we find that only a select group of organic molecules measure up to the standards dictated by compatibility.

A second approach to establishing a hospitable microenvironment for macromolecules is characteristic of marine cartilaginous fishes and the coelacanth (Table 10-1). In these organisms the intracellular fluids contain solutes which individually have strongly perturbing effects on macromolecular function and structure, but which cancel each other's effects at the concentration ratios found in the cells. This strategy is termed the *counteracting solute strategy*. It involves principally the counteraction of urea effects, which generally destabilize structure and inhibit function, by a family of methylamine solutes (Fig. 10-1) which stabilize protein structure and activate enzymes. In the context of counteracting solute mechanisms we will also investigate systems in which only urea is accumulated, with the not unexpected result of metabolic inhibition.

The Compatible Solute Strategy

The use of organic osmolytes which have no, or only negligible, effects on macromolecular function has several important advantages. First, large changes in the concentrations of compatible solutes during osmoregulation will not perturb metabolism. That is, enzymic activity will persist in the face of both high and varying levels of compatible solutes. A second major advantage of compatible solutes concerns the relative roles played by low molecular weight osmolytes and macromolecules in evolutionary adaptation to different salinities. Consider an evolutionary scenario in which an organism migrates to, and colonizes, an environment of much higher salinity than it previously was exposed to. If this hypothetical, salt-stressed organism merely opened the gates to ion influx and accumulated high concentrations of, for example, K^+ or Na^+ (plus Cl^- as counterion to maintain electroneutrality), it is likely that enzyme function would be seriously perturbed

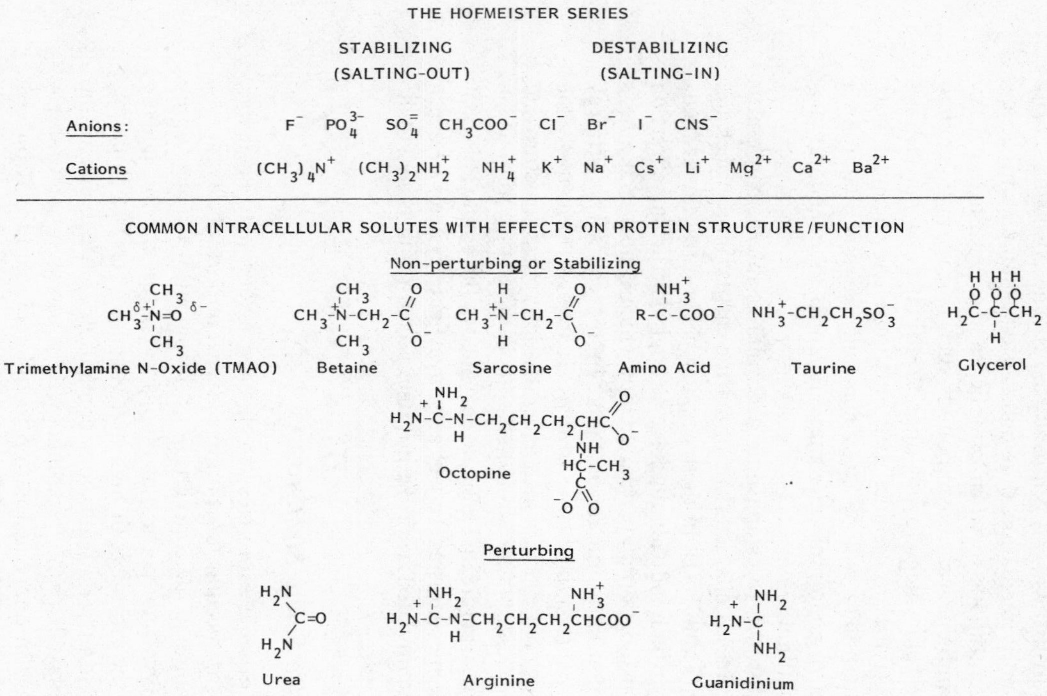

THE HOFMEISTER SERIES

	STABILIZING (SALTING-OUT)						DESTABILIZING (SALTING-IN)			
Anions:	F^-	PO_4^{3-}	$SO_4^{=}$	CH_3COO^-	Cl^-	Br^-	I^-	CNS^-		
Cations	$(CH_3)_4N^+$	$(CH_3)_2NH_2^+$	NH_4^+	K^+	Na^+	Cs^+	Li^+	Mg^{2+}	Ca^{2+}	Ba^{2+}

COMMON INTRACELLULAR SOLUTES WITH EFFECTS ON PROTEIN STRUCTURE/FUNCTION

Non-perturbing or Stabilizing

$(CH_3)_3N^{\delta+}{=}O^{\delta-}$ Trimethylamine N-Oxide (TMAO)

$(CH_3)_3N^+{-}CH_2{-}COO^-$ Betaine

$CH_3{-}N^+H_2{-}CH_2{-}COO^-$ Sarcosine

$R{-}CH(NH_3^+){-}COO^-$ Amino Acid

$NH_3^+{-}CH_2CH_2SO_3^-$ Taurine

$H_2C(OH){-}CH(OH){-}CH_2OH$ Glycerol

$H_2N^+{=}C(NH_2){-}NH{-}CH_2CH_2CH_2CH(COO^-){-}NH{-}CH(CH_3){-}COO^-$ Octopine

Perturbing

$(H_2N)_2C{=}O$ Urea

$H_2N^+{=}C(NH_2){-}NH{-}CH_2CH_2CH_2CH(NH_3^+)COO^-$ Arginine

$H_2N^+{=}C(NH_2)_2$ Guanidinium

Fig. 10-1. The Hofmeister (lyotropic) series of neutral salts, arranged in order of their tendencies to stabilize or destabilize macromolecular structure (the order from left to right is in the order of increasing destabilizing effect). The expressions "salting-in" and "salting-out" refer to the varied capacities of these salts to either increase or decrease, respectively, the solubilities of macromolecular structural components, e.g., amino acid residues. See von Hippel and Schleich (1969)

Commonly occurring organic solutes are shown in the lower half of the figure. Note the similarities in structure between the structure stabilizing methylamine solutes (TMAO, betaine, and sarcosine) and the highly stabilizing methylammonium ions of the Hofmeister series

in view of the sensitivities of many enzymic processes to salt concentration (see below). In this scenario, then, the accumulation of high internal salt concentrations would likely dictate the need for salt-tolerant enzymes. That is, by choosing a perturbing (noncompatible) solute, substantial changes in protein amino acid composition to offset salt perturbation may be needed. In contrast, if this hypothetical organism were to accumulate a compatible organic solute to reach osmotic balance, the development, over many generations, of salt-tolerant proteins would not be required. The relative rates of these two evolutionary courses (rebuilding many, perhaps most, proteins versus developing regulatory mechanisms for controlling the biosynthesis of compatible solutes) cannot be given with any rigor. However, we are inclined to view the latter evolutionary process, which would involve modifications of relatively few macromolecules, as likely to be far more rapid.

With these general advantages of the compatible solute strategy in mind, let us examine specific cases where organic osmolytes play dominant roles in the intracellular compartment. It is to be noted that, in multicellular animals, organic osmolytes may play relatively little role in osmotic balance of the extracellular fluids (Table 10-1). This marked difference between the intra- and extracellular fluids may be a reflection of the different metabolic complexities of the two body compartments. That is, virtually all metabolic transformations, e.g., of energy metabolism and biosynthesis, occur within the intracellular space, so the requirements for a hospitable microenvironment for macromolecules may be exceedingly high in this space.

The solute effects on enzymic activity illustrated in Figures 10-2 to 10-4 show several important facets of the design of compatible solute systems. Note that the readily available (and metabolically inexpensive) inorganic salts, NaCl and KCl, are highly perturbing of enzyme function. Thus, as argued above, an organism that were to osmoregulate using only NaCl or KCl would probably need to find ways of offsetting salt perturbation of enzyme function. Over evolutionary time courses this could entail large numbers of amino acid substitutions. Short-term, e.g., tidal cycle, changes in salinity might require the presence of a battery of isozyme forms for each salt-sensitive enzyme, wherein each single isozyme form possessed a different salt optimum; using compatible solutes, organisms avoid this salt perturbation problem.

The data given in Figures 10-2 to 10-4 indicate that some, but not all, commonly occurring organic compounds are compatible solutes. Among the free amino acids found in the intracellular fluids (Table 10-2), some are without significant effect on enzyme function, while

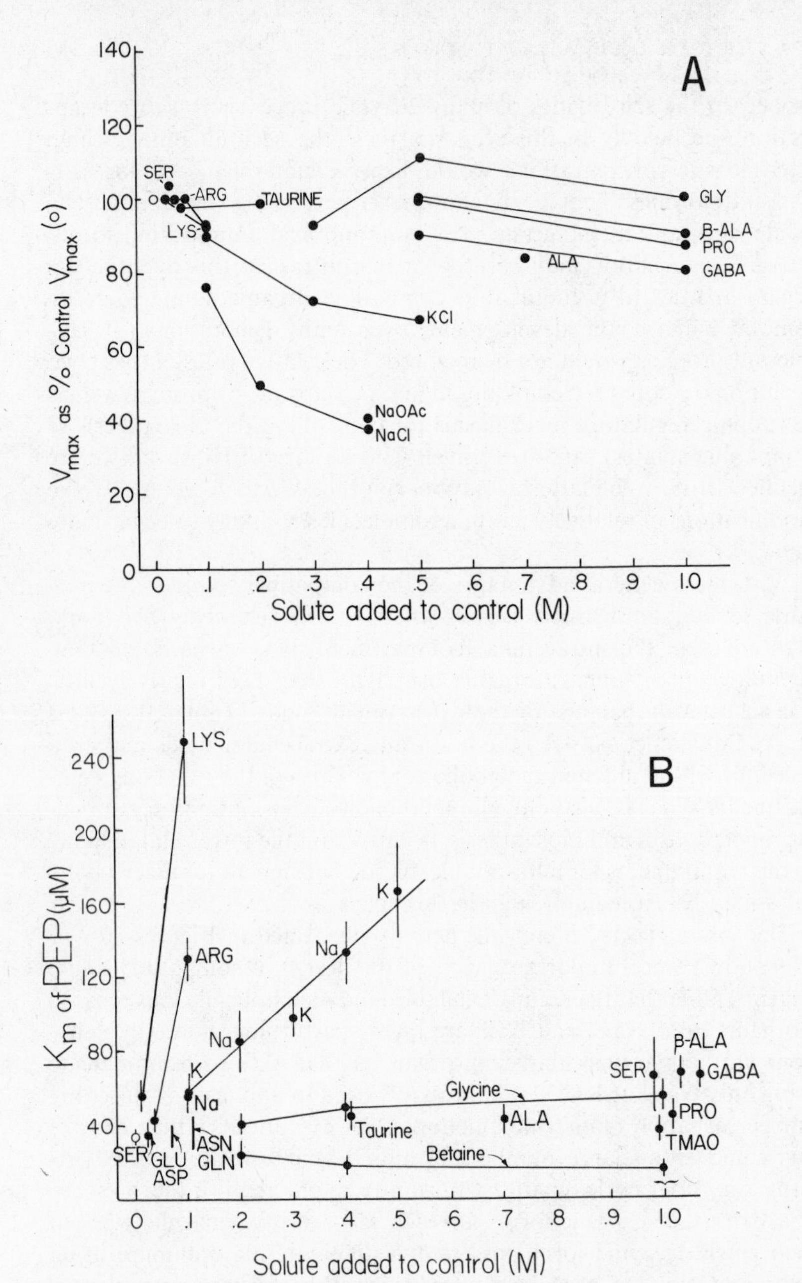

Fig. 10-2. A. The effects of different solutes on the maximal velocity (V_{max}) of the *Pachygrapsus crassipes* pyruvate kinase reaction. B. The effects of different solutes on the apparent Michaelis constant (K_m) of phosphoenolpyruvate (PEP) of the *P. crassipes* pyruvate kinase reaction. The vertical lines for each point represent 95% confidence intervals around the K_m values. The control (no additional solutes) K_m is indicated by the open symbol. Modified after Bowlus and Somero (1979)

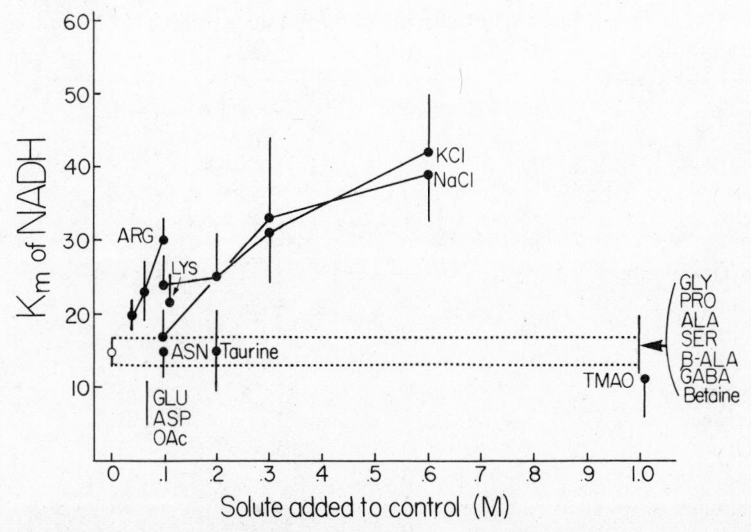

Fig. 10-3. The effects of different solutes on the apparent Michaelis constant (K_m) of NADH of tuna M_4-lactate dehydrogenase. The vertical lines for each point represent 95% confidence intervals around the K_m values. The control (no additional solutes) K_m is indicated by the open symbol. From Bowlus and Somero (1979)

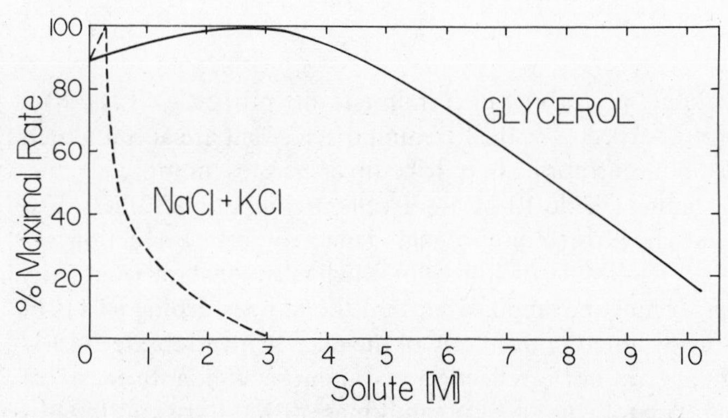

Fig. 10-4. The effects of salt (an equimolar mixture of NaCl and KCl) and glycerol on the activity of glucose-6-phosphate dehydrogenase from *Dunaliella tertiolecta* and *D. viridis* (enzymes of both species exhibited the same responses). Data from Borowitzka and Brown (1974)

Table 10-2. Free Amino Acid Contents in Body Fluids of Selected Marine Invertebrates

	Homarus vulgaris		*Eriocheir sinensis*	
Amino acid	muscle	serum	SW muscle	FW muscle
Alanine	23	1	72	18
Arginine	61	0	55	37
Aspartate (total)[a]	2	1	12	4
Cystine	—	—	—	—
Glutamate (total)[a]	34	0	28	10
Glycine	202	3	109	57
Histidine	1	0	—	—
Isoleucine	—	—	3	1
Leucine	6	0	5	2
Lysine	3	0	—	—
Methionine	1	0	—	—
Phenylalanine	1	0	—	—
Proline	104	1	24	5
Serine	—	—	6	3
Taurine	—	—	28	21
Threonine	1	0	15	4
Tyrosine	0	0	—	—
Valine	3	0	7	0

SOURCE: Data from the review of Schoffeniels (1976)
NOTE: Concentrations are expressed as mmol/kg wet wt. or mmol/L, rounded off to nearest mmol.
[a] "total" = aspartate + asparagine and glutamate + glutamine

others are strongly perturbing of certain enzyme properties (Figs. 10-2 and 10-3). Not unexpectedly, the free amino acids that are accumulated to the highest concentrations in marine invertebrates, notably glycine, proline and alanine (Table 10-2), are excellent compatible solutes. For the enzymes studied, these amino acids failed to perturb reaction velocities (V_{max}) (Fig. 10-2A) or apparent Michaelis constant (K_m) (Fig. 10-2B) values. It must be emphasized that the nonperturbing effects of these amino acids noted in the study of the marine invertebrate *Pachygrapsus crassipes* are not a reflection of "amino-acid-adaptiveness" of this particular homologue of pyruvate kinase (PK). Pyruvate kinases from a variety of marine and freshwater invertebrates and vertebrate species displayed similar insensitivities to glycine, proline, and alanine as well as similar sensitivities to noncompatible amino acids. The generality of these and other solute effects on proteins is further illustrated by the data given in Figure 10-3 from a study of tuna muscle M_4-lactate dehydrogenase. For this enzyme, as well as for PK, glycine, proline, alanine, and other amino acids contributing importantly to the free amino acid pools of marine invertebrates are without perturbing effects on enzyme function, but both NaCl and KCl are strongly perturbing. We argue, therefore, that the compatible amino acids do not owe their non-effects on enzyme function to adaptations of the enzymes per se, but rather that the nonperturbing traits of these osmolytes reflect general properties of protein-solute-water interactions common to all biological solutions (see discussion later in this chapter). These ubiquitous interactions are such that a solute which is compatible with one type of protein is very likely to be compatible with most, or all, other types of proteins. This is clearly an important attribute for a compatible solute.

Among the twenty or so free amino acids found in the intracellular fluids there also exist noncompatible species. Arginine (Fig. 10-1) and lysine are strongly inhibitory of the PK and LDH reactions (sharply increase K_m values). Whereas lysine concentrations are generally very low in cells, most marine invertebrates employ arginine-phosphate (Fig. 10-1) as the major muscle phosphagen (a role played by creatine-phosphate in other invertebrates and all vertebrates). It thus is appropriate to consider how invertebrate species using arginine-phosphate are able to cope with rising free arginine levels during periods of muscular activity when the arginine-phosphate stores are depleted, and free arginine concentrations could rise to levels of up to 20-40 mM (Table 10-2). In some instances, rising arginine concentrations, like falling pH, may serve as a signal to the energy metabolism reactions of the muscle to reduce their activity to prevent further acidification. In other instances, however, there may be mechanisms for further processing of arginine to compounds that are nonperturbing of metabolism. That is, arginine may be further metabolized to a compatible solute. One such compatible solute is the molecule octopine (Fig. 10-1), which is formed in a condensation reaction involving arginine and pyruvate. The octopine dehydrogenase (ODH) reaction takes the place of the LDH reaction in a variety of invertebrates, including bivalve and cephalopod molluscs, sipunculids and sea anemones (albeit not all members of each group have ODH). The reaction: pyruvate + arginine + NADH + $H^+ \rightleftharpoons$ octopine + NAD^+, leads to the regeneration of oxidized nucleotide, thereby facilitating a continuation of anaerobic glycolysis, and the conversion of arginine to a compound which has been shown to fit the criteria of solute compatibility (Bowlus and Somero, 1979). Thus, octopine is found to be nonperturbing of K_m values of the PK and LDH reactions. Also, octopine had no destabilizing effects on protein structural stability, unlike arginine (see below).

The use of selected free amino acids for osmoregulation is not restricted to the marine invertebrates, but is also characteristic of certain halotolerant (but nonhalophilic) bacteria and plants exposed to severe water stress (Table 10-1). Proline accumulation plays an especially critical role in these other halotolerant species as well as in marine invertebrates. In some bacteria, glutamate is the major amino acid osmolyte (Measures, 1975), and this amino acid has been found to be relatively non-perturbing of animal enzymes (Figs. 10-2 and 10-3). One difference between glutamate, on the one hand, and proline, glycine, and alanine, on the other hand, is that glutamate is negatively charged at physiological pH values. Thus, for each glutamate accumulated, one counterion (generally K^+) must be added to the cell as well. Measures (1975) argues that the development by Gram-positive bacteria of mechanisms to convert glutamate into proline and γ-aminobutyric acid (GABA) (note the compatibility of the latter solute shown in Figs. 10-2 and 10-3) can be viewed as an adaptation to minimize increases in monovalent cation accumulation during periods when free amino acid concentrations are rising.

The widespread use of selected free amino acids as osmolytes must not be taken to indicate that these are necessarily the only compatible solutes. In a variety of unicellular plants and fungi, glycerol serves as the major osmolyte, reaching concentrations of 4.4 molal in the unicellular green alga *Dunaliella* (Borowitzka and Brown, 1974; Brown and Borowitzka, 1979) and 2.7 M in the yeast, *Debaryomyces hansenii* (Gustafsson and Norkrans, 1976) (Table 10-1). Even at these high concentrations, glycerol is found to have little inhibiting effect on the enzymes which have been studied (Fig. 10-4). For glucose-6-phosphate dehydrogenase from *Dunaliella tertiolecta* and *D. viridis*, enzymic activity is virtually unaffected by increases in glycerol concentration up to almost 5 M. In contrast, equimolar mixtures of KCl and NaCl strongly inhibit the enzyme (after an initial activation at low ionic strength) (Fig. 10-4). An additional attribute of glycerol which makes it an excellent compound for rapid adjustments in intracellular osmolarity is the potential for storage of glycerol in an osmotically inactive form, a point we return to below when we review some of the regulatory mechanisms involved in controlling the concentrations of organic osmolytes.

A variety of other organic compounds have been found to play important osmotic roles in both green and blue-green algae (Table 10-1). In almost all cases, however, the effects of these compounds on enzymic activity have not yet been studied. Based on the effects—or non-effects—of osmolytes like proline, glycine, and glycerol, it is reasonable to expect that compounds such as isofloridoside, cyclohexanetetrol, and the recently discovered osmolyte 2-0-α-D-glucopyranosylglycerol of the blue-green alga, *Synechococcus* sp. (Borowitzka et al., 1980), will also display solute compatibility.

Regulation of Compatible Solute Concentrations

Considering the heterogeneous assemblage of organic osmolytes found in bacteria, plants, and animals, it is not surprising to find that a variety of metabolic routes are traversed to govern their rates and levels of synthesis. We will review briefly several of the best understood regulatory mechanisms to illustrate the diversity of control processes that are involved in attaining a common end result—the accumulation of an adequate concentration of organic osmolytes to restore osmotic equilibrium between the cell and the surrounding fluids.

1. *The glutamate dehydrogenase (GDH) reaction and amino acid synthesis.* In those species, including bacteria and marine invertebrates, which accumulate high concentrations of specific amino acids during adjustment to increased ambient salinity, a pivotal role has been suggested for the GDH reaction. In the GDH reaction:

$$\alpha\text{-ketoglutarate} + NH_4^+ \rightleftharpoons \text{glutamate} \quad (NAD(P)H \rightarrow NAD(P)^+)$$

the net addition of amino groups to α-keto acids takes place, and these glutamate-associated amino groups can then be transferred via transaminase reactions to other α-keto acids. For example, pyruvate may accept an amino group from glutamate to form alanine, a transamination reaction which draws on a glycolytic intermediate for production of a common organic osmolyte (see Chapter 2).

A key feature of the GDH reaction, noted in both animals and bacteria, is its pronounced stimulation by KCl and NaCl. For example, GDHs of several bacterial species were stimulated roughly twofold when K^+ concentration was increased from 250 to 500 mM (Measures, 1975). Salt activation is also reported for GDHs from a variety of marine invertebrates (see Schoffeniels, 1976). The significance of this salt activation is the generation of amino groups for transfer to α-keto acids such as pyruvate and oxaloacetate and the production of glutamate per se, since this amino acid is a precursor of proline, one of the most important amino acid osmolytes (Tables 10-1 and 10-2).

The contribution of free amino acids synthesized via transaminase reactions to the total free amino acid pool is not presently understood for any species, and the GDH plus transaminase control scheme outlined above must be viewed as but one component of the total free amino acid regulatory mechanism. Compelling evidence that other routes of free amino acid production contribute importantly to osmotic regulation includes the observations of Greenwalt and Bishop (1980) who demonstrated that specific inhibitors of transaminase enzymes were not able to prevent many of the increases in amino acid concentrations that accompanied hyperosmotic stress of isolated hearts of the mussel *Modiolus demissus*. Whereas the transaminase inhibitors aminooxyacetic acid and L-cycloserine did significantly block the buildup of proline and alanine, increases in other amino acids were not affected. These authors conclude that significant contributions to the free amino acid pool in hyperosmotically regulating animals are made by (a) protein catabolism and (b) arginine catabolism. Schoffeniels (1976) and co-workers have also conjectured that protein catabolism may play an important role in generating free amino acids during hyperosmotic regulation, albeit the fact that the amino acids selectively accumulated do not reflect the average amino acid compositions of proteins (see the section on halophilic bacterial proteins later in his chapter for a discussion of "typical" and "atypical" amino acid compositions), argues against the idea that general protein catabolism is the major source of free amino acids in those organisms employing these agents as the principal intracellular osmolytes.

2. *Regulation of glycerol concentrations in Dunaliella.* The concept of an osmotically inactive storage form of low molecular weight organic osmolytes is especially attractive in the case of those osmolytes that can be generated via a few metabolic transformations from a high molecular weight storage compound such as starch. Glycerol is one such osmolyte, and its storage in starch polymers represents one instance of maintaining an accessible source of osmolyte in a osmotically inactive state. (Recall that osmotic properties reflect the number of particles in solution—a colligative effect, in other words—so that one large starch polymer represents a potential for a vast number of low molecular weight osmolytes while confronting the cell with only a minimal demand on the cell's water supply.) In their review of glycerol regulation in *Dunaliella*, Brown and Borowitzka (1979) emphasize that the large and rapid (complete adaptation in sixty to ninety minutes) increases in glycerol concentration in hyperosmotically regulating *Dunaliella* occur equally well in the light or in the dark—provided that the cells contain adequate reserves of starch. The manner in which increases in ambient salinity trigger activation of glycerol synthesis is not presently known (albeit it is tempting to conjecture that the mech-

anism found in control of isofloridoside accumulation might also operate in *Dunaliella*; see below). The removal of glycerol from *Dunaliella* undergoing hypoosmotic regulation, while also not well established, involves metabolic utilization of glycerol, not extrusion of this energy-rich compound to the external medium. Perhaps the interconversion between starch and glycerol is reversible, and when high internal glycerol concentrations are no longer needed, glycerol moieties are returned to starch polymers. Whatever the precise regulatory events which are involved in glycerol metabolism in *Dunaliella*, these would appear to involve the activities of pre-existing enzymes, since inhibitors of protein synthesis do not block the glycerol-based osmoregulation.

3. *Potential control sites in isofloridoside accumulation.* Direct salt modulation of pre-existing enzymes has been discussed already in the case of the GDH reactions of bacteria and marine invertebrates. A regulatory system which combines such direct salt activation of enzymes and the recruitment of osmotically active monomers from a reserve polymer is found in the alga *Poterioochromonas malhamensis* (Kauss et al., 1979). In this species, a time lag of only one to three minutes after salinity increase was found before isofloridoside accumulation began (Kauss, 1979, for review). Approximately 70 to 80 percent of the osmotic adjustment during hyperosmotic regulation was due to this one organic compound. The rapidity of isofloridoside synthesis and the high concentrations of this osmolyte that were generated rapidly suggest that a storage form of this osmolyte exists in this alga and, furthermore, that the enzymes involved in producing isofloridoside are pre-existing in the cells and respond to rising salt concentrations. The model proposed by Kauss and colleagues involves both of these factors. A storage polymer termed chrysolaminarin may be the major reserve form of isofloridoside; studies using ^{14}C-labeled isofloridoside showed that, during hypoosmotic stress, this osmolyte was not released to the medium, but instead was incorporated into chrysolaminarin. Further regulation of isofloridoside concentrations involves an interesting cascade effect: increasing salt concentration is proposed to activate one or more proteolytic enzymes which, in turn, activate the enzyme isofloridoside-phosphate synthase via a covalent (proteolytic) modification step. (Recall that Ca^{++} activation of glycogen breakdown in muscle also involves an ion signal calling for covalent modification, phosphorylation in that particular instance.) Once activated, isofloridoside-phosphate synthase contributes to the production of high titres of isofloridoside. Through these types of enzymic transformations, a signal (rising osmolarity) is transduced into the catalytic activities necessary to generate large increases in the major osmolyte,

a regulatory mechanism that is at once highly sensitive and extremely rapid in response.

The Counteracting Solute Strategy

The independent discovery of a variety of compatible organic osmolytes during the evolutionary histories of diverse bacteria, plants, and marine animals would seem to suggest that, wherever organic solutes contribute predominantly to osmotic balance, these solutes will lack perturbing effects on macromolecules. It may seem surprising, therefore, that marine cartilaginous fishes and the coelacanth have adopted an osmotic strategy that is fundamentally different from the compatible solute strategy (Yancey and Somero, 1979, 1980). The key feature of the counteracting solute strategy found in cartilaginous fishes and the coelacanth is, as its name indicates, a balancing of opposing influences. In general, the effects of the perturbing solute, urea, are counteracted by the stabilizing or stimulating methylamine solutes, TMAO, sarcosine, and betaine (Table 10-1, Fig. 10-1). Compatibility of this complex suite of nitrogenous osmolytes is a property only of the whole solute system, not of the individual solutes themselves.

Before reviewing some of the counteracting effects of urea and the methylamines, certain important features of the elasmobranch-type osmotic system merit our attention. First, this type of solute system contains concentrations of urea that are high enough to perturb significantly enzyme function and structure. This fact suggests that these urea-rich organisms must either find ways of offsetting urea's perturbing effects via other solute influences or must accumulate special urea-adapted proteins. Whereas both types of adaptation are noted, as discussed below, protein modifications may be of limited extent. A second important property of elasmobranch-type solute systems is the characteristic 2:1 concentration ratio of [urea]:summed [methylamines] found in these animals. This approximate ratio is found among species and among individuals of a single species exposed to different salinities and containing different absolute amounts of the nitrogenous osmolytes. Not surprisingly, then, at this concentration ratio of urea: methylamines, the counteracting influences of these solutes are near their maximum.

Illustrations of urea-methylamine counteractions on a variety of enzymic reactions are found in Figure 10-5. For reactions catalyzed by creatine kinase (CPK) and pyruvate kinase (PK), urea is found to increase the apparent K_m of ADP (Fig. 10-5). All of the methylamines are found to decrease K_m of ADP. The key finding, however, is that when both urea and the methylamines are present at a 2:1 ratio of

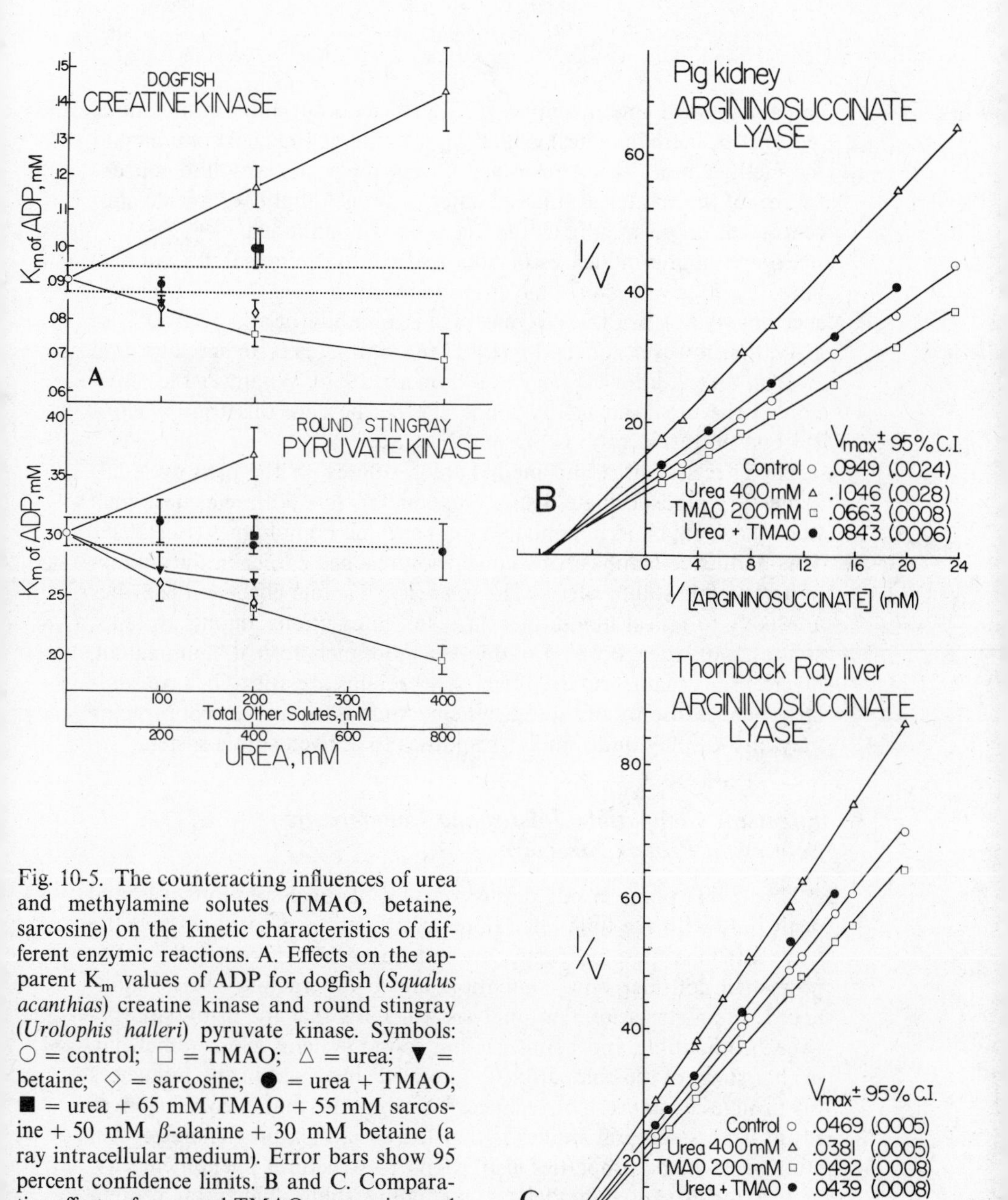

Fig. 10-5. The counteracting influences of urea and methylamine solutes (TMAO, betaine, sarcosine) on the kinetic characteristics of different enzymic reactions. A. Effects on the apparent K_m values of ADP for dogfish (*Squalus acanthias*) creatine kinase and round stingray (*Urolophis halleri*) pyruvate kinase. Symbols: ○ = control; □ = TMAO; △ = urea; ▼ = betaine; ◇ = sarcosine; ● = urea + TMAO; ■ = urea + 65 mM TMAO + 55 mM sarcosine + 50 mM β-alanine + 30 mM betaine (a ray intracellular medium). Error bars show 95 percent confidence limits. B and C. Comparative effects of urea and TMAO on porcine and thornback ray (*Platyrhinoidis triseriata*) arginosuccinate lyase V_{max} values. Figures modified after Yancey and Somero (1980)

[urea]:summed [methylamines], K_m values do not differ from control (no urea or methylamines) values. The counteracting effect of urea and the methylamines is not restricted to any particular absolute concentration of the solutes. Rather, counteraction is found over a wide concentration range so long as the 2:1 ratio is maintained.

Counteracting influences of urea and methylamine solutes are also noted on V_{max} values of some enzymic reactions (Fig. 10-5—argininosuccinate lyase). For this enzyme, urea is an inhibitor and TMAO is an activator; however, for the M_4-LDH reaction, urea is an activator and TMAO an inhibitor (Yancey and Somero, 1980). Whatever the direction of the urea and methylamine effects, they are offsetting when a 2:1 concentration ratio is present.

The effects of urea and methylamine solutes on the urea cycle enzyme, argininosuccinate lyase, are similar for both elasmobranch (thornback ray) and mammalian (pig) enzyme homologues (Fig. 10-5). This finding re-emphasizes a conclusion reached earlier in the context of compatible solute effects. The majority of solute effects on enzymes are likely to reflect ubiquitous classes of interactions among proteins, solutes and water. Because of this, elasmobranch, teleost, mammalian, invertebrate, plant, and even bacterial proteins are apt to behave similarly in response to any solute system. And, different types of proteins may also display quite similar responses to a given solute system.

Effects of Compatible Solutes and Counteracting Solutes on Protein Structure

Up to this point in our discussion of biological solutions we have dealt only with the functional (kinetic) properties of enzymes and the ways in which these properties are influenced by different solutes. We now shall consider how compatible and counteracting solutes influence protein structure. Our analysis will show that the distinction between compatible and counteracting solute systems blurs somewhat, in that some of the compatible amino acids have stabilizing influences on protein structure. On balance, however, the major conclusion to result from structural studies is that the design features manifested in studies of kinetic properties also extend to structural phenomena.

A frequently used model for examining solute effects on protein structural stability is the heat inactivation of ribonuclease (RNase) (von Hippel and Schleich, 1969). Solutes which stabilize protein structure characteristically increase the denaturation temperature (T_m) of RNase, while structure-destabilizing solutes lower T_m. Figure 10-6 compiles most of the available information on organic solute effects on

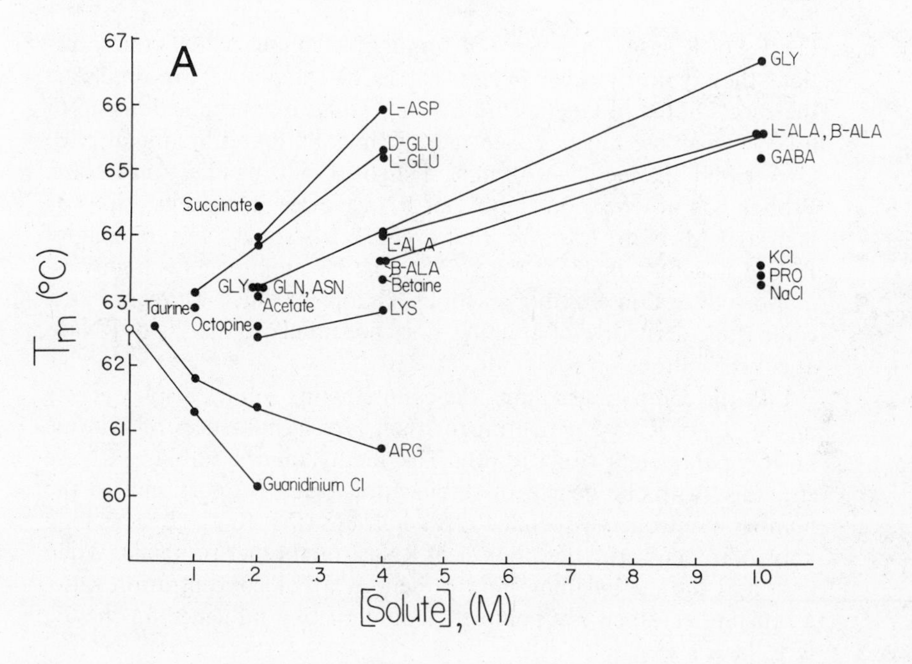

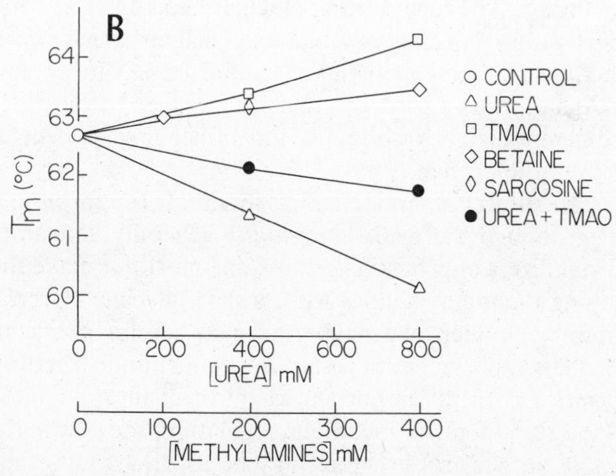

Fig. 10-6. The effects of different solutes on the thermal transition temperature ("melting temperature" = T_m) of bovine pancreatic ribonuclease. A. The effects of neutral salts, free amino acids, and various free amino acid-derived metabolites, e.g., octopine, on T_m. Modified after Bowlus and Somero (1979). B. The counteracting effects of urea and methylamine solutes on the T_m of ribonuclease. Modified after Yancey and Somero (1979)

the T_m of RNase. The effects of compatible solutes at the concentrations they typically occur *in situ* (Tables 10-1 and 10-2) are similar to the effects noted in kinetic studies—nil. Thus, at concentrations up to 0.4 M (which are higher conentrations than are found in animal cells) glycine and alanine have, at most, a slight stabilizing effect on RNase. Proline has no effect on the T_m of RNase even at concentrations as high as 1 M. Note, however, that arginine is strongly destabilizing of RNase structure, much as this noncompatible amino acid is highly inhibitory of certain enzymic reactions. Octopine, which we regard as a compatible derivative of arginine, does not influence the T_m of RNase at concentrations up to 0.2 M.

Like the compatible solutes, the counteracting solutes display effects on the T_m of RNase which mirror their typical influences on enzyme kinetic parameters (Fig. 10-6B). The methylamines stabilize RNase structure, with the degree of stabilization being proportional to the number of nitrogen-associated methyl groups (Fig. 10-1). Urea, as long known, is a powerful destabilizer of RNase (and other proteins). When urea and the methylamines are present at a 2:1 concentration ratio, a substantial, albeit not complete, counteracting influence on the T_m of RNase is seen.

The counteracting influences of urea and the methylamine solutes on RNase structural stability are not the only example of this type of structural balancing act. The renaturation of denatured LDHs and the rate of sulfhydryl group labeling of glutamate dehydrogenase also reflect counteracting urea and methylamine influences (Yancey and Somero, 1979). Again, these varied solute effects appear to be widespread among different classes of proteins and rather independent of the species source of the protein.

Before leaving the subject of counteracting solutes, it is appropriate to stress that even though the methylamines are generally activators of enzymes and stabilizers of protein structure, one must not make the mistake of grouping the methylamines with proline, glycine, glycerol, and other compatible solutes. We have argued at several points in this volume that the establishment of certain K_m values and structural features of enzymes represents important adaptive changes in these proteins. From the standpoint of metabolic regulation, too low a K_m value or too stable an enzyme structure may be as disadvantageous as elevated K_ms and reduced enzyme stability, as can be caused by the presence of urea. Thus, we feel it is as appropriate to regard urea as antagonizing methylamine effects as it is to view the methylamines as agents to offset urea's influences. Only the combined urea plus methylamine system is truly a compatible system.

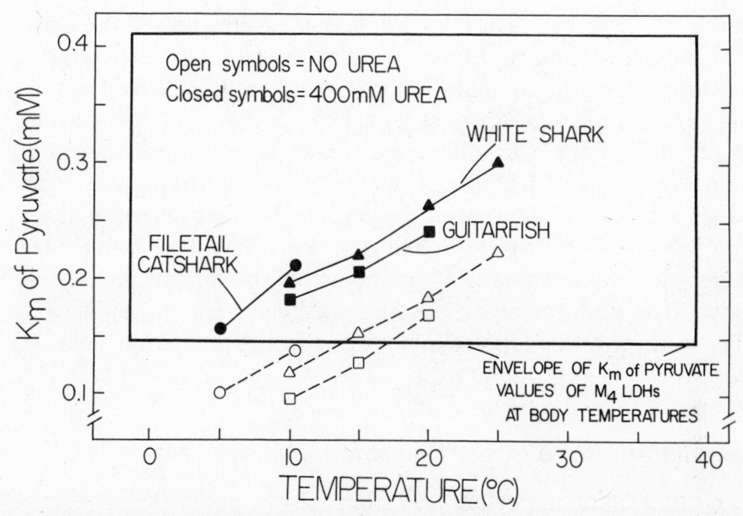

Fig. 10-7. The requirements of elasmobranch M_4-lactate dehydrogenases for physiological (approximately 400 mM) concentrations of urea to achieve a K_m of pyruvate typical of other vertebrate M_4-LDHs at body temperatures. The envelope of K_m of pyruvate values at physiological temperatures is based on data from over twenty vertebrate species (Yancey and Somero, 1978a,b; Graves and Somero, 1982; Somero et al., 1983). Elasmobranch data are from Yancey and Somero (1978b)

Urea-adapted Proteins

We stated earlier that organisms like elasmobranch fishes which contain high levels of urea could, in theory, adapt either by solute counteraction, as just discussed, or by developing urea-tolerant proteins. To round out our story of protein adaptation in species having high urea concentrations, we will consider one case in which methylamine counteraction of urea inhibition does not occur, and where a consequent change in the inherent properties of the enzyme serve to offset the inhibitory influences of urea.

Figure 10-7 illustrates the effects of temperature on the apparent K_m of pyruvate of M_4-LDHs from three elasmobranch fishes (symbols) and several other vertebrates (envelope of values; see Fig. 11-10). Unlike the counteracting urea-methylamine effects noted on K_ms of ADP for CK and PK, the increase in K_m of pyruvate caused by urea is not offset by any of the methylamine solutes (Yancey and Somero, 1978b). Conservation of the K_m of pyruvate, which is an important facet of enzyme adaptation and evolution, is achieved in this case by adjusting the set-point of the K_m. Thus, in the absence of urea, the K_m of pyruvate of the elasmobranch LDHs is lower than the K_m of LDHs of teleost fishes living at comparable body temperatures. When physiological concentrations of urea are added to the *in vitro* assay mixture, the K_m values of the elasmobranch enzymes increase into the range noted for the teleost LDH's K_m values. Here is one case, then, where tolerance of high urea concentrations has involved evolutionary adaptation of the urea-perturbed enzyme. The requirement for this adjustment may arise when urea is a specific active site binder and acts as a competitive inhibitor of the enzyme; methylamines are unlikely to be able to reverse this type of inhibitory effect (Yancey and Somero, 1978b).

Regulation of Urea and Methylamine Concentrations

Urea differs in an important way from the compatible solute osmolytes discussed earlier. Insofar as the metabolic abilities of animals are concerned, urea is a dead-end product and cannot be metabolized further to gain energy or to provide skeletons for biosynthesis. Urea is, therefore, a metabolite that need not be recycled, e.g., reincorporated into a polymer when demands are for a lowering of internal osmotic concentration. When excess urea is present, it is likely that its fate is excretion; this is not the fate of excess free amino acids, glycerol, or isofloridoside, as these osmolytes are too metabolically valuable for excretion.

The control of urea and methylamine levels in elasmobranch fishes is only partially understood. While it is clear that a complete urea cycle is present in elasmobranchs, generating urea via the same reactions found in mammals, the pathways of TMAO synthesis are relatively uncharted. Most likely TMAO is also a terminal metabolite, formed from the degradation of choline and other compounds. For both urea and the methylamines, therefore, control of intracellular and blood concentrations may be principally a close regulation of how much of each so-called waste product is allowed to pass from the body into seawater; synthesis rates per se may not be strongly dependent on external salinity. It is interesting in this regard that, at least in the case of urea, the gills of elasmobranch fishes are highly impermeable to osmolyte passage; this is not the case in teleost gills, which have a high permeability to urea. The mechanism of this urea-impermeability in elasmobranchs is not understood. Urea typically passes freely through cell membranes, so novel adaptations in elasmobranch gills appear probable. A second line of defense in urea retention is erected at the kidney level in elasmobranchs. Urea resorption occurs in these fishes, with the amount of urea returning to the body fluids from the formative urine likely to be a function of the osmotic gradient between the fish and the external milieu. A hallmark of the osmotic regulatory abilities of elasmobranchs is the preservation of a 2:1 ratio of urea: methylamines regardless of the total body fluid osmolarity (Forster and Goldstein, 1976). Thus in euryhaline (=tolerant of wide changes in ambient salinity) elasmobranchs like the ray, *Dasyatis americana*, acclimation to 50 percent seawater is marked by a large reduction in total [urea] plus [methylamines], but with the conservation of the critical ratio needed for counteracting effects on protein function and structure (Table 10-1).

Urea and Methylamine Effects: Some Unanswered Questions

In addition to the uncertainties remaining about the regulation of urea and methylamine concentrations in elasmobranchs and other urea-rich organisms, several other questions remain topics for further analysis of the effects of these nitrogenous solutes. One question concerns the roles of TMAO and other methylamines in organisms that do not accumulate high urea concentrations. For example, many marine invertebrates contain methylamines at concentrations up to approximately 0.1 M (Table 10-2). In the absence of a perturbant like urea, how do these methylamines affect intracellular macromolecules? Might these methylamines be antagonistic to inorganic ion effects on macromolecules? Evidence that one methylamine, betaine, is capable of offsetting salt (NaCl) inhibition of enzymes has been presented by Pollard and Wyn Jones (1979) for barley malate dehydrogenase. The abilities of methylamines to offset salt effects on proteins thus merits additional study.

Another point requiring investigation concerns the possible roles of methylamines in urea-rich amphibians like the crab-eating frog, *Rana cancrivora* (Gordon and Tucker, 1968) and the estivating toad, *Scaphiopus couchi* (McClanahan, 1967; Jones, 1980). Do these species accumulate TMAO or other urea antagonists in their intracellular fluids? Or, particularly in the case of *Scaphiopus* which is dormant during the times when urea concentrations are high, is urea an important mediator of metabolic depression? The same question can be raised about estivating lungfish, which also accumulate urea to concentrations of several tenths M. As discussed elsewhere in this volume, urea is in fact very likely to act as a potent inhibitor of glycolytic flux under the conditions of temperature, pH, and urea concentration characteristic of certain hibernating and, perhaps, estivating organisms. The roles of intracellular urea thus include both the prevention of desiccation (as seen in marine cartilaginous fishes and estivating species facing a desiccating environment, e.g., the burrowing spadefoot toad *Scaphiopus*) and the regulation of metabolism (in organisms lacking the urea antagonists such as TMAO).

The Halophilic Bacteria Strategy

The central thesis that was developed in the above discussion of compatible solutes and counteracting solute systems is that, by accumulating specific types of organic osmolytes rather than inorganic ions such as NaCl and KCl, osmotically concentrated organisms avoid the necessity of having to "redesign" their proteins to be salt-tolerant. We stated this conclusion in a slightly different way as well: the use of compatible solutes and counteracting solute systems shifts the evolutionary changes entailed in osmotic regulation from proteins in general to the specific proteins involved in regulating organic osmolyte concentrations. As a result of this shifting of sites of selective pressure solute-compensatory amino acid substitutions in proteins probably are not widespread in those organisms using compatible solutes or counteracting solutes.

There is a fascinating exception to the osmotic adaptation strategies discussed above that proves the central rule developed in the analysis of compatible and counteracting solute systems. One group of organisms has followed a radically different evolutionary route from the one traversed by halotolerant plants, cartilaginous fishes, and marine invertebrates. The halophilic bacteria such as *Halobacterium halobium* use inorganic ions, notably K^+, to attain osmotic balance with their environments. The environments in question are varied (salt-preserved meats, brine pools, and other highly saline bodies of water like the Dead Sea). What is notable about these environments is that their salt concentrations (mostly NaCl) approach 3–5 M, i.e., saturating levels. The halobacteria not only tolerate these high salt concentrations and exist in an isosmotic relationship to them, but they actually require 1–2 M NaCl in the external medium for growth (Lanyi, 1974; 1978). The halophiles appear to be relatively simple in terms of metabolic pathway complexity (Lanyi, 1974), a condition that may reflect solvent capacity limitations within these cells (see below) or, perhaps, the inabilities of certain classes of enzymes to acquire the capacity to function in the presence of multimolar salt concentrations. Halophilic bacteria are extremely effective in utilizing solar energy, however. Their cellular membranes contain a bacteriorhodopsin pigment system which is able to utilize photons to translocate protons and thereby establish energy gradients that can be coupled to useful work, notably the outwards transport of Na^+. The entry of Na^+ down Na^+ activity gradients can, in turn, drive the uptake of amino acids into the halobacterial cell. Much of the energy metabolism of these species appears to be based on amino acid catabolism (Lanyi, 1974; 1978; Stoeckenius et al., 1979).

The Nature of Proteins of Halophilic Bacteria

A particularly intriguing aspect of the halophilic bacteria is the set of adaptations in their proteins that enable the enzymic and ribosomal proteins to function at multimolar salt concentrations. Significantly, these same adaptations establish an obligatory requirement for high internal salt concentrations. This requirement contrasts with the situation found with compatible solutes and counteracting solute systems in which high solute concentrations are readily tolerated, but not required.

To understand how the proteins of halophilic bacteria differ from those of all other organisms studied to date, we must consider first the unique amino acid compositions of halophilic proteins and, then, the forces that maintain proteins in their native, functional conformations. The proteins of halophilic bacteria contain remarkably high percentages of acidic amino acid residues (glutamyl and aspartyl) and low percentages of strongly hydrophobic residues (leucyl, isoleucyl, and valyl) (Table 10-3; Lanyi, 1974). Halophilic proteins possess relatively large numbers of weakly hydrophobic residues. The bases for these differences between halophilic bacteria proteins and so-called "normal" proteins lie in the effects of high salt concentrations on protein stability. The principal effect of high KCl concentrations is a strong stabilization of protein structure. KCl is a structure-stabilizing, precipitating salt (see Fig. 10-1), and at concentrations in the molar range, a "normal" protein would be exceedingly rigid—perhaps too rigid to function—and might even be insoluble. The effects of structure-stabilizing salts like KCl are mediated largely through the salts' abilities to stabilize hydrophobic interactions. Thus, KCl and other salting-out salts of the Hofmeister series strongly favor the burial of hydrophobic amino acid residues in the interior of proteins (von Hippel and Schleich, 1969).

Table 10-3. A Comparison of Acidic and Basic Amino Acids and the Hydrophobicity of Cytoplasmic Soluble Proteins and Ribosomal Proteins from Halophilic and Nonhalophilic Bacteria

Species[a]	% Acidic[b]	% Basic[b]	% Excess acidic[b]	Hydrophobicity[c]
Cytoplasmic protein				
(hp) *Halobacterium salinarium*	26.8	9.7	17.1	0.958
(hp) *Halococcus* no. 24	27.8	9.9	17.9	0.936
(hp) *Halococcus* no. 46	26.8	10.3	16.5	0.956
(nh) *Pseudomonas fluorescens*	20.9	13.8	7.1	1.054
(nh) *Sarcina lutea*	21.2	12.6	8.6	1.025
(nh) *Escherichia coli*	18.4	18.2	0.2	1.114
Ribosomal 70s subunit				
(hp) *Halobacterium cutirubrum*	28.2	13.7	14.5	0.943
(nh) *Escherichia coli*	21.6	12.5	9.1	1.114

SOURCE: Modified after Lanyi (1974)

[a]The designations "hp" and "nh" refer to halophilic and nonhalophilic species, respectively

[b]Compositions expressed as mole percentage (moles per 100 moles of amino acid residues). Acidic residues are glutamic and aspartic acids; basic residues are lysine and arginine

[c]Hydrophobicity index (kcal/residue) is a measure of the average nonpolar character of the proteins' amino acid residues

Viewed from this standpoint, the unusual amino acid compositions of proteins from halophilic bacteria can be understood. The fundamental design feature of these proteins is *the establishment of a highly unstable structure which is converted to an appropriately stable structure only in the presence of physiological, i.e., multimolar, K^+ concentrations.* Two distinct effects are involved in this group of proteins. First, as the high percentage compositions of glutamyl and aspartyl residues indicate, charge repulsion among residues of a halophilic protein is very high in the absence of high concentrations of a counterion (K^+). Thus, part of the built-in instability of halophilic proteins is due to a high density of negatively charged groups, and the initial activating effects of K^+ on many halophilic bacteria enzymes (Fig. 10-8) is probably a reflection of K^+ titration of negatively charged sites, with

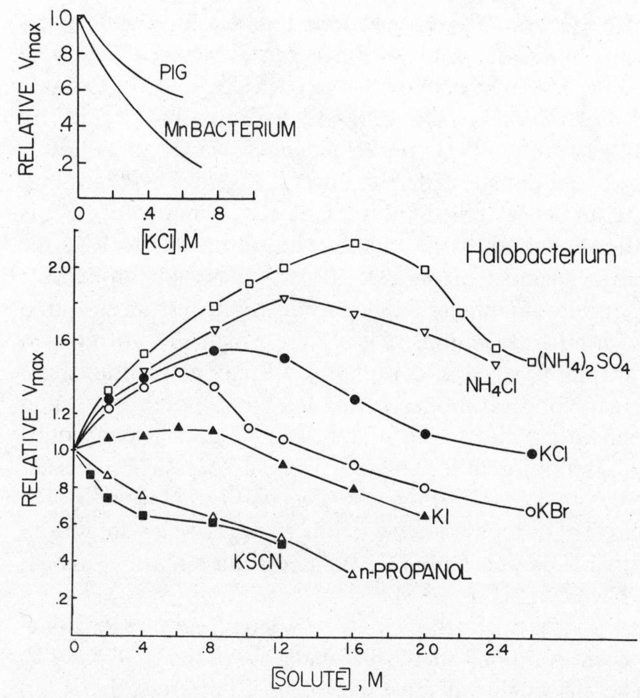

Fig. 10-8. The effects of Hofmeister series salts (plus n-propanol) on the V_{max} of the malate dehydrogenase (MDH) reaction of the extreme halophilic bacterium, *Halobacterium halobium* (unpublished data of Somero and Borowitzka). Note the consistent ranking of Hofmeister salts as illustrated in Figure 10-1. n-propanol is a strong destabilizer (solubilizes protein side-chains) like the neutral salt KSCN. For comparison, KCl effects on MDH of pig heart and a marine manganese-oxidizing bacterium (strain 52B, French, 1981) are shown. These two MDHs are strongly inhibited by KCl concentrations lower than those at which the halobacterium enzyme is maximally active

a concomitant folding of the enzyme into a more active conformation (Lanyi, 1974). The second source of structural instability is the reduction in strongly hydrophobic residues. The ability of high K^+ concentrations to stabilize hydrophobic interactions is offset by reducing the inherent hydrophobicity of the proteins. Weakly hydrophobic amino acids, e.g., glycine and alanine, take the place of valine, isoleucine, and leucine. As a result, Lanyi (1974) suggests, the *in situ* structural stabilities of halophilic bacteria proteins may resemble those of "normal" proteins in the latter molecules' relatively dilute microenvironment.

The salt effects on *Halobacterium halobium* malate dehydrogenase (MDH) shown in Figure 10-8 illustrate the stabilizing effects of high

K^+ mentioned above and the pronounced differences in salt effects based on the salts' positions in the Hofmeister series (Fig. 10-1). Note that, except for the strongly denaturing salt KSCN, and the organic solvent n-propanol, which would solubilize hydrophobic groups, all salts tested had activating effects on *H. halobium* MDH at the lower range of salt concentrations tested. This effect is taken as evidence that charge-shielding influences are the dominant activating factor at low salt concentrations, and that the charge neutralization by K^+, the monovalent cation of most of the salts tested, outweighs any effects of the various anions. At higher salt concentrations, however, anion effects become dominant. Anions, e.g., $SO_4^=$, which are effective at salting-out hydrophobic groups, continue to display activating effects up to very high salt concentrations. Salting-in anions, notably Br^- and I^-, have less activating effect and, at sufficiently high salt concentrations, lead to salt inhibition of the enzyme. These are classic Hofmeister series effects. Note, in contrast to the salt effects on *H. halobium* MDH, salt concentrations of more than a few tenths M are strongly inhibitory of both pig heart MDH and an MDH isolated from a marine manganese oxidizing bacterium (Fig. 10-8).

Salt effects on a variety of other soluble enzymes and on ribosomal proteins have been examined in the halobacteria (Lanyi, 1974, 1978, for reviews). The effects shown for MDH (Fig. 10-8) are rather typical of the salt responses of other halobacterial proteins. Ribosome stability requires high K^+ concentrations, and many enzymes attain their highest velocities at salt concentrations near or above 1 M. Some halobacterial enzymes are not salt-activated, but are resistant to salt inhibition. Other enzymes of halophiles are salt-inhibited, but less so than the corresponding enzymes of mesophiles.

This brief overview of halophilic bacteria proteins is adequate to show that these organisms are the exception proving the rule elaborated in the discussion of the other two osmotic solute strategies. The amount of evolutionary adaptation required to rebuild the proteins of halophiles must have been vast in view of the pervasive changes in amino acid composition noted in these proteins. It seems reasonable to argue that only a procaryote, with its relative simplicity and short generation time (compared to most eucaryotes), could effect this type of reconstruction effort. The halophile strategy is plainly a sword that cuts in two ways. Whereas the proteins of these bacteria have a unique ability to function in the presence of high salt concentrations, they also are in many cases absolutely dependent on the presence of high salt for function. Thus, halotolerance via this strategy would seem to lock a species into a concentrated environment. Users of compatible solutes such as *Dunaliella* show a much broader environmental tolerance. Even though *Dunaliella* can grow in saturated brines, it is also able to flourish in much more dilute solutions, due to the availability of an osmotic solute strategy that permits metabolic function in the presence of widely different internal osmotic concentrations.

Physical-Chemical Attributes of Osmolyte-Protein-Water Interactions

What are the general physical-chemical principles involved in the design of osmolyte systems? An answer, albeit a tentative one, can be developed starting with the scheme outlined above for the salting-out and salting-in influences of inorganic ions on proteins. As Clark and Zounes (1977) were the first to point out, the organic osmolytes commonly found in marine animals and certain halotolerant plants and bacteria bear structural similarities to the stabilizing (precipitating) ions of the Hofmeister series (Fig. 10-1). Most striking is the similarity between the methylamines and the strongly stabilizing methylammonium ion family; the latter ions are among the most powerful protein stabilizers and salting-out agents known. In contrast, urea is a potent destabilizer of protein structure. The commonly used free amino acid osmolytes are relatively weak stabilizers or may lack a stabilizing influence at physiological concentrations (see the RNase T_m data of Fig. 10-6). Glycerol may have stabilizing effects on protein structure, especially at low temperatures, where it is thought to impede cold denaturation by stabilizing hydrophobic interactions.

What do these resemblances between Hofmeister ions and certain of the organic osmolytes suggest in terms of fundamental mechanisms of solute effects? Perhaps the situation is clearest in the case of the elasmobranch type of system. One is tempted to suggest that the original discoverer of the Hofmeister series, especially its additivity property (such that a destabilizing ion's effects can be negated by the influences of a stabilizing ion), was the primitive ancestor to the modern cartilaginous fish! Thus, use of a balanced mixture of urea and methylamines is akin to employing a mixture of, say, sulfate ion and bromide ion or of tetramethylammonium ion and thiocyanate. In each case, the solution contains a pair of solutes from opposite ends of the Hofmeister series, and the effects of each solute are "titrated" or offset by the other solute of the system. We consider, therefore, that a major design principle in biological solutions is the establishment of conditions in which the solubilities of amino acid sidechains are optimal for protein function. Solutions which are either too salting-out or too salting-in, i.e., too stabilizing or too destabilizing, respectively, are avoided. This is perhaps the reason "why" only TMAO or only urea is not used in elasmobranchs; neither solute by itself will create an environment in which macromolecular solubility and stability are optimal. The compatible solutes may have a negligible or modest influence on amino acid sidechain solubility, as suggested by the lack of large-scale effects of most free amino acids tested on RNase thermal stability and by the finding that glycerol interacts only very weakly with proteins (Brown and Borowitzka, 1979). In a system where the salting-in/-out properties of the intracellular solution are grossly altered, as in the case of the halobacteria, the proper balance between protein stability and flexibility can only be achieved via changes in the protein, namely an extensive rebuilding of amino acid sequences. Thus, the cost of moving away from a solution having the correct balance between salting-in and salting-out influences is an intensive effort in protein evolution.

Monosaccharide Reactions with Proteins: Advantages of Glucose as a Metabolic fuel

Selection for low-molecular-weight organic molecules that lack perturbing effects on macromolecules is not restricted to cases of osmolytes. Bunn and Higgins (1981) have presented evidence that the selection of glucose as the major metabolic fuel monosaccharide is based on the relatively low tendency of glucose, compared to other monosaccharides, to undergo nonenzymatic glycosylation of proteins. When monosaccharides exist in the open (carbonyl) form of structure they are capable of reacting nonenzymatically with the terminal amino groups of proteins to form Schiff base linkages. The Schiff base (aldimine) so formed can then undergo rearrangement to form a ketoamine. These modified proteins may have perturbed functional properties, so there would appear to be an advantage to reducing this type of sugar-protein reaction.

One mechanism for reducing the rate of protein glycosylation would be the accumulation of monosaccharides that have a strong tendency to exist in the ring (hemiacetal or hemiketal) structure, since only the open form of the monosaccharide is capable of glycosylating terminal amino groups. As Bunn and Higgins argue, glucose has the strongest tendency to exist in the closed, ring form of any aldohexose. And, as their experiments involving glycosylation of hemoglobin demonstrated, glucose is the least reactive of all aldohexoses. The studies of Bunn and Higgins thus offer strong evidence in support of their hypothesis that the selection of glucose, rather than one of a number of other poten-

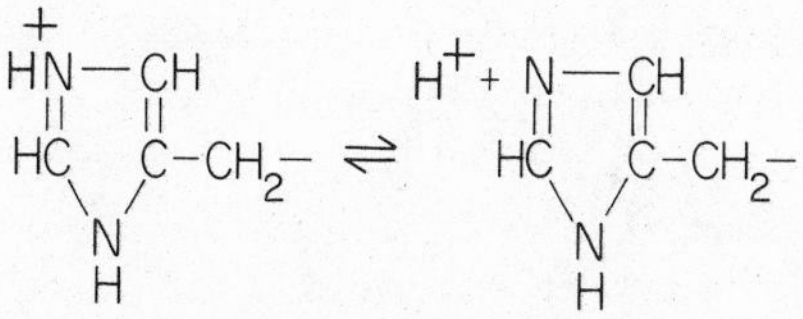

```
  +
HN—CH              H+ + N—CH
 ||  ||                 ||  ||
HC   C-CH2-   ⇌       HC   C-CH2-
  \ /                   \ /
   N                     N
   H                     H
```

Histidine Imidazole

Fig. 10-9. The imidazole group of histidine showing protonated (left) and nonprotonated (right) states

tially available stereoisomers of six-carbon sugars, might be traced to the stable ring structure of glucose and, therefore, its low rate of reaction with proteins and other molecules containing amino groups.

Optimal pH and Buffer System Design

Basic Considerations: Why Are pH Values What They Are? Up to this point in our treatment of the design and evolution of biological solutions we have neglected one solution component which, while present at only very low activities, has pervasive effects on biological structure and function. The proton (H^+) is the smallest solute of the cells, and its concentration generally falls into the range of 10^{-8} to 10^{-6} M, concentrations less than one-thousandth those of the dominant osmolytes. Yet, as any student of physiology and biochemistry is aware, slight changes in pH can have extraordinarily large influences on the rates of metabolic processes and the stabilities of proteins. Consequently, all organisms—procaryotes, plants, and animals—devote considerable effort to controlling the pH of the cell. To understand why this control is necessary in terms of basic molecular phenomena, it is appropriate to begin by asking, "Why are biological pH values what they are?"

One way of approaching this question is by considering the chemistry of one portion of one amino acid, the imidazole group of histidine (Fig. 10-9). An understanding of the acid-base chemistry associated with imidazole groups will help us understand some of the most fundamental principles involved in pH regulation, and will enable us to comprehend why it is not some unique pH value that is conserved among all organisms, under all conditions of temperature, but rather a stable protonation state of histidine imidazole groups (Reeves, 1972).

Some of the important properties of histidine imidazole groups are listed in Table 10-4. In the chapter dealing with temperature effects we discuss the significance for body fluid pH values of the enthalpy of

Table 10-4. Properties of Histidine Imidazole Groups

pK values	
imidazole	6.95
histidine-imidazole	6.00
histidyl-imidazole	
"typical"	6.5
adjacent to acidic (−) group	7–8
adjacent to basic (+) group	5–6
Enthalpy of ionization[a]	7 kcal/mol

NOTE: pK values determined at approximately 25°C. For data on the effects of neighboring groups on the pK values of histidyl-imidazoles, see Botelho and Gurd (1978), Botelho et al. (1978) [myoglobin], Matthew et al. (1979a,b) [hemoglobin], and Ohe and Kajita (1980) [hemoglobin]

[a] average value; actual value may vary with local environment of residue

imidazole ionization. In the present context, what is most important about imidazole is the pK values found at biological temperatures. Whereas the pK of imidazole is influenced by the neighboring amino acid residues of the protein (Table 10-4; see Matthew et al., 1979a,b; Ohe and Kajito, 1980), the majority of histidine imidazole groups probably have pK values close to 7 at 25°C. And it is no accident that the pH values defended by organisms are close to the pK values of histidine imidazole groups. This type of pH regulatory strategy is appropriately termed "alphastat regulation" (Reeves, 1972) to emphasize that the central goal of the regulatory mechanism is conservation of α-imidazole, defined as:

$$\alpha_{\text{imid}} = [\text{imid}]/([\text{imid}\cdot\text{H}^+] + [\text{imid}])$$

To appreciate why α_{imid} is conserved at a value approximating 0.55 in the intracellular fluids (a higher value, near 0.85, is found in blood), we must examine briefly the roles played by histidine imidazole groups. An important clue to the roles played by histidine imidazole groups is the fact emphasized above: at biological pH values, these groups are approximately half-protonated, a condition not found for any other amino acid sidechain at physiological pHs. Thus, histidine imidazole groups are uniquely prepared to function in a variety of roles where reversible protonation/deprotonation events are required. These events are often critical in a) enzyme-ligand interactions, b) catalytic steps in enzymic reaction sequences, c) structural transitions in proteins, and

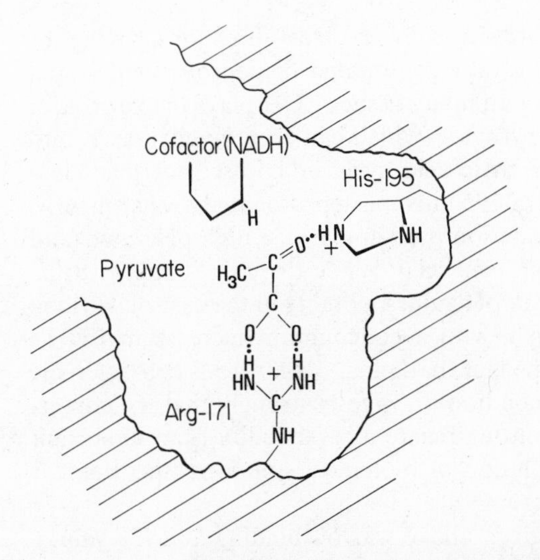

Fig. 10-10. The active center of lactate dehydrogenase, showing the key residues involved in pyruvate binding. Note the importance of histidyl-195.

d) buffering reactions. We will consider each of these roles of imidazole groups in turn, beginning with the functions of active site imidazoles of enzymes.

Binding Events and α_{imid}*: Conserving* K_m *Values and Reaction Reversibility.* To illustrate the importance of histidine imidazole groups—and the conservation of an approximately half-protonated state for these groups— in enzyme function we will consider the interactions between lactate dehydrogenase (LDH) and its substrates, pyruvate and lactate (Fig. 10-10). In vertebrate skeletal muscle, LDH has the key function of maintaining redox balance under conditions of limiting oxygen. Under these conditions LDH must be capable of binding pyruvate effectively. The ability to bind pyruvate is dependent on the protonation state of histidine residue number 195 (Fig. 10-10). When HIS-195 is protonated, LDH can bind pyruvate and reduce it to lactate. It would appear, therefore, that the "best" intracellular pH (pH_i) for the particular purposes of the pyruvate reductase reaction is a pH value low enough to ensure that HIS-195 is *always* protonated. However, the design of biochemical systems does not generally permit one characteristic to be optimized if, as a result, other features of the system are adversely affected. This rule holds true for LDH and other histidine-dependent enzymes. In the case of LDH and other enzymes which must function in both the "forwards" and "backwards" directions, having the active site histidine(s) either fully protonated or deprotonated would hinder reaction reversibility. In many tissues LDH plays the key role of oxidizing lactate to form pyruvate, which can then be channeled into aerobic pathways like the citric acid cycle or be used for gluconeogenesis. To bind lactate, HIS-195 must be deprotonated. Consequently, in tissues where lactate oxidation predominates, a high pH value (and very high α_{imid}) would seem beneficial. While there does appear to be some intertissue difference in pH values, notably in the case of working muscle, where pH_i may fall, with a concomitant increase in LDH's ability to bind pyruvate, the fact that an α_{imid} near 0.5 is characteristic of cytosols under normal conditions suggests strongly that the conservation of a pH value permitting reaction reversibility is an important design strategy in the evolution of biological solutions (this point is developed further in Somero, 1981).

Whereas the emphasis given above was on binding and K_m values, it is apparent that a parallel argument applies in the case of the catalytic step in the LDH reaction sequence. The proton transfers in the pyruvate reductase and lactate dehydrogenase reactions involve HIS-195, and the conservation of α_{imid} near 0.5 permits reversibility of the catalytic step as well as the binding step. Because of the widespread use of histidines in the active centers of enzymes catalyzing proton-transfer reactions (see Fersht, 1977), the advantages of maintaining α_{imid} near 0.5 apply to a very wide spectrum of enzymes.

Another result of α_{imid} conservation for enzymic catalysis involves the functions of enzymes, e.g., ribonuclease (RNase), which employ two histidines in their active centers, one of which acts as a base (and thus must be deprotonated at the start of the reaction), and one of which serves as an acid (and must be protonated at the start of the reaction). It is apparent that such enzymes will require a pH microenvironment permitting both base and acid functions for their active site histidines. Again, maintaining α_{imid} near 0.5 establishes the proper conditions for enzymic function.

In dwelling on the importance of histidine imidazole protonation state for enzymic catalysis and ligand binding, one must also consider enzymes which are geared to function unidirectionally, e.g., in proteolysis. These enzymes provide an example proving the rule elaborated above concerning the benefits of α_{imid} near 0.5 for most intracellular enzymes. For example, in the case of proteolytic enzymes, the highly acidic environments in which many of these digestive enzymes work ensure that any active site histidine imidazole groups will be protonated virtually all of the time, thus poising the hydrolytic reactions strongly in the direction of peptide bond cleavage. Extremes of pH, as occur in digestive organs, can be viewed as adaptations that favor essentially unidirectional catalysis via establishing an α_{imid} which is extreme, i.e., well removed from 0.5.

Another fluid compartment which must be examined in the context of α_{imid} values is the mitochondrial matrix. Albeit we currently know much less about intramitochondrial pH values than about blood or cytosol values, most available information suggests that mitochondrial matrix pH is several tenths of a pH unit more alkaline than cytosol pH (see Hazel et al., 1978; Somero, 1981). Does this higher pH indicate that mitochondrial matrix enzymes are faced with a less suitable pH for optimal ligand binding and catalysis? The answer seems to be negative. Hazel et al. (1978) report that the pH optima of selected mitochondrial matrix enzymes are shifted upwards by a few tenths of a pH unit relative to enzymes of the cytosol. Whereas many more data will be required to discern adequately the generality of this difference between cytosol- and mitochondrial matrix-localized enzymes, it seems on the basis of currently available data that the tracking of pH optima with microenvironmental pH values is characteristic of both cellular compartments. Do the higher pH optima of mitochondrial enzymes and the more alkaline pH of the matrix mean that histidine imidazole groups are no longer playing the key roles discussed for cytosol enzymes? Again, a negative answer seems to apply. Whereas the pK of histidine imidazole is, on the average, near 7 at 25°C, wide variations in this pK value are found among protein-bound histidine imidazole groups, depending on the polarity of the microenvironment offered by adjacent amino acid residues (Table 10-4). For example, by surrounding a histidine residue with negatively charged centers, e.g., aspartyl or glutamyl residues, the pK of the imidazole group can be elevated by one or two pH units (Table 10-4). In view of this fact, we conjecture that the evolution of mitochondrial enzymes may have entailed changes in the polarity of residues located close to key histidine residues, such that the pKs of the histidine imidazoles were elevated to values where an α_{imid} near 0.5 occurs at a pH value several tenths of a pH unit higher than cytosol pH. It is noteworthy that, according to this scheme, histidine not only remains the essential amino acid for protonation/deprotonation reactions where reversibility is required, but in addition histidine is the "tail wagging the dog," so to speak, in that other amino acids must be altered to establish the proper microenvironment (and pK values) for histidine imidazole groups.

Structural Transitions and α_{imid}. The structural properties of proteins also may be influenced by pH, as evidenced by the fact that most proteins are denatured by either low or high extremes of pH. Even within the biological pH_i range, however, important, pH-induced structural transitions may occur with large-scale effects on metabolism. And, again, α_{imid} appears as an important design principle. To illustrate these pH-protein structure relationships we focus on one enzyme, skeletal muscle phosphofructokinase (PFK), for which there is compelling evidence for metabolically significant structural and kinetic changes in response to small (one to a few tenths pH unit) alterations in pH_i.

Figure 10-11 shows the effects of changes in pH on the kinetics and subunit assembly of PFK. The structural and kinetic effects are closely linked and serve to regulate the activity of this important glycolytic "valve" enzyme. As pH is lowered into the pH_i range typical of mammalian muscle, PFK begins to exhibit the regulatory properties that are the hallmark of this enzyme's control characteristics (Fig. 10-11). Yet, while these small pH decreases effect the appearance of critical regulatory traits, they also make the enzyme highly metastable and susceptible to disaggregation into nonfunctional dimers. Further decreases in pH shift the tetramer/dimer equilibrium more toward the dimer, with a consequent loss of PFK activity and great reduction in capacity for glycolytic flux. This type of pH regulation may act to prevent large-scale decreases in pH_i during muscle metabolism. That is, the pH-induced structural transition may represent a "safety valve" for the cell, preventing pH decreases that could have serious and perhaps irreversible effects on a number of different processes and structures. It is noteworthy that the pH effects on the tetramer/dimer equilibrium are readily reversible; the dimers are not in any sense denatured, but instead can be viewed as remaining in a reserve pool of PFK molecules which will reassemble into functional tetramers when pH_i values return to higher levels.

Whereas the amino acid residues responsible for the pH effects on PFK structure have not been identified, the bulk of evidence strongly favors the involvement of histidine in the observed effects (Pettigrew and Frieden, 1978; Hand and Somero, 1982). As shown in Figure 10-11, a reasonable interpretation of the pH effects involves the titration of one or more histidine residues per subunit, with a concomitant structural change in the subunits which weakens the bonds holding the tetramer together. We see in this system yet another way in which α_{imid} values near 0.5 are important in metabolic design. In the muscle cytosol, PFK is exposed to pH values which can lead to titration of key histidine residues and, thereby, to changes in the enzyme's regulatory properties and structural features. The approximate pairing of pH_i with the pK values of certain imidazole groups enables a regulatory protein like PFK to alter dramatically its characteristics in response to slight, but physiologically significant, pH_i changes. To be thus sensitive, the

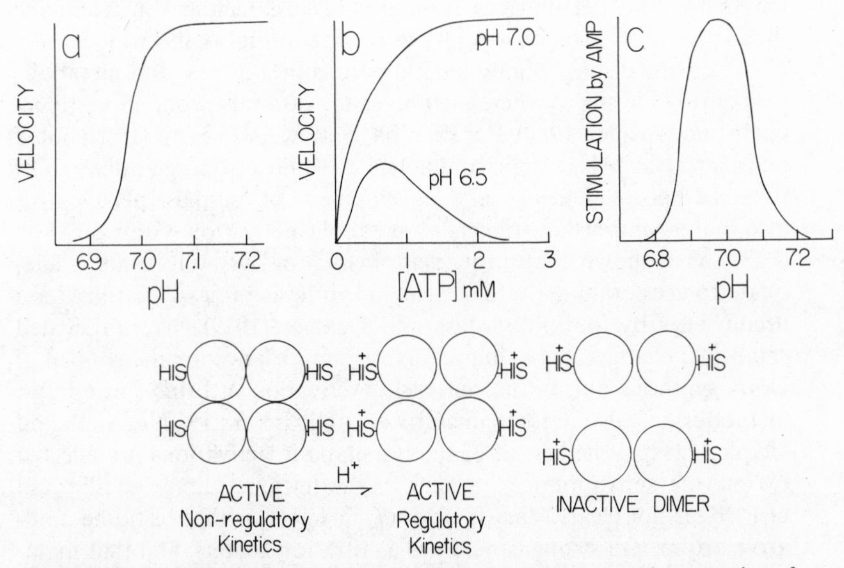

Fig. 10-11. The influences of pH on the kinetic and structural properties of phosphofructokinase (PFK). Kinetic effects are shown in frames a-c. (a) The effect of assay pH on the PFK reaction velocity (data are from an experiment by Trivedi and Danforth [1966] in which the [fructose-6-phosphate] was 0.43 mM and the [AMP] was 0.1 mM). Note the extremely sharp "titration" of PFK activity over a pH range where histidyl-imidazole titration is expected. (b) The effect of pH on the ATP-dependence of PFK activity (data from Uyeda and Racker, 1965). Note that ATP is both a substrate and a regulatory metabolite (allosteric effector) in the PFK reaction. At pH values in the range of working muscle, ATP displays both of these roles, while at higher pH values, ATP inhibition is not found. (c) The effect of pH on the regulation of PFK activity by 5′-AMP, an allosteric effector of the enzyme (data from Trivedi and Danforth, 1966).

In the lower portion of the figure a tentative model (see Bock and Frieden, 1976) for pH effects on PFK structure is given. Decreasing pH is proposed to lead to a titration of one or more histidyl residues per subunit in the tetramer. Protonation leads to the appearance of regulatory kinetics. The protonated tetramer dissociates relatively freely into nonfunctional dimers (which can reassemble into functional tetramers as pH rises).

protein must possess residues which are titratable over the range of pH_i values that characterize muscle in its resting and highly active states. Since titration curves are steepest near pK values of the species being titrated, the pairing of pH_i and pK of imidazole renders systems like PFK optimally sensitive to slight pH changes.

pH_i *and Metabolic Activation.* The metabolic regulation principles illustrated in the example of PFK suggest that alterations in pH_i of a few tenths of a pH unit can lead to very large changes in metabolic flux. In view of this fact, it is not surprising to find that a wide variety of organisms display highly correlated changes in pH_i and metabolic rate during life stages where metabolic activation is pronounced (Nuccitelli and Deamer, 1982). For example, Setlow and Setlow (1980) noted an increase in pH_i of approximately 1 pH unit during germination of bacterial spores. During early development of animals, pH changes also may be important triggers of metabolic activation. Grainger et al. (1979) have shown that an increase in pH_i of only 0.13 units is adequate to accelerate the rate of protein synthesis in newly fertilized sea urchin eggs by fourfold. Gillies and Deamer (1979) have implicated small pH_i changes as having regulatory importance for the control of DNA synthesis during the cell cycle. Activation of brine shrimp cyst metabolism is also accompanied by a small rise in pH_i (Nuccitelli and Heiple, 1982). Whether all of these metabolic activations are effected by enzyme activation, e.g., a shift in the tetramer/dimer equilibrium of PFK, is not clear. What is obvious, however, is that histidine imidazole groups are strong candidates as titration centers, and that maintaining pH_i at values allowing for imidazole titration is likely to be of significance for a wide spectrum of biological systems in which dormant-to-active metabolic shifts occur.

The Design of Intracellular Buffer Systems. In view of the importance of conserving α_{imid} within certain narrow ranges, it is pertinent to consider the buffers that are best suited for maintaining pH_i near pK_{imid}. In discussing buffer system design we must also keep in mind the principles of "solute compatibility" developed earlier in this chapter, for simply having an appropriate pK value may not be an adequate reason for using a particular buffer, at high concentrations, in the cellular fluid. We will see, in fact, that *buffer systems reflect selection for both appropriate pK values and compatibility with the metabolic apparatus of the cells.*

Because of the importance of maintaining pH_i near the pK of histidine imidazole groups, it is not surprising that histidine is usually the major source of intracellular buffering capacity (Burton, 1978; Somero, 1981). Histidine occurs in three distinct forms in the cell: free histidine, protein-bound histidine, and dipeptide histidine. In most tissues, free histidine is of minor significance to total buffering capacity (an exception is certain fish muscles; Lukton and Olcott, 1958). Protein-bound histidines contribute very importantly, and, in certain tissues, the dipeptide buffers play a dominant role in buffering (Davey, 1960; Crush, 1970; Burton, 1978; Somero, 1981). It is to these interesting di-

Table 10-5. Structure, Properties, and Sites of Occurrence of Dipeptide Buffers

Buffer	Chemical structure	pK value[a]	Occurrence (Concentration, mM)[b]
Carnosine	β-alanyl-L-histidine	6.83	Rabbit muscles
			l. dorsi (19.3) [3.2]
			psoas (21.1) [1.9]
			heart (0) [0]
			Blue whale muscles
			l. dorsi (47.8) [0.4]
			psoas (48.2) [2.3]
			Fin whale muscles
			l. dorsi (47.9) [4.3]
			psoas (50.0) [0.8]
			Chicken muscles
			pectoral (43.5) [12.3]
			leg (7.4) [2.2]
			Rat muscles
			l. dorsi (8.9) [3.0]
			gastrocnemius (6.7) [0]
			heart (0) [0]
Anserine	β-alanyl-L-1-methylhistidine	7.04	[see above values in brackets]
Ophidine	β-alanyl-L-3-methylhistidine		[see Crush, 1970]

SOURCES: Data assembled from Davey (1960) and Crush (1970)

[a] Assumed to be measured near 25°C. pK refers to the pK_2 of the histidyl imidazole group

[b] See the tabulation of occurrences and concentrations in Crush (1970) for an extensive treatment of the occurrence of dipeptide buffers

peptides that we will give the most attention, as they are a fascinating group of compounds to study from the perspective of molecular design.

The dipeptide buffers, carnosine, anserine, and ophidine, all contain histidine (albeit at times in a modified form) and a second amino acid (which also may be in a modified form) (Table 10-5). We can approach the design of these dipeptides by first asking why organisms bother to use dipeptides rather than free histidine in building up the buffering capacity of certain tissue. (As discussed in Chapter 4, buffering capacity is highest in muscle tissues which experience reduced oxygen availability during short-term burst activity, as in sprinting fish, or long-term, low-level muscle activity, as in diving mammals; dipeptides are known to supply most, or all, of this enhanced buffering ability [see Davey, 1960].) One answer to this question is based on metabolic reactivity. Free histidine interacts with a variety of enzymes involved in histidine synthesis and histidine incorporation into proteins. By incorporating histidine into a dipeptide, much, perhaps all, of this reactivity is lost. In this context it is interesting to note that one or both of the two amino acid residues of each dipeptide is a modified amino acid (Table 10-5; anserine, for example, contains β-alanine and 1-methylhistidine). These modified amino acid residues may further reduce the reactivity of the dipeptides with other amino-acid recognizing sites in the cell. Consequently, the dipeptide buffers are likely to be compatible solutes, being even less reactive with enzymes than the free amino acids accumulated in osmotically concentrated organisms.

The principal basis for utilizing the dipeptides, of course, is the pairing of their imidazole pK with the requisite pH of the intracellular fluid. It is obviously advantageous to buffer the cytosol with a compound having a pK near the optimal pH. The nonreactive dipeptide buffers thus appear to be an ideal means for establishing the pH_i values which we found to be appropriate for enzymic function and structure.

Imidazole is not the only compound that is a potentially good buffer, of course. Inorganic phosphate would seem to be an appropriate candidate for use in the cells and, in fact, phosphate buffering ranks second to imidazole buffering in most tissues (Burton, 1978). It is important to realize, however, that inorganic phosphate is likely to be an incompatible solute in many contexts. Thus, phosphate participates in a wide variety of enzymic transformations, including phosphotransferase reactions and reversible protein phosphorylations. It appears likely that very high phosphate concentrations could upset these equilibria and, further, that phosphate interactions with other ions could also alter the chemistry of the cell in maladaptive ways. The use of dipeptides as the major means for increasing buffer capacity thus can be regarded as an indication of their compatibility and as a suggestion that phosphate lacks this relationship to other components of the cellular chemistry.

Intracellular pH and "The Importance of Being Ionized"

Whereas the foregoing discussion has emphasized the importance of histidine imidazole groups and the necessity of keeping these groups approximately half-protonated, the astute reader may have already sensed that the entire histidine-based argument has started at the

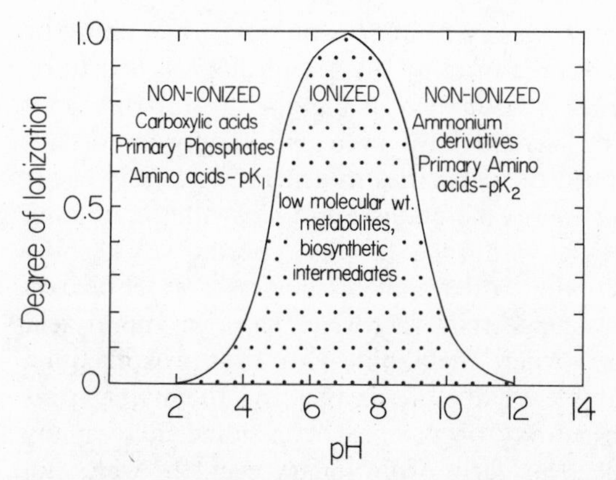

Fig. 10-12. "The importance of being ionized" (Davis, 1958): the influence of pH on the ionization state of key biomolecules. The area under the curve (dotted area) represents the region where most low-molecular weight metabolites, biosynthetic intermediates, and histidyl imidazole groups are ionized (charged). See text for a discussion of "the importance of being ionized." The figure and discussion are after Rahn et al. (1975)

second, not the first, step in the logic of biological solution evolution. Thus, it is entirely reasonable to ask, "Why wasn't some other amino acid selected at the dawn of cellular evolution for use as the key intracellular titratable group?" For instance, would an amino acid with an acidic sidechain (glutamate or aspartate) have been equally suitable for the diverse roles of histidine imidazole? Is there any reason why the entire solution and protein design system discussed above could not have been developed at a pH value of, say, 3–5? These are very important questions, and the best answer to them that we are aware of was suggested by Bernard Davis (1958) in his paper entitled, "On the importance of being ionized." Extension of Davis's idea into the realm of pH evolution and α_{imid} regulation was made by Rahn, Howell, and Reeves (1975). The gist of their argument is illustrated in Figure 10-12. Virtually all of the intermediates of metabolism are ionized at biological pH values. In contrast, if the original "decision" had been to use glutamate or aspartate as the cells' titratable residue many of these intermediates would be uncharged. What is the alleged "importance" of having a net charge? Davis makes the reasonable conjecture that charged species can be retained within the cell more readily than uncharged species, a conjecture that seems well supported by experimental evidence on a wide variety of metabolites (for example, phosphorylation of glucose by hexokinase "traps" glucose inside of liver cells). At the very earliest stages of cellular evolution, when many or all of the organic molecules required by the primordial cell had to be obtained from the external environment, being able to trap and hold these valuable compounds would have been exceedingly important. Similarly, as cells developed the capacities to generate a variety of organic molecules useful for energy metabolism and biosynthesis, holding these "homemade" molecules within the cell was important. By keeping the cellular pH at values where intermediates possessed a net charge, the trapping process was made relatively easy. The one amino acid having a pK near the pH where metabolite ionization was maximal was selected as the titratable residue of enzymes and the major intracellular buffer. The present-day occurrence of histidine in so many functions thus may reflect an early evolutionary event in which an advantageous ionization state was selected.

Solvent Capacity and Metabolic Evolution

Most of the discussion of this chapter has centered on the contents rather than the container, i.e., we have focused chiefly on osmolytes, protons, buffers, and proteins, while generally neglecting the solvent, water, that contains these diverse cellular components. Indeed, biochemists frequently neglect the medium in which virtually all of the cell's metabolic transformations occur, and focus only on the chemistry of the compounds dissolved in the cellular water. When we begin to list all of the many different types of compounds present in the small volume of water found in the cell, an important problem presents itself. This can be termed the "solvent capacity problem." How does the cell succeed in packaging so many different metabolites, enzymes, osmolytes, nucleic acids, etc. in such a small volume of water? There is no single answer to this question. Rather, several facets of cellular design appear to be related to the solvent capacity problem, and we shall briefly consider certain of the potentially more important adaptations which might play important roles in helping the cell resolve this problem.

Why K^+? An appropriate starting point in our analysis of the criteria involved in designing a biological solution that has adequate solvent capacity for thousands of different enzymic systems is to inquire whether different types of solutes place different demands on solvent water. As a case in point, we can ask whether the almost universal preference for K^+ over Na^+ as the dominant intracellular monovalent cation may reflect, at least in part, different hydration or water-structuring effects of the two ions. Wiggins (1971, 1979) has developed arguments that this is the case. Her reasoning is essentially as follows. The potassium ion is relatively large compared to the sodium ion and, therefore, has a lower charge density. The bulky potassium ion will have a different effect on water structure from the sodium ion. The latter is a water-structure-maker, while K^+ is a water-structure-breaker.

There are a number of implications arising from this difference between K^+ and Na^+. First, it may be advantageous for the cell to take up K^+ selectively and exclude (or pump out) Na^+ in order to reduce as much as possible the demands of the major inorganic ions for solvent water. Thus, when K^+ is accumulated instead of Na^+, the solvent capacity of the cellular water for other solutes may be higher. A second implication of the K^+ versus Na^+ difference in water structuring in effect turns the previous hypothesis around. Several workers (see discussions in Hazlewood, 1979; Ling, 1979; Wiggins, 1971, 1979) have proposed that the different demands of K^+ and Na^+ on the solvent capacity of water determine the ease with which these two ions can be accumulated in the already highly structured water of the cell. This hypothesis begins with the assumption that the intracellular water is highly structured due to the presence of proteins, membranes, nucleic acids, and thousands of organic metabolites. For an inorganic cation to be accommodated in such a structured solvent, especially at concentrations near 0.1 M (Table 10-1), it would appear necessary for the ion to have relatively low demands on solvent capacity. The selective accumulation of K^+ instead of Na^+ thus can be viewed as a physical-chemical consequence of incorporating ions into already densely packed solutions. Some investigators, in fact, propose that specific ion pumps, driven by splitting of ATP, are not even needed to maintain high intracellular K^+ and low Na^+ (see Ling, 1979). Whereas this is an extreme view not shared by the majority of workers in the field of ion transport, the different influences of K^+ and Na^+ on water structure would seem to play some role in the overall design of biological solutions.

Adaptations for Keeping Substrate Concentrations Low. Another important consideration in the design of biological solutions is the need to package enormous numbers of different enzymic reactions, each of which requires an enzyme plus one or more substrates and cofactors, into a minute amount of water. Atkinson (1969) appears to have been the first biochemist to view this problem in a comprehensive manner, and several of his ideas merit consideration. His analysis indicates that many of the common features of cellular architecture, enzyme kinetic properties, and even substrate chemistry can be attributed to selection for properties reducing the demands on the cell's limited water supply.

1. Enzyme-substrate affinities. Enzymes not only are able to drive metabolic reactions at rapid rates at low biological temperatures, but they accomplish these feats in the presence of very low concentrations of substrates and cofactors. Most metabolic intermediates are present within cells at concentrations of 1 mM and less (see the table on page 256 in Fersht, 1977). By having extremely high affinities for substrates and cofactors, enzymes do not require large, solvent-capacity-taxing concentrations of ligands. We can view the high affinities of enzymes for ligands as a crucial adaptation for permitting a high level of metabolic complexity, and the conservation of these ligand binding abilities in the face of the perturbing influences of temperature, pressure, and solute effects has been a major task for protein evolution, as we argue at several points in this volume.

It is important to realize, however, that high binding affinity is not a sufficient adaptation in and of itself. Atkinson (1969) makes this point very clearly when he shows what will happen when enzyme-substrate affinity is so high that, under normal cellular circumstances, the enzyme is saturated with substrate (functioning at V_{max}). Further increases in substrate entry into the pathway in which this enzyme is found will lead to an exponential rise in the concentration of this enzyme's substrate (Fig. 10-13). Suffice it to say that enzymes must work below saturation (V_{max}) if they are to have an adequate "reserve capacity" to deal with increased metabolic flux. For this reason one characteristically finds that substrate concentrations in cells are below the K_m values of enzymes (Fersht, 1977, p. 256). The evolution of enzyme-substrate affinities is thus marked by two important design principles. First, affinities must be very high to permit catalysis in the face of low substrate concentrations. Second, affinities must be adjusted such that enzymes are not saturated with substrates (i.e., enzymes must work below V_{max}), so that increases in metabolic flux can occur without exponential buildups of substrate. To achieve the latter end, regulatory processes function to keep physiological substrate concentrations near or below K_m levels.

2. Metabolite evolution. One of the dangers in allowing substrate concentrations to reach high values is that many substrates are highly reactive and/or will have strongly perturbing effects on cellular acid-base balance. Maintaining substrate concentrations at or below K_m

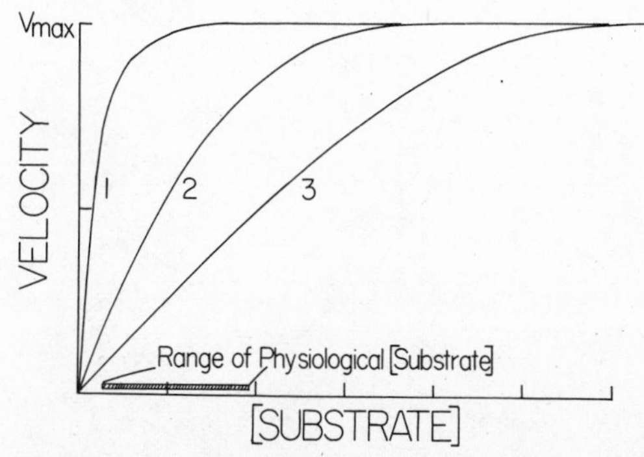

Fig. 10-13. A diagrammatic representation of three forms of a given type of enzyme, all of which have the same catalytic capacity (as shown by identical V_{max} values), but which differ in apparent K_m of substrate. Enzyme #1 exhibits an extremely low K_m, which enables the enzyme to reach V_{max} at the lowest substrate concentrations of all three enzymes. However, enzyme #1 is so rapidly saturated with substrates that, at the upper regions of the physiological substrate concentration range, the enzyme is functioning at V_{max} and thereby permitting substrate concentrations to rise rapidly. Enzyme #3 never saturates under physiological conditions, but it never reaches a significant fraction of its inherent catalytic potential. Enzyme #2 is viewed as the "best" enzyme of the three due to its relatively low K_m value, which permits an adequate rate of catalysis at physiological substrate concentrations, and its "reserve capacity," i.e., its ability to increase its rate of function significantly as substrate levels rise, yet not become saturated with substrate.

values not only minimizes solvent capacity problems, but also prevents this second type of negative effect of high metabolite concentrations. There are, however, many metabolic reactions which, on the basis of equilibrium considerations, would seem to require extremely high substrate concentrations with concomitant solvent capacity and toxicity problems for the cell.

While having extraordinary abilities to accelerate the velocities of chemical reactions, enzymes cannot influence the equilibrium positions (overall free energy changes) of these reactions. This fundamental and inviolable limitation for enzymes has required considerable evolutionary change in the design of key intermediates of metabolism. As Atkinson (1969) has stressed, adequate activity of a metabolic reaction having a highly unfavorable free energy change could necessitate exceedingly (and intolerably) high concentrations of highly reactive intermediates (Fig. 10-14). To enable the reaction, $A \rightarrow B$ to take place under conditions close to chemical equilibrium, concentrations of A vastly higher than can be tolerated by the cell could be required.

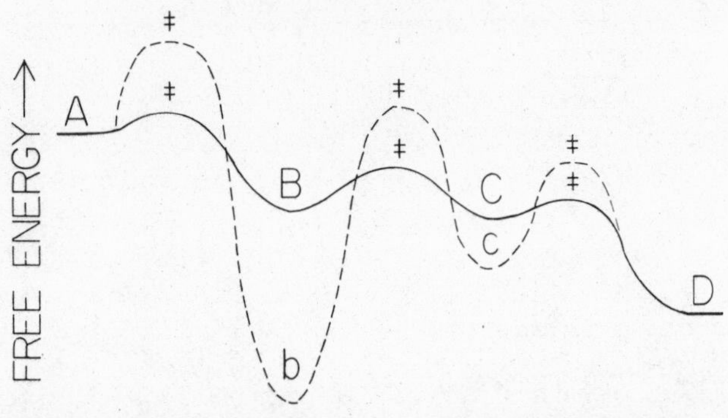

Fig. 10-14. Energy profiles for uncatalyzed (dashed line) and enzyme-catalyzed (solid line) reaction sequences in which the starting material A is converted to the final product, D, via a pair of intermediates, B (b) and C (c), where the upper case letters indicate the intermediates of the enzyme-catalyzed sequence, and the lower case letters stand for intermediates of the uncatalyzed reaction sequence. Enzymes have a major effect on the energy profile: they lower the energies of the activated complexes (transition states), symbolized by the double-dagger symbols (‡). The evolution of activated, high-energy forms of the substrates, B and C, prevents free energy "pits" from developing along the sequence; the latter would necessitate extremely high concentrations of intermediates, especially b, if the reactions were functioning close to equilibrium. See text for further discussion. Figure modified after Atkinson (1969)

The manner in which biological systems have resolved this problem entails the use of activated forms of metabolites such as A. A familiar example is the activated form of acetic acid, the thiolester derivative acetyl-coenzyme A. By using this activated form of acetic acid, which has a very large negative free energy of hydrolysis, the cell can conduct acetate-requiring reactions in the presence of very low acetate, i.e., acetyl-CoA, concentrations. As shown in Figure 10-14, the development of activated metabolites having high free energies of hydrolysis prevents the occurrence of deep energy "pits" along metabolic pathways, "pits" which require very high concentrations of low free energy reactants. Phosphorylated intermediates can be viewed as playing the same role as thiolester intermediates in this regard.

3. Enzyme regulation: allosteric control. The large increases in metabolic flux through pathways of intermediary metabolism, e.g., during vigorous muscular activity, do not require large buildups of intermediary metabolites (albeit anaerobic end products may accumulate significantly, a problem discussed elsewhere in this volume). A key to the abilities of metabolic pathways to increase flow-through without large changes in metabolite concentrations is the sensitive regulatory

function of "valve" enzymes. When signals appear calling for increased flow through a pathway, there is a highly coordinated activation of the regulatory enzymes, such that bottlenecks do not become established along the pathway. While transient rises in certain substrates may occur, these are short in duration and minor in size. Thus, the abilities of regulatory enzymes to respond rapidly to allosteric modulators must be viewed as beneficial in terms of facilitating the correct metabolic response to a particular circumstance, e.g., the need for a quick burst of speed, and in terms of allowing this increase in activity without a buildup of potentially toxic and solvent-capacity-taxing intermediates.

4. Other enzymic adaptations reducing solvent capacity requirements. Several other features of enzyme systems reduce demands for solvent capacity, albeit whether this is the primary raison d'être for these properties usually cannot be stated with any degree of certainty. For example, when a signal, e.g., the appearance of a new and metabolically useful foodstuff in the environment, calls for the appearance of new metabolic reactions, the enzymes of the new pathway are synthesized in a coordinated manner. By adding the entire pathway at once, buildup of intermediate products is minimized. Another property of many enzymic pathways which must play some role in solvent capacity adaptations is the assembly of many enzymes of a reaction sequence into one structural unit. The pyruvate dehydrogenase complex is a familiar example of this type of enzyme architecture. The reactants of the enzymes of such complexes may be largely restricted to the local microenvironment of the complex. Were the same enzymes to be distributed more or less randomly throughout the cellular water, requirements for higher substrate concentrations can be visualized.

Concluding Comments

Throughout this chapter we have developed the theme that biological solutions reflect types of evolutionary adaptation that often are overlooked by biologists. Certain pH values were shown to be optimal. Specific types and combinations of osmotic agents were shown to be most conducive of proper macromolecular function and structure. The properties of cellular water are intimately linked to the latter adaptations, albeit the full scope of water-solute-macromolecule interactions remains to be elucidated.

We have said relatively little about the problems associated with water and salt balance at the physiological level. Our neglect here is based on our belief that, whereas many good, recent sources of information are available on, for example, osmotic and ionic regulation in animals (Gilles, 1979; Maloiy, 1979; Yancey et al., 1982); few discussions exist on why certain types of solutes are the focal points of ionic and osmotic regulation. In a sense, then, our discussion is intended to provide the reader with a rationale for appreciating why organisms expend the vast amounts of work they do in regulating the compositions of their intra- and extracellular fluids.

CHAPTER ELEVEN

Temperature Adaptation

Introduction: Basic Effects of Temperature

Every student of biology is apt to be aware of the critical role of temperature relationships in establishing the distribution limits, the rates of function and, indeed, the very survival of organisms. Students of zoogeography, for example, have long been familiar with the patterns by which faunal compositions change along environmental temperature gradients. Physiologists have devoted a great deal of study to the effects of temperature on metabolic rates and other critical processes. Biochemists have devoted considerable attention to the influences of temperature on the catalytic and regulatory properties of enzymes and on the ways in which changes in temperature alter the structures of enzymic and other proteins. Thus, at all levels of biological organization, one finds that the influence of temperature is pervasive and often of dominant importance in establishing the environmental relationships of organisms. In this chapter we will explore several facets of temperature relationships, selecting for analysis phenomena which provide especially clear insights into the major biochemical adaptation strategies serving as the themes of this volume. Before considering these examples of temperature adaptation, however, we must briefly review the basic effects of temperature on biological systems. The perspective acquired from this review of basic effects will enable us to appreciate more fully the variety and severity of temperature effects on living systems and to comprehend the possible avenues of adaptive response open to organisms encountering changes in habitat temperature.

In order to understand the effects of temperature on the diverse types of processes and structures present in living systems, we must first clearly define what we mean by "temperature," and distinguish this concept from the related phenomenon of "heat." A concise way of distinguishing between these two concepts is to refer to temperature as the intensity of the heat energy present in the system. Temperature is a measure of the intensity of the kinetic energy of the constituents of a system (at biological temperatures, this movement is primarily translational and, to a lesser extent, rotational). Temperature is expressed in units of degrees Celsius (°C) or degrees absolute (°K). In contrast to temperature, heat is an expression about the total energy in a given system and is expressed in calories or joules (1 joule = 4.18 calories). For any given substance, one can, of course, relate the temperature of the substance to the heat present in the system. Thus, for water, 2.6 kcal/mole of thermal energy are present at 0°C (273°K) and 3.8 kcal/mole of thermal energy are present at 60°C (333°K).

These somewhat abstract definitions, in and of themselves, may give the reader little insight into the nature and magnitudes of temperature effects on biological systems. To appreciate how pervasively temperature affects organisms, it is appropriate to dwell in some detail on two categories of temperature effects that lie at the base of most of the thermal relationships of organisms; these are temperature effects on a) reaction rates, and b) reaction equilibria, especially those equilibria involving the formation/rupture of noncovalent ("weak") chemical bonds.

Rate Effects of Temperature. We are all aware that an increase in temperature speeds up the rate at which a chemical transformation occurs. What most readers may not be aware of, however, is the large effect which very small temperature changes have on reaction rates, and the fundamental mechanisms responsible for this relationship. In biological jargon it is common to find the expression Q_{10} used to refer to the effect of temperature changes on reaction rates, where Q_{10} is defined as the ratio of reaction velocities at temperatures (T°) ten degrees C (or K, since a °C and a °K measure the same temperature interval) apart. That is,

$$Q_{10} = \frac{\text{Velocity}\ (\text{T}^\circ + 10^\circ)^*}{\text{Velocity}\ (\text{T}^\circ)}$$

Intuitively, one might suppose that a doubling of reaction velocity for a 10°C (K) rise in temperature, i.e., a Q_{10} of 2.0, is associated with a large relative change in temperature. However, Q_{10} values near 2, which are common Q_{10}s for many physiological and biochemical processes, correspond to an increase in absolute temperature of only about 3 percent near room temperature (10/(273 + 25)). What are the fundamental mechanisms that allow a small percent change in absolute

* A more general expression for Q_{10}, which allows a Q_{10} to be computed for any temperature interval, is:

$$Q_{10} = \left(\frac{k_2}{k_1}\right)^{10/(T_2^\circ - T_1^\circ)}$$

where k_1 and k_2 are the velocity constants at temperatures T_1° and T_2°, respectively.

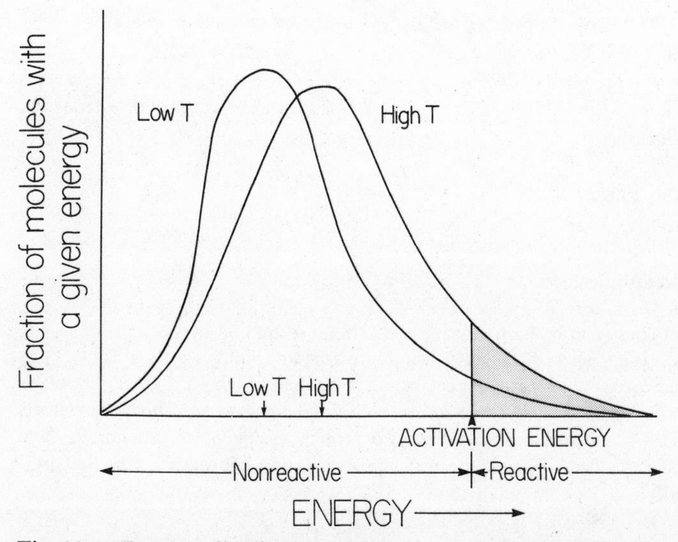

Fig. 11-1. Energy distribution curves, as determined by the Maxwell-Boltzmann equations, for a population of molecules at two different temperatures. The average energy of the molecules equals the temperature of the system in degrees Kelvin. Only those molecules having energies equal to or greater than the activation energy are reactive

temperature to be translated into an approximately 100 percent change in reaction velocity?

The question was answered by Svante Arrhenius during the 1880s. In his analysis, which has subsequently been adopted as part of the dogma of physical chemistry and thermodynamics, emphasis is placed on the fraction of a population of molecules which is reactive at any given temperature. As illustrated in Figure 11-1, only a small fraction of the total population of molecules possesses enough energy to react at a certain temperature, and changes in temperature lead to marked increases in the size of this reactive population. The reactive molecules are referred to as having energies equal to or greater than the minimal *activation energy* required for the reaction. In the present context involving temperature effects on reaction velocities, the activation energy we are referring to is the *activation enthalpy* ($\Delta H^{\ddagger}$) expressed in kcal/mol. The formal expression for the temperature dependence of reaction velocity includes the $\Delta H^{\ddagger}$ term:

$$\frac{d \ln k}{dT} = \frac{\Delta H^{\ddagger}}{RT}$$

Table 11-1. The Approximate Enthalpy Changes Associated with the Formation of Weak Bonds

Class of weak bond	Approximate enthalpy of formation (kcal/mol)
Van der Waals interactions	−1
Hydrogen bonds	−[3 − 7]
Electrostatic (ionic) bonds	−5
Hydrophobic interactions	+[1 − 3]

NOTE: The net change in enthalpy during weak bond formation between groups on a protein (or between nucleic acid constituents, proteins and lipids, etc.; see Table 11-2) also is apt to involve the rupture of weak bonds with the surrounding water. Thus, the net change in enthalpy is likely to be quite small since, for example, the heat released during hydrogen bond formation between two protein groups in effect helps to "pay back" the heat expended to break the water-protein group hydrogen bonds that previously existed.

where k is the rate constant of the reaction, T is absolute temperature and R is the universal gas constant. In subsequent sections of this chapter we will also consider the entropy term ($\Delta S^{\ddagger}$) in the overall expression for the free energy of activation ($\Delta G^{\ddagger}$). It is $\Delta H^{\ddagger}$, however, that underlies the sharp changes in reaction velocity resulting from small percentage changes in absolute temperature. And it is the relationship among reaction velocities, temperature change, and $\Delta H^{\ddagger}$ that can be viewed as one of the critical focal points of biochemical adaptation to temperature.

Temperature Effects on Reaction Equilibria. Much as temperature has sharp effects on the reaction velocity constants of chemical reactions, changes in temperature frequently have substantial effects on the equilibrium constants of biochemical reactions, especially those involving the reversible formation of noncovalent ("weak") chemical bonds (Table 11-1). The covalent bonds that link the amino acid residues of a protein or the nucleotide bases of DNA and RNA are relatively strong, and in the absence of enzymes catalyzing the hydrolyses of these bonds, covalent bonds are not very vulnerable to thermal perturbation at biological temperatures. In contrast, the hydrogen bonds, charge-charge interactions, van der Waals interactions, and hydrophobic interactions responsible for a broad suit of biological functions (Table 11-2) are readily disrupted by small temperature changes. These are

Table 11-2. Biological Structures and Processes which are Dependent on Weak Chemical Bonds[a] for Their Precision and Efficiency

Protein structure
Secondary structure (alpha helices and beta pleated sheets)
Tertiary structure ("conformation")
Quaternary structure (subunit assembly)
Protein compartmentalization, e.g., protein-membrane interactions
Nucleic acid structure, e.g., double helix structures and tRNA structures
Enzyme-ligand complexes
Lipid physical state ("viscosity")
Hormone-receptor binding
Water structure

[a] See Table 11-1

the equilibrium effects which will occupy our attention, for weak-bond-based equilibria at once are essential for a diverse family of biological structures and processes and are highly sensitive to thermal perturbation.

It behooves us to consider in a bit more detail the roles played by, and the energetic characteristics of, weak bonds. This analysis will allow us to glean insights into an exceedingly important property of macromolecular systems, including enzymes, nucleic acids and membranes. This is the property which can be termed "semistability," using the appropriate term of Alexandrov (1977). All of the biological structures stabilized by weak bonds share a common property: they are structures that are not rigid and invariant, but rather they change during the performance of their activities. Consequently, these structures must have an inherent flexibility, and this property is conferred by the use of weak bonds in maintaining the higher orders of structure (Table 11-2). For example, enzymes often undergo changes in conformation during binding of ligands and catalysis. These conformational changes may be essential for the precise alignment of reactive centers in the enzyme active site with the reactive groups in the substrates and cofactors. An enzyme that is highly rigid and unable to undergo these conformational transitions may be an extremely stable, long-lived enzyme; however, it is apt to be functionally useless. We shall examine this relationship between enzyme flexibility and functional capacity in detail later in this chapter (and in Chapter 12, dealing with pressure effects). For the moment, it will suffice to re-emphasize that a broad spectrum of biological structures, especially those of enzymes and functional membranes, display their optimal functional properties only when they exist in a fine balance between lability and stability.

One additional aspect of weak bonds is of special significance in the context of temperature effects. This is the difference between the various types of weak bonds in the *sign* of the enthalpy change associated with their formation (Table 11-1). Note that hydrogen bonds, charge-charge interactions, and van der Waals interactions form with a negative enthalpy change, whereas hydrophobic interactions form with a positive enthalpy change. Thus, an increase in temperature will destabilize the first three types of weak bonds, while hydrophobic interactions will be stabilized. Due to the different directions of temperature effect on the different types of weak bonds, the influence of a temperature change on a weak-bond-dependent system is seen to depend in part on which class of weak bond plays the dominant role in stabilizing the system.

The Nature of the Problems Posed by Temperature Changes. Recalling three of the essential, conservative goals of biochemical adaptation elaborated in the first chapter of this book, the conservation of appropriate macromolecular structural parameters, the maintenance of adequate metabolic flux, and the close regulation of metabolism, the full scope of temperature "problems" is apparent. Thus, the reliance on weak chemical bonds dictates a sharp temperature dependence of macromolecular structure, and this dependence in turn makes the metabolic apparatus and the regulation thereof highly sensitive to temperature change. Likewise, the rate effects of temperature will impact on the velocities with which metabolic flux occurs. For these reasons evolutionary changes directed toward coping with the effects of temperature are pervasive, as are phenotypic changes observed in organisms confronted with short-term fluctuations in temperature.

In responding to the problems generated by temperature effects, organisms have taken two very different adaptational approaches. On the one hand, many species have developed the ability to defend their body temperatures within a narrow range. Birds and mammals are the prime examples of this strategy, but we also find homeothermy in certain invertebrates. The benefits of homeothermy are clear. However, the costs of maintaining a stable body temperature are considerable, and for reasons of such costs (as well as reasons of basic anatomical geometry in some cases), the majority of organisms conform closely in body temperature with the ambient temperature. These organisms are termed "ectotherms." It is in the ectotherms that we find the most pervasive restructuring of biochemical systems to offset (or exploit) the effects of temperature changes. In homeothermic species, notably those animals which are capable of generating their own heat for use in temperature regulation ("endotherms"), we find that biochemical adaptations are much less pervasive, and are centered on enhancing the organism's abilities to a) produce metabolic heat in response to thermal stress, and b) modulate the flow of body heat in concert with the needs to warm or cool the body. We now turn to a consideration of endothermy and homeothermy, with emphasis on the novel ways in which enzymic systems predating endothermy have been remodeled to provide endotherms with a high degree of control of body temperature and, thereby, an escape from the temperature stresses discussed above.

Endothermy and the Regulation of Body Temperature

The Benefits and Costs of Endothermy. In view of the pervasive influences of temperature changes on the structures and functions of organisms, many benefits accrue from maintaining the body temperature independent of variations in ambient temperature. Organisms with the ability to regulate their body temperature via heat generated within their cells (endotherms) can maintain the rates of their vital processes at relatively stable levels even in the face of large changes in ambient temperature. The stability of neural function may be an especially important advantage to the endothermic homeotherm. A stable brain temperature may preclude periods of torpor as observed, for instance, in most reptiles faced with low temperatures. The ecological consequences of stable body temperatures in mammals are profound. For instance, the ability to remain active at night, when ambient temperatures are generally much lower than daytime temperatures, allows mammals to conduct important activities such as food acquisition when competing species—and potential predators—are quiescent. Perhaps partly for this reason, most mammals are nocturnal or crepuscular in habit.

A related advantage of endothermic control of body temperature is the ability of such an organism to select its habitat on other than thermal criteria. Dramatic examples of this type of ecological advantage are found in high latitude seas. The rich krill populations in the Antarctic serve as a major dietary source for certain whales and seals; endothermy enables these mammals to function in an environment where water temperatures are below the freezing point of mammalian blood. A similar example is offered by warm-bodied fishes like the bluefin tuna (*Thunnus thynnus*). Because of its high metabolic rate and large body mass, this fish has a substantial capacity to hold its body temperature above ambient temperature, at least in the case of large, adult specimens. Consequently, bluefin adults (while originating in the tropics, where spawning occurs) are able to spend much of their lives in cold, food-rich waters at high latitudes. As in the cases of mammals, birds, and even some insects, endothermy may largely free the organism from having to select a habitat on strictly thermal criteria.

The benefits of endothermy do not arise without substantial costs, however. Even for well-insulated endotherms like many birds and mammals, the basal metabolic rate necessary for supplying the heat needed for body temperature control is roughly ten times the basal metabolic rate of a similar-sized ectotherm (Bennett and Ruben, 1979). The endotherm must acquire and process vastly larger amounts of food than an ectotherm of similar size. Consequently, the population densities of endotherms cannot be as large as those of ectotherms. This, of course, is apparent from even a casual observation of natural food chains.

If, on balance, endothermy confers important advantages, why are there not more endothermic species? The answer to this question involves a broad suite of anatomical, physiological, and ecological factors. Many organisms have too large a surface-to-volume aspect to permit them to retain significant amounts of metabolically generated heat. Thus, very small animals may not be capable of endothermic body temperature regulation, although their access to this strategy will also depend on ambient temperature (Tracy, 1977). When animal-to-environment thermal gradients are low, control of body temperature via endothermy obviously is facilitated. The respiratory mode of an animal also may influence its access to endothermy. Air-breathers have a distinct advantage over water-breathers in this regard, since air has a much higher oxygen concentration than water, and water is such an effective conductor of body heat. Because of these factors, we find few examples of water-breathing endotherms. The general activity level of an animal, as often determined by its foraging or predatory strategy, may also influence its access to endothermy. Consider the striking contrasts between a tuna and a deep-sea angler fish. Both of these water-breathers occur in cold marine environments, yet the high metabolic rate of the tuna, which is associated with its need to swim continuously at high velocity for ram ventilation of its gills and for buoyancy, provides adequate heat for thermoregulation, while the metabolic rate of the sluggish angler fish, which is likely to be two or three orders of magnitude lower than that of the tuna (Chapter 12), could not begin to supply the energy needed for endothermy.

As in the case of most categorizations imposed by humans on the natural world, the division between endothermy and ectothermy or between homeothermy and poikilothermy (variable body temperature) is not hard and fast. Many animals are endothermic homeotherms when possible, e.g., when ambient temperatures are permissive of this strategy and/or when food supplies are high, yet allow their body temperatures to fall drastically when conditions preclude homeothermy. Hibernators are a familiar example; a squirrel with a normothermic temperature near 37–39°C may allow its body temperature to fall to within a few degrees of 0°C during hibernation. Other endothermic animals undergo substantial diurnal body temperature changes, e.g., hummingbirds, which may allow their body temperatures to track the ambient temperature at night. These small birds have the advantage of rapid heating abilities, due to their spectacularly high capacities for aerobic respiration. In contrast, very large hibernators like bears, which warm only very slowly, do not allow their body temperatures to fall to ambient levels. This strategy of maintaining a stable body temperature when possible, but allowing body temperature to fall when environmental conditions preclude homeothermy, is generally termed "heterothermy." It is a common strategy among mammalian and avian species, especially those of high latitude habitats.

The "Raw Material" for Endothermy. Even though endothermy is of infrequent occurrence in the lower vertebrates (fishes, amphibians, and reptiles), the fundamental "raw material" for fabricating endothermy is present in these animals. Indeed, this "raw material" is ubiquitous in living organisms. The principal engineering feat accomplished by endotherms is the exploitation of the basic heat-generating capacities associated with all metabolic systems in such a way as to force these systems to work at the higher rates appropriate for the endothermic conditions. (The second requirement for an endothermic homeotherm is the close regulation of heat exchange with the environment; we will discuss examples of the heat-flow engineering of endotherms later in this chapter.)

The most important fact to keep in mind in this analysis of heat generation mechanisms in endotherms is this: *regardless of the organism, the food source being catabolized, or the relative balance between aerobic and anaerobic processes, approximately 75 percent of the enthalpy content of the chemical bonds in foodstuffs is dissipated as heat during catabolism.* Table 11-3 presents convincing data on this point (Prusiner and Poe, 1968). What this fact tells us is that no new type of metabolic scheme needs to be elaborated to permit the endothermic condition. Instead, the task of the endotherm is to exploit pre-existing

Table 11-3. Enthalpy Balances during Substrate Oxidations

Reaction	ΔH of combustion/ oxidation (kcal/mol)	ΔH after ATP synthesis (kcal/mol)	% heat[a]
NADH + $\frac{1}{2}O_2$ = NAD^+ + H_2O	−62	−48	77
Succinate + $\frac{1}{2}O_2$ = fumarate + H_2O	−36	−27	74
Oleic acid + $25\frac{1}{2}O_2$ = $18CO_2$ + $17H_2O$	−2,657	−1,994	75
Acetic acid + $2O_2$ = $2CO_2$ + $2H_2O$	−209	−162	78
Lactic acid + $3O_2$ = $3CO_2$ + $3H_2O$	−326	−264	81
Glucose + $6O_2$ = $6CO_2$ + $6H_2O$	−673	−519	77

[a] % heat refers to the heat *not* trapped in "high energy" bonds of ATP, and equals the ratio of the enthalpy change after ATP synthesis divided by the overall enthalpy change associated with complete combustion or oxidation of the substrates in question (for the first two reactions, complete oxidation is assumed; for the other reactions total combustion to CO_2 and H_2O is assumed). The % heat values are from calculations of Prusiner and Poe (1968), who assumed that the enthalpy change of ATP hydrolysis (synthesis) is approximately 4.7 kcal/mole.

metabolic reactions, with their inherent releases of heat energy. Stated most simply, what an endotherm needs to do is to find ways of increasing the flux through existing catabolic pathways and, if possible, to find mechanisms whereby the 25 percent or so of heat energy trapped in chemical bonds, e.g., those of ATP, also can be released as heat under circumstances where the need to generate additional heat is overriding. It is within this relatively simple framework that all known endothermic systems are designed, and we will examine selected examples of endothermic systems in vertebrates and invertebrates to illustrate how widely different organisms have utilized common approaches to a single adaptational phenomenon.

Brown Adipose Tissue (*BAT*). We begin our analysis of mechanisms of endothermy with a consideration of a tissue which is the sine qua non of thermogenic function. Indeed, brown adipose tissue (BAT) is the only tissue of animals known to have thermogenesis as its sole function. Nonetheless, BAT follows the principle elaborated above; namely that the exploitation of pre-existing biochemical systems is the basic engineering principle used in endothermic mechanisms. To date, in fact, only a single new type of protein has been found to distinguish the key heat-generating component of BAT, the mitochondrion, from the mitochondria of other tissues. Not unexpectedly, this is a protein involved in regulating metabolic flux through channels common to most, if not all, mitochondrial systems.

Before considering the details of BAT biochemistry, it is important to review the general roles and species distribution patterns of this unique thermogenic tissue. BAT is a mammalian invention; it is not known in other endothermic species such as birds or warm-bodied fishes. Not all mammals possess BAT, however. Typically, BAT is restricted to 1) hibernators, who utilize BAT-produced heat during arousal; 2) neonatal mammals, which frequently lack adequate insulation or non-BAT capacities for heat production; and 3) some cold-stressed mammals. The BAT content of an organism is not fixed, but varies with ontogeny and adaptation to cold. Thus, in some mammals BAT is absent during most of ontogeny, and appears only shortly before birth (Brooks et al., 1980). In some species, e.g., rats, cold acclimation leads to marked increases in BAT (and its attendant mechanisms for heat generation and control thereof).

What particularly distinguishes the metabolism of BAT from that of other tissues which generate heat as a metabolic byproduct is the total commitment of foodstuff bond enthalpy to heat production. That is, BAT possesses the means for "wasting" 100 percent of chemical bond energy directly as heat. No biologically useful storage form of this energy, e.g., as ATP or reducing power, is generated in the mitochondria of BAT in its fully thermogenic mode. How this feat is accomplished has only recently been discovered (Nicholls, 1976, 1979), and the complete regulatory mechanisms governing the thermogenic poise of BAT mitochondria remain to be established. The key to the thermogenic capacities of BAT is, not surprisingly, a short-circuit of mitochondrial ATP synthesis, and it is worth considering the nature of this short-circuit, and putative mechanisms for controlling this break in normal mitochondrial energy flux, to understand not only the design of BAT, but also the basic features of mitochondrial ATP synthesis.

Figure 11-2 illustrates the chemiosmotic mechanism of ATP synthesis (Mitchell, 1979) in a normal mitochondrion and the short-circuited proton leakage pattern that characterizes BAT functioning in a fully thermogenic mode (model after Nicholls, 1976). In a typical mitochondrion actively synthesizing ATP, the driving force for ATP synthesis by mitochondrial ATP synthetase (also known as the F_1 ATPase) is a

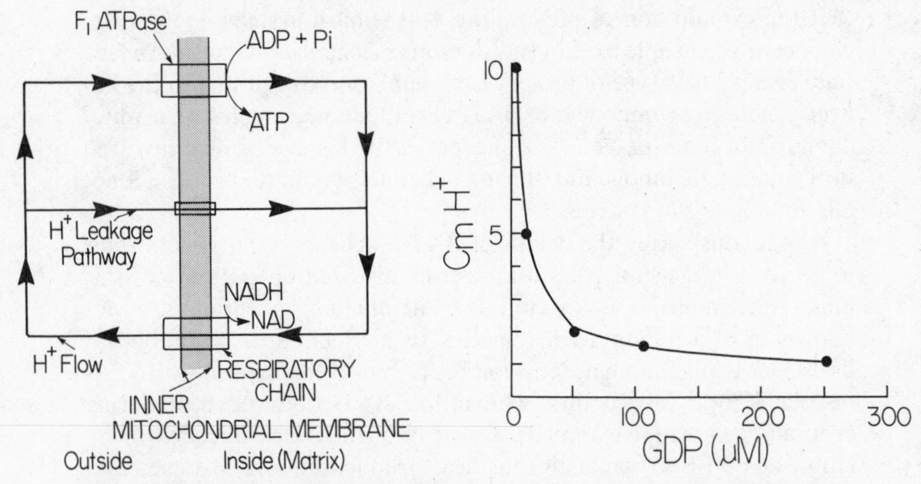

Fig. 11-2. Left: Proton flow circuitry across the inner mitochondrial membrane, showing ATP-synthesizing (F_1 ATPase) and BAT H^+ leakage pathways. The oxidation of NADH involving the respiratory chain drives the transport of protons from the inside (matrix) of the mitochondria to the outside compartment. The negative free energy change that results from the inward flow of protons back into the matrix can drive the synthesis of ATP via activity of the F_1 ATPase system or, as in BAT under thermogenic conditions, can lead to the dissipation of energy as heat.

Right: The influence of GDP on the proton conductance (C_mH^+) of the inner mitochondrial membrane. Figure after Nicholls (1976)

gradient in proton activity established across the inner mitochondrial membrane. As protons flow down their activity gradient, the negative free energy change accompanying these movements is coupled to the endergonic synthesis of ATP (Mitchell, 1979). As shown in the upper panel of Figure 11-2, a critical feature of this chemiosmotic ATP synthesis mechanism is a closed proton circuit.

In BAT this circuit is open. A short-circuit channel is present in the inner mitochondrial membrane which permits protons to bypass the ATP synthetase pathway. Thus all of the energy accompanying proton translocation between mitochondrial compartments is dissipated as heat. BAT is, therefore, not faced with the problem of having to find a way to split large quantities of ATP in order to keep a high rate of heat generation occurring (other thermogenic mechanisms typically must face the ATP-splitting problem, as discussed below). In BAT the only products of foodstuff, e.g., fat, degradation are water, CO_2, and heat.

As one would expect, BAT is a highly aerobic tissue, possessing large concentrations of triglycerides for fuel plus the enzymes of the β-oxidation spiral needed to degrade this energy-rich substrate. BAT therefore generates relatively nontoxic endproducts, as one would expect for a tissue having an exceptionally high metabolic flux which is frequently maintained for periods of hours to days.

From the standpoint of regulation, BAT would appear to offer the same types of threats to an organism that a nuclear fission reactor presents to its operators. Both "reactors" possess enormous capacities for heat generation, and in both cases the unchecked activity of the reactor could lead to serious consequences for the system. For BAT, unregulated function could possibly overheat the organism and could certainly waste valuable energy when BAT-supplied heat is no longer needed for endothermic regulation. It is not surprising, therefore, that BAT metabolism is under tight control. This control system appears to be the major evolutionary addition to the mitochondria of BAT, distinguishing these mitochondria from those of other tissues. The one new protein discovered to date in BAT mitochondria is a 32,000 dalton protein which binds purine nucleotides (GDP is most effective, ADP less so) and which is localized on the outer face of the inner mitochondrial membrane. The role of this protein is to modulate the proton conductance of the inner mitochondrial membrane, specifically the conductances of the short-circuit channel through which protons flow during thermogenesis. The binding of GDP to this protein leads to a sharp decrease in proton conductance (lower panel of Fig. 11-2). The concentration of this regulatory protein in BAT is found to vary with developmental stage and cold exposure (Brooks et al., 1980), a finding which demonstrates that both the heat-generating capacity of BAT and the ability to regulate this capacity vary concomitantly—as one would expect in a properly designed metabolic system.

The complete mechanism of BAT regulation at the systemic and cellular levels is not yet fully understood. It appears probable that a basic feature of the total control mechanism involves a coupling of an external signal, e.g., a change in ambient temperature, to a change in inner membrane conductance. Interfacing these two events may be a) hormonal signals (norepinephrine release can elicit a large increase in lipolysis and oxygen consumption of BAT within a minute after hormone binding to the tissue) and b) modulation of a battery of enzymes, including lipases, protein kinases, and enzymes associated with purine nucleotide metabolism (Nicholls, 1976; 1979).

In summary, BAT is unique among thermogenic systems in that a) its sole function is heat production, and b) it is the only thermogenic mechanism known in which the entire enthalpy content of the covalent bonds of foodstuff molecules is directly exhausted as heat, without an intermediate synthesis and splitting of ATP. In the systems examined below, we find that ATP synthesis and splitting are common phenomena during the thermogenic response, and this additional component of thermogenesis establishes the requirements for novel types of enzyme regulation.

Shivering Thermogenesis. Many organisms face the requirement for warm-up prior to activity. This objective usually is met by utilizing the enzymic machinery available in the muscle cells to split ATP to ADP and, thereby, boost the rate of substrate catabolism and heat production. The commonest form of heat generation in muscles is shivering. This involves neurally controlled function of the ATPase enzymes of the contractile system (actomyosin ATPase). The splitting of ATP by the contractile ATPases does not lead to coordinated muscular activity, however, but rather to a pattern of muscle contraction in which little net movement is observed. Shivering thermogenesis is common in mammals and certain insects, e.g., bumblebees (Heinrich, 1974a, 1979). The latter are frequently observed to shiver while stationary, either to keep their thoracic muscles (flight muscles) near operating temperature or to warm these muscles if they are cooler than approximately 30°C, the minimal muscle temperature permissive of flight.

Nonshivering Thermogenesis: Futile Cycles as Heat Generators. Nonshivering thermogenesis can involve a variety of mechanisms in a number of different tissues (liver, muscle, adipose tissue). Control of non-shivering thermogenesis may be hormonal, e.g., via thyroxine effects on ion transport ATPases, or ionic, e.g., in futile cycle mechanisms. Futile cycles, as their name suggests, involve the simultaneous function of enzymes poised in opposite directions, with the net effect of their activities being the splitting of ATP. Futile cycles may involve the simultaneous synthesis and hydrolysis of triglycerides, the catabolism and resynthesis of glycogen and glucose, and, as we now discuss in detail, the simultaneous operation of PFK and FBPase.

In bumblebees (genus *Bombus*) the preflight heating of thoracic muscles involves a net hydrolysis of ATP via the simultaneous functioning of PFK and FBPase (Fig. 11-3). PFK is a key regulatory site in glycolysis, controlling the rate of glycogen (or glucose) catabolism. FBPase, on the other hand, is typically involved in gluconeogenesis. Because of these contrasting metabolic roles, PFK and FBPase gen-

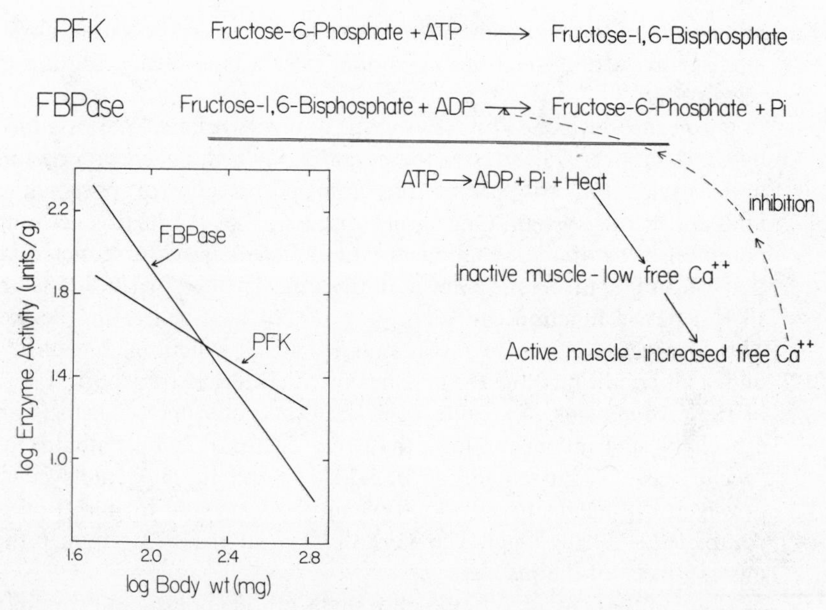

Fig. 11-3. The PFK-FBPase futile cycle found in the thoracic (flight) muscle of bumblebees (genus *Bombus*). The simultaneous operation of PFK and FBPase during preflight warm-up leads to a net hydrolysis of ATP and evolution of heat. Control of the system is mediated, at least partially, by Ca^{++} signals. When the muscle reaches a temperature permissive of flight (approximately 30°C), contraction is initiated with the concomitant release of Ca^{++}. The calcium ion strongly inhibits FBPase, but has no influence on PFK. Thus, Ca^{++} shuts off the futile cycle and allows a unidirectional flux of carbon through the glycolytic sequence. The inset to the figure shows the scaling relationship reported by Newsholme et al. (1972) for PFK and FBPase activities in bumblebee muscle. Note that the capacity for futile cycling increases in smaller individuals which have a surface to volume aspect less conducive for heat retention. The regulatory model is after Clark et al. (1973)

erally do not both occur in high titres in the same cell. PFK activity is typically highest in glycolytic tissues like skeletal muscle, while FBPase activity is highest in gluconeogenic tissues such as liver and kidney cortex. The discovery of extremely high FBPase levels in bumblebee flight muscle by Newsholme and co-workers (Newsholme et al., 1972) thus seemed to present a paradox. FBPase was some thirty times higher in flight muscle than in other animal tissues, including mammalian liver; the enzyme was not subject to inhibition by AMP, a common property of other species' FBPases; and, interestingly, the amount of FBPase (and PFK as well) was inversely correlated with body mass of the bumblebees (inset to Fig. 11-3). To the astute physiologist, this scaling relationship should offer a clue to the solution of this mystery.

The model developed by Newsholme and co-workers, later extended by Clark et al. (1973), stresses the potential for heat generation by this futile cycle, and emphasizes the appropriate control possibilities inherent in this system. Our earlier discussion of the heat-generating potential of catabolic metabolism is a sufficient basis for concluding that an ATP-splitting mechanism of the sort provided by simultaneous PFK-FBPase function can serve as a potent heat-generating device. This is especially true in insect muscle, which functions aerobically under all conditions due to the efficient tracheal gas transport system of these organisms. As Table 11-3 shows, aerobic metabolism is an especially efficient means for exhausting chemical bond enthalpy as "waste" heat. What remains to be added to this thermogenic system, however, is a means to activate it when necessary and to shut if off—completely—during flight. The work of Clark et al. (1973) supplied the missing piece of the puzzle.

These authors studied bumblebee flight muscle *in vivo* and carefully examined the possible regulatory effectors that might govern the on/off function of FBPase. They found that during flight, no futile cycle activity occurred; that is, FBPase was totally inactive. During stationary periods, however, especially at low temperatures, substrate cycling reached levels twice those of *net* glycolytic flux. The key regulatory signal for bumblebee FBPase is free calcium ion. The logic underlying this regulatory property can easily be appreciated when it is recalled that calcium ions are released from the muscle sarcoplasmic reticulum system during muscle contraction. The following regulatory scheme thus can be proposed: during stationary periods when muscle temperatures are below approximately 30°C, the lack of muscle contractile activity is paired with low free calcium concentrations in the myoplasm. The futile cycle involving FBPase and PFK function can then occur. As the muscle warms to its minimal operating temperature, contractile activity becomes possible, and as contraction commences, free Ca^{2+} is released into the myoplasm and FBPase is inhibited. Glycolytic activity then becomes unidirectional, providing adequate ATP for muscle contraction during flight. This control circuitry is consistent with the currently available data on bumblebee FBPase and PFK; the latter enzyme is not inhibited by physiological concentrations of Ca^{2+}. Additional control components almost certainly remain to be discovered, however. For instance, how is the balance between shivering thermogenesis and futile cycle thermogenesis controlled? The Ca^{2+}

released during shivering could prevent futile cycle activity. What are the temperature characteristics of the interaction between FBPase and Ca^{2+}? Does the enzyme bind calcium better at higher temperatures, insuring tighter regulatory control? Whatever the remaining details of this interesting thermogenesis mechanism, it is clear that the futile cycle established by the presence of high concentrations of both PFK and FBPase in bumblebee flight muscle may serve an important thermogenic function. Moreover, this futile cycle shares with BAT an engineering principle common in thermogenic systems: to elevate the thermogenic capacity of a tissue, two basic types of biochemical adaptation are most important, one being an increase in the amount of enzymic machinery for catabolism, and the other being subtle regulatory modifications (the 32,000 dalton purine binding protein in BAT and the Ca^{2+} sensitivity of bumblebee FBPase) added into the preexisting metabolic system. In the case of shivering thermogenesis, regulation is typically systemic, involving neural modulation of ATP splitting, with Ca^{2+} again serving as an important mediator of the regulatory response.

Lastly, the scaling relationship noted in Figure 11-3, which shows that smaller bees have higher concentrations of both PFK and FBPase in their flight muscles, can be interpreted as an indication that bees having a small mass face greater heat flux problems than larger bees. To compensate for a greater tendency for loss of heat, due to their larger surface-to-volume aspects, the smaller bees maintain higher titres of the two enzymes involved in the futile-cycle heat generation scheme.

Control of Heat Flow in Endotherms: Circulatory and Insulation Adaptations. To function effectively as an endotherm, an adequate capacity for heat generation must be paired with an ability to control the flow of heat between body and environment. While heat flow regulation is, strictly speaking, an anatomical and physiological phenomenon, rather than a biochemical adaptation, we shall briefly treat some examples of heat flow mechanisms which illustrate the integration of anatomical, physiological, biochemical, and behavioral levels of biological organization in the endothermic adaptations of organisms. Biochemical reactions provide the raw material for endothermy—heat—and it falls on higher-level systems to make use of this potential.

Insulation. The vast majority of endotherms—birds, mammals, and certain insects—employ external insulation to aid in the retention of metabolically generated heat. Keratin-based structures, hair and feathers, are characteristic of birds and mammals. Subcutaneous lipid layers (blubber) are found in marine mammals, for which the insulation provided by dead air spaces among erect hairs is not possible during submersion. Insects may also employ superficial insulation, e.g., in the form of scales in the case of the endothermic sphinx moth, *Manduca sexta* (Heinrich, 1971). Insulation per se, however, is a largely passive mechanism for controlling heat flow. Although the control of piloerection may provide a mammal with an effective means for altering the insulation offered by hair, in mammals and most other endotherms the critical factor in governing heat exchange with the environment is circulatory control.

Circulatory Flow: Variable Heat Dissipation. In the majority of endotherms the attainment of homeothermy is highly dependent on controlled dissipation of heat to the environment via the blood (or, in insects, the hemolymph). Circulatory control of heat flux is marked by several common traits. First, certain body sites, e.g., the abdomen in endothermic insects like *Manduca sexta* and bumblebees, the flippers of seals, and the relatively hairless stomachs of canine mammals, are largely devoid of insulation. These "bare" regions provide an effective radiating surface to the organism. Second, the control of blood flow through certain of these radiators is precisely modulated. For example, in endothermic insects the flow of heat from the thorax (where the flight muscles generate large quantities of heat both at rest and during flight) to the abdomen (where little heat is produced) is modulated by variations in hemolymph circulation. When the insect is in cool surroundings, most of the heat generated in the thorax is retained there. When vigorous flight activity takes place in a warm enviroment, heat is ducted to the abdomen via increased hemolymph flow. The ability to regulate heat flow between the thorax and the abdomen is illustrated especially clearly during brooding periods by bumblebees (Heinrich, 1974b). Under brooding conditions the queen bee maintains the temperatures of eggs, larvae, and pupae within a remarkably narrow range in the face of large fluctuations in ambient temperature. Brooding is done via the abdominal regions of the body, which receive considerable heat of thoracic origin. In *Bombus vosnesenskii*, thoracic and abdominal temperatures are held between 35-38°C at ambient temperatures between 3 and 33°C (Heinrich, 1974b). Comparable homeothermy of the flight muscle (but not the abdominal region) is noted during the flight of endothermic insects. *Manduca sexta* is capable of holding its thoracic temperatures between 32–33°C over the ambient temperature range of 9–24°C (Heinrich, 1974a).

In mammals an analogous control system is common, and this involves variations in blood flow, dilation of capillaries, and supplemental sweat production/evaporation in many cases.

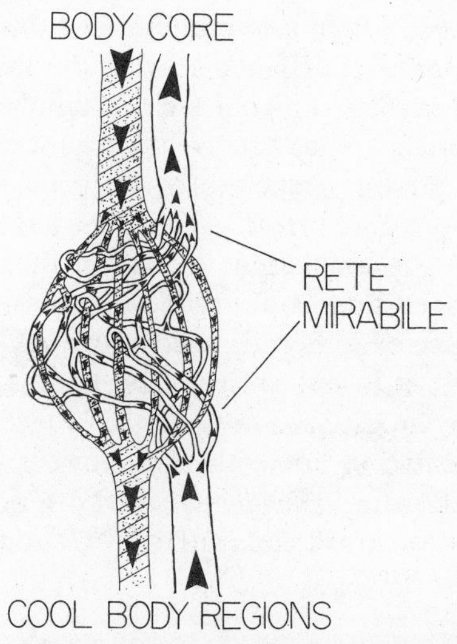

Fig. 11-4. A diagrammatic representation of the circulatory pattern in a countercurrent heat exchanger. The dense network of capillaries termed the *rete mirabile* is where the warm blood passing, e.g., from deep-body tissues, exchanges heat with cool blood entering, e.g., from the gills in tunas or the cool periphery in seal flippers or shore bird feet

Circulatory Flow: Countercurrent Heat Exchangers. The control of heat flow between body core and the surroundings is often a two-way street in terms of circulatory functions. Above we have emphasized the mechanisms controlling heat dissipation when the organism attains a thermal state dictating loss of heat to the environment. In many other cases, especially in aquatic or amphibious species, elaborate mechanisms are found which restrict heat loss to the medium. Of great importance in this regard are countercurrent heat exchangers (Fig. 11-4).

The design principles reflected in heat exchangers are the same whether the exchanger is found in a seal flipper, a bird leg, or the locomotory muscle of a tuna fish—to name three important contexts where countercurrent exchangers operate. As shown in Figure 11-4, warmed blood passes into intimate contact with cooled blood in the *rete mirabile.* In the case of the leg exchanger of a shore bird and the flipper exchanger of a seal, the effect of the counter-current exchanger system is to retain metabolically generated heat within the body core. The blood leaving the core is cooled in the exchanger en route to the periphery (bare portion of the bird's leg or the non-insulated end of the flipper). In tunas, on the other hand, the blood leaving the heart is cooled to ambient temperature (due to thermal equilibration at the gills, which takes place much more rapidly than gas equilibration); this arterial blood is warmed by the heated venous blood leaving the locomotory musculature. Via this combination of high muscle metabolic rate and effective heat trapping through a countercurrent exchanger system, tunas may hold their deep muscle temperatures more than 10–15°C above ambient water temperature. Heat exchangers have also been reported in the brain and gut circulations of certain tuna species (e.g., Stevens and Fry, 1971; see Carey et al., 1971 for review). It remains unclear how effectively these endothermic teleosts can regulate their body temperatures; that is, it is still unclear how homeothermic these fishes are. Data on the largest tuna species, the bluefin (*Thunnus thynnus*) are at least suggestive of homeothermy, however, and even smaller species like the yellowfin (*Thunnus albacares*) may possess at least some capacity for endothermic homeothermy (Dizon and Brill, 1979; Graham and Dickson, 1982).

The Origin of Endothermic Homeothermy. Who were the first endothermic homeotherms? Did endothermy as we view it in contemporary organisms arise because of the advantages of high or stable body temperatures, or did endothermic capacities appear as a side benefit of a more fundamental type of adaptation? Were the dinosaurs "hot-blooded?" Perhaps because of the curious types of animals involved in these questions, and certainly because biologists and laypersons alike may enjoy speculating about major evolutionary events, the questions above have provided a great deal of stimulation in recent years. We shall not answer them, for we view such a task as empirically, if not logically, impossible. It is worth our while, however, to discuss the issues contained in these questions, particularly for the purposes of preparing a world-view that will be useful in helping us understand the problems and potentials associated with ectothermy.

We must first consider what is needed to fabricate an endotherm from an ectothermic ancestor. Perhaps the major biochemical embellishment which must be added to the ectothermic biochemical apparatus is a capacity for *more* aerobic catabolism. Since aerobic respiration leads to such large-scale production of heat (Prusiner and Poe, 1968; Table 11-3), any change that significantly raised the capacity of an organism for aerobic metabolism would provide the key element for supporting endothermy. Is it likely, then, that selection for higher aerobic capacities was due to the advantages of endothermy? Bennett and Ruben (1979) argue otherwise. Their argument, which is based on the energetic costs of maintenance and locomotion in ectotherms and endotherms, states that the primary selective advantage of the increased aerobic capacities of mammals and birds is an increased ability for sustained, aerobically supported locomotory function. The lower vertebrates rely very strongly on anaerobic glycolysis for vigorous bouts of locomotory activity, as might be needed in prey capture and predator avoidance. Mammals and birds, in contrast, power their active locomotory bouts largely by aerobic processes (mammals and birds may possess impressive glycolytic abilities too, of course). Bennett and Ruben thus conclude that selection for aerobically based locomotory performance was the primary factor leading to the approximately ten-fold higher capacities for oxygen consumption in endotherms relative to ectotherms. Interestingly, mitochondrial densities differ by approximately an order of magnitude between reptiles, on the one hand, and mammals, on the other ((Bennett, 1972). Once this high capacity for aerobic metabolism was built into the ancestors of modern avian and mammalian endotherms (each contemporary group derives from a different ectothermic ancestor in all likelihood), the use of this aerobic machinery for endothermy could be developed.

Although the Bennett and Ruben scenario is logical and consistent with most available information on the evolutionary histories, and contemporary properties, of ectotherms and endotherms, several additional questions cannot be answered entirely by the discussion they present. For instance, given that an ectotherm with high aerobic capacity has the potential for endothermy, what selective factors will favor endothermy? What selective factors will favor the curious coincidence of body temperatures near 37–39°C for the large majority of avian and mammalian endotherms? Answers to these questions, again, will contain a high degree of conjecture. We can envision some of the reasons favoring the acquisition of endothermy, based on earlier discussion. For instance, the capacity to function effectively at night could lead to selection for a means of preserving the typical daytime temperature of the body. Shivering, known in both invertebrate and mammalian endotherms, could provide an early and simple start toward endothermy, using nervous control of ubiquitous muscle ATPases. Over longer time periods a similar selective basis for endothermy can be envisioned. Colonization of a cooler habitat or global cooling trends could both favor selection for conservation of a body temperature to which the biochemical apparatuses of the potential endotherm were adapted. The latter consideration, involving genetic adaptations, raises two issues. Firstly, the retention of a body temperature near 37–39°C by most birds and mammals (and some endothermic insects during activity bouts) may be a reflection of typical *ectothermic* body temperatures at some time in the past. When these warm temperatures were replaced by cooler temperatures during global cooling, a need to retain the 37–39° body temperature at which biochemical systems such as enzymes and membranes were adapted to function optimally may have first arisen. The second point following from this scenario concerns what we mean by "optimal" in the context of temperature effects on biochemical systems. To treat adequately this critical issue concerning temperature effects on macromolecular structures and functions, we must look in detail at the ways in which temperature affects macromolecules and, then, at the manners in which macromolecules are adapted to establish optimal functional and structural characteristics at—and often only at—their normal temperature ranges.

Ectothermy

Basic Considerations. In ectotherms, organisms whose body temperatures are determined by the ambient temperature, there typically are no physiological or biochemical defenses to prevent a close tracking of ambient temperature by the body temperature. Many ectotherms do have pronounced behavioral thermoregulatory capacities, and seek appropriate microenvironments for holding the body temperature within a preferred range. Behavioral regulation of body temperature is usually of most importance during short-term fluctuations in ambient temperature, as might occur on a diurnal cycle. Seasonal temperature changes and, especially, variations in temperature associated with long-term climatic changes and the colonization of new habitats generally require more than behavioral thermal regulation. These long-duration temperature changes provide organisms with enough time for fabricating extensive modifications in their biochemical systems, e.g., enzymes and membranes. These modifications will involve a) phenotypic regulation of gene expression and, thereby, changes in the quantities and types of proteins in the cells; and b) genetic changes over many generations, which allow the evolution of new macromolecular variants that are adapted to the new thermal regime. In the sections below we will examine both the phenotypic changes of acclimation (acclimatization) and the genetic changes occurring in the course of the long evolutionary histories of species. Several questions will serve as the focal points for our analysis. Firstly, what features of biochemical systems are instrumental in establishing the *thermal optima* and *thermal tolerances* of organisms? Secondly, what types of biochemical adjustments are most important in setting these optima and tolerances? That is, how do organisms defend the parameters most demanding of conservation? Thirdly, to what extent are temperature-related adaptations achieved by macromolecules such as enzymes, and

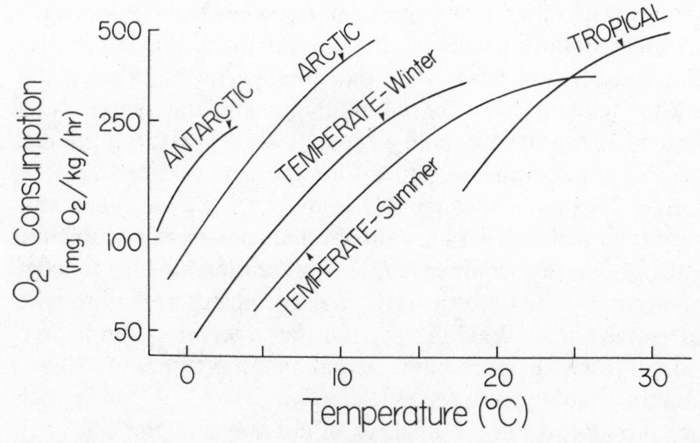

Fig. 11-5. Temperature compensation of oxygen consumption rates in fishes from different thermal regimes. Note the close conservation of metabolic rate, at appropriate physiological temperatures, in the differently adapted and differently acclimated fishes. Note that comparisons of this sort must involve careful control of activity of the specimens, and comparisons must involve fishes with generally similar locomotory habits and swimming abilities. Figure modified after Brett and Groves (1979)

to what degree can so-called "micromolecules" contribute to the adaptation response? Lastly, how do the processes of acclimation (acclimatization) and evolutionary (genetic) adaptation resemble each other? Are the phenotypic adjustments made during an ectotherm's lifetime comparable to those which different species make as a result of changes in available genetic information? To approach these questions it is convenient to "dissect" the biochemistry of the cell into separate components, notably the proteins and the lipid-based systems such as membranes, and to examine initially each component separately. This mode of analysis will teach us that, while each component displays its own characteristic responses to temperature, certain common patterns of temperature perturbation and adaptive reaction are ubiquitous among diverse biochemical systems.

Protein Adaptations

Temperature Compensation of Catalytic Rates. The Primordial "Rate Problem." Discussions of temperature effects on ectothermic organisms typically begin with a consideration of how ectotherms having widely different body temperatures succeed in maintaining relatively similar metabolic rates (Fig. 11-5). This "temperature compensation of metabolism" is observed during acclimation (acclimatization) and evolutionary time courses (Fig. 11-5). It is, therefore, pertinent to ask if the adjustments in metabolic capacity reflect fundamental differences in the enzyme systems of warm- and cold-adapted (-acclimated) ectotherms, and, if enzymic adaptations are crucial for the temperature compensation process, are similar mechanisms found over, e.g., seasonal and evolutionary time scales? Two enzymic mechanisms seem possible for achieving metabolic rate compensation: 1) alterations in the concentrations of enzymes present in the cells, and 2) changes in the catalytic efficiencies* of enzymes. We will examine the latter mechanism first; it will be found to be the more fundamental and of greater importance in evolutionary adaptation to temperature.

It is appropriate to begin an analysis of enzyme catalytic efficiency by considering what can be regarded as the primordial rate compensation problem. At biological temperatures, which fall into the approximate range of -1.9°C (the freezing point of seawater) to 100°C (the boiling point of water, a temperature tolerated by thermophilic procaryotes), the rates of chemical reactions involving the formation and rupture of *covalent* chemical bonds are extremely low relative to the rates of enzymic transformations in organisms. The primordial rate compensation problem, that involving the establishment of chemical reaction rates of the high level characteristic of life, was "solved" by the development of protein catalysts. Enzymes are generally thought to be able to increase chemical reaction rates by *six to nine orders of magnitude*, a testimony to the extreme catalytic powers of these molecules. A brief review of how this catalytic feat was achieved will

* "Catalytic efficiency" is defined in several different ways. In all definitions, however, the common criterion is how rapidly the enzyme can convert substrate to product, as expressed on a per enzyme molecule basis. One expression for "catalytic efficiency" is the substrate turnover number, often measured as moles substrates converted to product per mole enzyme per unit time (generally one minute). An alternative measure is the "catalytic constant" (k_{cat}) which expresses the turnover per enzyme active site. Both of these measurements entail function of the enzyme at maximal possible velocity (V_{max}). A third way of estimating the catalytic power of an enzyme involves the ratio of k_{cat} to K_m; this first-order rate constant provides a more realistic estimate of catalytic activity under *in vivo* substrate concentrations, which typically are below saturation. However, for estimating the inherent efficiency of the enzyme under optimal conditions (V_{max}), the former two definitions are preferable, since they follow directly from the free energy of activation of the reaction which is the true thermodynamic measurement of how effectively an enzyme can reduce the free energy "barrier" to its reaction.

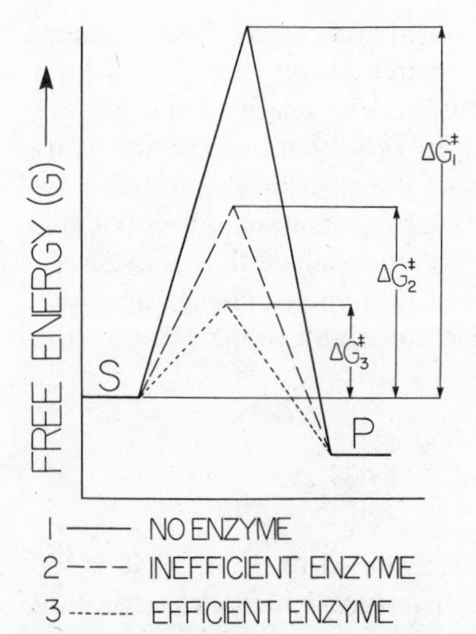

Fig. 11-6. Free energy profiles for three versions of the same chemical reaction, the conversion of S to P. The solid line (1) represents the uncatalyzed reaction. The line formed of long dashes (2) illustrates the reaction as catalyzed by a relatively inefficient enzyme, while the line formed of short dashes (3) portrays the reaction as catalyzed by a highly efficient enzyme. The activation free energies, $\Delta G^\ddagger$, corresponding to each form of the reaction are shown by the vertical lines on the right side of the figure

enable us to understand some of the key features of enzyme catalysts and to anticipate how further evolutionary adaptation to different temperatures may be achieved.

The reason why chemical reactions occur so slowly at biological temperatures was hinted at earlier, when we discussed the concept of activation enthalpy. We stressed that only a small fraction of the molecules in a system possess kinetic energy equal to or greater than the threshold activation energy (activation enthalpy) which is obligatory for the reaction to occur. We shall now expand on the phenomenon of activation energy, and consider in some detail its components and its role in adjustment of catalytic rates. Figure 11-6 presents a simple model of activation energy relationships in a system where the same chemical transformation (S is converted to P) occurs 1) in the absence of catalysis, 2) in the presence of an inefficient enzyme, and 3) in the presence of a highly efficient enzyme. The ordinate of this figure measures the free energy ($\Delta G^\ddagger$) of activation, a term defined as:

$$\Delta G^\ddagger = \Delta H^\ddagger - T\Delta S^\ddagger$$

As in the case of free energy changes which accompany the attainment of chemical equilibrium, the activation free energy change has both an enthalpy and an entropy contribution. The relative contributions of $\Delta H^{\ddagger}$ and $\Delta S^{\ddagger}$ to $\Delta G^{\ddagger}$ will be found to vary from one enzymic reaction to another and, interestingly, from one species to another for a given class of enzymic reaction (the latter observation will provide us with some clues about the mechanisms used to adjust the catalytic efficiencies of enzyme homologues). The critical energy change is always $\Delta G^{\ddagger}$, however, for the rate of the reaction is an exponential function of this energy change:

$$\text{Velocity} = \frac{kKT}{h} e^{-\Delta G^{\ddagger}/RT}$$

where k is a transition factor (usually assumed to equal 1), K is the Boltzmann constant, T is absolute temperature, h is Planck's constant and R is the universal gas constant.

Our analysis of interspecific differences in $\Delta G^{\ddagger}$ values for a given type of enzymic reaction will focus on three questions. First, how do the activation free energy changes that accompany a given type of enzymic reaction vary among species? Are these differences temperature compensatory in that cold-adapted species possess the most efficient enzyme homologues? Second, what mechanisms are available for modifying $\Delta G^{\ddagger}$ values in temperature-compensatory fashion? Third, what are the costs associated with the development of highly efficient enzyme catalysts? We can rephrase this question as follows: Why aren't all enzymes as catalytically efficient as would seem to be possible? In suggesting an answer to this last question we will discover that an intimate relationship exists between the functional and structural features of enzymes, such that optimization of one property may be precluded by the need to retain another characteristic at a suitable level. As in the case of anatomical evolution, the selective modification of one trait cannot occur in a vacuum, but instead must be performed in the broad context of the interrelatedness of most, or all, features of the system.

Enzyme catalytic efficiency and adaptation temperature. How do homologous forms of an enzyme differ in their abilities to reduce the activation free energy barriers to a particular type of metabolic reaction? Table 11-4 lists data for several enzymes that have been compared in species having broadly different cell temperatures. A consistent trend is noted among all sets of enzyme homologues: enzymes of cold-adapted species have higher catalytic efficiencies than

Table 11-4. Activation Parameters for Sets of Homologous Enzymic Reactions of Species Differing in Adaptation Temperature

	$\Delta H^{\ddagger a}$	$\Delta S^{\ddagger b}$	$\Delta G^{\ddagger a}$	Rate[c]
Lactate dehydrogenase[d]				
Pagothenia borchgrevinki (fish) (−1.9°C)	10,467	−12.7	14,000	1.00
Sebastolobus alascanus (fish) (4–12°C)	10,515	−12.6	14,009	0.98
Thunnus thynnus (fish) (15–30°C)	11,384	−10.0	14,152	0.76
Rabbit (37°C)	12,550	−6.4	14,342	0.54
Mg^{++}–Ca^{++} myofibrillar ATPase[e]				
Champsocephalus gunnari (fish) (−1–+2°C)	7,400	−31.0	15,870	1.00
Notothenia neglecta (fish) (0–2°C)	11,300	−17.8	16,130	0.62
Cottus bubalis (fish) (3–12°C)	13,550	−9.9	16,290	0.45
Dascyllus carneus (fish) (18–26°C)	23,350	+25.1	17,540	0.05
Pomatocentrus uniocellatus (fish) (18–26°C)	26,500	+32.5	17,620	0.04
Carassius auratus (goldfish)				
2°-acclimated	11,804			
31°-acclimated	20,454			
Glutamate dehydrogenase[f]				
Urticinopsis antarcticus (anemone) (−1.9°C)	13,990			
Metridium senile (anemone) (northern: 8–13°C)	14,661			
Metridium senile (southern: 12–18°C)	16,361			
Anthopleura elegantissima (anemone) (15–35°C)	19,479			
D-glyceraldehyde-3-phosphate dehydrogenase[g]				
Rabbit	18,450	11.4	15,300	
Lobster	13,950	−2.2	14,550	
Cod	13,900	−2.6	14,700	
Muscle glycogen phosphorylase-b[h]				
Rabbit	20,650	17.2	15,950	
Lobster	15,350	1.1	15,050	

[a] kcal/mol [b] Entropy units (cal/mol/°)
[c] Rates are relative to the velocity of the most efficient variant of the enzyme
[d] Data from Somero and Siebenaller (1979) computed for 5°C
[e] Data from Johnston and Walesby (1977) obtained at 0°C. Goldfish data from Johnston (1979)
[f] Data from Walsh (1981) computed for 0°C [g] Values computed from data of Cowey (1967)
[h] Values computed from data of Assaf and Graves (1969)

those of warm-adapted species. This trend is noted not only in broad comparisons of ectotherms, on the one hand, and endotherms, on the other, but it is also evident in comparisons of enzymes from ectotherms having different body temperatures. Thus, body temperature, not phylogenetic status, appears to be the key factor in dictating the efficiency of an enzyme. Note that the high-body-temperature bluefin tuna (*Thunnus thynnus*), whose muscle temperatures may reach approximately 30°C has an M_4-LDH of relatively low efficiency compared to the Antarctic teleost fish, *Pagothenia borchgrevinki*, which has the most efficient M_4-LDH. Similarly large variations in catalytic efficiency among enzymes of ectotherms are found for the actomyosin ATPase homologues (Table 11-4).

These results would seem to provide unequivocal evidence that, over evolutionary time spans during which rebuilding of the amino acid sequences of enzymes is possible, temperature-compensatory adjustments in catalytic efficiency are fabricated in a variety of types of enzymes. In fact, no exception to this pattern of temperature compensation has been found. We thus conclude that the activation free energy barrier to a given type of metabolic reaction is not in any sense fixed, but can be raised or lowered according to the thermal relationships of the species. This conclusion is not without its difficulties, however, for it raises the perplexing question, "Why are enzymes of warm-adapted species less efficient than they apparently could be?"

CONFORMATIONAL BASES FOR DIFFERENCES IN ENZYME EFFICIENCY. To approach this question, it is necessary to consider how $\Delta G^{\ddagger}$ values could be adjusted in temperature-compensatory ways. Certain types of adjustment mechanisms would seem to be precluded. Substrate and cofactor structures are obviously the same in all species, so the requisite adaptations must be attributed to portions of the enzyme itself. Of the various regions of the enzyme, the active site residues which interact with the substrate and cofactor are usually strongly conserved in all interspecific homologues of an enzyme. Thus, we conclude that the actual chemistry of the reaction, by which we refer to the types of enzyme-ligand interactions, the nature of the covalent bonds formed or broken, and the classes of proton donor/acceptor species, is likely to be invariable among interspecific homologues. What *is* modifiable, however, is the change in enzyme conformation which accompanies catalysis, and to understand the $\Delta G^{\ddagger}$ differences among homologues we must, therefore, explore the ways in which the energy changes resulting from alterations in protein conformation can contribute to elevating or reducing the free energy barrier to catalysis.

Early theoretical treatments of enzymic catalysis tended to view enzymes and their substrates as "locks" and "keys," respectively. The enzyme was viewed as a rigid body, containing a site (the "keyhole") into which the substrate fit as a "key." This concept is now regarded as misleading, if not incorrect. Whereas there must be a precise geometrical complementarity between binding site and ligand (as well as the correct charge and polarity characteristics), this complementarity may not be wholly pre-existing, but may be induced by the binding of ligand. Koshland (1973) has coined the appropriate expression of "induced fit" to describe these interactions. One can consider such interactions as "hand" and "glove" effects, where the final form of the "glove" (the enzyme) is established only after insertion of the "hand" (the ligand). Flexibility in enzyme structure is thus seen as an obligatory part of many—perhaps all—binding events. Furthermore, during the steps in the catalytic reaction sequence subsequent to binding, e.g., during the formation of the activated complex, conformational changes may also occur (Fersht, 1977). Because changes in enzyme conformation entail the breaking and formation of weak chemical bonds, they will be accompanied by needs for energy inputs or by releases of energy. And it is these energy changes associated with conformational changes that may be responsible for the differences in $\Delta G^{\ddagger}$ discussed above. The question thus becomes one of analyzing the bases for different conformational change energy costs in different species.

Two closely related models can be proposed to account for different conformational energy expenditures during catalysis in enzyme systems from organisms adapted to different temperatures. One model is quite general, and emphasizes the role played by general protein stability considerations in temperature adaptation. Let us suppose that adaptation to higher temperatures necessitates increases in protein structural stability (a conjecture that will be discussed in detail in a later section of this chapter). Increasing structural stability could be achieved by additional covalent bonds in the protein structure, notably disulfide links (—S—S— bridges). However, interspecific comparisons of proteins, including those of thermophilic bacteria living near the boiling point of water, have not found evidence for widespread utilization of this mechanism for structural stabilization. Recall that proteins must remain flexible, and the type of stability conferred by additional covalent interactions may preclude the type of flexibility that reliance on weak (noncovalent) bonds makes possible. The alternative strategy then involves the use of different numbers (or types) of

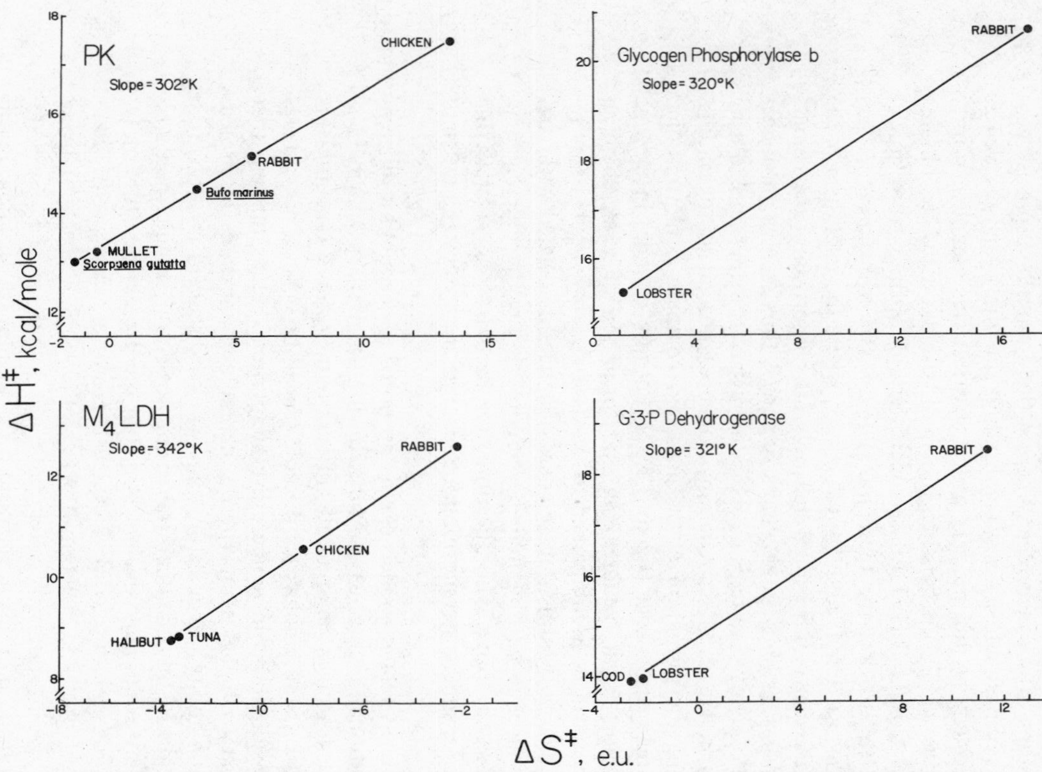

Fig. 11-7. Compensation plots ($\Delta H^{\ddagger}$ versus $\Delta S^{\ddagger}$) for the activation enthalpies and entropies of reactions catalyzed by interspecific homologues of different enzymes. The slopes of the regression lines are in degrees Kelvin (°K). The enzymes are: skeletal muscle PK, M_4-lactate dehydrogenase, glyceraldehyde-3-phosphate dehydrogenase, and glycogen phosphorylase. Figure modified after Somero and Low (1976)

weak interactions to stabilize protein conformation. What would be the consequence of adding one or more additional weak bonds to a protein that must function at elevated temperatures? One consequence could be a paired increase in thermal stability and decrease in catalytic efficiency (increase in $\Delta G^{\ddagger}$). If during the conformational change associated with the activation event this additional weak bond had to be broken to permit the enzyme to attain the conformation needed for catalysis, an addition to the energy barrier to the reaction would occur. According to this model, the more inherently rigid a protein, the more likely is the protein to display a high $\Delta G^{\ddagger}$ in its reaction. As discussed later, there is, in fact, a strong correlation between body temperature and protein thermal stability, just as body temperature and $\Delta G^{\ddagger}$ vary directly. The analysis just offered suggests that there may exist causal relationships between these trends.

Another line of evidence favoring a tight relationship between catalytic efficiency and the number of weak bonds formed or broken during catalysis is the covariation in activation enthalpy and activation entropy found in comparisons of interspecific homologues (Table 11-4, Fig. 11-7). In all enzyme families compared, increases in activation enthalpy are highly correlated with increases in activation entropy. When $\Delta H^{\ddagger}$ and $\Delta S^{\ddagger}$ are plotted as a "compensation plot" as in Figure 11-7, this correlation is demonstrated particularly clearly. The expression "compensation plot" is used to emphasize that simultaneous increases in $\Delta H^{\ddagger}$ and $\Delta S^{\ddagger}$ have offsetting, compensating, influences on $\Delta G^{\ddagger}$ ($\Delta G^{\ddagger} = \Delta H^{\ddagger} - T\Delta S^{\ddagger}$). The slope of a compensation plot is in units of °K, and is termed the "compensation temperature." What is revealing about the compensation temperatures found in analyses like those of Figure 11-7 is a consistent and small temperature range into which values fall—and the values observed are those characteristic of the formation and rupture of weak bonds in aqueous solution (Lumry and Biltonin, 1969).* The compensation phenomenon found in comparisons of enzyme homologues thus is interpreted as further evidence

* Lumry and colleagues (see Lumry and Biltonin, 1969) have proposed that there is something "magic" about the temperature of mammalian and avian cells: near 37°C, changes in enthalpy and entropy have almost fully counterbalancing effects on free energy changes for those reactions involving the formation/rupture of weak bonds in aqueous solutions. These investigators propose, therefore, that the body temperatures of birds and mammals are not an evolutionary "accident" due, for example, to a particular ambient temperature that happened to be common during the origin of endothermic vertebrates. Rather, a temperature was selected at which the formation and rupture of weak chemical bonds could occur with minimal net free energy change.

for different amounts of weak bond formation/rupture during catalysis in different homologues of a given type of enzyme (Low and Somero, 1974). Thermodynamic arguments do not, of course, allow one to specify details of mechanism. For example, the differences in activation energy parameters could reflect different amounts of bond breakage or bond formation during the activation event. Thus, a high activation free energy could result from a relatively large energy input into bond breakage or a small energy yield from bond formation. While there is evidence in support of both sorts of bond energy changes during activation events (see Greaney and Somero, 1979), an important determinant in establishing the energetics of activation is the need to hold ligands to the binding sites. This "anchoring" may be energetically costly and may sharply influence the subsequent energy changes during catalysis.

RELATIONSHIPS BETWEEN BINDING BNERGY AND ACTIVATION ENERGY. We have repeatedly emphasized that biochemical adaptations must be analyzed in a holistic context; changes in one trait often effect modifications in other characteristics of the system, so evolutionary processes must be analyzed on a systemwide basis. In the present context, we must consider two separate, yet energetically related steps in an enzymic reaction sequence: binding and the activation step of catalysis. The analyses of Jencks, Fersht, and other enzyme chemists (see Fersht, 1977) have stressed that the stability of the enzyme-substrate complex, as determined by the ΔG of substrate binding, can have a marked effect on the activation free energy barrier. The same consideration of course applies for cofactor binding. As illustrated in Figure 11-8, a large free energy decrease during substrate or cofactor binding places the enzyme-ligand complex into a deep energy pit. Consequently, a much greater climb is necessary to surmount the activation free energy barrier. This relationship shows us that tightness of ligand binding is apt to correlate inversely with catalytic efficiency. How do these relationships apply to the adaptation of enzymes to temperature?

A key fact offering insights into the relationship of binding to activation energies is that increases in temperature usually perturb enzyme-ligand complexes. That is, the binding of substrates, cofactors, and modulators to enzymes generally takes place with a negative enthalpy change. That is understandable in light of the importance of ionic interactions and hydrogen bonds in stabilizing enzyme-ligand complexes. It follows that enzymes that function at relatively high temperatures may need to invest more energy in stabilizing ligand complexes than do enzymes of low-body-temperature species. Consequently, the model shown in Figure 11-8 may describe the differences between the enzymic

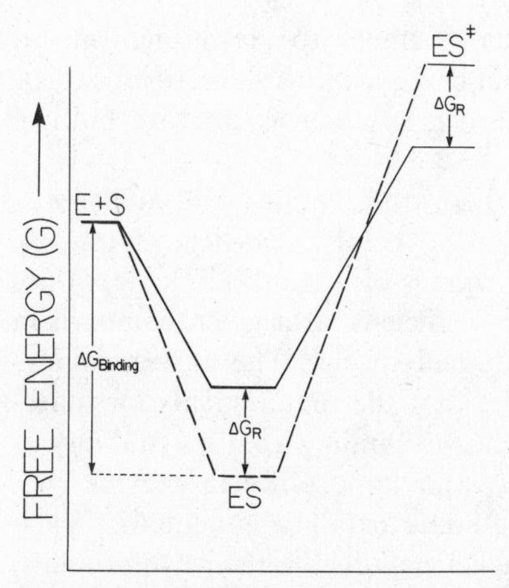

Fig. 11-8. Energy profiles for the reactions catalyzed by two forms of an enzyme which differ in the energy changes that occur during binding and the activation event. In the reaction represented by the dashed line, the enzyme possesses a substrate (S) binding site highly complementary to the shape and charge properties of the substrate, and a maximal amount of energy is released on binding. Thus, the ES complex sits in a deep free energy "pit." The reaction indicated by the solid line involves an enzyme having a lesser complementarity to the shape/charge properties of the substrate and, as a result, less free energy is released during binding, and the ES complex sits at a higher energy level. The amount of possible binding energy not released during substrate binding in the looser-binding enzyme (ΔG_R) is released during the activation event. At this time, the shape of the binding site changes such that the shape/charge properties of the transition state form of the substrate are complementary to the enzyme binding site. The energy released, i.e., ΔG_R, lowers the free energy state of the transition state complex ($ES^{\ddagger}$) and, therefore, facilitates the reaction velocity. It is seen from this analysis that the commitment of large quantities of energy during the initial binding event can increase the activation free energy "barrier" to the reaction. Thus, the differential allocation of the intrinsic binding energy ($\Delta G_{Binding}$)(the total binding energy released during enzyme interactions with the ground state and transition state forms of the substrate) between stabilizing the ES complex and reducing $\Delta G^{\ddagger}$, will determine whether a highly stable ES complex or a rapid rate of catalysis, respectively, is achieved by the enzyme. Figure modified after Somero and Yancey (1978)

reactions of high- and low-body-temperature species. The former species stabilize enzyme-ligand complexes with one or more additional weak bonds during catalysis and therefore a higher energy hill must be surmounted. Borgmann and Moon (1975) have presented compelling evidence that this type of difference is responsible for the activation free energy differences noted among M_4-LDH homologues (Table

11-4). This mechanism is conceptually similar to the mechanism involving differences in conformational energy changes described earlier, and the compensation phenomenon again is consistent with both of these models.

An important conclusion from these observations and analyses is that structural factors may set the limits of catalytic efficiency for an enzyme. We earlier inquired "why" enzymes of warm-body-temperature species like mammals should be less efficient, at a given comparison temperature, than those of cold-adapted species. The answer following from the analysis given above is that the requirements for inherent structural stability and proper ligand binding abilities (stability of enzyme-ligand complexes) are of greater importance in enzyme evolution than the development of ultimate catalytic potentials. When birds and mammals lost enzymic efficiency during their evolutionary development from ectothermic ancestors, this loss was not without a compensating gain. We will now consider what forms this "gain" has taken, and this analysis will enable us to begin to appreciate some of the molecular determinants of organisms' (and enzymes') thermal optima.

CONSERVATION OF APPARENT K_m. Even the simplest enzymic reaction involves at least three distinct chemical events: the initial binding of substrate (and cofactors, if these are part of the reaction), the chemical conversion of substrate to product, and the release of bound product to generate the free enzyme, which is then capable of initiating another round of catalysis. The second or third step in this simplified sequence is usually the rate-governing step at saturating substrate concentrations. For the M_4-LDH reaction functioning in the direction of pyruvate reduction, release of oxidized cofactor (NAD^+) at the end of the reaction appears to be the rate-limiting step (Everse and Kaplan, 1973). The modification of catalytic activity thus may be restricted to altering the energy changes associated with one event in the catalytic sequence, as discussed above. However, any change in the enzyme that might influence the energetics of product release, as in adaptations of LDHs, might also affect the energetics of substrate (or cofactor) binding. Thus, as suggested above, one component of the enzyme machinery cannot be rebuilt without pervasive influences on a number of enzyme properties. It follows from this reasoning that priorities must be established among the different properties of enzymes, such that those traits which are of greatest importance are conserved—even if this necessitates suboptimal values for less critical parameters.

One of the most critical aspects of enzyme function is the capacity of the enzyme to maintain its apparent Michaelis constant (K_m) of sub-

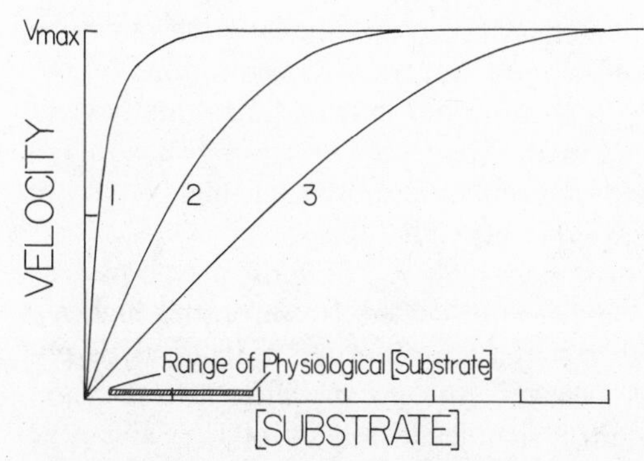

Fig. 11-9. Substrate saturation curves for three variants of a given type of enzyme having different apparent K_m values. Enzyme 1 has a very low K_m value, and functions near or at V_{max} under all physiological substrate concentrations. This variant has a minimal reserve capacity. Enzyme 3 has a very high K_m of substrate, and never reaches an appreciable fraction of its potential activity under physiological substrate concentrations. Enzyme 2 displays the optimal relationship between K_m and physiological substrate concentration. Increases in substrate concentration lead to a large increase in catalytic rate, yet the enzyme is never saturated with substrate, a condition which could lead to uncontrolled buildup of metabolic intermediates (see text)

strate within the range appropriate for the proper catalytic rate and regulatory sensitivity. These catalytic and regulatory relationships are illustrated in Figure 11-9. An enzyme having a very low K_m such as the enzyme whose saturation curve is farthest to the left in Figure 11-9 will function at or near maximal velocity under basal metabolic conditions. Any need to increase metabolic flux through this point in metabolism is unlikely to be met by such an enzyme. Since the enzyme is already functioning at or near V_{max}, increases in substrate concentration will be without effect on the enzyme (in fact, many enzymes are inhibited by substrate concentrations several times the K_m level; LDH is one such enzyme). Activation of the enzyme via regulatory signals leading to decreases in K_m will also be without effect. For example, decreases in muscle pH during vigorous exercise can activate M_4-LDH by lowering the K_m of pyruvate. If a particular M_4-LDH is already functioning at V_{max}, this important regulatory mechanism cannot function to increase the rate of pyruvate reduction and regeneration of oxidized cofactor. Considerations such as these indicate that a "good" enzyme will not function at or near its potential V_{max}, but instead will maintain a *reserve capacity* which enables it to increase its rate of function in response to regulatory signals (changes in pH or changes in concentrations of metabolite modulators) and to increases in substrate concentration. Atkinson (1969) has argued that the latter effect can be viewed as an important buffering of substrate concentrations by enzymes. If enzymes reach V_{max}, then further influx of substrate beyond the saturating levels will lead to an exponential buildup of substrate—and this can be disastrous for the cell if the substrate is highly reactive chemically or interferes with other enzymic processes.

An enzyme having an extremely high K_m of substrate allows the cell to circumvent the problems just discussed. However, this high-K_m enzyme is apt never to realize a significant portion of its catalytic potential. That is, this high-K_m enzyme (the enzyme farthest to the right in Fig. 11-9) will never attain a significant fraction of V_{max}. Even the addition of regulatory signals which lower K_m may not adequately activate the enzyme.

What we view as the ideal compromise situation is the middle enzyme in Figure 11-9, the enzyme having an adequate reserve capacity yet which attains a significant fraction of V_{max} under cellular substrate and modulator concentrations. To be accurate, one must emphasize not only the enzyme's properties, but also the intracellular substrate concentrations available to the enzyme, for the correct pairing of K_m values with intracellular substrate concentrations is the real adaptational requirement. How is this pairing met in species adapted to different temperatures? Are enzymes the same and substrate concentrations variable? Or, conversely, are substrate concentrations conserved and enzymes modified accordingly to keep K_m values near or slightly above (Fig. 11-9) intracellular substrate concentrations?

The M_4 isozyme of LDH will again serve as a model enzyme in our analysis, since the most data on K_m values and substrate (pyruvate) concentrations in different vertebrates are available for this system. Figure 11-10 shows the effect of measurement temperatures on the K_m of pyruvate for M_4-LDHs of a broad suite of vertebrates, including an endotherm (rabbit) and ectotherms adapted to temperatures between approximately -1.86°C (the Antarctic teleost, *Pagothenia borchgrevinki*) and 35–47°C (*Dipsosaurus dorsalis*, a desert lizard). For each LDH homologue temperature increases raise the K_m of pyruvate, and the slope of this relationship varies somewhat among the LDHs. However, at the physiological temperatures of the species, indicated by solid lines connecting the measured K_m values, a very strong conservation of K_m is noted. All of the M_4-LDHs examined display K_ms of pyruvate in the range of 0.15–0.35 mM at normal body temperatures. The biological significance of this conservative pattern is indicated by the data in Table 11-5. Pyruvate concentrations are highly similar among different vertebrates under resting conditions (pyruvate levels

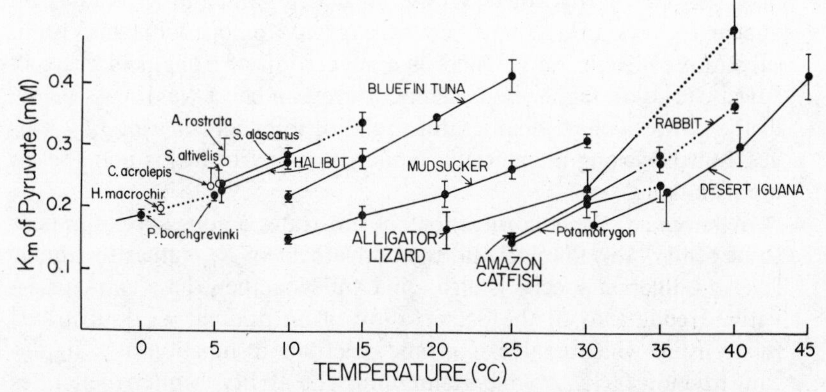

Fig. 11-10. Apparent K_m of pyruvate versus temperature for M_4-LDHs of vertebrates having different adaptation temperatures. The dark lines connecting the K_m values indicate physiological temperatures for the different species. Ninety-five percent confidence intervals around the K_m values are indicated. The species studied include a mammal (rabbit); a reptile, the desert iguana, *Dipsosauris dorsalis*; and a variety of fishes (all other species shown). Figure modified after Somero et al. (1983)

Table 11-5. Pyruvate Concentrations in Skeletal Muscle of Different Vertebrates

Species (condition)	[pyruvate] (mmol/kg tissue)
Dog (resting muscle)	0.33
Rat (resting muscle)	0.18
Frog (resting muscle; 20°C)	0.11
Trout (resting muscle; 12°C)	0.11
Carp (resting muscle; 8–14°C)	0.12
Eels (resting muscle; 15°C)	0.14
Goldfish (resting muscle; 5°C)	0.06
Goldfish (resting muscle; 25°C)	0.33
Goby fish (resting muscle; 15°C)	0.02
Goby fish (resting muscle; 25°C)	0.04
Goby fish (exhausted animal; 25°C)	0.33

SOURCES: Data for the goby, *Gillichthys mirabilis*, are from Walsh and Somero (1982). Other data are from various sources, reviewed by Yancey and Somero (1978b)

may rise during vigorous exercise, of course, and the M_4-LDHs all appear to have an adequate reserve capacity to cope with this rise in substrate concentration). There is a suggestion of a slight increase in [pyruvate] with higher body temperatures, an effect we discuss below in the context of K_m-temperature relationships and *in vivo* Q_{10} values, but the strong interspecific similarity in [pyruvate] is nonetheless unmistakable.

In summary, the concentrations of substrates among species appear to be remarkably similar, and the adjustment of K_m values to similar levels in different species is also apparent. Together, these two conservative trends lead to the preservation of an optimal K_m-[substrate] relationship which enables enzymic reactions to function at a significant fraction of V_{max}, while maintaining the ability to alter their rates significantly in response to modulator signals.

TEMPERATURE-K_m-*pH* RELATIONSHIPS: K_m CONSERVATION THROUGH *pH* VARIATION. Up to this point in our discussion of K_m conservation we have not indicated whether the observed interspecific similarity in K_m values noted for M_4-LDHs and all other enzymes so examined (Somero, 1978) is a function of amino acid substitutions in the enzymes or, alternatively, due to microenvironmental influences in the solution bathing the enzymes. We will see, in fact, that both types of adaptations play important roles in maintaining the K_m values of M_4-LDHs near levels optimal for correct function. Microenvironmental changes are important over all time courses of temperature adaptation as well, an exceedingly important point.

The micromolecule responsible for adaptively adjusting K_m of pyruvate values is the proton. We must examine in detail the relationships among pH, body temperature, and enzyme functional properties, for these relationships are pervasive among biochemical systems, and are not restricted to enzyme-substrate interactions. Figure 11-11 illustrates the variations in blood and cytosol pH with temperature noted for a wide range of animals, vertebrate and invertebrate, ectotherm and endotherm (see Reeves, 1977, for review). As a general rule, albeit a rule with interesting exceptions, as discussed later, the pH values of blood and cytosol decrease with rising temperature, with the approximate slope of -0.017 pH units per degree C rise in temperature. This temperature-dependent variation in blood pH and cytosol pH (pH_i) is observed a) during acclimation of individual organisms, and b) in differently adapted species. Thus, to cite extreme cases, the pH of resting muscle cytosol of a mammal at 37°C is near 6.7–6.9, while the pH_i of muscle of a -1.9°C Antarctic fish should fall in the range of approximately 7.3–7.5. What are the consequences of these large (almost order

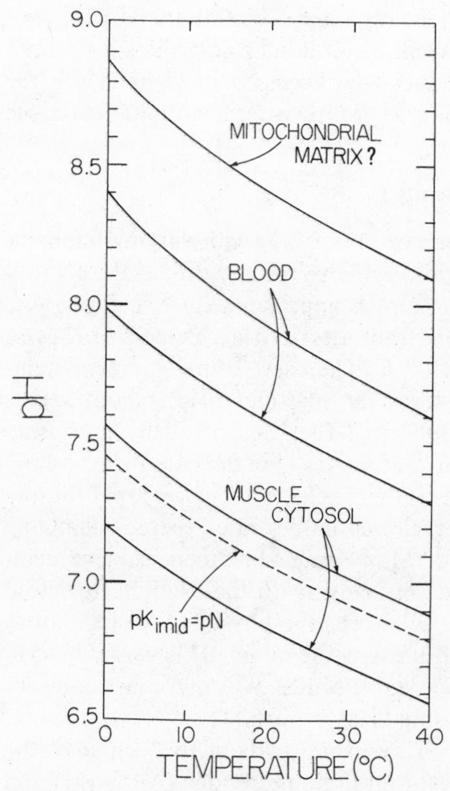

Fig. 11-11. The relationships of blood, muscle cytosol, and mitochondrial matrix pH to body temperature in animals. The effect of temperature on the pK of imidazole (pK_{imid}) is also shown. The effect of temperature on the neutral pH of water (pN) is seen to be the same as the pK_{imid} function. Ranges of measured blood and muscle pH are indicated. The mitochondrial matrix pH values are conjectural, as they have not been measured as a function of temperature. Figure modified after Somero (1981)

of magnitude) differences in proton activity among different species? And, what are the effects of short-term variation in pH for a eurythermal ectotherm?

CONSERVATION OF IMIDAZOLE PROTONATION STATE. We can glean some appreciation of the benefits of this temperature-pH relationship through further study of LDH-pyruvate interactions, for these are appropriate examples for illustrating the phenomenon that lies at the heart of the temperature-pH pattern of Figure 11-11: the conservation of imidazole charge state. Reeves (1972) has suggested that the critical parameter that is *conserved* by the observed variation in pH with temperature is the fractional dissociation of imidazole groups e.g., those of protein-bound histidyl residues. As discussed in Chapter 10, the fractional dissociation of imidazole groups is termed alpha-imidazole (α-imidazole), and is defined as:

$$\alpha_{imid} = [imid]/([imid] + [imid \cdot H^+])$$

The physical basis for the conservation of α_{imid} with varying temperature and pH is the dissociation enthalpy of the imidazole protonation step. The dissociation enthalpy is approximately 7 kcal/mol and, as shown in Figure 11-11, this translates into a change in pK with temperature of -0.017 pH units/°C. The curve of pK_{imid} versus temperature thus parallels the curves for pH_i and pH_B (blood) versus temperature. It should be emphasized that this parallel relationship among these three curves is not "automatic," but depends on active pH regulation by organisms. Air-breathers achieve the correct pH for any given temperature by altering their ventilatory rates, thereby changing the CO_2 titres of their blood. Many water-breathers achieve acid-base regulation via bicarbonate exchange with the medium (reviewed by White and Somero, 1982). Whatever the physiological regulatory mechanisms employed, the end result is the same: pH is varied in such a manner as to keep α_{imid} at a stable value. We now must consider why this conservative relationship is so important.

For M_4-LDH, the binding of pyruvate to the active center of the enzyme entails coordination with an arginine residue (ARG-171) and a histidyl imidazole group (HIS-195) (see Fig. 11-12). The arginine residue (specifically its guanidino group) will always bear a positive charge at physiological pH values, so pyruvate binding will always be facilitated by ARG-171. A different situation pertains in the case of pyruvate-imidazole interactions, however, and what transpires in the interaction of substrate with HIS-195 serves as an excellent illustration of the importance of α_{imid} regulation (alphastat regulation; Reeves, 1972). At physiological pH values, histidyl imidazole groups are approximately half-protonated, regardless of body temperature. That is, α_{imid} is near 0.5 in the cytosol. For LDH function this is critical in at least two ways, as discussed in Chapter 10. Firstly, binding of pyruvate and lactate is highly sensitive to pH. For pyruvate to bind to the active center, HIS-195 must be protonated (see the inset to Fig. 11-12); for lactate to bind, HIS-195 must be deprotonated. As a first approximation, therefore, holding α_{imid} near 0.5 *allows reaction reversibility*. Were pH values to be extremely low or high relative to the pK of imidazole, enzymic reactions would be strongly poised in one direction. Secondly, by having the set-point of HIS-195 protonation near

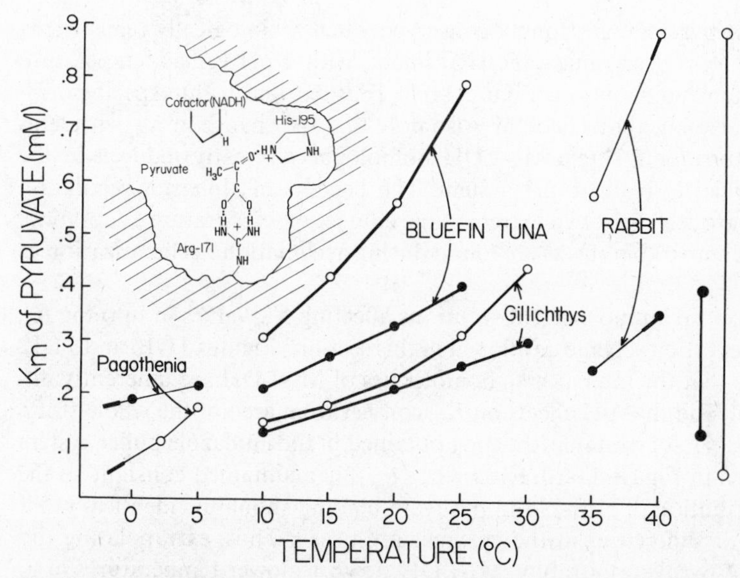

Fig. 11-12. The effect of experimental temperature on the apparent K_m of pyruvate for M_4-LDHs of three vertebrates adapted to different temperatures. Assays were performed under two pH regimens: temperature-dependent pH in an imidazole buffer system (closed symbols; assay conditions identical to those for Fig. 11-9), and a constant pH (phosphate) system (open symbols). For the latter system, the pH was maintained at 7.4 at all temperatures. The vertical lines to the right of the figure illustrate the range of K_m values at physiological temperatures found under the two assay systems. The inset to the figure shows key amino acid residues of the LDH binding site. Note the stabilization of pyruvate binding by the fixed charges of arginine-171 and the protonated form of histidine-195. Figure from Somero (1981)

0.5, decreases in pH can have the effect of enhancing pyruvate binding, and decreasing lactate binding. During exercise, when the working skeletal muscle may depend strongly on anaerobic glycolysis, the fall in pH during muscle activity will activate the pyruvate reductase function of LDH through enhancing pyruvate binding. We again see the benefits of maintaining a reserve capacity for increased rates of function in response to modulator signals (H^+ in this case).

Returning now to the question of K_m conservation, we find that the pH-temperature relationship just discussed leads to significant stabilization of K_m values in the face of short-term and evolutionary time course changes in body temperature (Fig. 11-12). Under the nonphysiological (but experimentally common!) conditions of constant pH, the variation in K_m of pyruvate among interspecific homologues of M_4-LDH is approximately tenfold (open symbols of Fig. 11-12). However,

K_m of pyruvate is strongly conserved when a biologically realistic pH regimen is used (imidazole/HCl buffer, with a pH versus temperature relationship shown in Figure 11-11 [dashed line]). Interspecific variation is only about twofold. Also note that the change in K_m with temperature for a single M_4-LDH homologue is greatly reduced in the biologically realistic pH system. The benefits of alphastat regulation thus are seen over two extreme time courses of temperature adaptation, and comparable effects are found during acclimation/acclimatization as well.

How do amino acid substitutions affecting K_m values fit into the K_m conservation scheme developed in this section? Figures 11-10 and 11-12 show that the interspecific homologues of M_4-LDH are inherently different, and that pH effects on K_m conservation are not the whole story. Consider, for example, the data obtained in the imidazole buffer system shown in Figure 11-10. Even when α_{imid} is maintained constant so the contribution of HIS-195 to pyruvate binding should be identical in all species, differences in the enzymes are noted. Thus, extrapolating the K_m of pyruvate of rabbit M_4-LDH down to lower temperatures, e.g., those of the Antarctic fish, *Pagothenia*, we would find that the mammalian LDH has a greatly lower K_m value, which may reflect a much higher inherent affinity for substrate. Modifications in the amino acid compositions of LDHs (and other enzymes) are needed to ensure that K_m conservation is maintained. pH changes can achieve only part of the needed K_m stabilization, a conclusion which follows directly from the fact that interactions besides those involving imidazole groups are involved in stabilizing the LDH-pyruvate complex and, indeed, most other enzyme-substrate complexes. For enzymes lacking imidazole-substrate interactions, the alphastat regulatory scheme may have little if any effect on K_m values (or K_m stabilization in the face of temperature changes). For these enzymes K_m stabilization may have to be achieved entirely through changes in amino acid sequence. The evolutionary modification in enzyme primary structures to produce correct K_m values may have an important side effect on catalytic efficiency. As discussed above, activation free energy barriers may be influenced by the amount of energy involved in stabilizing enzyme-ligand interactions. The additional energy commitment for stabilizing enzyme-ligand complexes of high-body-temperature species may exact the cost of reduced catalytic efficiency.

THRESHOLDS OF TEMPERATURE STRESS: AT WHAT POINT DO BODY TEMPERATURE CHANGES "START TO MATTER"? There seems little doubt that species living in widely different thermal environments, e.g., polar and tropical seas, have enzymes whose properties are adapted for the temperature range at which catalysis and regulation must take place. Indeed, the conservation of K_m values found for M_4-LDHs and other families of enzyme homologues (Somero, 1978) provides us with firm grounds for speaking of "optimal temperatures" for enzyme function. These are the temperatures at which appropriate K_m values (or, more exactly, appropriate relationships between K_m and substrate concentration) are found. The question we now will address concerns the minimal difference in average body temperature that is needed to select for temperature-adaptive changes in enzymes.

A particularly appropriate experimental approach to this question about threshold effects involves the use of closely related congeneric species which are generally similar in anatomical and ecological attributes, but which live in environments having slightly different temperatures. Congeners offer the advantage of minimizing the dangers of "apple to orange" (*Pagothenia* to rabbit!) comparisons, since these broad comparisons could lead to confusion about which of the observed effects are related strictly to differences in body temperature, and which are due to "fishness" or "rabbitness," for example.

We will consider a quartet of closely related species living in habitats which differ by, at most, 6–8°C in temperature. Barracuda species (genus *Sphyraena*) of the eastern Pacific are found in cool temperate zones on both sides of the equator, in slightly warmer waters of the Gulf of California, and in warm tropical waters (Fig. 11-13, upper left panel). All *Sphyraena* congeners are pelagic schooling fishes with similar body forms and general ecologies (Graves and Somero, 1982). Thus, there is a strong basis for arguing that any differences in the kinetic properties of homologous enzymes from these four species, adapted to three different thermal regimens (Fig. 11-13; Table 11-6), are reflections of temperature adaptation.

All of the data obtained in comparative studies of purified M_4-LDH homologues from these four congeners bolster the arguments raised above concerning the role of activation free energy differences in temperature adaptation, and the importance of conserving K_m values. It is especially significant that differences are found among species which experience habitat temperatures differing by *only a few degrees C* (Fig. 11-13; Table 11-6). It is also interesting that the two temperate zone species, the north temperate *S. argentea* and the south temperate *S. idiastes*, display identical kinetic properties. These two M_4-LDHs also are identical by standard electrophoretic analysis, while the homologues of the other two species display different electrophoretic mobilities (Fig. 11-13). The two temperate species have been separated for the past 3–4 million years, and the finding that their M_4-LDHs are apparently identical offers a compelling case for the role of maintaining enzyme kinetic parameters within specified narrow ranges.

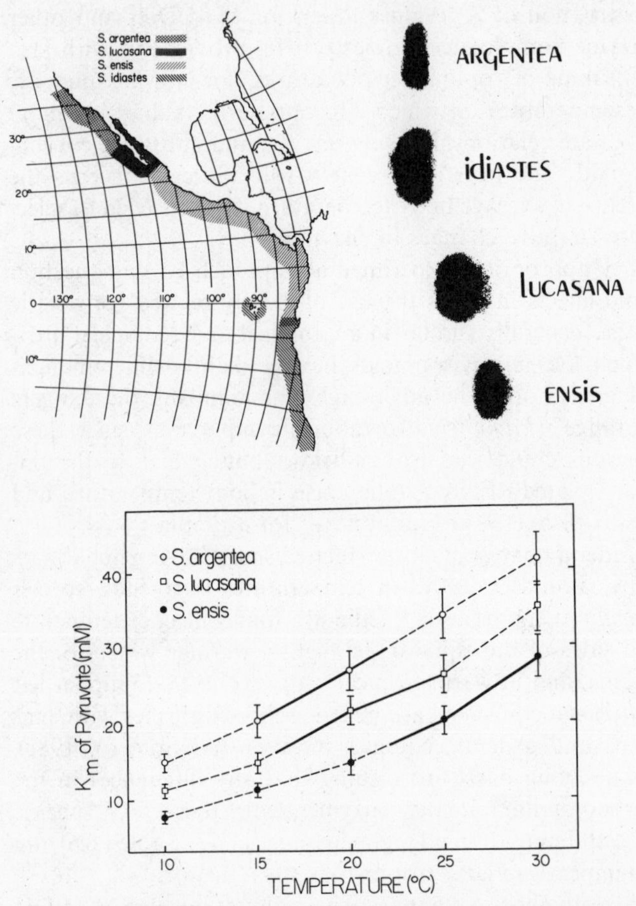

Fig. 11-13. Distribution patterns and electrophoretic and kinetic properties of M_4-LDHs of four species of eastern Pacific barracudas (genus *Sphyraena*). Upper left panel: The distribution patterns of the four *Sphyraena* congeners. Upper right panel: Electrophoretic patterns of the M_4-LDHs of the four species. Note the identical migration patterns of the enzymes from the two temperate zone species, *S. argentea* (north temperate) and *S. idiastes* (south temperate). Lower panel: The effect of assay temperature on the K_m of pyruvate for the M_4-LDHs of *S. argentea*, *S. lucasana*, and *S. ensis*. The K_m values for *S. idiastes* are not plotted, but are indistinguishable from the values for the enzyme from *S. argentea*. Figures from Graves and Somero (1982)

Table 11-6. Kinetic Parameters for the Lactate Dehydrogenase Reactions of Three Barracuda Congeners: *Sphyraena argentea* (temperate), *S. lucasana* (subtropical), *S. ensis* (tropical)

	S. argentea	*S. lucasana*	*S. ensis*
K_m of pyruvate at 25°	0.34 ± 0.03 mM	0.26 ± 0.02 mM	0.20 ± 0.02
k_{cat} at 25°	893 ± 54 sec.$^{-1}$	730 ± 37 sec.$^{-1}$	658 ± 19 sec.$^{-1}$
Temperature midrange (TM)	18°	23°	26°
K_m of pyruvate at TM	0.24 mM	0.24 mM	0.23 mM
k_{cat} at TM	667 sec.$^{-1}$	682 sec.$^{-1}$	700 sec.$^{-1}$

SOURCE: Graves and Somero (1982)

For all four congeners, K_m of pyruvate values at average body temperatures are virtually identical, and k_{cat} values at body temperature are also highly conserved. This comparison of closely related congeners from different temperature regimens is taken as a strong argument for the conclusion that the interspecific differences noted in the broad comparisons of mammals and differently adapted ectotherms do, in fact, represent important facets of biochemical adaptation to temperature. Body temperature, not phylogenetic status, is the dominant component in establishing the relationships of enzymes to temperature.

TEMPERATURE-K_m RELATIONSHIPS AND Q_{10} VALUES. For virtually all enzymes that have been studied, increases in temperature cause increases in K_m values. This relationship between temperature and K_m could have pronounced effects on the Q_{10} values of enzymic reactions in the cell. Because substrate concentrations are below saturating, changes in K_m will affect reaction velocity, other things being equal. Thus, as temperature rises, the K_m value can move further away from physiological substrate levels—assuming these are not changing as well—with the result that Q_{10} values at subsaturating concentrations of substrate are much lower than those found at high or saturating substrate concentrations. Table 11-7 documents this effect. However, some of the data of Table 11-5 suggest that an assumption often made in discussions of K_m-Q_{10} effects may be invalid: physiological substrate concentrations do not always remain stable as temperature changes, either during short-term fluctuations occurring over periods of minutes or during acclimation. In fact, for the LDH-pyruvate system of the

Table 11-7. The Effects of Substrate Concentrations on Q_{10} Values for Enzymic Reactions in which the K_m of Substrate Increases with Rising Temperature

	M₄-lactate dehydrogenase[a]				Pyruvate kinase[b]
[pyruvate] (mM)	*Dipsosaurus dorsalis* (lizard) Q_{10}	*Nezumia bairdii* (fish) Q_{10}	*Coryphaenoides carapinus* (fish) Q_{10}	[PEP] (mM)	(Alaskan king crab; *Paralithoides camtschatica*) Q_{10}
0.100	1.49	1.28	1.21	0.02	1.6
0.125	1.58	1.40	1.31	0.05	1.9
0.250	1.71	1.42	1.36	0.20	2.2
0.500	1.87	1.50	1.41	0.50	3.0
1.000	1.85	1.63	1.52		
V_{max}	2.04	1.70	1.72		

NOTE: See Fig. 11-10 for M_4-LDH. V_{max} Q_{10}s are for theoretical V_{max}
[a] Q_{10}s for the temperature range 20–30°C. Unpublished data of Somero
[b] Q_{10}s for the temperature range 5–15°C. From Somero (1969)

estuarine teleost fish, *Gillichthys mirabilis*, the concentration of pyruvate in skeletal muscle shows a temperature dependence that actually exceeds the variation in K_m of pyruvate with temperature. Consequently, for this particular enzymic reaction, not only is the rise in K_m with temperature ineffective in lowering the Q_{10} value at nonsaturating [pyruvate], but the slightly greater increase in [pyruvate] relative to the rise in K_m of pyruvate leads to a higher Q_{10} (2.1) at physiological [pyruvate] than the Q_{10} computed for V_{max} (Fig. 11-14). We propose that the retention of a relatively stable ratio of [pyruvate] to K_m of pyruvate is of greater adaptive value than the Q_{10} reduction that would occur if K_m rose considerably faster than [pyruvate]. How general this tracking of K_m and [substrate] is among different enzymic reactions remains to be established. Likewise, the extent to which temperature-dependent increases in K_m values can contribute to reductions of biological Q_{10} values (termed "positive thermal modulation," Hochachka and Somero, 1973) is yet unclear. What is more apparent, however, is the adverse effect that rises in K_m with *decreasing* temperature can have on metabolism. Sharp rises in K_m with falling temperature ("negative thermal modulation") lead to dramatically high Q_{10} values at nonsaturating substrate concentrations, due to the combined effects of a) lower kinetic energy and, therefore, a slowing of catalysis, and b) reduced

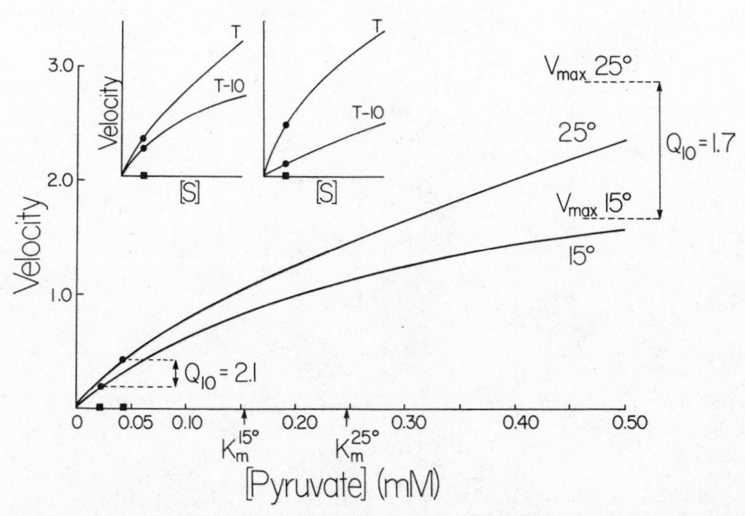

Fig. 11-14. The effects of temperature-dependent changes in K_m and [substrate] on the temperature coefficients (Q_{10}) of enzymic reactions. The major figure illustrates the effects of changes in the K_m of pyruvate and [pyruvate] on the M_4-LDH reaction of *Gillichthys mirabilis*. Data are from Walsh and Somero (1982). The solid squares on the [pyruvate] axis illustrate the measured skeletal muscle pyruvate concentrations in 15°- and 25°C-acclimated specimens, and the solid circles on the saturation curves denote the rates of LDH activity at 15° and 25°C at these pyruvate concentrations. Since both the K_m of pyruvate and [pyruvate] rise with acclimation (experimental) temperature, no Q_{10} reduction at nonsaturating concentrations of pyruvate is observed. Indeed, the Q_{10} at physiological [pyruvate] is slightly higher than the V_{max} Q_{10} (the Q_{10} value which would be found at *all* pyruvate concentrations if K_m were independent of temperature)

The left inset shows the Q_{10}-reducing effects of K_m decreases with falling temperature when physiological substrate concentrations do not change with temperature. This pattern corresponds to "positive thermal modulation" (Hochachka and Somero, 1973). The right inset shows the Q_{10}-increasing effects of K_m increases with falling temperature, under conditions where the substrate concentration does not change with temperature. This pattern corresponds to "negative thermal modulation" (Hochachka and Somero, 1973). Physiological [substrate] ([S]) is again indicated by the closed square, and physiological reaction rates are shown by closed circles

ability to function at low substrate concentrations. Effects of this sort have been noted for several enzymes, including acetylcholinesterase (Fig. 11-15). The nature of isozyme variants of acetylcholinesterase (AChE) is such as to suggest the selective advantage of avoiding negative thermal modulation—high Q_{10} values at low temperatures, and it is worth considering how AChEs differ among species and, in the case

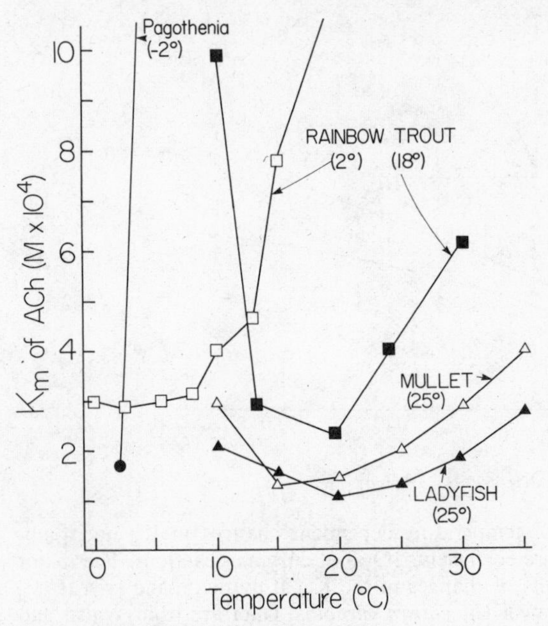

Fig. 11-15. The effect of temperature on the apparent Michaelis constant (K_m) of acetylcholine (ACh) for acetylcholinesterases from fishes adapted or acclimated (rainbow trout) to different temperatures. The adaptation or acclimation temperature of the organism is given in parentheses beneath the species' name. The two isozyme forms of acetylcholinesterase found in cold (2°) and warm (18°) acclimated trout display different electrophoretic mobilities. Data from Baldwin (1971), and Baldwin and Hochachka (1970). Reprinted by permission from Baldwin (1971). Copyright 1971 Pergamon Press, Ltd., and Baldwin and Hochachka (1970). Copyright 1970 The Biochemical Society.

of the rainbow trout (*Salmo gairdneri*), in differently acclimated specimens of the same species.

Isozymes and temperature acclimation. In discussing the M_4-LDHs of the barracuda congeners, we learned that differences in average habitat temperature of only 6–8°C are sufficient to favor the evolution of distinct variants of enzymes having kinetic properties suitable for the different temperatures encountered by the species. In view of the fact that many eurythermal ectotherms experience annual temperature changes and even diurnal temperature changes in excess of 6–8°C, it is very likely that one form of a particular class of enzyme may not be capable of functioning satisfactorily over the entire body temperature range of the species. Even though alphastat regulation is found to dampen the temperature-dependent changes in K_m for some systems, e.g., LDH-pyruvate interactions, other enzymes may not be aided by this microenvironmental adjustment. In the latter cases, two or more variants, either multiple locus isozymes or allelic isozymes (allozymes) may be required to enable the organism to cope successfully with a broad temperature range.

The acetylcholinesterases (AChEs) of fishes provide an example of the role which multiple locus isozymes can play in temperature acclimation. Figure 11-15 illustrates the effect of temperature on the K_m of substrate (acetylcholine-ACh) for AChE variants of different species and differently acclimated rainbow trout (*Salmo gairdneri*). The interspecific pattern of K_m conservation is similar to that noted previously for LDHs. For the trout, however, a similar pattern is found intraspecifically; 2°C-acclimated fish differ from 18°C-acclimated fish. The AChEs of the differently acclimated trout display similar K_m values at their respective acclimation temperatures, but the temperature-dependent variation in K_m of ACh is different for the cold- and warm-acclimated trouts' enzymes. Most important is the rapid increase in K_m of ACh below approximately 15°C for the enzyme from the warm-acclimated specimens. The putative biological significance of this sharp rise in K_m with falling temperature derives from the synergism between rising K_m and falling temperature on reaction velocity. The combined effects of reduced substrate binding (assuming that K_m serves as a measure of this capacity) and reduced kinetic energy for driving the reaction over its activation enthalpy "barrier" can lead to Q_{10} values greater than 10 at physiological substrate concentrations (Hochachka and Somero, 1973). Thus, in addition to the importance of maintaining the appropriate K_m of substrate over a range of body temperatures, an organism must prevent rapid increases in K_m with falling temperature. In the rainbow trout, the use of two isozyme forms of AChE solves both problems. How the differential adjustment of isozyme levels is achieved remains to be discerned. Season-specific patterns of isozyme synthesis may be important, although in another system, trout isocitrate dehydrogenase (IDH), Moon and Hochachka (1971) found that a complex family of isozymes having different K_m-versus-temperature patterns was present at all seasons. The IDH system seems to resemble the "backup" hemoglobin systems discussed in Chapter 9. In these complex systems there exist reserve forms of molecules for function under altered environmental conditions.

How common is the "multiple isozyme" strategy of enzymic adaptation to temperature in the case of eurythermal ectotherms? Presently available evidence suggests that most species do not synthesize distinct "cold" and "warm" isozyme variants like those found in the AChE system of rainbow trout (see Somero, 1975; Shaklee et al., 1977).

Why is the "multiple isozyme" strategy not used more widely? A partial answer to this question can be developed on the basis of an important characteristic of certain species in which the "multiple isozyme" strategy occurs. For rainbow trout it is apparent that a tetraploid condition is characteristic of the genome (Ohno, 1970). That is, in contrast to a normal, diploid organism, the rainbow trout and many of its relatives have essentially twice as much genetic information to utilize in coping with different environments. The luxury of having this much genetic information may enable the rainbow trout and other tetraploid species to synthesize season-specific forms of enzymes (Somero, 1975). In contrast, for species lacking this redundant genetic information, only a single gene locus may be available for coping with the full range of habitat conditions. Thus, the occurrence of the "multiple isozyme" strategy may be restricted to organisms, like rainbow trout, which have enough genetic loci to permit the evolution of environment-specific forms of many types of proteins.

In the context of environment-specific isozymes it is important to realize that different isozyme forms often are found on tissue-specific and even organelle-specific bases. These multiple isozyme systems derive from gene duplication and subsequent evolutionary divergence of the two (or more) loci. In the rainbow trout and other tetraploid animals, the duplication of a genome which already contained a full battery of tissue-specific isozymes appears to have occurred. Thus, the additional genetic information acquired by these tetraploids did not have to be used to develop tissue- or organelle-specific isozymes, but rather could be used to generate isozyme forms specific for different *external* environments.

Allozymic variants and temperature adaptation. The uncertain status of multiple locus isozyme variants in temperature acclimation/acclimatization is shared by allelic variants (allozymes). Even though many workers have examined the role of allozymes in environmental adaptation (reviewed by Hedrick et al., 1976), there are very few clear-cut examples of allozymes possessing adaptively different functional or structural properties that correlate with environmental differences. To some extent the dearth of good examples of allozymic adaptations to temperature may reflect the experimental protocols commonly used in allozyme studies. Many such studies involve relatively crude techniques such as measurement of heat stabilities using tissue homogenates.

Few studies have involved fully purified enzymes which were assayed under biologically realistic *in vitro* conditions, e.g., of pH.

One study which satisfies the criteria of good experimental design and which provides a convincing example of adaptively different allozymes is the investigation of LDH isozymes in the teleost fish, *Fundulus heteroclitus*, by Place and Powers (1979). The heart-type (B) isozyme of LDH in liver was examined in populations of *Fundulus* inhabiting different environments along the east coast of the United States, from Maine to Florida. Two allozymes of LDH-B were found (LDH-B^a and LDH-B^b). The (b) gene was dominant in northern populations, but was progressively replaced by the (a) gene along a north to south gradient (along this coastal region, a 1°C increase in average water temperature occurs for each degree change in latitude). The kinetic characteristics of the purified LDH-B allozymes differed in a pattern that reflected the temperature gradient in the fishes' habitats. Using the ratio of k_{cat}/K_m (a measure of catalytic rate at physiological substrate concentrations) Place and Powers found that the (b) allozyme had a higher rate at 10°C than the (a) allozyme, while the (a) allozyme was superior at temperatures above 25°C. The maximal value of k_{cat}/K_m was attained at 20°C for the "cold" (b) allozyme, and near 30°C for the "warm" (a) allozyme. This study, while limited to one enzyme of one species, offers strong support for the hypothesis that allelic polymorphism may "fine tune" enzyme systems in different populations of a given species which inhabit environments differing in temperature.

There are other, somewhat less definitive studies, which argue for the same conclusion. Studies of esterases (Koehn, 1969), lactate dehydrogenases (Merritt, 1972), phosphoglucoseisomerases (Watt, 1977), and alcohol dehydrogenases (see Malpica and Vassallo, 1980) are in general agreement with the findings of Place and Powers. The study of Malpica and Vassallo (1980) is of particular interest in that they extended the examination of adaptive allozyme differences to the level of the genetic background effects that determine not the genotype (allozyme) which is present, but rather its actual *in situ* activity. Malpica and Vassallo studied the activities of the "fast" (F) allozyme of alcohol dehydrogenase (ADH) in populatons of *Drosophila melanogaster* (fruit fly) living in different thermal regimes. The F allozyme of ADH has a higher specific activity than the "slow" (S) allozyme, and the F form of the enzyme tends to be found in colder habitats. Since differences in enzymic activity between individuals can be due to variations in catalytic efficiency (k_{cat}), i.e., to structural gene differences, and to genetic background effects, e.g., factors that may determine enzyme concentration,

Malpica and Vassallo reasoned that differences in genetic background, as well as in allozyme type per se, could play a role in temperature adaptation. Moreover, they argued that a correlation of genetic background effects with temperature would provide strong evidence that the differences in allozyme type were adaptively important. Their results show that, indeed, background effects fostering high levels of ADH activity were more prevalent in cold than in warm habitats. Their findings demonstrate, therefore, that allozyme variants may have their activities significantly modified by the genetic background in which they are placed. Perhaps the correct way to view the results of Malpica and Vassallo, in terms of the preceding discussion of adaptation in kinetic properties, is to view changes in enzyme primary structure as being responsible for establishing optimal kinetic properties, while total tissue activity is adjusted in large measure by varying enzyme concentration (total activity = $k_{cat} \cdot$[enzyme]). This conclusion leads us to consider another aspect of enzymic adaptation to temperature—the adjustments in enzyme concentration that play important roles in the process of metabolic temperature compensation on an intraspecific basis.

TEMPERATURE COMPENSATION VIA CHANGES IN ENZYME CONCENTRATION. The discussion of enzymic mechanisms of temperature adaptation up to this point has centered on what can be termed "qualitative" changes in enzyme systems. Emphasis has been on isozyme or allozyme variants of enzymes having adaptively different kinetic properties, notably K_m and k_{cat} characteristics. Another way of achieving temperature-adaptive changes in metabolic systems is via "quantitative" adaptations—changes in the concentrations of the same type of enzyme. Adjustments of this nature seem particularly important in seasonal adjustments of metabolic rates in a temperature-compensatory fashion (Figs. 11-5 and 11-7). Ectothermic species, animals (Fig. 11-5) and plants (Fig. 11-17), frequently exhibit a capacity to offset the effects of decreasing temperature on metabolic processes and thereby maintain a relatively stable metabolic rate at different seasons. In most of the cases studied the organisms did not appear to utilize season-specific isozyme forms, but rather altered the concentrations of many enzymes, especially enzymes likely to be rate-limiting to different metabolic pathways (Hazel and Prosser, 1974; Shaklee et al., 1977). In animals the enzymes of aerobic metabolism often show particularly strong temperature compensation (Hazel and Prosser, 1974; Sidell et al., 1973), while glycolytic enzymes like LDH may show no compensation or even an inverse compensation. The disparity between aerobic and anaerobic enzymic systems in water-breathers like fishes may be

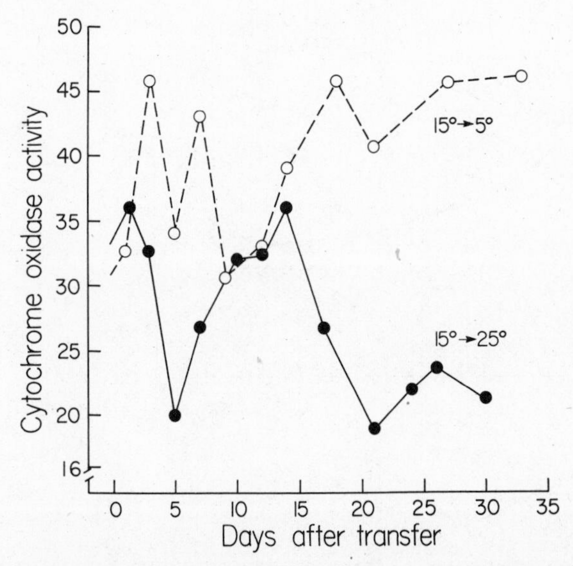

Fig. 11-16. Cytochrome oxidase activity in goldfish skeletal muscle as a function of time at different acclimation temperatures. Fish were first acclimated to 15°C and then transferred to either 5°C (open circles) or 25°C (closed circles). Cytochrome oxidase activity is expressed as micromoles of cytochrome-C oxidized per second per mg of protein in the muscle homogenate. Data from Sidell et al. (1973)

reflection of the role played by dissolved oxygen concentrations in determining the metabolic balance of the organism. Thus, in addition to compensation for temperature effects on metabolic rates, the metabolic systems may need to adjust to different oxygen availabilities. The increased oxygen available at low temperatures would facilitate aerobic metabolism, while warm-acclimated individuals may need an enhanced glycolytic capacity, especially during burst locomotion. Nonetheless, at least the adjustments in aerobically poised enzymes like cytochrome oxidase (Fig. 11-16) and succinic dehydrogenase (Sidell et al., 1973) appear interpretable as mechanisms for ensuring that changes in body temperature do not establish disadvantageously low or high metabolic rates. Certainly the changes in aerobically poised enzymes provide a good mechanistic explanation for the concomitant changes noted in oxygen consumption (Fig. 11-5).

In the examination of enzyme concentration changes during acclimation it is important to consider both the kinetics of the process and the final, equilibrium state. Figures 11-16 and 11-17 provide both types of

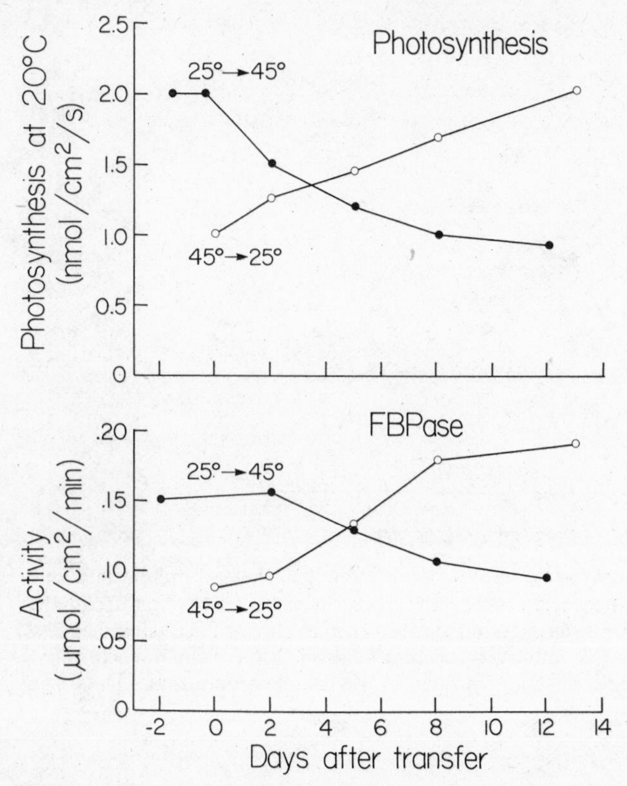

Fig. 11-17. The time course of change in light-saturated photosynthetic capacity measured at 20°C (nmoles net CO_2 uptake/cm^2 of leaf area/second) [upper panel] and fructose-1,6-biphosphatase (FBPase) activity [lower panel] in mature leaves of *Nerium oleander* grown under different acclimation regimens. Plants were first acclimated to either 25°C or 45°C and then transferred to the other temperature. Figure modified after Bjorkman et al. (1980)

information. In the goldfish system studied by Sidell et al. (1973), a considerable amount of fluctuation in enzyme levels occurred before a new steady-state was reached. This effect was noted both in 15° to 5° transfers and 15° to 25° transfers. The basis and significance of these oscillations are not known. In the plant study (Fig. 11-17), a more regular change in the Calvin-Benson cycle enzyme fructosebisphosphatase was observed. Activity of this enzyme exhibited a smooth change in temperature step-up and step-down experiments, and the similarity in kinetics of change in enzymic activity and photosynthetic capacity is striking. Bjorkman et al. (1980) suggest that activity of fructosebisphosphatase may be the critical rate-limiting factor involved in temperature compensation of photosynthesis in this plant (*Nerium oleander*). The *N. oleander* system exhibited a much more rapid adjustment in enzymic activity than that noted in the goldfish study.

It should be emphasized that in most cases where enzymic activity changes during acclimation have been found, clear-cut evidence in the form of actual enzyme concentration measurements has not been obtained. Rather, most studies have measured enzymic activities using tissue homogenates. In a few cases, however, actual enzyme concentrations have been obtained (Sidell, 1977). In other cases in which activity changes are accompanied by no change in electrophoretic pattern of the enzymes studied (e.g., Shaklee et al., 1977), it is reasonable to conclude that concentration changes, not isozyme changes, are responsible for the alterations in enzymic activity.

To what extent do adjustments in enzyme concentration contribute to temperature compensation noted on an *interspecific* basis? This is an extremely difficult question to answer because interspecific differences in metabolism due, for example, to variations in locomotory habit, may obscure metabolic differences due to temperature per se (Sullivan and Somero, 1980). As we have argued at several points, comparisons of closely related congeners may offer the best avenue of attack on such questions. In this regard the study of Wilson et al. (1974) is appropriate to consider. Wilson and co-workers studied two congeners of the genus *Sebastes*. *Sebastes miniatus* lives below the thermocline in summer and experiences relatively cool temperatures (10–12°C) compared to *S. auriculatus*, which lives above the thermocline (temperature range of approximately 12–22°C). Cytochrome oxidase activities in *S. miniatus* muscle were 70 percent higher than in *S. auriculatus* in comparisons involving freshly collected specimens. Cross-acclimation experiments which involved transfer of *S. miniatus* to 20°C and *S. auriculatus* to 10°C led to compensatory shifts in cytochrome oxidase activity; enzymic activity rose in *S. auriculatus* and fell in *S. miniatus*. In contrast to the results obtained in study of cytochrome oxidase, the levels of LDH and PK activity did not differ between species and exhibited no acclimation effect. These results suggest two conclusions. First, these two congeners appear to be acclimatized to their different temperature regimes via holding different concentrations of aerobically poised enzymes in their muscles. The concentration difference clearly is not fixed, as cross-acclimation can largely remove the differences noted in freshly collected specimens. Second, aerobically poised and anaerobically poised enzymes may exhibit different patterns of acclimation.

A final consideration concerning the role played by enzyme concentration changes in temperature compensation on an interspecific scale is that, when changes in temperature perturb the kinetic properties of an enzyme to the extent that these properties are no longer within the optimal range, e.g., in the case of temperature-K_m interactions, achieving temperature compensation via quantitative mechanisms does not appear to be a suitable strategy. There appears little sense in synthesizing high levels of a poorly functional enzyme during cold adaptation, for example. Instead, given sufficient time for changes in primary structure to be effected, the development of new homologues of an enzyme with K_m and k_{cat} characteristics appropriate for the new thermal regimen represents a much more favorable solution to the problem of enzymic adaptation to temperature.

PROTEIN THERMAL STABILITY. Why do organisms die at extremes of temperature? Attempts to answer this question have frequently involved study of the thermal stability of proteins (see Alexandrov, 1977, for a thorough review of this literature). A wide variety of proteins from animals, plants, and prokaryotes have been investigated to determine whether the temperatures at which proteins denature reflect the upper thermal tolerance limits of the organism. In almost all cases an interesting correlation between denaturation temperature (or melting temperature [T_m]) and adaptation temperature has been found: proteins of organisms adapted to high temperatures denature at higher temperatures than proteins from cold-adapted species, yet the T_m of the protein is invariably many degrees higher than the upper lethal temperature of the organism (collagen monomers may represent one exception, as discussed below). This relationship is illustrated by the data of Table 11-8. Does this correlation between body temperature and T_m

Table 11-8. Thermal Stabilities of Homologous Proteins from Organisms Adapted to Different Temperatures

Species (body temperature)	
Skeletal muscle actin[a]	Rate constant of denaturation, k ($\times 10^3$/min)[a]
Dipsosaurus dorsalis (30–47°C) (lizard)	0.00
Rabbit (37°C)	1.76
Chicken (39°C)	1.60
Cyprinodon macularius (10–40°C) (teleost)	4.45
Thunnus alalunga (15–25°C) (teleost)	5.74

Table 11-8. (*continued*)

Species (body temperature)	
Sebastolobus alascanus (3–10°C) (teleost)	4.78
Sebastolobus altivelis (3–10°C) (teleost)	4.70
Pagothenia borchgrevinki (−1.9°C) (teleost)	9.20
Gymnodraco acuticeps (−1.9°C) (teleost)	4.66
Pyruvate kinase[b]	Heat inactivation temperature (°C)[b]
Chicken	60
Rabbit	62
Mugil cephalus (18–30°C) (teleost)	58
Bufo marinus (25–32°C) (amphibian)	57
Sebastes paucispinus (8–17°C) (teleost)	55
Sebastes serriceps (8–17°C)	55
Scorpaena guttata (8–17°C) (teleost)	55
Salmo gairdneri (4–15°C) (teleost)	56
Pagothenia borchgrevinki (−1.9°C) (teleost)	42
Collagens[c]	Thermal transition temperature (°C)[c]
Cod skin	12
Cod swim bladder	16
Dogfish skin	16
Earthworm cuticle	22
Carp swim bladder	29
Perch swim bladder	31
Calf skin	39
Ascaris cuticle	52

[a] Data from Swezey and Somero (1982a). The rate constant of denaturation was computed from the rate of loss of native actin during incubation at 38°C. The lower the rate constant, the more heat stable the actin. For example, the actin of the desert iguana, *D. dorsalis*, exhibited no denaturation during the time course of the incubation (up to three hours).

[b] Data from Low and Somero (1976). The heat inactivation temperature was taken as the temperature at which catalytic activity first began to decrease for aliquots of enzyme that were incubated for three minutes at a series of high temperatures.

[c] Data after Bailey (1968). The thermal transition temperature is the temperature at which the helix to coil transition occurs.

have any causal significance? Or are proteins sufficiently strong to remain always in the proper structural and, hence, functional state over the full range of temperatures experienced by the organism (protein)?

An important component of our answer to these questions is the fact that gross denaturation, i.e., the dissociation of subunits and unfolding of tertiary structure and much of secondary structure, is too crude an index of protein state to serve as a reliable guide to the thermal stability of protein-based processes like enzymic catalysis. The proper catalytic and regulatory functions of proteins are apt to be lost at temperatures well below those at which gross denaturation occurs. For instance, we have observed that apparent K_m values generally rise with increasing temperature, such that at temperatures well above physiological temperatures for a particular species the K_m value for a particular enzyme-substrate pair is apt to be much higher than the optimal value (see Figs. 11-10 and 11-15). There is no question but that the structure of the enzyme under these conditions is "native" in the sense that the enzyme retains its subunit aggregation state and tertiary structure. Nonetheless, the reaction catalyzed by the enzyme is not able to retain its optimal properties, e.g., in terms of regulation. For this reason we conclude that serious and perhaps lethal effects on enzymic reactions can be effected by temperatures that are too low to denature the enzyme. The key point to remember is that it is the metabolic reaction catalyzed and regulated by the enzyme that matters to the organism, and a reaction can be severely impaired long before the enzyme itself is irreversibly damaged by high temperatures.

The foregoing analysis still does not explain the correlation between adaptation temperature and T_m, however. One possible basis for this correlation has been alluded to earlier, when we discussed the relationship between catalytic efficiency and adaptation temperature. We argued that enzymes of low-body-temperature species gain some, or all, of their heightened catalytic efficiency via possession of a more flexible structure, a structure that allows catalytic conformational changes to occur with less energy input. If the primary adaptation in enzymes related to temperature is adjustment of k_{cat} values and conservation of K_m, then the correlation between body temperature and T_m of proteins can be rationalized. More flexible structures are associated with a) increased catalytic efficiency and b) a reduced need for large amounts of stabilization energy for holding enzyme-substrate or enzyme-cofactor complexes together. The T_m differences are not, per se, significant temperature adaptations in this view, at least in cases for which the thermal stabilities of all homologues of a particular type of enzyme are greater than the highest body temperature of any species examined.

There is a group of organisms for which enhanced protein thermal stability clearly is necessary to permit survival at their extreme environmental temperatures. These are the thermophilic bacteria, which may thrive in waters with temperatures approaching 100°C. Proteins of non-thermophiles typically denature at the normal growth temperatures of thermophiles (Table 11-8), so the amino acid compositions of proteins of thermophiles might be expected to display striking differences from the sequences of homologous or analogous proteins of mesophilic and psychrophilic ("cold-loving") species. Alternatively, the cytoplasm of a thermophilic bacterium might contain substances for protecting proteins from thermal denaturation.

Studies of a variety of proteins from thermophilic bacteria have shown that, in fact, these proteins are extremely resistant to thermal denaturation (Singleton and Amelunxen, 1973; Singleton et al., 1977). Some of these proteins can tolerate prolonged heating *in vitro* at temperatures near 90°C. Not surprisingly, considerable attention has been directed to the mechanisms that facilitate this high thermal stability. On the basis of available information—and data are quite sparse due to the lack of information on the three-dimensional structures of homologous or analogous proteins of mesophiles and thermophiles (see Argos et al., 1979)—several types of structural adaptations seem of importance in stabilizing the structures of proteins from thermophiles. First, as shown by comparisons of D-glyceraldehyde-3-phosphate dehydrogenase from *Bacillus stearothermophilus* (enzyme structure and activity maintained up to 60–70°C) and *Thermus aquaticus* (enzyme structure and activity maintained up to 95°C), only one or two additional electrostatic interactions are needed to provide the required increase in thermal stability, some 5–10 kcal/mole of net stabilization free energy (Biesecker et al., 1977). Increased numbers of electrostatic interactions (salt bridges) have been proposed to be of general significance in increased thermal stability (see Zuber, 1979).

Additional mechanisms for enhancing the thermal stabilities of proteins have been discovered by Argos et al. (1979) who compared data on three-dimensional structures of LDH, D-glyceraldehyde-3-phosphate dehydrogenase, and ferredoxin, purified from thermophiles and mesophiles. These authors conclude that thermal stability in these enzymes is achieved by a substantial number of very small stability-enhancing changes throughout the proteins. The proteins from thermophiles had reduced surface hydrophobicity and enhanced internal hydrophobicity. Because hydrophobic interactions tend to be stabilized by increases in temperature (see Brandts, 1967), these shifts in hydrophobicity seem reasonable. The proteins of the thermophiles also had amino acid substitutions that favored stability of the alpha-helical regions of the molecules. Sheet-forming abilities of amino acid residues in β-sheets also were slightly enhanced. Lastly, for some of the aliphatic sidechains of amino acids buried within the structures of the proteins, an increase in sidechain bulkiness was found; this change could also stabilize hydrophobic interactions.

In considering mechanisms for enhancing the thermal stabilities of proteins from thermophilic bacteria, it is important to emphasize that the net stabilization free energy that maintains the native structures of proteins is only of the order of 30–50 kcal/mole (Brandts, 1967). The net stabilization free energy is the algebraic sum of a number of energy changes, including the decrease in entropy associated with formation of the regularly folded protein structure, the favorable increase in entropy associated with the formation of hydrophobic interactions among buried residues, and the energy changes that accompany the formation of protein-protein hydrogen bonds at the expense of protein-water hydrogen bonds. When all of these energy changes accompanying the folding of a protein into its compact native conformation are added together, the net stabilization free energy is of the order of only several weak bonds, despite the fact that hundreds of weak bonds must be formed and broken during the folding of the protein into its final shape. The low net stabilization free energy of a protein indicates that relatively few additional weak bonds—or relatively little strengthening of existing weak bonds—is needed to create a thermophilic protein from a mesophilic protein. Indeed, the structural studies of Argos et al. (1979) show very clearly how subtle are the changes involved in the building of a thermophilic protein. The analysis of Argos et al. also shows that, for the large majority of the amino acid substitutions involved in transforming a mesophilic protein into a thermophilic protein, only a single base change in the triplet code is needed. Thus, in terms of genetic structure as well as amino acid composition, the building of thermophilic proteins appears a relatively simple evolutionary task.

While the tolerance of high temperatures by proteins of thermophiles is most likely due to modified primary structures in most cases, there may be systems whose thermal stability is established by protective substances. Oshima (1979) has found that four novel polyamines in the thermophile, *Thermus thermophilus*, protect the protein synthetic apparatus of this bacterium from thermal denaturation. Even though the individual components of translation were thermally stable, protein synthesis was essentially blocked at temperatures above 50°C unless polyamines were present. The protective polyamines facilitated translation at high growth temperatures, and the composition and concentration of the polyamines was dependent on culture temperature.

Protein synthesis is an especially sensitive locus to thermal perturbation (reviewed by Bernstam, 1978), so novel mechanisms such as the protective polyamines may be needed to facilitate the proper interactions between messenger RNA, the ribosomes, transfer RNAs, and the protein components of translation.

In this context it is worth discussing modifications in transfer RNA (tRNA) structure in thermophilic bacteria. In *T. thermophilus* two mechanisms have been found to enhance the thermal stability of tRNAs (Oshima, 1979). As in the case of ribosomal RNAs and tRNAs from other thermophiles, the tRNA of *T. thermophilus* has an increased G + C content relative to the homologous RNAs of mesophiles. G:C base pairs are more thermally stable than A:U pairs, so this change in nucleic acid primary structure can easily be rationalized. In addition, the tRNAs of *T. thermophilus* and several other extreme thermophiles contain 5-methyl-2-thiouridine (S). The 2-thiouridine derivatives are known to strengthen the interactions involved in base-stacking in the helical nucleic acid conformation. Thiolation was found to depend on growth temperature; the S content was directly proportional to the temperature at which the cells were cultured. Most likely, the thiolation reactions occurred after transcription. In common with the conclusion reached earlier concerning the number of additional weak bonds needed to convert a mesophilic protein into a thermophilic protein, it is noteworthy that the thiolation mechanism of enhancing tRNA stability involves the addition of only three more weak bonds to a system which is stabilized by a total of more than sixty weak bonds (Oshima, 1979).

While the proteins of thermophilic bacteria may function within a few degrees of their denaturation temperatures in many cases, among animal proteins there is only one example known of a protein whose denaturation temperature is close to the average body temperature of the species. Collagen, the most abundant protein in most animals, comprising up to one-quarter of total protein, exhibits melting temperatures that correlate with average body temperature *and* which lie very close to the upper lethal temperature of a species (Rigby, 1968). The form of collagen in question is tropocollagen, which is a helical coil formed from three linear polypeptide chains. This building block for collagenous structures is subsequently modified through formation of covalent cross-links such as those between modified lysine sidechains. When stabilized by these cross-links, the thermal stability of collagen is much higher than that of tropocollagen. The melting of tropocollagen to form gelatin, a random coil, typically occurs at temperatures very close to the upper lethal limit of the species and is strongly dependent on the proline and hydroxyproline content of the molecule. Low-body-temperature species have lower hydroxyproline contents than warm-adapted species. The placement of hydroxyproline residues within the collagen helix is also thought to affect the thermal stability of the molecule.

Why should a protein be so thermally labile? One possible explanation for the metastability of tropocollagen is that the assembly of more complex collagenous structures requires a flexible building block to work with. Formation of the triple helix (tropocollagen) and the subsequent modification of tropocollagen via removal of peptides from both ends of the molecule by proteolytic enzymes, followed by assembly of collagen fibers and covalent cross-link formations, may dictate that the basic structure of tropocollagen be highly flexible. Once the mature form of collagen is built, however, the system has more than adequate thermal stability to withstand temperatures well in excess of normal body temperatures. Thus, the threat of heat denaturation to collagen-containing systems may be only at the assembly phase; mature, assembled collagenous structures are not subject to this stress.

The treatment of protein structural stability just presented may have given the reader the impression that selection invariably favors proteins with very long life spans. In fact, the half-lives of proteins range from several minutes to several weeks, with an average protein half-life of 3.5 days characterizing rat liver (reviewed by Goldberg and Dice, 1974; Goldberg and St. John, 1976). The most rapid turnover is noted for enzymes that are the rate-limiting valves in their respective pathways (Goldberg and St. John, 1976). The short half-lives of rate-limiting enzymes have been viewed as an indication that protein turnover can play a major role in regulating the activities of multi-enzyme pathways. For instance, when the synthesis of a rate-limiting enzyme ceases, a short half-life for the enzyme ensures a rapid decrease in the pathway's activity.

The variation in half-life among different proteins has been shown to correlate strongly with the resistance of the proteins to thermal denaturation (Goldberg and Dice, 1974). A highly flexible protein structure may favor an enhanced susceptibility to proteolytic attack due to a greater tendency to unfold at normal body temperatures. In the context of selection for thermal stability, we must take into account the possibility that a close conservation of half-life, at physiological temperatures, is adaptive. For example, if low body temperatures favor maladaptively long half-lives for rate-limiting enzymes, then selection could favor reduced thermal stability for proteins of low-body-temperature species to ensure an adequate rate of protein turnover. Whether this type of adaptive change in protein structure is important remains to be determined. Temperature effects on protein turnover are not well understood and appear to be very complex (Somero and Doyle, 1973; Goldberg and Dice, 1974).

REVERSIBLE TEMPERATURE AND pH EFFECTS ON SUBUNIT ASSEMBLY. As the foregoing discussion of temperature effects on protein turnover suggests, temperature induced changes in protein structure, while disadvantageous to an organism in many cases, also have the potential for contributing to the precise regulation of metabolic processes. In addition, the effects of temperature on a protein may amplify the effects of low molecular weight modulators of the protein's activity. Physiologically significant relationships between protein structure, temperature, and the effects of low molecular weight modulators are evident in the regulation of an important metabolic regulatory enzyme we have already met in a variety of contexts, phosphofructokinase (PFK).

Figure 11-18 illustrates the interacting effects of temperature, pH, and [urea] on the activity and structure of PFK from the skeletal muscle of the squirrel, *Citellus beecheyi*, a species that can hibernate. As discussed in Chapter 7, transitions between euthermic (37°C) and hibernating states are marked by large reductions in metabolism and a cessation of thermogenesis, e.g., of shivering. The properties of *C. beecheyi* muscle PFK appear appropriate for facilitating rapid—and reversible—transitions between active and inactive states of muscle glycolysis. Under simulated hibernating conditions of low body temperature (6°C) and acidotic pH (the extra- and intracellular fluids of hibernators are between 0.2 and 0.4 pH units more acidic than would be predicted according to the temperature versus pH curve in Fig. 11-11), PFK tetramers rapidly disassemble into dimers. This change is shown by a loss in catalytic activity (Fig. 11-18A) and a decrease in light scattering (Fig. 11-18B). Light scattering is a sensitive measure of the size of a molecular aggregate, and this technique provides a good index of the extent to which the functional tetrameric state of PFK has dissociated to form the inactive dimer. Over a period of approximately one hour under these *in vitro* conditions, approximately 90 percent of PFK activity is lost. This inactivation of PFK under *in situ* conditions would lead to a virtually complete block to muscle glycolysis and, thereby, to the elimination of shivering thermogenesis and to the sparing of muscle glycogen reserves for use in thermogenesis during recovery of the euthermic state and in locomotory activity following arousal.

The temperature and pH effects on PFK structure and function would not be well suited for regulatory roles were it not for the fact that these effects are rapidly reversible, as shown by the data in Figure

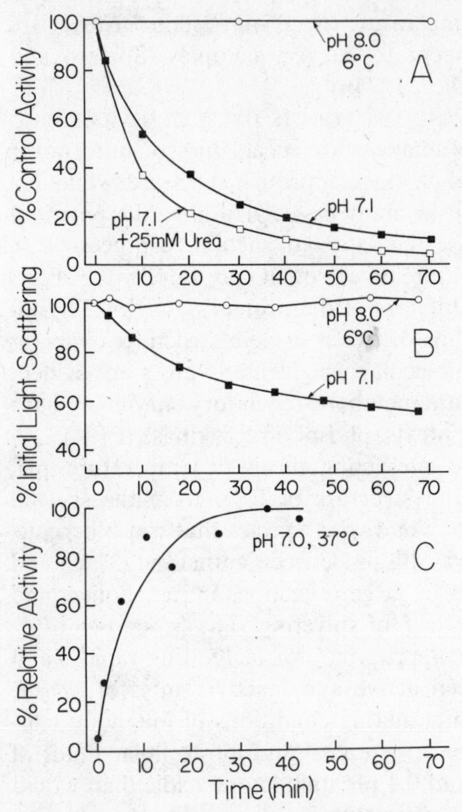

Fig. 11-18. The effects of temperature, pH, and [urea] on the catalytic stability and subunit assembly state of muscle phosphofructokinase purified from the ground squirrel, *Citellus beecheyi*. A. Loss of PFK activity during incubation at different pH values and in the presence or absence of 25 mM urea. The control activity refers to the initial activity of the preparation; activity is completely stable at alkaline pH values for long time periods. B. Change in light-scattering ability of PFK as a function of time and incubation conditions. The decrease in light scattering at pH 7.1 reflects disassembly of the tetramer to the dimer. Since the dimer retains some light-scattering ability, the percentage loss of activity (A) differs from the percentage loss in light scattering. C. Reactivation of PFK that had been inactivated by dialysis at low temperature and low pH. Data from Hand and Somero (1983a), *Physiological Zoology* 56: 380–388. By permission of the University of Chicago Press. © 1983 by the University of Chicago.

11-18C. When PFK that has been fully inactivated by dialysis at low pH and low temperature is placed in a buffer that simulates the pH and temperature conditions of euthermic mammalian muscle (37°C, pH near 7.0), the dissociated PFK rapidly reassembles to restore full activity. This rapidly reversible equilibrium between tetrameric and dimeric states of PFK thus appears to provide a hibernating mammal with a highly responsive on-off switch for muscle glycolysis.

Viewing these temperature and pH effects in the context of the thermal relationships of ectothermic species, we can understand more completely the selective advantages of the body temperature-versus-pH relationship shown in Figure 11-11. The rise in pH with falling body temperature noted in virtually all organisms so examined except hibernating species is adequate to prevent establishment of pH/temperature conditions that are conducive to the loss of PFK tetrameric structure (=activity).

The lability introduced into protein structure by changes in temperature may make the protein increasingly sensitive to other structural perturbants, and these effects may also have regulatory significance. For example, as shown in Figure 11-18A, very low (25 mmoles/L) concentrations of urea enhance the rate of activity loss for PFK. This urea effect appears to be ubiquitous among PFKs from a variety of species (Hand and Somero, 1982, 1983a), and in urea-rich species like cartilaginous fishes and certain estivating teleosts and amphibians, these urea effects may be of physiological significance. In the present context, it is appropriate to consider whether urea could serve as a metabolic inhibitor under conditions that dictate an inhibition of glycolysis. In mammalian hibernators, urea concentrations appear in general not to increase during hibernation, so urea potentiation of low temperature/low pH effects probably is not common. However, in certain estivating species that accumulate high concentrations of urea during estivation (200–400 mmoles/L in the spadefoot toad, *Scaphiopus couchi*; McClanahan, 1967), urea could be an important inhibitor of glycolytic flux. Similar effects could occur in lungfishes that also buildup high urea concentrations during estivation.

The low-temperature-induced disassembly of PFK is but one example of cold-inactivation of enzymes (see Beyer, 1972). It is not clear, however, how important temperature-pH interactions are in many of these cases, since the data typically were gathered without consideration for physiological values of temperature and pH. One good example of a significant physiological low-temperature inactivation can be given, however. In a number of C4 plants, the enzyme pyruvate orthophosphate dikinase, which is important in regenerating the primary CO_2 acceptor, PEP, dissociates into an inactive form at reduced temperature (Sugiyama et al., 1979). The kinetics of *in vitro* inactivation differ according to the plant species from which the enzyme is isolated; cold-adapted species possess a more cold-stable enzyme than heat-adapted species. Similar correlations were noted in *in vivo* studies in which intact leaves were subjected to cold stress and, following this treatment, the amount of extractable enzyme activity was measured.

In summary, reversible temperature effects on enzyme subunit assembly may in some cases pose threats to metabolic integrity; yet, in other cases, they may serve as critical features of regulatory mechanisms geared to facilitating rapid and reversible changes in flux through metabolic pathways.

TEMPERATURE ADAPTATION OF ASSEMBLY REACTIONS: MUSCLE ACTIN. In the preceding discussion of temperature effects on subunit assembly processes, no mention was made of possible interspecific differences in subunit binding properties. Are the energy changes that accompany the binding of subunits the same in all species, or do adaptive differences exist among interspecific homologues of multimeric proteins? Only one protein system, skeletal muscle actin, has been examined from this perspective (Swezey and Somero, 1982a). The data shown in Figure 11-19 demonstrate that large temperature-related differences in binding energy parameters (binding enthalpy and entropy) are present in actins, and the thermal stabilities of the homologues of actin also vary according to the species' adaptation temperatures. With the interesting exception of deep-living fishes (see Chapter 12), both the binding enthalpy and binding entropy of the actin polymerization reaction increase with rising adaptation temperature. Thus, the highest ΔH and ΔS of polymerization are found for the desert iguana, *Dipsosaurus dorsalis*, and the lowest values are found for highly cold-adapted fishes like *Pagothenia borchgrevinki* and *Gymnodraco acuticeps*. The interspecific differences in ΔH and ΔS of polymerization mirror the differences in $\Delta H^{\ddagger}$ and $\Delta S^{\ddagger}$ observed in comparative studies of activation energy changes during enzymic catalysis; in both assembly and catalytic processes the enthalpy and entropy changes rise with increasing adaptation temperature and, in both cases, a strong compensation relationship is found between the enthalpy and entropy changes. In view of the suggested bases for the interspecific differences in $\Delta H^{\ddagger}$ and $\Delta S^{\ddagger}$ given earlier in this chapter, can we provide a mechanistic explanation for the differences in the ΔH and ΔS of actin polymerization that involves alterations in protein structure and/or in water organization?

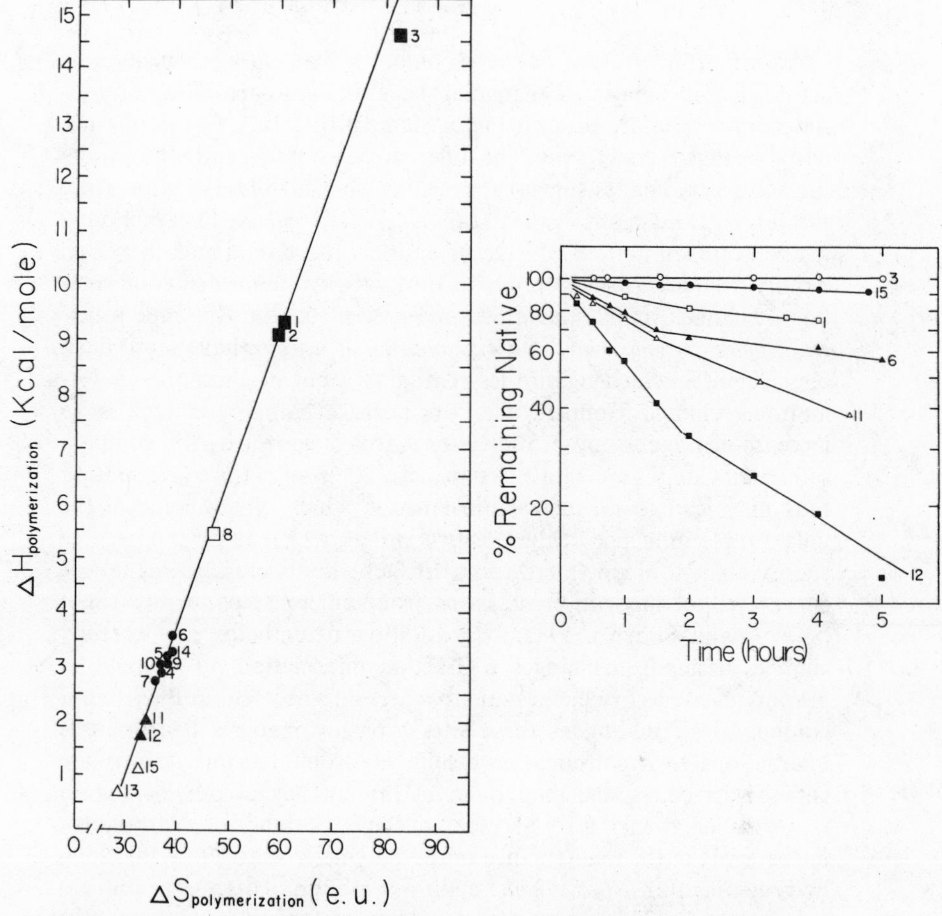

Fig. 11-19. Polymerization thermodynamic parameters (left panel) and heat denaturation (lower panel) for skeletal muscle actins purified from animals having widely different body temperatures. Left panel: An enthalpy-entropy compensation plot for the enthalpy ($\Delta H_{polymerization}$) and entropy ($\Delta S_{polymerization}$) changes occurring as monomeric (G) actin assembles into filamentous (F) actin. Right panel: Loss of native (G) actin as a function of time of incubation at 38°C. Rate constants for the denaturation processes are given in Table 11-8. Reprinted by permission from Swezey and Somero, 1982a. Copyright 1982 American Chemical Society.

Species names (code numbers): rabbit 37° (1), chicken 39° (2), desert iguana, *Dipsosaurus dorsalis* 30-47° (3), *Sebastolobus alascanus* 3-10° (teleost fish, 4), *Sebastolobus altivelis* 3-10° (5), *Caranx hippos* 25-28° (teleost, 6), *Cyprinodon macularius* 10-40° (teleost, 7), *Rhinobatos productus* 10-20° (elasmobranch fish, 8), *Thunnus alalunga* 15-25° white muscle (teleost, 9), *T. alalunga* red muscle 15-25° (10), *Gymnodraco acuticeps* −1.9° (teleost, 11), *Pagothenia borchgrevinki* −1.9° (teleost, 12), *Coryphaenoides armatus* 2° (teleost, 13), *Coryphaenoides acrolepis* 2° (teleost, 14) *Halosauropsis macrochir* 2° (teleost, 15)

An important feature of the globular to filamentous transition in actin is a conformational change of the actin monomer (G actin) associated with its addition to the actin filament (F actin). This conformational change is endergonic. The differences in binding enthalpy among the actin homologues suggest that actins from high-body-temperature species, e.g., the desert iguana (*Dipsosaurus dorsalis*), which has body temperatures of up to 47°C, are structurally more rigid and, therefore, require a larger enthalpy input to bring about the needed conformational change. Actins of cold-adapted species like the Antarctic fishes, *Pagothenia borchgrevinki* and *Gymnodraco acuticeps*, have more flexible structures which require less enthalpy input to effect the conformational change. Coupled with the positive enthalpy change is an increase in the entropy of the system. This is shown by the compensation plot in Figure 11-19 (left panel). Altering actin conformation thus may lead to an increase in entropy which offsets some of the unfavorable influence of the positive enthalpy change on the net free energy change of polymerization. In fact, the differences in the free energy of polymerization of actins from different species are slight (Swezey and Somero, 1982a). In addition to enthalpy and entropy changes arising from changes in G-actin conformation, polymerization also involves energy changes due to water displacement at the subunit contact sites. Binding at these sites typically involves hydrophobic interactions to a considerable extent. Hydrophobic interactions are entropy-driven, so the magnitude of the entropy of polymerization may provide at least a rough index of the importance of hydrophobic interactions in the maintenance of the polymeric state. Note that low-body-temperature species have actin assembly processes with low entropy changes. This may reflect a reduced reliance on hydrophobic interactions at subunit contact sites in low-body-temperature species, a change which appears adaptive in view of the fact that hydrophobic interactions are destabilized by low temperature. The relative roles of energy changes due to 1) alterations in G actin conformation, and 2) water displacement from subunit contact sites in actin polymerization remain to be established.

Temperature-Lipid Interactions:

GENERAL FEATURES. An important theme developed in the discussion of temperature-protein interactions is that a protein must exist in a "semistable" state if it is to have the correct abilities to undergo the changes in shape (or assembly state) that catalysis and regulation demand. In lipid-based systems like membranes, a similar "semistability" of structure is found. Many lipid-based systems exist in a "liquid-crystalline" state which is poised between extreme fluidity, on the one hand, and a rigid, gel-like state, on the other hand. In this section we deal with the biological significance of this conservation of lipid physical state, termed "homeoviscous" adaptation (Sinensky, 1974), and with the mechanisms whereby homeoviscosity is achieved.

In analysis of temperature-lipid interactions, it is essential to establish first the basic working materials of the system in question. Several classes of lipids are shown in Figure 11-20. Of particular importance for this discussion, which focuses on the lipids of functional membranes, are the phospholipids. These consist of a molecule of glycerol to which two fatty acid chains are bound and to one end of which is attached a strongly polar group. Phospholipids thus are amphiphilic; the polar end orients into the aqueous phase, while the nonpolar fatty acid chains aggregate away from water (hydrophobic effect). This dual nature of phospholipid solubility helps establish the membrane bilayer (Fig. 11-21). Membrane proteins may be plated onto the polar surface of the membrane or may be fully or partially buried within the nonpolar interior of the membrane lipid bilayer. The former proteins are termed "peripheral" or "extrinsic"; the latter proteins are termed "integral" or "intrinsic."

The fluidity of a membrane is determined by several physical and chemical variables. As detailed in Table 11-9, fatty acids differ in their melting temperatures according to chain length and double-bond content. Shorter fatty acids are more liquid, and the addition of double bonds greatly decreases melting temperature. The basis of the latter effect is shown in Figure 11-20; the kinks in the fatty acid chain introduced by a *cis* double bond prevent tight alignment of the chains and reduce the stabilization energy arising from van der Waals interactions. Membrane fluidity is also influenced by cholesterol content and by the ionic microenvironment offered by the bathing solution. Membrane fluidity is also strongly influenced by temperature and hydrostatic pressure, and it is the interplay between chemical and physical determinants of fluidity that will be the primary focus of our discussion.

HOMEOVISCOUS ADAPTATION: BASIC MECHANISMS. We will shortly consider numerous facets of the functional correlates of changes in membrane fluidity. For the present it will suffice for the reader to accept the fact that a conservation of some particular state of membrane viscosity or fluidity may be critical for a wide spectrum of membrane-based functions, including enzymic catalysis, ion transport, and synaptic conduction. We will first consider how this "homeoviscosity" is conserved,

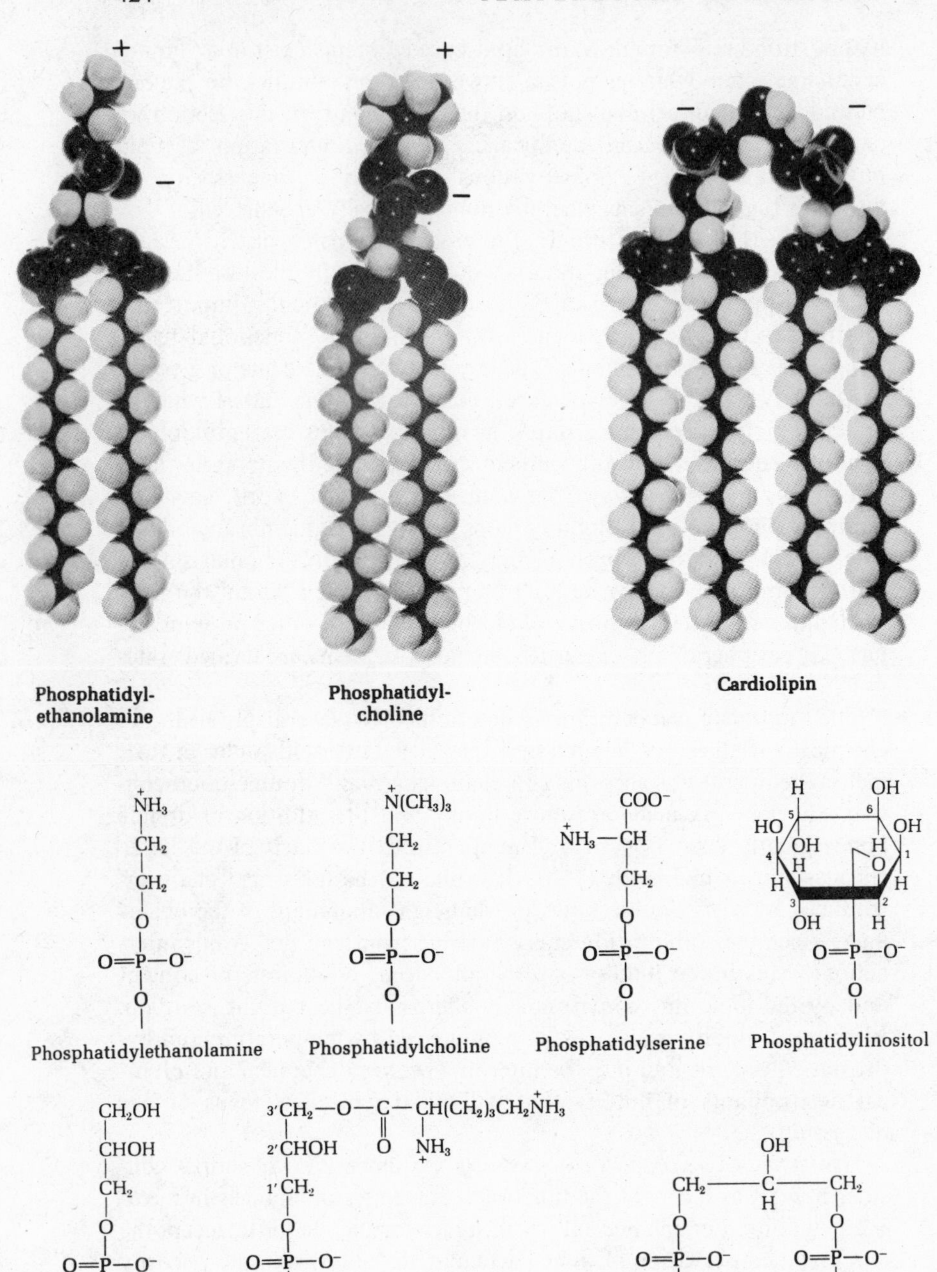
+
−
+
−
−
−
Phosphatidyl-
ethanolamine
Phosphatidyl-
choline
Cardiolipin
Phosphatidylethanolamine
Phosphatidylcholine
Phosphatidylserine
Phosphatidylinositol
Phosphatidylglycerol
3′-O-Lysylphosphatidylglycerol
Cardiolipin (diphosphatidylglycerol)

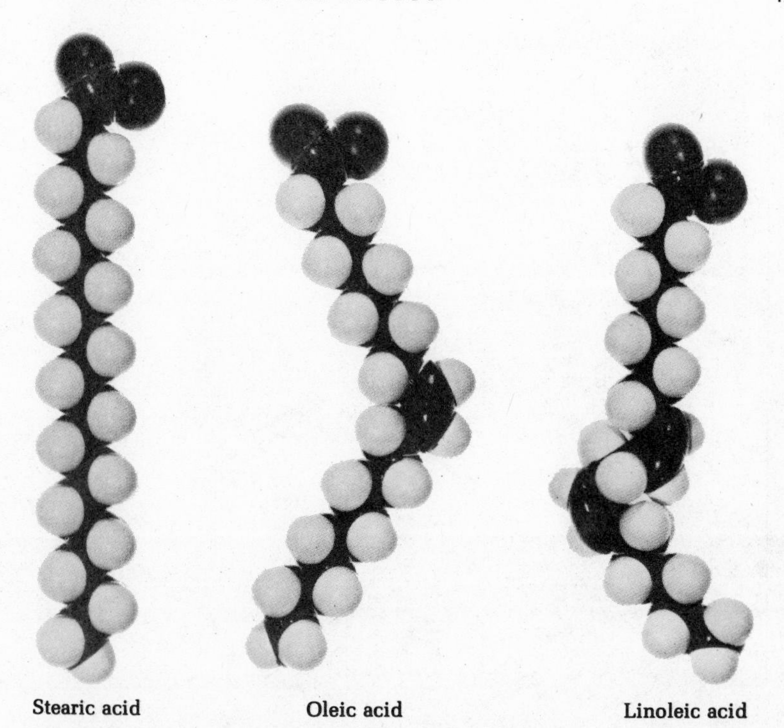

Fig. 11-20 (above and on facing page). Some commonly occurring lipids in biological membranes. Note the effects of double bond(s) on the configuration of the fatty acid chain, e.g., compare stearic acid with oleic acid

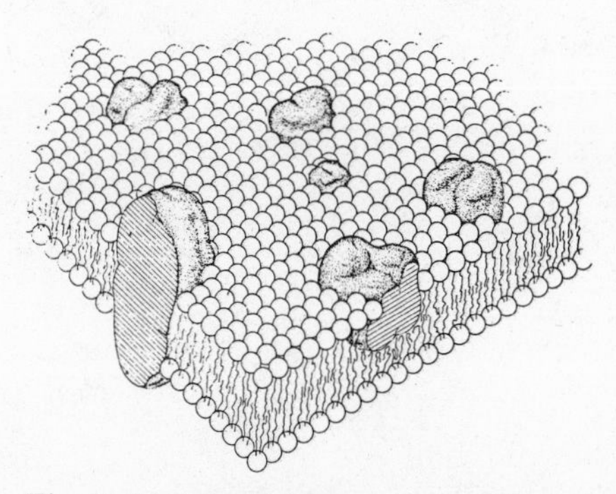

Fig. 11-21. The fluid-mosaic model of a biological membrane, showing the varied locations of membrane proteins within (=intrinsic proteins) or plated on (=extrinsic proteins) the phospholipid bilayer. Model after Singer and Nicholson (1972) (Copyright 1972 by the American Association for the Advancement of Science.)

Table 11-9. Structures and Melting Temperatures for Commonly Occurring Fatty acids

Symbol	Structure	Common name	Melting temperature
Saturated fatty acids			
12:0	$CH_3(CH_2)_{10}COOH$	Lauric	44.2
14:0	$CH_3(CH_2)_{12}COOH$	Myristic	53.9
16:0	$CH_3(CH_2)_{14}COOH$	Palmitic	63.1
18:0	$CH_3(CH_2)_{16}COOH$	Stearic	69.6
20:0	$CH_3(CH_2)_{18}COOH$	Arachidic	76.5
Unsaturated fatty acids			
$16{:}1^{9}$	$CH_3(CH_2)_5CH{=}CH(CH_2)_7COOH$	Palmitoleic	−0.5
$18{:}1^{9}$	$CH_3(CH_2)_7CH{=}CH(CH_2)_7COOH$	Oleic	13.4
$18{:}2^{9,12}$	$CH_3(CH_2)_4CH{=}CHCH_2CH{=}CH(CH_2)_7COOH$	Linoleic	−5.0
$18{:}3^{9,12,15}$	$CH_3CH_2CH{=}CHCH_2CH{=}CHCH_2CH{=}CH(CH_2)_7COOH$	Linolenic	−11.0
$20{:}4^{5,8,11,14}$	$CH_3(CH_2)_4(CH{=}CHCH_2)_3CH{=}CH(CH_2)_3COOH$	Arachidonic	−49.5

SOURCE: After Lehninger (1975)

and then examine in detail certain of the consequences of this important adaptive mechanism.

From the lipid taxonomy information presented in Table 11-9, the basic strategy for achieving homeoviscous adaptation should be clear. To offset the solidifying influences of temperature decreases, the organism may incorporate relatively high percentages of unsaturated lipids. The result of this increased double-bond content is a lowering of the transition temperature where the fluid-gel transformation takes place. This mechanism for conserving membrane fluidity has been observed in bacteria, plants, and animals (reviewed by White and Somero, 1982; Hazel, 1983). The saturation levels of different classes of phosphoglycerides of brain synaptosomal membranes isolated from species adapted to different temperatures and from differently acclimated individuals of the same species are given in Table 11-10. Note the similarity between inter- and intraspecific effects. The same types of substitutions characterize cold- and warm-adapted species, e.g., the arctic sculpin and desert pupfish, and differently acclimated goldfish. The data in Table 11-10 for brain synaptosomal membranes are comparable to those obtained in studies of most other membrane systems, e.g., mitochondria (Caldwell and Vernberg, 1970), various membranes of the unicell, *Tetrahymena pyriformis* (Kasai et al., 1976), and thylakoid membranes of the chloroplast (Raison et al., 1982). Homeoviscous adaptations therefore appear to be a common feature of membrane systems.

The regulation of membrane viscosity via changing the fatty acid composition of polar lipids like phospholipids entails several regulatory events of fatty acid synthesis and the subsequent incorporation

Table 11-10. Fatty Acid Composition of Phosphoglycerides from Brain Synaptosomes Isolated from Animals Acclimated to or Adapted to Different Temperatures

	Phosphoglyceride class	Arctic sculpin 0°C	Goldfish 5°C	Goldfish 25°C	Desert pupfish 34°C	Rat 37°C
Ratio of saturated to unsaturated fatty acids	Choline	0.593	0.659	0.817	0.990	1.218
	Ethanolamine	0.260	0.340	0.506	0.568	0.651
	Serine/inositol	0.477	0.459	0.633	0.616	0.664

SOURCE: From Cossins and Prosser (1978)

of fatty acids into polar lipids. Different species display varied regulatory mechanisms. Some species appear to effect control primarily by regulating the double-bond content of fatty acids; other species achieve homeoviscosity via regulating the types of fatty acids added during phospholipid synthesis, and in some species the key regulatory event appears to be the activation or deactivation, as needed, of the enzymes responsible for desaturase activity. The latter type of regulatory effect in which desaturase activity is rapidly modulated to suit the thermal regimen of the organism merits our attention, as this type of regulatory effect illustrates how direct temperature effects on enzymes can lead to important regulatory changes.

Two different types of temperature effects on desaturase systems have been observed. In the bacterium *Bacillus megaterium*, the desaturase enzyme normally present at low growth temperatures is rapidly denatured when culture temperatures are stepped up (Fujii and Fulco, 1977). Thus an enzyme whose activities would be detrimental to the cell at high temperatures appears to have a built-in thermal lability which ensures that the enzyme is inactivated at higher growth temperatures. In temperature step-down experiments, rapid induction of the desaturase was observed.

A second mechanism for regulating desaturase activity has been described in *Tetrahymena pyriformis* (Kasai et al., 1976; Kitajima and Thompson, 1977). In *Tetrahymena* the desaturase enzyme is proposed to undergo a change in membrane localization with shifts in culture temperature, and this movement leads to changes in desaturase activity. When culture temperatures are stepped down, the endoplasmic reticulum undergoes a fluid-to-gel transition, and the desaturase enzyme which normally is buried within the plane of the bilayer (and is inactive when so situated) is "squeezed" out of the bilayer into contact with the aqueous phase (cytosol). Thus exposed to substrates and cofactors, the desaturase becomes activated and begins synthesizing lipids with reduced saturation. These lipids are incorporated into the different membrane systems of the unicell to achieve homeoviscosity. Interestingly, the endoplasmic reticulum does not complete its homeoviscous adjustment as rapidly as peripheral membranes, e.g., those of the pellicle. This difference in kinetics among different membrane systems makes eminent sense, for if the endoplasmic reticulum were to complete its lipid readjustments first, the desaturase might sink into the now fluid bilayer, cease its activity, and leave the rest of the cell's membranes with inappropriate fatty acid composition. Figure 11-22 illustrates the topological shifts proposed to underlie the regulation of desaturase activity in *Tetrahymena*.

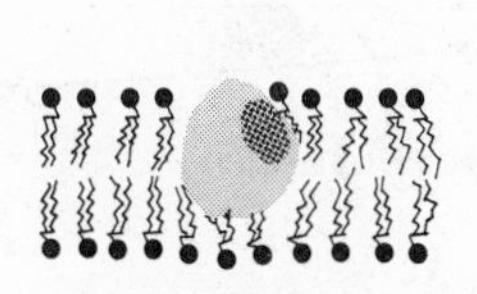

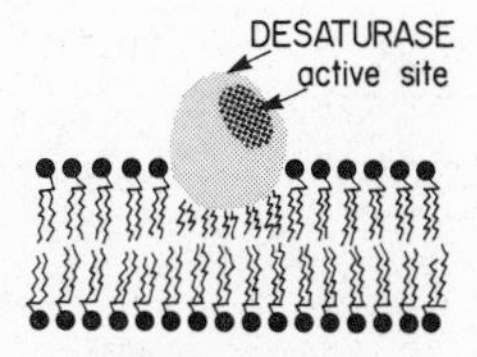

Fig. 11-22. A model for the fluidity-dependent regulation of desaturase activity in the endoplasmic reticulum membranes of *Tetrahymena*. When the membrane is highly fluid (high temperature, high degree of fatty acid unsaturation), the desaturase enzyme lies embedded in the lipid bilayer. When the fluidity of the membrane is low (low temperature, high degree of fatty acid saturation), the desaturase is extruded from the plane of the bilayer, into contact with the surrounding solution. In the exposed state, catalysis (fatty acid desaturation) is possible; in the embedded state, the enzyme is inactive. This model is hypothetical and based on the model proposed by Kasai et al. (1976) and Kitajima and Thompson (1977)

FUNCTIONAL CORRELATES OF HOMEOVISCOUS ADAPTATION.

1. Behavioral correlates. Among the membrane systems likely to play important roles in establishing the thermal relationships of animals are those involved in neural function, e.g., synaptic membranes. The study of Cossins et al. (1977) illustrated in Figure 11-23 shows that changes in the fatty acid composition of synaptosomal phospholipids closely parallel behavioral changes during acclimation. As an index of membrane fluidity, the polarization of a fluorescent label embedded in the synaptosomal lipids was utilized. The more fluid the lipid phase, the lower the amount of polarization. This effect is due to the greater freedom for movement of the fluorescent probe in a more liquid microenvironment. The polarization parameter is known to depend strongly on membrane phospholipid fatty acid composition (Cossins and Prosser, 1978).

In the acclimation study of Cossins et al. goldfish were shifted from one acclimation temperature, 5° or 25°C, to the opposite temperature, and the change in membrane fluidity (polarization) and alterations in chill coma, equilibrium loss, and hyperexcitability temperatures were monitored as a function of time of acclimation. Several findings of this study are significant for our discussion. First, the changes in lipid fluidity and behavioral traits exhibit similar kinetics. This similarity in time courses argues for a causal relationship between synaptic membrane changes and behavioral shifts. Second, the time courses of cold (5°C) and warm (25°C) acclimation are markedly different. Acclimation to 25°C was completed much more rapidly. Third, regardless of

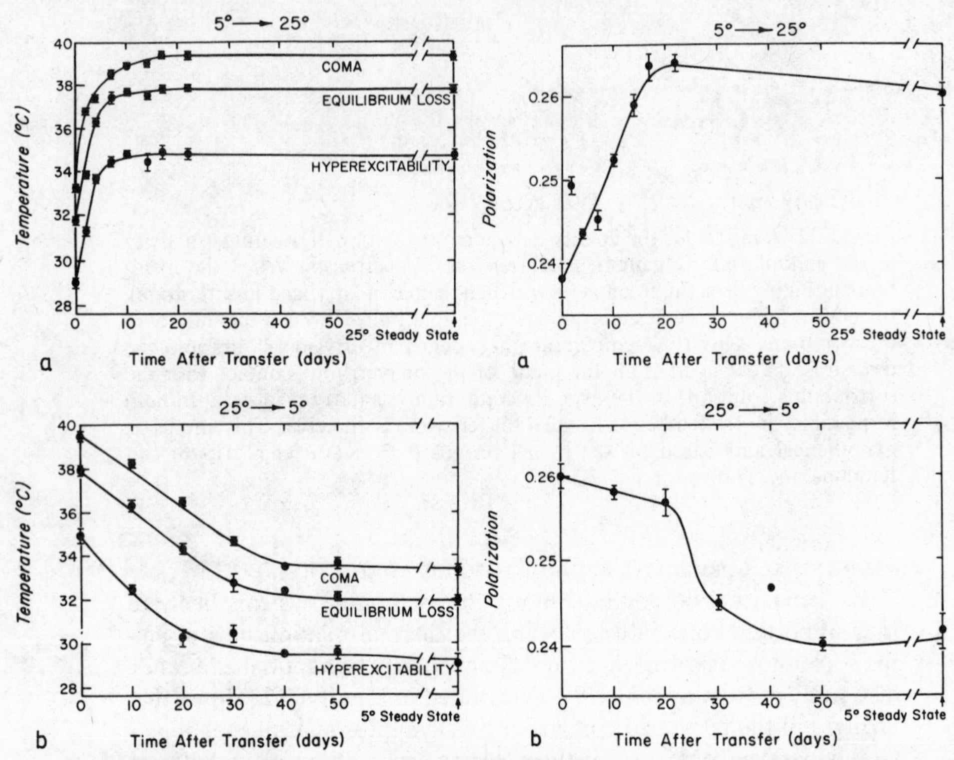

Fig. 11-23. Time courses of acclimation for three behavioral traits (coma temperature, equilibrium loss, and hyperexcitability) and for synaptosome lipid fluidity (as measured by polarization) in goldfish. Specimens were first acclimated to either 25°C or 5°C and then transferred to the other temperature. Figure from Cossins et al. (1977)

acclimation temperature, the time periods to complete the membrane restructuring were large relative to the times likely to be needed to alter the activities of enzymes involved in desaturation of fatty acids and the building of appropriate phospholipids for the temperature in question. The fact that from two to six weeks were needed to complete the acclimations suggests that the rate-limiting event in homeoviscous adaptation in this multicellular species is not the synthesis of the correct phospholipids—these changes are quite rapid, at least in unicells (see above)—but rather is the turnover of the membranes. Particularly at 5°C, membrane turnover may be so slow that even the capacity to generate appropriate phospholipids is not sufficient to bring about a rapid homeoviscous adjustment.

2. Photosynthetic Adaptations. Parallel changes in membrane lipid properties and the performance of membrane-localized systems are well-known in plants as well as in animals. In plants the thylakoid membranes of chloroplasts, the sites of many of the energy-transfer reactions associated with light harvesting and CO_2 fixation, are apt to be an especially crucial membrane component in determining the thermal optima and thermal tolerance limits of photosynthesis.

As in the case of animal respiration (Figure 11-5), green plant photosynthesis displays temperature compensatory shifts in capacity and tolerance, and this is seen both on evolutionary (interspecific) and acclimatory time scales (Fig. 11-24). Two important classes of adaptation are associated with these adjustments in the temperature responses of photosynthesis. As discussed earlier, the elevated photosynthetic capacities of cold-adapted (-acclimated) plants, especially at low growth temperatures, appear to be due to increased concentrations of one or more of the enzymes associated with the Calvin-Benson cycle. The increased tolerance of high temperature, however, appears to be the result of homeoviscous lipid adaptations (Raison et al., 1980; Pike and Berry, 1980). The critical achievement of this particular type of homeoviscous adaptation is the stabilization of the interactions between light harvesting and energy transferring components of the thylakoid membrane. In simplest terms, thermal disruption of the thylakoid membrane places a "short-circuit" into the photosynthetic system via disruption of the organization of the chloroplast energy transfer system. Specifically, heat damage involves a block to excitation energy transfer from chlorophyll b to chlorophyll a, and changes in the allocation of excitation energy between photosystems II and I (Armond et al., 1978). The perturbation by high temperatures of the interactions among the pigment-protein complexes of the chloroplast can be monitored by measuring the fluorescence of chlorophyll (see Raison et al., 1980, p. 270). When free in solution, chlorophyll fluoresces very strongly. When in the intact chloroplast, however, the fluorescence of chlorophyll is strongly quenched because the excitation energy is trapped by the reaction centers. Increases in chlorophyll fluorescence with rising temperature thus are indicative of disruption of the thylakoid membrane system (Fig. 11-24), and these fluorescence increases are closely correlated with decreased photosynthetic performance as measured by quantum yield. Desert plants such as *Tidestromia oblongifolia* exhibit great heat tolerance in photosynthetic performance and a correlated thermal stability of the thylakoid membrane (Fig. 11-24). The lipid changes responsible for these capacities appear to be much the same as those found in differently adapted (acclimated) animals: heat-

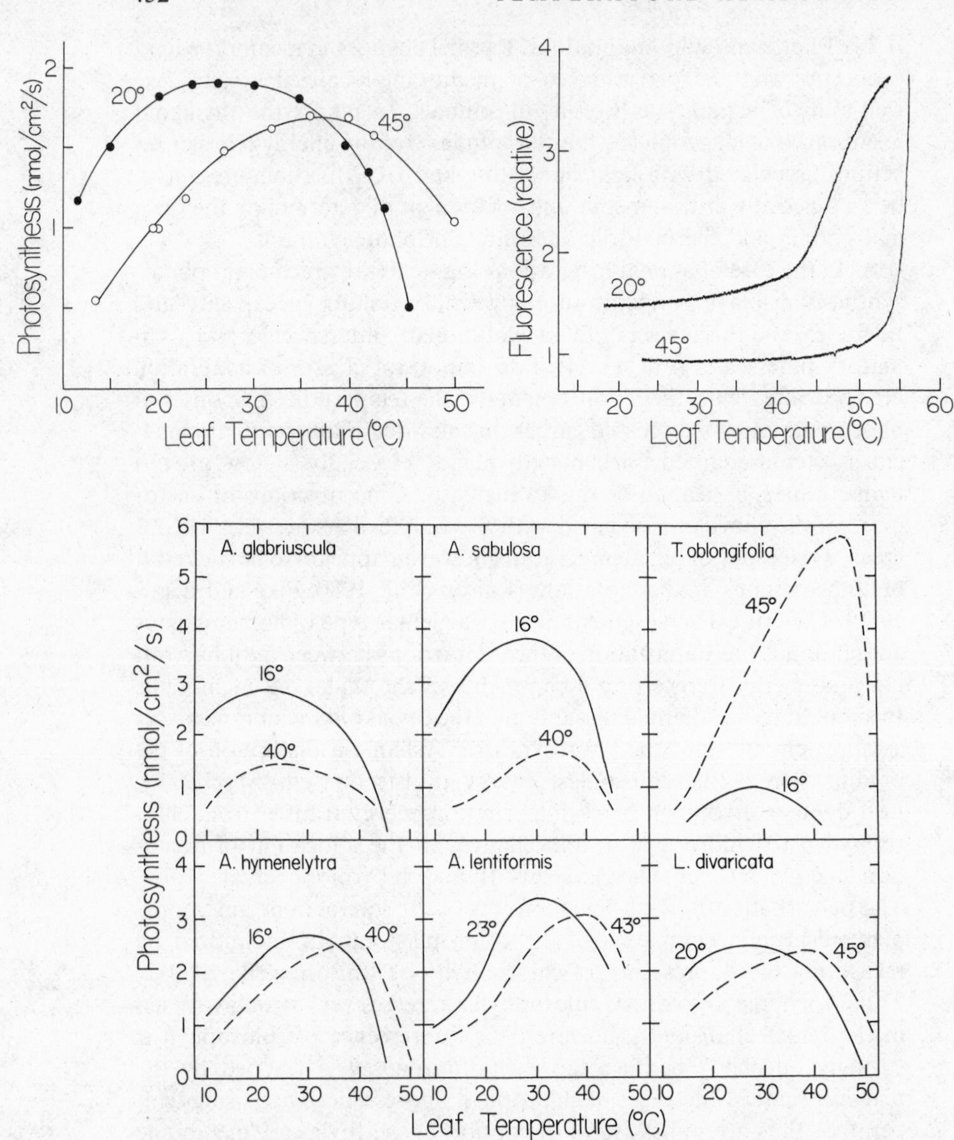

Fig. 11-24. The effects of growth and assay temperatures on photosynthetic rates (nmol CO_2 fixed/cm^2 of leaf/second) in a variety of plants adapted to different habitat temperatures. Upper left panel: The temperature dependence of light-saturated CO_2 uptake for *Nerium oleander*, grown under two different thermal regimens (20° day/15° night; and 45° day/32° night). Figure after Bjorkman et al. (1980). Upper right panel: The change in relative intensity of chlorophyll fluorescence in whole *Nerium oleander* leaves as a function of temperature. Figure after Raison et al. (1980)

tolerant species (populations) contain higher percentages of saturated fatty acids in their phospholipids and galactolipids (Raison et al., 1982).

3. Activation energies of membrane-associated enzymic reactions. In much the same way as changes in lipid fluidity can influence the functions of multiprotein complexes and protein-pigment complexes, the behavior of individual membrane-associated enzymes is often sharply dependent on the local viscosity of the enzyme's environment. Figure 11-25 portrays schematically the effects of temperature change and homeoviscous adaptation on the thermodynamic activation parameters of a membrane-bound enzyme, in this case an ion-activated ATPase. At some characteristic "break" temperature, the slope of the Arrhenius plot (log rate versus reciprocal of absolute temperature) changes sharply. Below the "break" temperature the activation enthalpy of the reaction is much higher than above the break temperature. There is, however, a corresponding rise in activation entropy as activation enthalpy increases (inset to figure). Thus the net change in the free energy of activation is small relative to the enthalpy effect, but nonetheless is large enough to retard the reaction velocity.

The basis of these effects on the activation parameters is likely to be the necessity for the enzyme protein to "move"; i.e., to change conformation or to move in or through the plane of the membrane during catalysis. Movement would require displacement of the vicinal lipids. If these lipids are relatively fluid, movement of the enzyme is energetically low; if the lipids are rigid, then extensive localized melting of the lipids may be necessary. This would require a substantial enthalpy input,

---

Lower six panels: The effect of growth temperature on the rate and temperature dependence of light-saturated net CO_2 uptake for several plants from contrasting thermal regimens. Growth temperatures (=acclimation temperatures) are given for each species. *Atriplex glabriuscula* and *Atriplex sabulosa* are coastal species of California and are incapable of acclimating to 40°C daytime temperatures. Although the upward shift in thermal optimum for photosynthesis occurs during 40°C growth, the rate of photosynthetic activity does not increase, unlike the response observed for the thermophile, *Tidestromia oblongifolia*. The latter species fails to cold-acclimate. The two evergreen species native to Death Valley, *Larrea divaricata* (creosote bush) and *Atriplex lentiformis*, remain active throughout the year and are capable of a high rate of photosynthesis over an especially wide temperature range. Note the shifts in thermal optima as a function of acclimation temperature. *Atriplex hymenelytra*, a highly eurythermal species found in both coastal and desert habitats, likewise displays a broad thermal range for photosynthesis and a pronounced capacity for acclimation. Figure modified after Bjorkman et al. (1980)

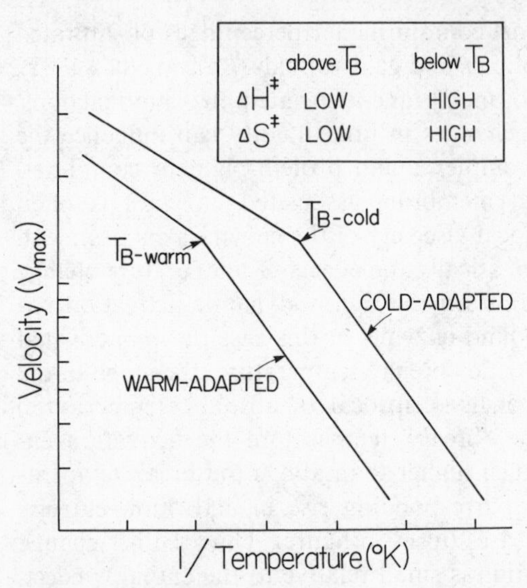

Fig. 11-25. Arrhenius plots (log V_{max} versus 1/absolute temperature) for an ion-dependent ATPase from membranes of warm- and cold-adapted (-acclimated) organisms. The process of adaptation (acclimation) leads to a shift in the "break" temperature (T_B), the temperature at which the slope of the plot changes. Alterations in lipid composition, e.g., saturation, effect these shifts in T_B. A high degree of enthalpy/entropy compensation is found in these effects. The high enthalpy of activation ($\Delta H^{\ddagger}$) noted below the break temperature is associated with a high entropy of activation ($\Delta S^{\ddagger}$), and low values of both activation parameters are found above T_B. These compensating changes in $\Delta H^{\ddagger}$ and $\Delta S^{\ddagger}$ reflect localized disorganization of lipids surrounding the enzyme protein during catalysis. Model is based on data in G. Inesi, M. Millman, and S. Eletr. *J. Mol. Biol.* 81: 483–504.

but would also lead to a marked increase in the entropy of the system (the melted lipids obviously would possess less order than the frozen lipids). Through homeoviscous adjustments in the membrane, the phase transition temperature of the membrane phospholipids can be adjusted to facilitate the maintenance of a relatively fluid environment for enzyme function at physiological temperature. Indeed, evidence gathered by Hazel (1972) and Vik and Capaldi (1977) suggests that catalysis by lipoprotein enzymes (enzymes requiring both a protein [catalytic] component and a lipid component for full activity) is most efficient when the protein component is bound to a highly fluid lipid. This correlation between lipid fluidity and catalytic efficiency can be interpreted as a reflection of low energy barriers to protein movement during catalysis when the lipids surrounding the protein have low

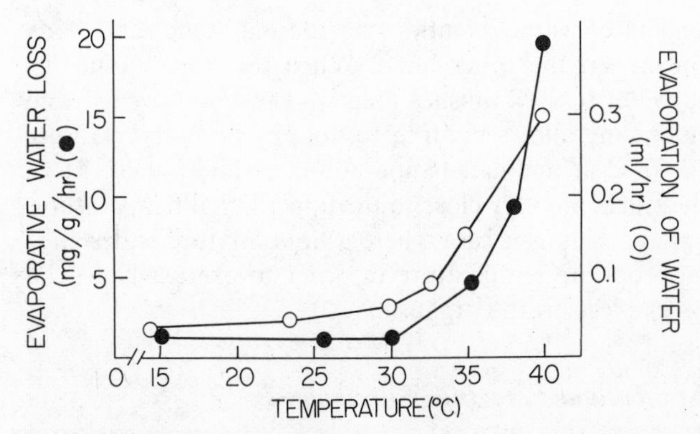

Fig. 11-26. The effects of temperature on: 1) evaporative water loss in *Phyllomedusa sauvagei* (closed circles); and 2) evaporative water loss from the surface of water covered with a film of lipid from skin secretions of this frog (open circles). Data from McClanahan et al. (1978) *Physiological Zoology* 51: 179–187 by permission of the University of Chicago Press. © 1978 by the University of Chicago.

melting temperatures. It must be recalled, however, that constraints operate on the degree of fluidity that is tolerable. Even though a highly fluid milieu might favor the highest catalytic efficiency for membrane-associated enzymes, too fluid a lipid bilayer could have the unfavorable effects discussed in the preceding section—a loss of integrity of multi-component systems such as those found in the thylakoid membranes.

4. Skin waxes and temperature regulation. Our treatment of lipid phase transitions with changing temperature has been presented in a negative light in that we have emphasized the disadvantageous effects of alterations in lipid fluidity at extremes of temperature. As we argued in the case of temperature and pH effects on PFK assembly in hibernators, however, what is an adverse effect for some circumstances may be highly advantageous under other conditions. An example of this type of phenomenon in the context of lipid phase changes is found in the water balance and thermoregulatory strategies of an arboreal frog from Argentina, *Phyllomedusa sauvagei* (McClanahan et al., 1978). This heat-tolerant frog, which may experience ambient temperatures over 40°C in the summer, retards water loss by waxing its skin with a lipid substance secreted by glands in the skin. Up to ambient temperatures of approximately 35°C the frog is a thermal conformer, and the wax layer on its skin is highly impermeable to water because the wax is in a gel-like state. However, somewhere between 34 and 38°C the wax begins to melt (Fig. 11-26). This phase transition marks a trade-off

between the benefits of water retention, on the one hand, and body temperature control, on the other hand. When the wax liquifies its ability to retard water loss diminishes greatly. The frog "sweats" and this evaporative cooling allows the frog to hold its body temperature between 36 and 37°C at ambient temperatures as high as 40–41°C. The latter temperatures are very close to the upper lethal temperatures of this species. Here, then, is a case where a lipid mixture is designed to melt at an appropriate temperature in order to exert a beneficial (thermoregulatory) effect on the organism.

Freezing Resistance and Freezing Tolerance

Much as organisms must closely regulate the physical states of their lipid-based systems, for many cold climate species regulating the physical state of the extra- and intracellular fluids assumes critical importance during much or all of the year. With rare exceptions, intracellular ice formation is lethal to cells. While it is true that cryopreservation methodologies do allow long-term storage of frozen cells and tissues (Ashwood-Smith and Farrant, 1980), the conditions necessary to achieve this feat, e.g., the addition of high concentrations of cryoprotectant substances like dimethylsulfoxide, are not accessible to animals in nature. Most organisms are likely to die even if ice formation is confined to the extracellular spaces, although there are striking examples of species that not only tolerate freezing of the extracellular fluids, but even nucleate ice formation. In the majority of species that experience the threat of freezing, however, mechanisms are employed that prevent ice formation in all body compartments. The unusual macromolecular entities, biological "antifreeze" compounds, which retard ice formation and confer freezing resistance to a varied suite of fishes and invertebrates, serve as the primary focus of this final section of the chapter on thermal relationships.

The choice of strategy for coping with ambient temperatures that are below the body fluid freezing points of organisms not adapted to resist or tolerate freezing depends on several factors. One important consideration relates to the ability of the organism to remain active at these low temperatures. If an ecological factor such as lack of food dictates that dormancy is desirable during the cold season, then the extracellular body fluids of the animal may be modified to facilitate the formation of ice at temperatures close to the freezing point of the blood or hemolymph. In such cases supercooling is largely avoided. These types of animals are termed "*freeze-tolerant*" because they can withstand ice formation in the extra-, but not the intracellular fluids. It may seem maladaptive to produce ice-nucleating agents to trigger ice formation at relatively high temperatures, a feat common in freeze-tolerant species (see Zachariassen, 1980), for at first glance there appears to be little basis for concluding that ice formation in the extracellular fluids, which is followed by withdrawal of water from the intracellular fluids, could be beneficial to the organism. It has long been thought that dehydration of the intracellular space, with concomitant increases in inorganic ion concentrations and distortion of intracellular structures, is a major cause of low temperature-induced lethality. However, the point of using nucleating agents to foster ice formation at relatively high freezing temperatures is that, by avoiding supercooling, ice formation will occur in the extracellular spaces at temperatures well above those at which spontaneous ice formation can occur intracellularly. Potential damage from dehydration may still exist, albeit freeze-tolerant animals have marked capacities to withstand dehydration (Kanwisher, 1955; Murphy and Pierce, 1975). Ice formation in the intracellular space is prevented, however, and this achievement may be most critical for a freeze-tolerant organism. The potential damage caused by ice crystal growth within the cell appears much greater than damage due to dehydration.

A second major strategy for dealing with low body temperatures is found in *freeze-resistant* species, organisms that employ biochemical mechanisms to prevent ice formation in both the extra- and intracellular fluids. In many cases the species that utilize this strategy remain active at potentially freezing temperatures. Sustained activity and the presence of ice-containing extracellular fluids and dehydrated intracellular spaces are conditions that seem imcompatible. Polar fishes offer especially good examples of freezing resistance, and the peptide and glycopeptide antifreezes in the body fluids of these species allow them to remain active in the presence of ice at seawater temperatures of −1.86°C (DeVries, 1980, 1982).

While many fishes do erect biochemical defenses against freezing, it is appropriate to note that many species employ seasonal migrations which lead to removal of the threat of freezing. For example, the long horn sculpin, *Myoxocephalus octodecemspinosus*, migrates out of nearshore waters in the winter where ice formation occurs, and seeks a deeper, ice-free habitat during these cold months. This behavior eliminates the danger of freezing via the seeding of the body fluids (freezing points are more than 1°C above the freezing point of seawater for fishes lacking antifreezes). However, migration into deeper water would appear to present this sculpin with a reduced food supply as well as exposure to another set of predators. Thus, behavioral avoidance of freezing may carry the costs of existence in a suboptimal habitat for at least part of the year. The development of antifreeze molecules has allowed organisms to select habitats on criteria other than the presence or absence of temperatures that are lethal to freeze-susceptible species.

EXAMPLES OF FREEZE-TOLERANT ORGANISMS. Prior to reviewing the biochemistry of antifreeze molecules, we shall consider briefly a variety of organisms that tolerate, and often induce, ice formation in their extracellular fluids. These organisms typically are terrestrial species that are dormant in winter. Numerous examples of insects are known that contain ice-nucleating agents that effectively prevent supercooling of the hemolymph (Duman, 1980; Zachariassen, 1980). As mentioned above, this relatively high-temperature freezing prevents more deleterious low-temperature freezing of the cytosol.

A somewhat similar adaptive strategy has recently been reported in the plant, *Lobelia telekii*, which is native to an Afro-alpine environment where temperatures near −10°C may occur at night throughout the year (Krog et al., 1979). This plant, unlike an Arctic beetle, cannot go dormant for weeks to months at a time, but instead must find ways to achieve tolerance of freezing at night while remaining metabolically active and growing during warmer periods of the day. The key factor of this plant's diurnal thermal regulatory strategy is a large, fluid-filled compartment in the inflorescence of the plant. This cavity is filled with a viscous fluid that serves at least two thermally related functions. First, the aqueous fluid has a high heat capacity and thermal inertia, so some of the changes in ambient temperature will be buffered by this fluid. Second, this fluid behaves like the blood of freeze-tolerant insects in that it contains an ice-nucleating agent that triggers freezing near 0°C. This ice formation can be beneficial in two ways. As in the case of freeze-tolerant animals, the prevention of supercooling acts as a means for reducing the dangers of intracellular ice formation which could occur if the body temperature of the plant fell well below 0°C. In addition, at higher subzero temperatures the heat of fusion released during ice formation can actually warm the plant. Krog et al. (1979) found that as the ambient temperature decreased from about 12°C during daytime highs to nearly −8°C during the late night, the temperature of the central part of the plant did not fall below 0°C. Ice was generated at a sufficient rate to buffer the central temperature. These authors calculated that approximately 2 percent of the central cavity froze under these conditions, suggesting that a significant "heat reservoir" remained should more extreme temperature conditions be experienced.

It seems highly unlikely that similar heat-generating functions of ice-nucleating agents occur in animal body fluids. The relative fluid volumes of *L. telekii* and animals are grossly different, and the difficulties of controlling heat flow in small animals are obvious. Moreover, as stated above, terrestrial invertebrates that tolerate ice formation in their extracellular fluids generally are in a state of months-long hibernation.

Among aquatic animals the intertidal invertebrates of high latitudes may display impressive degrees of freezing tolerance. For instance, Kanwisher (1955) showed that two intertidal molluscs, *Mytilus edulis* and *Littorina rudis*, withstand freezing of approximately 70 percent of their total body water at an ambient temperature of −20°C, a common winter temperature for the North Atlantic intertidal regions where these specimens were obtained. All freezing was in the extracellular space. Although the cells appeared shrunken and distorted under microscopic observation, they contained no ice crystals (Kanwisher, 1955). In a related study, Murphy and Pierce (1975) examined the ability of another intertidal mollusc, *Modiolus demissus demissus*, to acclimate to low temperatures and, thereby, to increase its tolerance of freezing. They were able to demonstrate that a pronounced increase in the mollusc's capacity to withstand tissue desiccation accompanied low temperature acclimation. Specimens acclimated to 23°C died when 35 percent of the tissue water was removed via extracellular ice formation; 0°C-acclimated individuals could tolerate the loss of 41 percent of tissue water. An important conclusion from this study is that low temperature acclimation does not reduce the amount of tissue water lost during extracellular freezing, but instead leads to a greatly increased tolerance of partial dehydration. The molecular basis of this acclimatory effect is not known, albeit the involvement of macromolecular antifreezes in certain intertidal molluscs (Theede et al., 1976) raises the possibility that these molecules may play some role in stabilizing biochemical structures and functions in the face of reduced water activities, a suggestion that has also been made in the case of xerotolerant insects such as *Tenebrio molitor* (Patterson and Duman, 1979; Schneppenheim and Theede, 1980).

A final group of compounds merits consideration in the context of freezing tolerance. These are polyhydroxyl alcohols (polyols) like glycerol, sorbitol, and mannitol, which may accumulate to high concentrations in many insects that are either freeze-tolerant or freeze-resistant. Polyols lower the supercooling point of a solution approximately twice as much as they reduce the true freezing point, so these compounds may be of great significance in the prevention of ice formation. The potential disadvantages of polyols are at least twofold. First, because they function in a strictly colligative manner, unlike the peptide and glycopeptide antifreezes discussed later, polyols will sharply increase the osmotic concentration of the body fluids. If polyols are uniformly

Table 11-11. "Antifreeze" Proteins and Glycoproteins of Fishes and Invertebrates

Group of organisms	Type of "antifreeze"	Molecular weight (daltons)	Thermal hysteresis[a] (°C)	Reference
Antarctic fishes				
Nototheniidae (e.g., *Pagothenia borchgrevinki*)	Glycoproteins	8 size classes 2,600–33,700	1.27	De Vries (1974)
Zoarcidae (*Rhigophila dearborni*)	Glycoproteins		0.76	De Vries (1974)
Arctic Fishes				
flounder (*Pseudopleuronectes americanus*)	Proteins	3 size clases	0.62 (winter)	Duman and DeVries (1974, 1976)
cod (*Gadus ogac*)	Glycoproteins	7 size classes same as Antarctic nototheniids	1.18	Van Voorhies et al. (1978)
saffron cod (*Eleginus gracilis*)	Glycoproteins		1.0	Raymond et al. (1975)
sculpin (*Myoxocephalus verrucosus*)	Protein		1.4	Raymond et al. (1975)
Insects				
Tenebrio molitor	Proteins	several size classes		Patterson and Duman (1979) Schneppenheim and Theede (1980)
spider (*Philodromus* sp.)	Proteins		2.44 (February) 0 (June)	Duman (1979)
spider (*Clubiona* sp.)	Proteins		1.88 (January) 0 (warm-acclimated)	Duman (1979)
beetle (*Dendroides canadensis*)	Proteins		3.62 (cold-acclimated)	Duman (1980)
beetle (*Meracantha contracta*)	Proteins		3.71 (February)	Duman (1977)

[1] Thermal hysteresis = melting temperature − freezing temperature. Thermal hysteresis values are measured with serum of animals adapted to their normal habitat temperatures or acclimated to different temperatures as indicated

distributed throughout all body fluid compartments, however, this osmotic effect may not be much of a problem, since redistribution of water among compartments would not occur. Second, the polyols increase the viscosity of body fluids greatly, a problem that again is avoided by the peptide and glycopeptide antifreezes, which work in noncolligative manners.

One important feature of polyols like glycerol is that they may serve a variety of functions in organisms that experience freezing or desiccating conditions. Glycerol functions as a cryoprotectant, as a supercooling agent, and has been shown to aid in desiccation resistance in cysts of *Artemia* (Clegg, 1962). As noted in Chapter 10, glycerol is a compatible solute that lacks perturbing effects on proteins. Common inorganic ions, e.g., K^+, Na^+, and Cl^-, do not exhibit compatibility at high ionic strengths. Thus, glycerol may be an appropriate solute to use at low water activities, whether these low activities are associated with high external osmotic concentrations, with encystment and dormancy, or with freezing tolerance. It is necessary to realize, however, that glycerol is not a ubiquitous component of the body fluids of freeze-tolerant and freeze-resistant animals (Duman, 1980; Zachariassen, 1980). The full chemical armament of such animals usually contains other compounds whose contributions are critical for survival at low temperatures. These compounds are the peptide and glycopeptide antifreeze molecules, which have now been described in a variety of high latitude fishes as well as in several terrestrial invertebrates.

FREEZING RESISTANCE AND MACROMOLECULAR ANTIFREEZES. One of the most intriguing stories about molecular evolution to have been developed during the past two decades involves the macromolecular antifreezes (peptides and glycopeptides), first discovered in larvae of the insect, *Tenebrio molitor* (Ramsay, 1964) and blood sera of Antarctic fishes (reviewed by DeVries, 1980, 1982). The currently identified family of antifreezes listed in Table 11-11 reveals that these peptides and glycopeptides are found in a taxonomically diverse assemblage of animals, differ substantially in chemical structure and molecular weight, and can vary in concentration as a function of acclimation or acclimatization. All of these antifreezes share one important common property, however, which is a diagnostic trait for these molecules: they lower the freezing point (temperature of ice crystal growth) of a solution more than the melting point (temperature of shrinkage of ice crystals). This "thermal hysteresis" is used to detect the presence of antifreezes in solutions and, moreover, may provide an important clue to the mechanism of antifreeze action, as discussed later. We will see, in fact, that in spite of the diversity of antifreeze primary structures (Table 11-11), a common mechanism of antifreeze action is likely to apply in all cases.

The glycopeptide and peptide antifreezes are common in high latitude fishes that inhabit the upper regions of the water column where ice is present (Table 11-11). Fishes of high latitude seas lacking these antifreezes must seek out ice-free niches, and many species are likely to remain in a supercooled state throughout their lives (DeVries, 1980). The antifreezes present in high latitude fishes are a fascinating example of convergent evolution at the molecular level. The evolutionary development of glycopeptides and peptides capable of causing thermal hysteresis has occurred several times. Distantly related fishes have independently "discovered" the same antifreeze primary structures (Fig. 11-27). The repeat tripeptide unit, "alanyl-alanyl-threonyl" (with a carbohydrate moiety attached to the threonyl residue), is found in the Antarctic nototheniid fishes, e.g., of the genus *Trematomus*, and in the Arctic rock cod, *Gadus ogac* (Van Voorhies et al., 1978). The peptide antifreezes of the Arctic sculpin, *Myoxocephalus verrucosus*, and flounder, *Pseudopleuronectes americanus*, likewise appear to have similar structures. Cysteine-rich antifreezes have been described in both fish (the sea raven, *Hemitrepterus americanus*; Slaughter et al., 1981) and insects (*Tenebrio molitor*; Schneppenheim and Theede, 1980). Thus, there appears to be a considerable variety of molecular structures that can function as antifreezes, and the independent evolution of these different structures has occurred again and again, in a variety of fishes and invertebrates.

The structures of two types of antifreezes are shown in Figure 11-27. The glycopeptide antifreezes were the first to be sequenced, and they characteristically exhibit a simple structure in terms of amino acid sequence. Alanyl residues contribute a high percentage of total residues, and threonyl residues serve as the attachment sites for the carbohydrate components of the antifreezes. The basic repeated unit shown in Figure 11-27 is found in antifreezes of different molecular weights. The smallest glycopeptide is 2,600 daltons, and the largest is approximately 33,000 daltons (DeVries, 1980). In the smallest class of glycopeptide, some of the alanyl residues are replaced by prolines.

The partial sequence of the peptide antifreeze from the winter flounder, *P. americanus*, is shown in the lower panel of Figure 11-27. Alanyl residues again constitute a major fraction of the amino acid chain. Threonyl and aspartyl residues are the other dominant components of the peptide antifreezes. Like the galactose and galactose-amine residues of the glycopeptide antifreezes, the polar or charged threonyl and

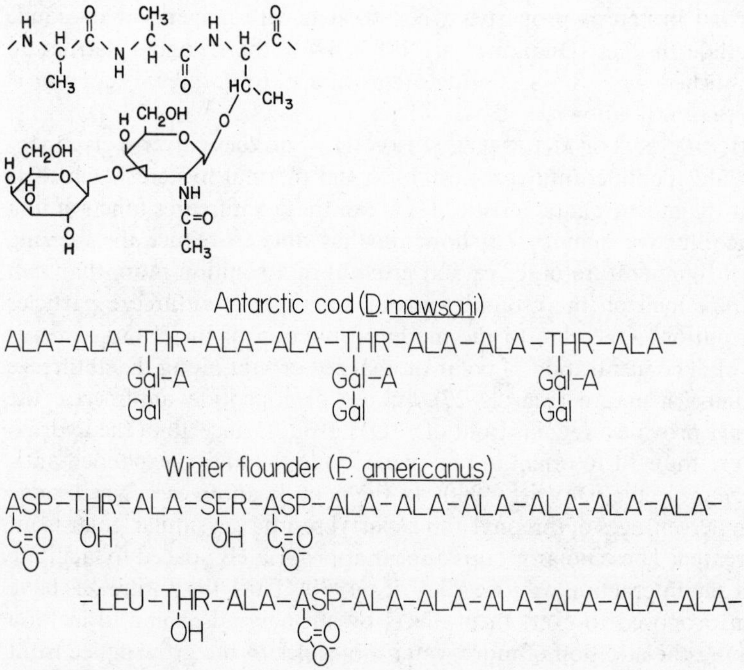

Antarctic cod (D. mawsoni)

ALA–ALA–THR–ALA–ALA–THR–ALA–ALA–THR–ALA–
Gal-A Gal-A Gal-A
Gal Gal Gal

Winter flounder (P. americanus)

ASP–THR–ALA–SER–ASP–ALA–ALA–ALA–ALA–ALA–ALA–
C=O OH OH C=O
O⁻ O⁻

–LEU–THR–ALA–ASP–ALA–ALA–ALA–ALA–ALA–ALA–
OH C=O
O⁻

Fig. 11-27. Structures of the glycopeptide and peptide antifreezes of polar fishes. Upper panel: The basic repeating structural unit of a glycopeptide with antifreeze properties. The peptide chain contains only two types of residue (alanyl and threonyl). To each threonine residue is joined a disaccharide (β-D-galactopyranosyl-(1→3)-2-acetamido-2-deoxy-α-D-galactopyranose. Lower panel: Primary structures of the glycopeptide antifreezes of the Antarctic nototheniid fish, *Dissostichus mawsoni* (observe the ALA-ALA-THR- repeat unit, shown in the upper panel), and the peptide antifreeze of the winter flounder, *Pseudopleuronectes americanus*. Figure after DeVries (1980)

aspartyl residues provide a “front” of polar groups which may facilitate strong adsorption to ice crystals, as discussed below.

The insect antifreezes have not been as well characterized at the structural level. None of the insect antifreezes has been found to contain carbohydrate components, i.e., all of these antifreezes appear to be peptides (Duman et al., 1982). The percentage contribution of different amino acid residues differs from the fish antifreezes. The insect peptide antifreezes characteristically have low alanyl contents, for example. The antifreeze peptide from the milkweed bug, *Oncopeltus fasciatus*, has a high serine content (30.5 percent; Patterson et al., 1981). Other insect antifreeze peptides contain high cysteine contents and lose their thermal hysteresis properties when treated with reagents that reduce disulfide bridges (Duman et al., 1982). How the secondary structure established by —S—S— bridges contributes to antifreeze function is not presently known.

MECHANISMS OF ANTIFREEZE ACTION. How do these diverse glycopeptide and peptide antifreezes establish the thermal hysteresis which is their diagnostic characteristic? How can these antifreezes function in a noncolligative manner, i.e., how are they able to reduce the freezing point (temperature of ice crystal growth) of a solution more than can be explained on the grounds of total numbers of antifreeze particles in solution? One clue to the mode of function of these antifreezes is the high concentration of polar or charged groups along the antifreeze primary structure (Fig. 11-27). In the glycopeptide antifreezes, the sugars provide a regular front of —OH groups, since all of the hydroxyls are thought to orient in one plane along the linear, expanded antifreeze molecule (DeVries, 1980). In the peptide antifreezes, which have high percentages of threonyl and aspartyl residues, a similar polar front is created. These polar groups appear appropriately spaced to facilitate a strong interaction with ice (DeVries, 1980). Thus, the antifreezes have been proposed to exert their effects by strongly adsorbing to ice and making the addition of more water molecules to the growing ice front less thermodynamically favorable (a more detailed discussion of this "poisoning" of growth mechanism is given by Raymond and DeVries, 1977).

Evidence in support of this hypothesis has come from several types of experiments. For example, unlike solutes that obey colligative relationships, the antifreezes do not freeze out of a solution as ice formation occurs (Duman and DeVries, 1973). This finding suggests that a strong interaction takes place between the ice and the antifreeze. Using scanning electronmicroscopy, Raymond and DeVries (1977) provided striking visual proof of this interaction. They showed that the antifreezes did adsorb to ice crystals, inhibiting their growth and affecting the crystal form of the ice that did propagate. The thermal hystersis observed in the freezing-melting behavior of solutions containing antifreezes further suggests an unusual interaction between ice and these molecules. The reduced freezing point observed in antifreeze-containing solutions may reflect the inhibition of ice crystal growth that results when antifreeze molecules adsorb to the growing front of a small ice crystal. Once ice is formed, however, its melting temperature is that of normal ice. The dependence of antifreeze action on polar groups has been demonstrated by chemical modification studies, e.g., by blocking the hydroxyl groups on the galactose residues with sodium borohydride (DeVries, 1980). The removal of these polar groups may prevent efficient hydrogen bonding between antifreeze and ice, thereby eliminating the molecules' capacities to bind to ice and to inhibit further crystal growth.

It is important to point out that the unusual capacities for interaction with ice possessed by the peptide and glycopeptide antifreezes do not mean that these antifreezes have atypical interactions with liquid water (see DeVries, 1980). The quantity of water bound to the antifreezes is similar to the amount bound by proteins of similar size which lack the thermal hysteresis properties of the antifreeze molecules. The fraction of serum water that could be bound by the antifreezes is only about 1 percent (DeVries, 1980), and this minute amount of bound water cannot account significantly for the freezing-point depressing abilities of the glycopeptide and peptide antifreezes.

WHAT IS THE RELATIONSHIP BETWEEN IN VITRO AND IN SITU FUNCTION OF ANTIFREEZES? The elucidation of the mechanism of antifreeze action has, of course, involved in *vitro* experimentation. The careful studies of ice crystal growth *in vitro* may mislead us into thinking that a similar phenomenon occurs throughout the body fluids of antifreeze-containing animals. In reality, ice formation per se must be avoided in most freezing-resistant species, notably the Antarctic fishes which live continuously at $-1.86°C$ (DeVries, 1980). Thus, even if the antifreezes in these fishes were capable of preventing the growth of ice crystals in the blood or cytosol, there is apparently no way to remove these crystals once they are formed. That is, the fish never reach temperatures as high as the melting points of the antifreeze-containing body fluids (approximately $-1°C$).

These considerations have led to the hypothesis that the *in situ* function of antifreezes is one of inhibiting ice crystal propagation across integumental surfaces (DeVries, 1980). For example, in fishes the most likely site of ice propagation is across the gills, which lack a covering of protective scales and mucus. Antifreeze molecules in the gill membranes and circulating fluids may, therefore, be critical in preventing the passage of minute ice crystals from the seawater into the body fluids. Schneppenheim and Theede (1979) have observed that the integument of the Arctic sculpin, *Myoxocephalus scorpius*, contains antifreeze peptides, a finding in support of the hypothesis that one of, if not the key, function(s) of the antifreezes is a "peripheral defense" involving the blocking of the ice crystal penetration into the interior of the organism.

ANTIFREEZES, NUCLEATING AGENTS, AND POLYOLS IN INSECTS: WHAT ARE THEIR ROLES? Unlike polar fishes in which each species contains only one type of antifreeze compound in terms of basic chemical structure (but not molecular size), some insects contain a battery of compounds that all may be involved in establishing freezing resistance or freezing tolerance: freezing-point depressing peptide antifreezes, ice nucleating agents, and cryoprotective substances like glycerol. How do these multiple components interact to provide insects with the abilities either to avoid ice formation or to withstand freezing of the extracellular fluids?

Duman (1980) has summarized the different roles likely to be played by these three classes of molecules. It is fairly clear that the nucleating agents have a single role, that of preventing significant supercooling which could lead to spontaneous ice formation in the intracellular space. The full roles of polyols remain controversial despite years of study. They may aid in supercooling, they may be cryoprotectants, and they may aid in desiccation resistance. The contributions played by antifreeze peptides may also be multifaceted. Prior to the acquisition of freezing tolerance late in the fall by the beetle *Dendroides canadensis*, the antifreeze peptides present in its hemolymph may adequately protect against ice formation. The antifreeze peptides may also act as supercooling agents (Duman et al., 1982). Once the beetles acquire freezing tolerance, it is not clear what role the antifreeze peptides can play. Nonetheless, these peptides reach their maximum concentrations in this beetle when freezing tolerance is greatest. Duman (1980) conjectures that these proteins may serve as cryoprotectants under these latter conditions; this idea awaits experimental test.

REGULATION OF ANTIFREEZE SYNTHESIS AND DEGRADATION. Antifreeze peptides and glycoproteins may be present at relatively high concentrations in the body fluids (approximately 3 percent weight/volume in the serum of Antarctic fishes; DeVries, 1980), and their biosynthesis probably involves a substantial amount of metabolic energy. It is not unexpected, therefore, that antifreeze synthesis in many species occurs on a seasonal basis, beginning in the autumn and ceasing in the spring. Antarctic fishes are an exception, of course, as they continuously experience subzero temperatures and must maintain antifreeze titres during their entire lives.

A number of recent studies have addressed the control mechanisms involved in triggering changes in antifreeze synthesis/degradation rates. Much as antifreeze molecules differ in structure among animals, so are the control mechanisms varied, at least in terms of the environmental cues which are utilized to sense the needs for antifreeze production. Duman and DeVries (1974) found that a "fail-safe" control mechanism was operative in governing antifreeze synthesis and degradation in some northern fishes. Although the synthesis of antifreeze was triggered by holding fish at low temperatures, the disappearance of antifreezes required a combination of long photoperiod and warm temperatures. Thus, removal of antifreezes in response to an unusual warm spell in late winter or early spring would not occur due to the overriding influence of photoperiod. The kinetics of antifreeze synthesis required three to six weeks for peak antifreeze titres to be built up, and loss of antifreeze occurred over a similar time course.

The hormonal and molecular-genetic bases of antifreeze turnover have also been studied in some fishes. In the winter flounder, *Pseudopleuronectes americanus*, pituitary regulation of antifreeze degradation was shown by Hew and Fletcher (1979). Hypophysectomized fish continued to synthesize antifreeze in the summer, while control fish did not. The levels of control involved in regulating antifreeze synthesis appear to be complex. Lin (1979) and Lin and Long (1980) have shown that the specific messenger RNA (mRNA) for antifreeze in the winter flounder appears well before antifreeze synthesis begins, and disappears about a month prior to the disappearance of antifreeze in the spring. Thus, both transcriptional and translational control of antifreeze synthesis in fish liver appear to occur.

In terrestrial invertebrates, a variety of regulatory schemes exist. Not all species studied exhibit the "fail-safe" degradation scheme noted for certain fishes. Thus, while the darkling beetle, *Meracantha contracta*, required a combination of high temperature and long photoperiod to induce the loss of antifreeze (Duman, 1977), the spider *Philodromus sp.* lost antifreeze only in response to elevation in temperature. In the beetle *Dendroides canadensis*, Duman (1980) showed that a long photoperiod was necessary to trigger loss of antifreeze; high temperatures and short photoperiod were not effective in this regard. The neural and hormonal mechanisms effecting these responses to environmental cues remain to be elucidated.

Lessons from the Study of Temperature Adaptation—and Some Important, Unanswered Questions

In Chapter 1 we proposed that a major reason for doing comparative biochemistry, that is, for examining adaptive variations on different molecular themes, is to discover the crucial aspects of biochemical design. Through the study of molecular adaptations, one not only may discover how different organisms are biochemically adapted to their particular habitats, but also what characteristics of molecular systems are rigorously conserved in all species. The study of temperature adaptation offers an excellent means for elucidating these aspects of biochemical design, because temperature has such a pervasive influence on biochemical systems that all of the major chemical components of organisms—water, proteins, lipids, and nucleic acids—will have their structural and functional properties altered by changes in temperature.

Among the strongly conservative trends we noted in temperature adaptation was the defense of K_m values and catalytic rate potentials of enzymes at normal physiological temperatures. Temperature compensatory adjustments in enzyme kinetic properties that lead to the conservation of K_m and catalytic rate seem a ubiquitous feature of enzyme evolution. Only small changes in habitat (body) temperature are necessary to foster these types of evolutionary modifications of enzymes. Thus, in the case of the barracuda congeners, we found that differences in average body temperature of only a few degrees Celsius have been sufficient to favor selection for temperature-adapted M_4-LDH variants. The types of amino acid substitutions that effect these functional adaptations remain to be discovered. Through understanding the structural bases of these kinetic adaptations we may gain powerful new insights into the basic mechanisms of enzyme function; that is, we may be able to discern more clearly how changes in primary sequence effect adjustments in binding and catalytic properties. Through studies of enzymes from differently adapted ectotherms we may come to understand structure-function relationships in enzyme families, e.g., M_4-LDHs, at the detailed level that structure-function relationships in respiratory proteins, e.g., vertebrate hemoglobins, are now understood. Likewise, study of the structures of subunit interaction sites in multimeric proteins like filamentous actin may provide us with key insights into the types of bondings that stabilize the reversible assembly of many different classes of proteins.

In many ways the adaptations noted in lipid-based systems parallel those found in proteins. In both cases the need for reversible structural transformations necessitates a "semistability" of structure (Alexandrov, 1977). To achieve the correct level of structural flexibility for the given temperature of function, amino acid composition and lipid composition are found to be adjusted. Like the adaptations in enzyme structural and functional properties, the adaptations in lipid systems leading to homeoviscosity appear to be ubiquitous among different types of organisms. Even though a variety of different regulatory mechanisms for modifying the fatty acid and head group compositions of membrane lipids exist in procaryotes and eucaryotes, the end result of these regulatory events is the same in all cases.

Adaptive changes in the structures of large molecules, proteins, lipids, and nucleic acids, are complemented by adaptive adjustments in the microenvironment in which the large molecular ensembles function.

In particular, regulation of pH according to the alphastat scheme was shown to be critical for the retention of correct kinetic properties, e.g., the K_m of pyruvate for M_4-LDH, and assembly states of multimeric proteins, e.g., of PFK. The importance of histidine imidazole groups in temperature adaptation was strongly emphasized. How different organisms and different tissues within an organism regulate pH merits much additional study, for although we are beginning to appreciate the significance of alphastat regulation, the mechanisms employed to achieve the correct pH_i for a particular body temperature are not that clearly understood.

The study of temperature adaptation provides us with a good means for comparing the results of different time courses of molecular adaptation. We found that long-term, evolutionary changes often were similar to short-term acclimatory changes. Lipid adaptations offer a good case in point; homeoviscous adjustments were seen to be comparable in differently adapted species and in differently acclimated populations of a single species. Alphastat regulation also is noted on inter- and intraspecific bases. The role of protein variants in temperature acclimation remains unclear. The adaptive trends noted in interspecific comparisons, e.g., temperature adaptive modifications in K_m and k_{cat} values, are typically not observed in comparisons of differently acclimated (-acclimitized) populations of a single species. Nonetheless, there are enough good examples of acclimation or acclimatization effects, e.g., the acetylcholine esterase system of *Salmo gairdneri* and the H_4-LDH system of *Fundulus heteroclitus*, to encourage more studies of these changes.

Finally, our increasing understanding of the end-results of temperature adaptation processes is not yet matched by an appreciation of the kinetics of these processes. How rapidly do evolutionary changes in proteins occur? Do acclimatory adjustments in protein and lipid systems follow similar time courses? What regulatory mechanisms are rate governing in acclimation processes? Questions such as these may serve as useful focal points in our future attempts to appreciate how the diverse biochemical systems of organisms are adaptively modified over evolutionary time and during short-term acclimatory periods for function under diverse thermal regimes.

CHAPTER TWELVE

Adaptations to the Deep Sea

Introduction

Now that the fundamental characteristics of biochemical adaptation have been presented, largely through discussions involving one environmental stress at a time, it seems appropriate to conclude this volume by considering an environment in which all classes of environmental factors—physical, chemical, and biological—play conspicuous and interacting roles in shaping the biochemical design of organisms. The unique features of the deep sea provide us with an excellent study system for illustrating the basic concepts developed throughout this volume and, in particular, for showing how the basic biochemical designs of organisms are shaped by the combined selective influences of a complex set of environmental factors including: hydrostatic pressure, temperature, light, food supply (quantity, caloric content, distribution, source of primary production), currents, and the chemical composition of seawater.

Selectively Important Features of the Deep-Sea Environment

1. Physical characteristics. The deep sea is an environment noted for its extremes of hydrostatic pressure, its darkness (except for light due to bioluminescence), its low temperatures (except near the hydrothermal vents, where water temperatures may be as high as 360°C; Spiess et al., 1980), its great distance from primary productivity (again, the hydrothermal vents are an exception), and the general absence of strong currents. Each of these factors will be seen to influence the designs of deep-sea animals. These essential features of the deep sea are illustrated in Figure 12-1. Note that sunlight-driven primary productivity is restricted to the first 300 m of the water column; at greater depths light penetration is inadequate to support photosynthesis. Also note that the deep sea is typically a very cold environment. Average deep-sea temperatures are near 2–3°C. Thus, we can ask whether the types of temperature adaptations noted in cold-adapted ectotherms from shallow waters or terrestrial habitats are also characteristics of deep-living species. Since hydrostatic pressure increases by 1 atmo-

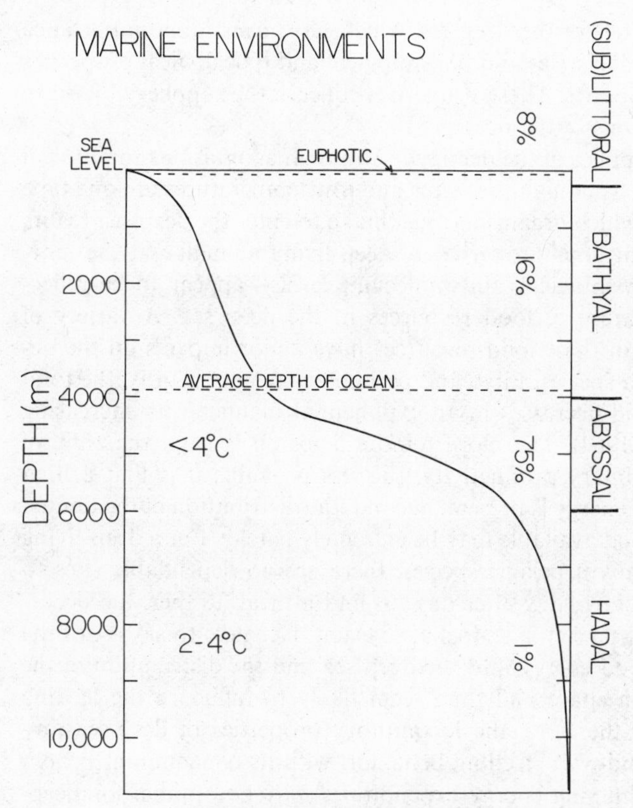

Fig. 12-1. Classification of the marine environments. The percentages refer to the area of the ocean floor included in each depth zone. Figure from Somero (1982)

sphere (atm) for each 10 m increase in depth, the pressures in the deepest regions of the oceans are near 1,100 atms. Average deep-sea pressures are near 380 atm. Darkness is the rule in the deep sea, except where biologically produced light is present. Darkness can have major effects on predator-prey interactions. Visual localization of prey is difficult, so feeding strategies may differ from those found in shallow waters. Choice of feeding strategy may have profound influences on the biochemical designs of deep-sea animals, as we argue in sections to follow. In summary, then, the physical features of the deep-sea environment would seem to present organisms with several distinct problems necessitating biochemical adaptations. How each of these physical features influences the properties of deep-sea animals cannot be appreciated in isolation, however; instead we must include in our analysis a second class of properties of the deep sea which have overriding importance in affecting the life styles and physiological and biochemical properties of deep-living species. These properties concern the supplies of food in the deeper regions of the oceans.

2. Food supplies in the deep sea. Although adaptations to physical stresses arising from high pressures and low temperatures are one type of adjustment which organisms must make to enter the deep-sea realm, the most conspicuous properties of deep-living animals—at the morphological, physiological, and molecular levels—appear to be reflections of the nature of food resources in the deep sea. A variety of characteristics of these food resources have major impacts on the design of deep-sea species. First, and perhaps most importantly, the total quantity of food decreases in an exponential manner with increasing depth (Banse, 1964). The more remote a region is from the zone of light-driven primary productivity, the less probable it is that a high density of food energy is present. Second, the distribution of the limited quantities of food available may be extremely patchy. For a deep-living animal, especially a pelagic species, there are no dependable sites to which it can return, day after day, to find a meal. Rather, the occurrence of a food packet is a stochastic event. Needless to say, both the total flow of food energy into the deep sea and the distribution of the food packets in space and time seem likely to influence the feeding strategies and, therefore, the locomotory properties of deep-sea animals. "Float-and-wait" feeding behavior, with its concomitant benefit of reduced swimming energy expenditure, may be optimal for deep-living forms. And, in view of the fact that burst swimming may lead to approximately 100-fold increases in power output over the basal level in fishes (see Brett and Groves, 1979), the energy saved by a "float-and-wait" predator may provide this species with a bonus in terms of energy that can be channeled into growth. The energy budget of a deep-sea animal may differ considerably from that of a shallow-living species, both in terms of total caloric flow through the animals and the relative parceling of energy between maintenance (including swimming costs) and growth.

Until very recently, the food supplies available to deep-sea animals were discussed entirely in terms of "energy limitation," following the basic treatment given above. The belief that all primary productivity on earth was driven by solar energy leads to the conclusion that the amount of food energy available is directly correlated with closeness to sites of sunlight-driven primary production. It follows that the deep sea must be an environment in which energy limitation is a dominant fact of life. In recent years a discovery was made that is causing biologists to re-examine critically food-chain relationships in the deep sea. This discovery is that primary productivity may be driven by sources other than solar energy. At the hydrothermal vent communities located along spreading centers (Corliss et al., 1979; Spiess et al., 1980), energy in the form of hydrogen sulfide (H_2S or HS^-) is a primary driving force of life. Sulfides generated through geological processes (Edmond et al., 1979), i.e., produced independently of solar energy, are used as a primary source of energy to drive ATP synthesis and the production of biological reducing power (NADPH) in the vent communities (Felbeck et al., 1981; Felbeck and Somero, 1982). Thus, when we speak about nutrient levels and trophic interactions in the deep sea, we must develop two distinct stories. One begins in the euphotic zone, and one originates in the magma beneath the ocean floor. The contrasts between the organisms of the "typical" deep-sea regions and the hydrothermal vent communities are such as to provide a convincing argument for the importance of the proximity to a site of primary production in establishing the key morphological, physiological, and biochemical attributes of animals. Studies of the dense, metabolically robust vent community animals allow us to appreciate better than was formerly possible how the food sources available to typical deep-sea creatures shape their unusual properties.

Adaptations to Hydrostatic Pressure

Prior to considering the trophic relationships of deep-sea animals, whether these be typical deep-sea creatures or members of the hydrothermal vent communities, it is appropriate to analyze first the types of molecular adaptations that allow life to exist, at any intensity of metabolism, under the high pressures found in the deep sea. Without the capacity to maintain an adequate rate of closely regulated metabolic flow in the face of high pressures, it would be moot to be concerned about relative food abundance in typical and hydrothermal vent habitats in the deep ocean. Pressure adaptations, notably of enzymic function, will be seen to play major roles in establishing the depth distribution patterns of marine animals—both in terms of restricting access of shallow-living animals to the deep sea and in terms of making deep-sea species less fit competitors with their shallow-living counterparts. To appreciate how these adaptations achieve their importance in setting depth distribution patterns, we must first examine the sources of pressure perturbation of biological systems.

Pressure effects on all types of processes arise from a common basis: volume changes. If a process takes place with no change in volume, then it will not be influenced by pressure. However, when a volume increase takes place during the process, pressure will inhibit the process. Conversely, if a volume decrease takes place, pressure will enhance the process. These are the basic ground rules involved in establishing pressure sensitivity and, as we argue below, in fabricating adaptive responses to pressure.

For purposes of this discussion it is necessary to examine volume changes in somewhat greater detail, inquiring, first, about their sources and, second, about the quantitative relationships between these changes in volume and their effects on both the equilibrium positions and rates of chemical reactions. At the outset of this discussion it is important to stress that pressure can alter the equilibrium of a reaction and the rate at which this equilibrium is attained. These relationships are given by equations (1) and (2):

$$\left(\frac{\delta \ln K_{eq}}{\delta P}\right)_T = \frac{-\Delta V}{RT} \tag{1}$$

$$k_p = k_0 \exp(-P\Delta V^{\ddagger}/RT) \tag{2}$$

where P is pressure (in atms), T is the absolute temperature, R is the gas constant (equal to 82 cm^3 atm° K^{-1} mol^{-1}), K_{eq} is the equilibrium constant for the reaction, k_0 and k_p are the rate constants for the reaction at pressures of 1 atm and P atms, respectively, ΔV is the volume change associated with the initial versus final states of the system, and $\Delta V^{\ddagger}$ is the "activation volume" of the reaction, the change in system volume occurring during the rate-limiting step in the transformation of reactants to products. Detailed accounts of these relationships are found in Johnson et al. (1974) and Laidler and Bunting (1973).

From these equations it is apparent that the volume change associated with the initial (reactant-containing) and final (product-containing) systems will determine the effect of pressure on K_{eq}, while the effect of pressure on reaction rate (k) is determined by the change in system volume as the ground-state complex is excited to form the transition state. As in the case of temperature effects, the symbol "double-dagger" (‡) is used to denote the activation step.

Having established a quantitative relationship between volume changes and the effects of pressure on reaction equilibria and reaction rates, it is now appropriate to consider the sources of volume changes in biochemical transformations. Which components of metabolic reactions can expand or be compressed? To develop a satisfactory answer to this question it is essential to keep in mind that the vast majority of chemical transformations in biological systems occur in an aqueous phase. Water is likely to be the major contributor to equilibrium and activation volume changes (Low and Somero, 1975). The basis for the pivotal role of water in establishing volume changes of biochemical reactions is that most constituents of the cellular fluids have effects on water structure. In very general terms, most metabolic intermediates and protein amino acid sidechains are surrounded by a layer, perhaps several water molecules in thickness, of relatively ordered water. This ordered or structured water is likely to have a smaller volume than would be occupied by the same number of water molecules in the so-called "bulk" phase of the solution. A consequence of the water-structuring abilities of metabolites, allosteric effectors, and amino acid sidechains is that whenever an enzymic reaction takes place, some change in water organization and, hence, in system volume is likely to occur. For instance, addition of a substrate or cofactor molecule to the active center of an enzyme may require at least a partial dehydration of both the ligand and the amino acid residues which interact with the ligand. Enzyme-ligand interactions typically are mediated through charge-charge interactions or through polar, but noncharged interactions. Carboxyl groups, carbonyl oxygens, guanidinium moieties (of arginine), phosphates, and histidyl residues all have strong water-structure-affecting properties. Thus, the reversible hydration/dehydration of these groups during binding events is likely to occur with substantial changes in system (water-protein-ligand) volume, whether or not the volume of the protein changes.

Proteins seem a second likely candidate for volume changes in view of the occurrence of substantial changes in enzyme conformation and/or subunit aggregation state during catalysis and regulation. For proteins, two distinct classes of volume changes must be examined. First, alterations in the packing efficiency of the amino acid residues of a protein could contribute to volume changes. These are changes in the inherent volume of the protein, as distinguished from hydration volume changes which occur at the protein-water interface. In fact, most proteins examined to date display extraordinary levels of packing efficiency; their amino acid sidechains, notably the nonpolar (hydrophobic) chains that fill most of the interior of the protein, are packed virtually as tightly as the molecules in an organic solution of a single hydrocarbon species (see discussion in Chotia, 1975). In the case of protein tertiary structure formation, Nature seems to abhor a vacuum, and there is little space left for improving the tightness or compactness of fit among amino acid residues. Thus, changes in the volume of a protein during catalysis and regulation, while they clearly do occur in certain cases (Low and Somero, 1975), are apt to contribute less to the

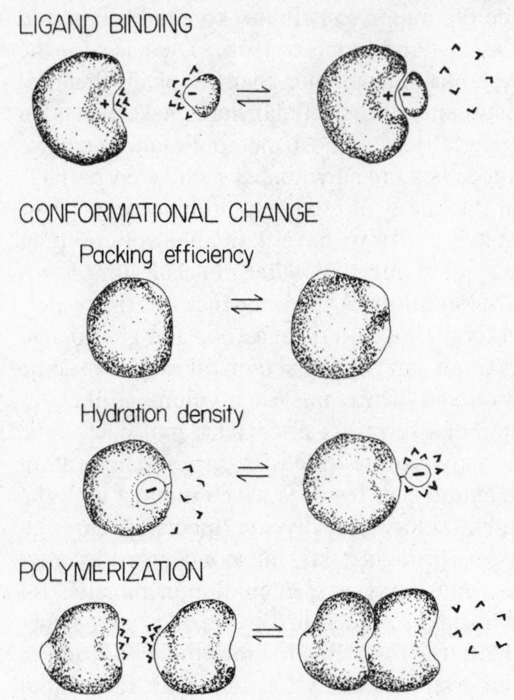

Fig. 12-2. Sources of volume changes during enzymic reactions and protein subunit aggregation events. Ligand Binding: For the binding event illustrated, a negatively charged ligand (substrate, cofactor, or regulatory modulator) complexes with a positively charged site on the enzyme (binding site). Densely ordered water around the ligand and the binding site is "squeezed" into the bulk water phase during the binding step. The expansion of this released water creates an increase in system (protein + ligand + water) volume. Conformational Changes: 1) "Packing Efficiency" effects. During a change in enzyme conformation the density of the protein itself changes due to alterations in the packing efficiencies of amino acid residues. 2) "Hydration Density" effects. Changes in protein conformation lead to alterations in the exposure to water of water-structure-modifying groups. Illustrated in this figure is the reversible transfer of a negatively charged group from a relatively "buried" (inaccessible to water) location to an environment where strong hydration is possible. The water organizing around the exposed charged group is denser than bulk water. Polymerization: When two subunits aggregate to form a polymer, organized water on the subunit contact sites is "squeezed" into the bulk phase, leading to an expansion of system volume, as in the case of the ligand binding example.

pressure-sensitivities of enzymic reactions than hydration-based volume changes. The latter can arise as a result of changes in protein conformation whether or not the volume of the protein itself changes during the reaction. For instance, alterations in exposure to water of amino acid sidechains which affect water structure (density) seem almost unavoidable during changes in enzyme conformation. In fact, these hydration changes may be an important contributor to the catalytic efficiencies of enzymes, since an exergonic hydration event which occurs during the activation step in catalysis will reduce the free energy of the activated complex and thereby enhance its probability of formation (Low and Somero, 1975; Greaney and Somero, 1979). As in the case of ligand binding, where volume changes seem unavoidable as water is removed from and added to ligands and binding sites residues, catalytic conformational changes may involve hydration reactions that confer a pressure-sensitivity to enzymic processes. It should also be noted that pressure-sensitivity can arise during assembly events involving proteins. The removal of water from subunit contact sites is likely to be a ubiquitous step in the assembly of multimeric proteins from their constituent subunits. Consequently, even without changes in the inherent volumes of the subunits during polymerization, subunit assembly equilibria may be perturbed by increases in pressure. The "threat" of pressure disruption of multimeric proteins seems especially great in the case of entropy-driven assemblies, where water removal is responsible for the large entropy increase that drives the assembly reaction.

The varied sources of volume changes during ligand binding, catalytic activation events, and subunit assembly processes are illustrated diagrammatically in Figure 12-2.

Pressure Adaptations of Enzyme Kinetic Properties

Do the various sources of volume changes during enzymic reactions actually establish high enough pressure-sensitivities of enzymic processes to necessitate pressure-adaptive changes in enzymes? An affirmative answer can be given to this question, even though it is not clear what particular types of volume changes establish the pressure response of the system in most cases. Our treatment of pressure effects on enzymes of shallow- and deep-living animals will, therefore, focus strongly on pressure effects on the kinetics of enzymic reactions, and only a limited and conjectural treatment of basic volume-change-adjusting mechanisms will be given.

In our discussions of temperature, osmolyte, and pH influences on enzyme kinetics, we found that certain key features of enzymes were at once highly perturbable and, as a result of adaptive changes in either the enzymes themselves or the solutions bathing them, strongly conserved. One of these enzymic traits is the apparent Michaelis constant (K_m) of substrates and cofactors, a parameter which in many cases reflects the effectiveness with which the enzyme-substrate or the enzyme-cofactor complex forms. Much as K_m values are conserved in the face of varying temperature, pH, or osmotic conditions, as detailed in earlier chapters, adaptation to the deep sea has necessitated a stabilization of K_m values in the face of high pressures and low temperatures. It is worth examining in some detail the role of pressure-adaptive changes in K_m values, for this analysis will teach us several important lessons about adaptation to deep-sea conditions. We will find that shallow-living, cold-adapted species' enzymes are not always pre-adapted for function at high pressures. Thus, access to the deeper regions of the water column necessitates overcoming an important biochemical barrier. We will also find that striking examples of convergent evolution are found in enzymes of deep-sea fishes; there appears to be a single "best solution" to maintaining controlled enzymic catalysis at depth. Also, we will discover that the acquisition of pressure-resistant enzymes has not been without a certain price, and the cost which accrues from tolerance of pressure may be reflected in poor competitive abilities of deep-living animals at shallower depths. Lastly, we will find that the necessity for pressure-adaptive changes in enzymes may arise at relatively shallow depths, in the range of 500 to 1,000 m (51 to 101 atms).

The enzyme which has received the most detailed examination from the standpoint of pressure adaptation is the M_4 isozyme of lactate dehydrogenase (LDH), an enzyme which by now should be a familiar acquaintance of the reader. Possibly for the reasons suggested at the end of this section, LDH is a markedly pressure-sensitive enzyme. All of the kinetic features examined to date, including K_m values for substrate (pyruvate) and cofactor (NADH) and substrate turnover number, exhibit pressure-adaptive differences between the LDHs of shallow- and deep-living fishes. Analysis of LDH's of shallow- and deep-living fishes will allow us to gain important insights into the role of protein adaptations in deep-sea animals, and into the effects of these adaptations on depth distribution patterns in the water column.

Figure 12-3 illustrates the effects of experimental pressure on the K_m values of NADH (upper panel) and pyruvate (lower panel) for M_4-LDHs from several shallow- and deep-living marine teleost fishes, all

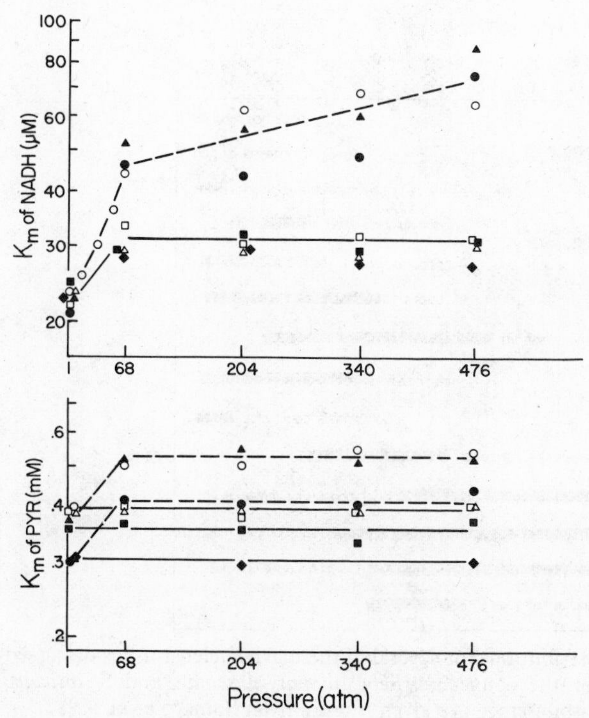

Fig. 12-3. The effects of hydrostatic pressure on the apparent K_m of NADH (upper panel) and pyruvate (lower panel) for purified M_4-lactate dehydrogenases of several deep- and shallow-living marine teleost fishes. Deep-living species: *Sebastolobus altivelis* (□), *Antimora rostrata* (△), *Coryphaenoides acrolepis* (■), and *Halosauropsis macrochir* (◆). Shallow-living species: *Pagothenia borchgrevinki* (●), *Scorpaena guttata* (▲), and *Sebastolobus alascanus* (○). All measurements were made at 5°C. Figure from Siebenaller and Somero (1979)

of which are adapted to roughly similar temperatures (approximately −2° to 8°C). As we would expect on the basis of K_m-conservation arguments raised in Chapter 11 dealing with temperature effects, the 1 atm K_m values for both NADH and pyruvate are extremely similar among the different species. For instance, the K_m of pyruvate values are in the range characteristic of highly cold-adapted ectotherms. The close similarity found among the K_m values at 1 atm pressure is not observed at elevated pressures, however. For the LDHs of the shallow-living fishes, increases in pressure cause the K_m values of both substrate and cofactor to increase, with an especially large effect occurring for the K_m of NADH. For the LDHs of the deep-living fishes, the K_m of

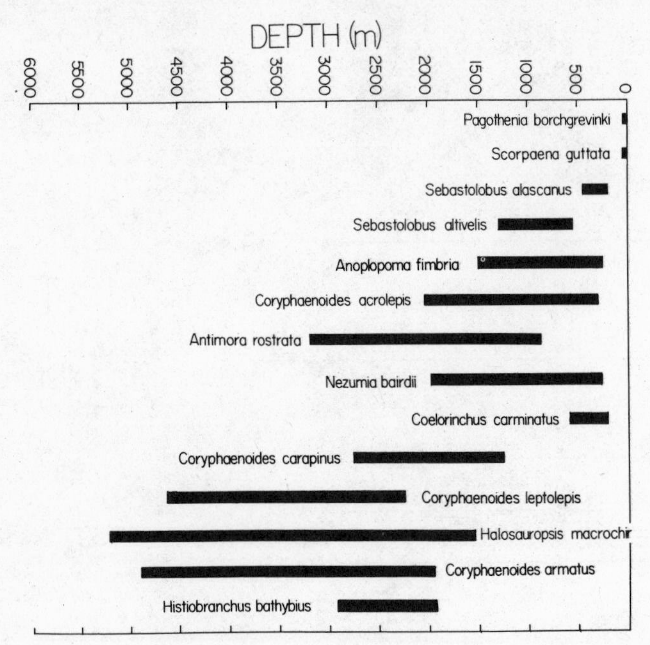

Fig. 12-4. Depth distributions for several of the marine teleost fishes discussed in this chapter. For the congeners, *Sebastolobus alascanus* and *S. altivelis*, depths of maximal abundance are given. Figure from Somero et al. (1983)

pyruvate is totally insensitive to pressure, while the K_m of NADH exhibits only a modest increase between 1 and 68 atms, and then remains stable up to pressures of 476 atms, the highest pressure used in this study, and a pressure close to the upper limits experienced by the deep-sea fishes examined (Fig. 12-4).

What are the implications of the kinetic differences noted between the M_4-LDHs of shallow- and deep-living fishes? First, it is apparent that the LDHs of the deep-sea fishes can maintain appropriate K_m values at elevated pressures, unlike the homologous enzymes from all of the shallow-living fishes examined. As discussed in detail in the context of temperature adaptation of enzyme function, maintaining K_m values within a range optimal for catalysis and regulation is a critical feature of enzymic adaptation. The enzymes of shallow-living fishes would be handicapped at depth by having abnormally high K_m values for both substrate and cofactor. The large increase in K_m of NADH could be particularly disadvantageous since the cofactor binding sites of dehydrogenases must bind cofactor extremely tightly to remain cofactor-saturated at all times. The direction of metabolic flow is thus established by the redox balance of the cell (ratio of NADH to NAD^+ in the case of LDH). Greatly reduced cofactor binding ability could have a strong rate-decreasing effect on glycolytic flux, since cofactor binding is the obligatory first step in the LDH reaction. For the LDHs of deep-sea fishes, substrate and cofactor binding is sufficiently pressure-insensitive to avoid these problems.

From the foregoing analysis it is clear that the M_4-LDHs of shallow-living, cold-adapted species are not pre-adapted for function at depth. Thus the acquisition of pressure-tolerant enzymes can be viewed as a prerequisite for penetration into the deeper regions of the water column. Two of the species examined in these studies allow an interesting scenario for the establishment of depth zonation patterns to be proposed (Siebenaller and Somero, 1978). *Sebastolobus alascanus* and *Sebastolobus altivelis* are very similar in their morphological features and general ecologies (Moser, 1974). They differ, however, in their depth distributions, as shown in Figure 12-4. They also display the deep-versus-shallow differences in LDH kinetic functions, as well as a number of other biochemical differences we discuss below. The different pressure responses of the M_4-LDHs are due to only very minor changes in the primary structures of the two LDH homologues. Peptide mapping studies and amino acid analysis reveal that only a single amino acid substitution appears to distinguish two LDHs (an asparagine residue in *S. altivelis* substitutes for a histidine in *S. alascanus*) (Joseph Siebenaller, unpublished observations). Thus, the amount of evolutionary change that is needed to convert a pressure-sensitive LDH to a pressure-insensitive molecule may be relatively small. In view of these considerations it is reasonable to conjecture that the deeper-living species, *S. altivelis*, may have arisen from an ancestral population of *S. alascanus* via acquisition of pressure-adapted LDHs (and, no doubt, other proteins). According to this hypothesis, occasional individuals of an earlier *S. alascanus* population (or a common ancestor to both current species) possessed enzymes having reduced pressure-sensitivity, and these individuals were able to colonize deeper regions of the water column. This model provides at least a partial account of how a shallow-living species could generate a second, deeper-living species. This model does not, however, explain why the pressure-tolerant species is restricted to greater depths and does not co-exist with—or replace—the shallow-living species. We return to the question of what keeps deep-sea species deep later in this chapter.

Another important conclusion arises from the comparison of *Sebastolobus alascanus* and *S. altivelis* lactate dehydrogenases. This is that pressures of only 50–100 atm seem adequate to favor selection for pressure-resistant enzymes. An organism need not experience extremes of abyssal pressures to need the types of adaptations noted for the M_4-LDH of *S. altivelis* and other deep-sea fishes. The similarities noted between the M_4-LDHs of *S. altivelis* and the much deeper occurring rattail (*Coryphaenoides*), halosaur (*Halosauropsis macrochir*), and *Antimora rostrata* suggest that the pressure adaptations needed to overcome the effects of hydrostatic pressures in the range of 50–100 atms are also adequate to permit enzyme function at several hundred atms pressure. In fact, the similarities in kinetic properties among the M_4-LDHs of all of the deep-sea fishes studied represent a clear example of convergent evolution at the molecular level. Each of the species studied arose from a different, shallow-water ancestor. It is pertinent to ask whether identical amino acid substitutions have taken place in all of the pressure-adapted LDHs; we hope future studies of the primary structures of these enzymes will reveal an answer to this question. We suggest a possible, common adaptation mechanism at the end of the next section of this chapter.

It is appropriate to emphasize that the M_4 isozyme of LDH represents one of several thousand different proteins of a fish, and generalizations based on LDH-pressure interactions do not rest on too strong an inductive base. Other enzymes may be more—or less—pressure-sensitive, and different steps in the catalytic reaction sequence may be perturbed in other enzymic reactions. Based on the limited amount of data available from studies of deep-sea animals (reviewed in Somero et al., 1983), it seems fair to state that some, but not all, proteins of deep-living species will display adaptations to pressure. In terms of pressure effects on apparent K_m values, the effects noted for M_4-LDHs may be fairly typical of the effects for a wide variety of enzymes. Data on acetylcholinesterases (Hochachka, 1974), citrate synthases (Hochachka et al., 1975), and Na-K ATPases (Pfeiler, 1978) all suggest that K_m values are frequently highly sensitive to elevated pressures and, as a result, are an important focal point of adaptation in deep-sea animals.

Pressure-related Differences in Catalytic Efficiency

The astute reader may have detected a key asymmetry in the arguments formulated above concerning the role of pressure adaptation of enzymes in establishing depth distribution patterns in the water column. We proposed that the acquisition of pressure-insensitive kinetic properties may allow members of a shallow-living species to move to deeper waters and, over time, to become reproductively isolated

Table 12-1. Activation Energy Parameters and Relative Absolute Velocities of M_4-Lactate Dehydrogenase Reactions of Species Adapted to Different Temperatures and Pressures

Species (depth/body temperature)	$\Delta H^{\ddagger}$ (cal/mol)	$\Delta S^{\ddagger}$ (cal/mol/°K)	$\Delta G^{\ddagger}$ (cal/mol)	Relative velocity[a]
Pagothenia borchgrevinki (surface, −2°C)	10,467	−12.7	14,000	1.00
Sebastolobus alascanus (180–440 m; 4–12°C)	10,515	−12.6	14,009	0.98
Sebastolobus altivelis (550–1,300 m; 4–12°C)	11,985	−8.1	14,249	0.64
Coryphaenoides acrolepis (1,460–1,840 m, 2–10°C)	11,813	−8.7	14,222	0.67
Halosauropsis macrochir (approx. 2,300 m; 2–5°C)	11,843	−8.6	14,227	0.66
Antimora rostrata (1,300–2,500 m; 2–5°C)	12,557	−6.4	14,343	0.54
Thunnus thynnus (bluefin tuna) (surface to 300 m; 15–30°C)	11,384	−10.0	14,152	0.76
Rabbit (37°C)	12,550	−6.4	14,342	0.54

SOURCE: Data from Somero and Siebenaller (1979)
[a] Measured at 5°C and 1 atm pressure.

and form a new species. The origin of *Sebastolobus altivelis* from *S. alascanus* through a speciation event of this type was proposed. However, this model has the following shortcoming: it cannot explain why the pressure-adapted species stays in deeper waters exclusively and does not distribute through both the deeper waters and the shallower waters where its non-pressure-adapted congener lives. While there is no single answer to questions about "what keeps deep-sea animals restricted to the deep sea," study of an additional characteristic of M_4-LDHs from shallow- and deep-living fishes has provided evidence that biochemical factors may be important in setting upper limits to depth distributions of deep-sea species, much as they may be instrumental in establishing the lower distribution limits of shallow-living forms.

The data in Table 12-1 show that the acquisition of pressure-insensitivity of function does not come without a price: the catalytic efficiencies of M_4-LDHs of deep-sea fishes are markedly lower than those of shallow-living, cold-adapted fishes. Under identical experimental conditions using saturating substrate and cofactor concentrations, a single M_4-LDH molecule of a deep-sea fish can convert pyruvate to lactate at only about one-half to two-thirds the rate characteristic of the M_4-LDH from a shallow-living fish. As in the case of the K_m responses to pressure, these differences in catalytic efficiency are found in all deep- and shallow-living fishes studied. The *Sebastolobus* congeners again display distinct shallow-versus-deep differences.

What are the implications of these differences in catalytic efficiency? First, it appears that a deep-sea fish would be at some metabolic disadvantage in shallow water—all other factors being equal—due to the possession of an enzyme having relatively low activity compared to the enzymes present in the shallow-living species it must compete with. Because of the critical role of M_4-LDH in locomotory function, especially "burst" swimming, a deep-sea fish attempting to eke out a living in shallow water might be relatively poorly prepared to catch prey, and to avoid serving as prey for a more robust swimmer. Of course, the deep-sea fish might counter the low efficiency of its M_4-LDHs by synthesizing and retaining higher concentrations of the enzyme in its locomotory musculature. This adaptive strategy bears an obvious cost in protein synthesis energy, however. Inefficient enzymes, therefore, may be one factor of importance in establishing the upper limits of distribution of high-pressure-adapted species. These differences may be especially important during the early stages of species differentiation, when most of the other characteristics of the shallow- and deep-living species pair are still extremely similar.

The implications of reduced catalytic efficiency of high-pressure-adapted enzymes must be considered in a cost-versus-benefit fashion under the *in situ* conditions faced by deep-sea organisms. There are at least two bases for rationalizing the putative benefits of less efficient enzymes in deep-living forms. On the one hand, the fact that most deep-sea animals have extremely low metabolic rates (see below) would seem to suggest that enzymes having low catalytic efficiencies could be a mechanism for slowing down metabolism in a cold, dark, food-poor environment. That is, reduced catalytic efficiency is directly advantageous in terms of establishing an appropriate level of metabolic flux for deep-living organisms. However, while this hypothesis may seem to have a ring of truth to it, we feel it is erroneous. If, in fact, the physical and biological features of the deep sea favor selection for reduced metabolic rates, as seems indisputable, it would seem a poor energetic strategy to use inefficient enzymes to accomplish this task. If energy is present in low quantities, then it would seem advantageous to use lower concentrations of efficient enzymes to achieve a given rate of metabolism, rather than to use higher concentrations of less efficient enzymes. In view of the fact that M_4-LDH is one of the major soluble proteins in skeletal muscle, clear energy savings from reduced protein synthesis seem possible if more efficient forms of LDH can be produced. An alternative explanation for the benefits of inefficient enzymes in deep-living organisms thus seems called for.

One potential advantage associated with inefficient enzymes has already been discussed in a different context. When comparing the catalytic efficiencies of enzyme homologues from animals having different body temperatures (but all living at approximately 1 atm pressure), we observed that the most efficient catalysts were found in low-body-temperature species like an Antarctic fish (Table 12-1). The least efficient homologues of an enzyme were those of warm-bodied animals like birds and mammals. However, whereas catalytic efficiency varies inversely with body temperature, enzyme thermal stability was found to be directly proportional to body temperature. We interpreted these relationships in terms of a "compromise" between rate of function, on the one hand, and structural stability, on the other hand. Enzymes functioning at high temperatures were proposed to benefit from enhanced thermal stability, even if this structural advantage occurs at the cost of reduced rate of function (the latter effect is, of course, largely offset by the thermal acceleration of reaction rates in high-body-temperature species). In the deep sea, where hydrostatic pressure may act as a perturbant of enzyme structure, selection again may lead to "compromises" between enzyme stability and enzyme catalytic efficiency. Pressure-resistant structure may be crucial for the establishment of the pressure-insensitive K_m values of NADH, for example. Thus, to ensure correct binding of cofactor and substrate, the enzyme may "pay a price" in catalytic efficiency.

Recent studies of C_4-LDH, the isozyme present in mammalian testes, suggest that an intimate relationship between catalytic efficiency and the conformational changes linked to cofactor binding may exist (Musick and Rossmann, 1979). This relationship may provide at least a partial explanation for the kinetic characteristics of the M_4-LDHs of deep-sea fishes. Musick and Rossman paid particular attention to the "loop" region of C_4-LDH, a segment of the enzyme that closes over the cofactor to form the final binding pocket of the enzyme. Ordinarily, in M_4-LDHs, the loop region does not fold over the rest of the active site until cofactor binds to the enzyme. However, in X-ray studies of the C_4-LDH isozyme, Musick and Rossmann found that the loop region existed in a closed (C) conformation even in the absence of co-factor and substrate. In other words, the collapsed loop configuration appears to be especially thermodynamically stable in the C_4 isozyme. A possible consequence of this is a reduced turnover rate. Since binding of NADH occurs only to the open (O) conformation of the enzyme loop region, the proportion of C to O forms of the enzyme will establish the fraction of C_4-LDH molecules competent to initiate the catalytic cycle. In an efficient M_4-LDH system, the O:C ratio may be extremely high, whereas the C_4 system will have a lower ratio and, thereby, lower rate of function for a given number of LDH molecules. Extending these considerations to pressure-adapted LDHs, we can envision a situation in which enhanced stabilization of the closed loop configuration is needed to maintain the integrity of the M_4-LDH-NADH complex in the face of high pressure. To ensure that elevated pressures do not disrupt the enzyme-cofactor complex, an inherently stronger interaction between the loop region and the remainder of the LDH molecule may be established in high-pressure-adapted M_4-LDHs. Consequently, the ratio of C:O configurations of the loop may be similar to that found for the C_4 isozyme. The similar catalytic efficiencies of the mammalian C_4 isozyme and the M_4-LDHs of deep-sea fishes are consistent with this model. An experiment of obvious interest would be to examine the pressure-sensitivity of the C_4 isozyme; is this isozyme "deep-sea-like" in its binding properties?

To fill out the cost-benefit-ratio balance sheet associated with the evolution of pressure-resistant enzymes like the M_4-LDHs of deep-sea fishes, it is appropriate to consider how the M_4-LDHs of both shallow- and deep-living species would compare in performance under *in vivo* conditions of substrate and cofactor concentration and deep-sea pressures. Although the M_4-LDHs of shallow-living fishes have a clear advantage at 1 atm and under conditions of saturating substrate concentrations, they will not maintain this rate advantage under the *in situ* conditions found in the muscle of a deep-sea fish. Intracellular pyruvate concentrations are below saturating, so the pressure perturbation of pyruvate K_m will be inhibitory for the shallow species' enzymes. More significantly, the large increases in the K_m of NADH will lead to non-saturation of the cofactor binding site. NADH binding is the obligatory first step in the LDH reaction, so this K_m perturbation will have a marked inhibiting effect on pyruvate reduction. Taking all of these factors into account, it is likely that the pressure-resistant LDHs of deep-sea species will be capable of reducing pyruvate at least as well as the LDH homologues of shallow-living species under conditions of high pressure and in the presence of physiological levels of substrates and cofactors. Equally important, the LDHs of the deep-sea species will retain K_m values optimal for regulation. An important lesson to be learned from comparing the M_4-LDH homologues of shallow and deep-living fishes is that adaptations to pressure, like adaptations to temperature, involve compromises in which the interplay between functional and structural traits necessitates that conservation of highly critical enzymic features, e.g., ligand binding ability, may entail some sacrifice in other properties, e.g., catalytic efficiency.

Pressure Effects on Protein Structure

At present we know very little about the classes of structural adaptations that facilitate protein function in deep-sea organisms. Whereas pressure-resistant structures seem necessary, as hypothesized above for the loop region of LDHs of deep-sea fishes, the nature of the amino acid substitutions underlying these adaptations are only now being studied in detail. What is clear about structural adaptations is that they have one common goal: the reduction of positive volume changes associated with the alterations in protein structure that accompany ligand binding, catalysis, and subunit assembly.

Figure 12-2 illustrates in a highly diagrammatic fashion the different sources of volume changes in protein-based systems, including both enzymes and contractile proteins such as actin. Volume changes are seen to arise from changes in protein volume (due to alterations in packing efficiency) and changes in water organization (density) around the protein. Available evidence indicates that the latter source of volume change is quantitatively the more important, and that of the various types of structural transformations undergone by proteins, subunit assembly processes are most apt to occur with large changes in water organization and, therefore, be particularly pressure-sensitive. Let us, then, examine subunit assembly processes to understand better how the proteins of deep-sea animals may need to be modified to permit, for example, multimeric enzyme structures and polymerized (filamentous [F]) forms of actin to exist under high pressure.

At the outset of this discussion it is appropriate to remind the reader that the pressure-sensitivity of assembly processes has long been recognized as a potentially critical problem for deep-sea organisms. For instance, Penniston (1971) found that multimeric proteins from a number of 1-atm-adapted organisms were inhibited by pressure, an effect he attributed to subunit dissociation leading to loss of catalytic activity (most multimeric enzymes become inactive when dissociated into their constituent subunits). Penniston thus proposed that deep-sea organisms either must rely on monomeric proteins or must stabilize the subunit interactions of multimeric proteins with unusually strong noncovalent bonds compared to 1-atm-adapted species. Has Penniston's hypothesis stood the test of time?

For the first of his two alternative hypotheses (monomeric enzymes replace multimeric enzymes in deep-sea species), enough information is now available to indicate that it is not correct. Deep-sea animals possess multimeric proteins at the same metabolic loci where these proteins are found in 1-atm-adapted organisms. Penniston's second hypothesis may have merit, however. A number of studies of hydrostatic pressure effects on multimeric enzymes, including mammalian M_4-LDHs, have presented evidence that maintaining the proper subunit aggregation state for enzyme function is not likely to be possible under deep-sea pressures. For example, Schade et al. (1980) found that pig M_4-LDH disassembled into monomers as pressure was increased above approximately 400 atm, with complete disassembly being achieved by 1,000–1,500 atms. Upon release of pressure, native (fully functional) tetramers reassembled, indicating that high pressure did not cause the tertiary structures of the subunits to unfold in an irreversible manner. Although the *in vitro* conditions used in this study (incubation at 20°C at nonphysiological pH) require one to refrain from hasty conclusions about the biological relevance of this finding, these results are at least suggestive of an inability of 1 atm-adapted multimeric proteins to maintain their aggregated form at typical deep-sea pressures. Further comparative studies are obviously needed.

If multimeric proteins of deep-sea organisms do have increased abilities to remain assembled at high pressures, what types of "stronger bonds" are apt to be involved in these adaptations? One possible avenue to follow in achieving pressure-resistant subunit assemblies is to alter the type of energy change used to drive the polymerization step. In terrestrial organisms, subunit assembly is often entropy-driven (see Fig. 12-2). In these entropy-driven processes, organized water at the subunit contact surfaces is shed during the assembly step, and the expansion of this water into the bulk solution produces an increase in system entropy that "pays for" the assembly reaction. For deep-sea species, then, we can ask if assembly processes are less entropy-driven, and, if they are, how might reductions in the entropic contribution to assembly energetics be achieved. The only assembly process which has been studied using homologous proteins from shallow- and deep-living species is the formation of filamentous (F) actin from globular (G) actin monomers. The results given in Figure 12-5 show that the enthalpy

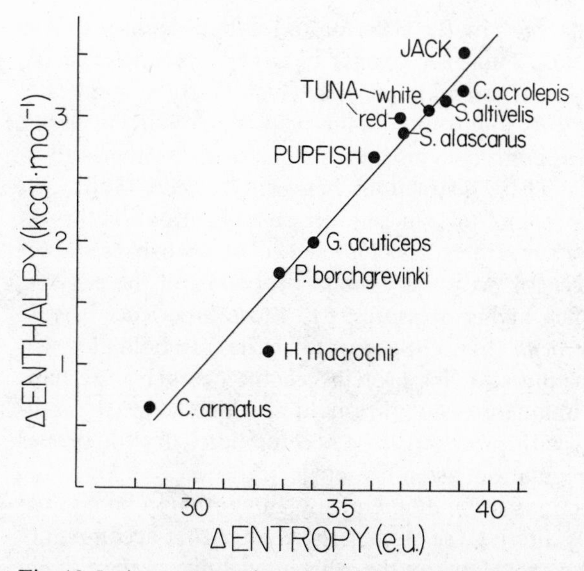

Fig. 12-5. A compensation plot of the enthalpy and entropy changes accompanying the polymerization of muscle actin in several species of teleost fishes. Note that the deepest-occurring species (*Halosauropsis macrochir* and *Coryphaenoides armatus*) have the lowest entropy and enthalpy changes. See Figure 11-19 for data on other species. Figure modified after Swezey and Somero (1982a)

and entropy changes accompanying the addition of a G-actin monomer to a growing F-actin filament are variable among species from different temperature and pressure regimens. As discussed in Chapter 11, the enthalpy and entropy changes accompanying the G to F transformation are larger in warm-adapted than in cold-adapted species. In addition, the deepest-living fishes studied, *Coryphaenoides armatus* and *Halosaropsis macrochir*, have even lower ΔH and ΔS values than those of shallower-living fishes which encounter similar habitat temperatures (Fig. 12-5). This finding suggests that selection does favor reductions in entropy changes and, concomitantly, in water expansion, in the protein assembly processes of deep-living fishes.

Notice that there is a distinct difference between the actin systems examined and the LDH homologues discussed earlier in terms of the grouping of "pressure-adapted" and "non-pressure-adapted" species. For the LDHs, all fishes having depths of occurrence greater than approximately 500–1,000 m were found to possess enzymes with reduced pressure-sensitivity. For muscle actin, however, many of the species showing pressure-adapted characteristics in their LDHs do not appear "pressure-adapted" by the criterion of reduced ΔS of the G to F actin transformation. Thus, a congener of *Coryphaenoides armatus*, *C. acrolepis*, has an actin assembly process which thermodynamically is similar to the process found in cold-adapted, shallow-living fishes. The two *Sebastolobus* congeners also are similar to each other and to shallow-living fishes. These distinctions between the LDH and actin systems reveal that proteins differ in their pressure perturbation thresholds. For LDHs, pressures in excess of a few tens of atmospheres seem high enough to select for pressure-resistant properties of the enzyme, while for actins, much higher pressures, e.g., those experienced by *C. armatus* and *H. macrochir* (Fig. 12-4) seem to be the threshold at which pressure-adaptive changes in this protein become necessary. Perhaps each protein has a unique pressure threshold at which adaptive modifications for coping with pressure stress become necessary for proper function (kinetics, regulation, assembly, etc.).

How might reductions in binding entropy and enthalpy be effected without significantly altering the free energy change that accompanies the polymerization event? How can the inherent stability of the F-actin polymer be maintained while reductions in particular types of bonding energy are achieved? One potential mechanism involves what can be thought of as energy and volume "titrations" (Somero and Low, 1977). To counterbalance a volume increase resulting from water expansion, for example, a water-constricting group could be transferred from the interior of the protein, e.g., the G-actin monomer, simultaneously with the occurrence of the volume-increasing event. In the G to F transformation of actin, therefore, as the subunits join together and water is forced away from the subunit contact sites, an accompanying change in subunit conformation could bring into contact with the solution a charged or polar group, which would have the effect of decreasing the system volume (see Fig. 12-2). This charged (or polar) group exposure would be exergonic due to the favorable enthalpy change that occurs during the group's hydration (the entropy change would be unfavorable, i.e., negative). It is important to realize that the coupled group transfers involved in these type of "titrations" need not occur at the subunit contact sites themselves. Rather, energy changes occurring simultaneously with the polymerization event can be due to structural changes at any locus on the protein molecule.

In conclusion, the subunit interactions present in high-pressure-adapted organisms may not be "stronger" in the sense that the total stabilization free energy involved in maintaining the aggregated state is increased. Rather, the type of energy (enthalpy versus entropy) change contributing to stabilization of the polymer may differ qualitatively from one environment to another, as shown by the data in Figure 12-5. Conditions of low temperature and high pressure may favor a reduced dependence on entropy-driven reactions, while reliance on favorable (negative) enthalpy changes may be increased.

A final point involving protein structural adaptations concerns the structural stability of globular proteins in deep-sea animals. We previously have considered a possible means for enhancing NADH binding in deep-sea species' dehydrogenases via stabilization of the loop region of the LDH molecule. In the case of skeletal muscle actins, there is evidence that the tertiary structure of G-actin is relatively rigid in the case of the two deepest-living fishes studied, *C. armatus* and *H. macrochir*. By the criterion of heat stability (refer to Fig. 11-11), the G-actins of these two deep-living fishes are as stable in structure as the G-actin of the thermophilic desert reptile, *Dipsosaurus dorsalis*, which has core temperatures of up to 47°C. Thus, elevated hydrostatic pressure, like high body temperatures, may favor selection for rigid protein structures.

Adaptations in Lipid-Based Systems

Lipid-based components of organisms, notably membrane systems, display responses to temperature and pressure which in certain ways are highly similar to the responses shown by protein systems. Like enzymic proteins in particular, lipid-based structures must persist in a type of "semistable" state to effect their biological roles. And, as has been shown to be true for proteins, this fine balance between stability and flexibility of structure is easily perturbed by changes in the physical environment. As discussed in Chapter 11, the lipid-based components of cells appear to function optimally when in a "liquid-crystalline" physical state. A reduction in temperature may lead to a disruption of the function of lipid-based systems, e.g., membrane transport proteins, due to a) a liquid crystalline-to-solid phase transition, or b) local "freezing-out" of clusters of membrane phospholipids.

Pressure effects on lipid-based systems, while less studied than temperature effects, seem of sufficient magnitude to create potential problems for deep-living organisms. The fact that increases in hydrostatic pressure favor solidification of lipids suggests that the combination of high pressure and low temperature faced by typical deep-sea organisms could create especially severe problems for cellular components dependent on a relatively fluid lipid microenvironment. The tendency of high pressure to solidify lipid systems, i.e., to raise the melting temperature of lipids, is easily understood on the basis of volume relationships.

Thus, the tighter packing of fatty acid chains in the "solid" state relative to the "fluid" state renders hydrostatic pressure a potent "freezing agent" of lipid systems.

The interacting effects of pressure and temperature on a membrane-localized, phospholipid-dependent enzyme, Na-K-ATPase (from pig kidney), are illustrated in Figure 12-6. In the upper panel of this figure the influences of pressure on the reaction velocity are illustrated. Above a certain pressure the slope of the Arrhenius plot changes, an effect reminiscent of temperature perturbations of membrane-localized enzymes. At pressures above the "break" pressure (P_b), the activation enthalpy of the reaction is increased, an effect which is probably due to the "solid" nature of the phospholipids in the enzyme's microenvironment. These "solid" lipids may have to be "melted" to allow the enzyme protein to undergo the required changes in conformation or intramembrane orientation associated with ion transport function. The P_b value of the system is highly dependent on temperature (lower panel of Fig. 12-6), showing the strong synergism between reduced temperature and elevated pressure on lipid-based systems. Note that for this mammalian enzyme system, the lipids would be in a "solid" form at deep-sea temperatures and pressures, for P_b reaches a value of 1 atm at a temperature of approximately 18°C. If we assume that many membrane-localized enzyme systems of deep-sea species must be surrounded by a "fluid" lipid microenvironment to achieve proper function, then adjustments in the fluidity of lipid components seem necessary. Theoretical P_b versus temperature curves for deep-living organisms are shown in the lower panel of Figure 12-6; the reality of these curves remains to be established, of course. Important evidence along these lines could come from studies of membrane phospholipid fatty acid composition of deep-sea organisms. Are the phospholipids of deep-living creatures built from fatty acids having unusually high double-bond contents, a trend noted in cold-adapted species?

Although currently there is little information on the lipid compositions of functional membranes of deep-living organisms (see Patton, 1975), behavioral and physiological studies of pressure effects on whole organisms have provided strong evidence for pressure adaptation at the membrane level. For example, Brauer et al. (1980a,b) found that shallow- and deep-living amphipods of Lake Baikal (1,400 m deep) differed significantly in behavioral and sodium-exchange-system responses to elevated pressure. The deeper-living amphipods displayed clear pressure adaptations, and these seem very likely to be membrane-based. Electrophysiological studies, using preparations from non-pressure-adapted animals, have typically shown strong pressure disruption

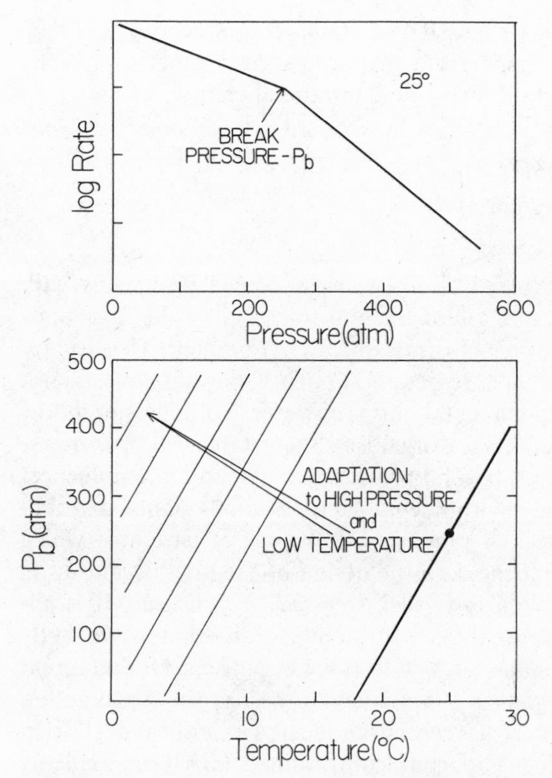

Fig. 12-6. The effects of measurement pressure and temperature on the activity of the Na-K-ATPase of pig kidney. Upper panel: influence of pressure on reaction velocity at 25°C. Note the "break" in the slope of the rate-versus-pressure function near 250 atms. This "break pressure" (P_b) is thought to be due to a change in the structure of the lipid microenvironment surrounding the Na-K-ATPase protein. Phospholipid "solidification," which is enhanced by increases in pressure and decreases in temperature, leads to a higher activation energy for the reaction and a larger activation volume as well. Lower panel: the relationship between P_b and measurement temperature. The heavy line with the P_b point from the upper frame (25°C P_b) represents the relationship between P_b and temperature found for the Na-K-ATPase of pig kidney. The parallel series of lines rising in the left half of the figure illustrates the effects which pressure (and temperature) adaptations could have on the P_b-versus-temperature relationship. For example, adaptation to deep-sea conditions of low temperature (near 2°C) and high pressure would be facilitated by a leftwards and upwards shifts of the P_b-versus-temperature plots, such that membrane phospholipids would remain in a "liquid-crystalline" state even under deep-sea conditions. The Na-K-ATPase data for pig kidney are from de Smedt et al. (1979)

of membrane function (reviewed by Wann and Macdonald, 1980). Thus, as in the case of temperature adaptation, the cell membrane may be an extremely important locus of adaptational change.

Metabolic Rates in the Deep Sea: The Interplay of Physical and Biological Factors

The adaptations discussed above were concerned principally with giving deep-sea organisms a fundamental tolerance of the high pressures and low temperatures of the deep-sea environment. These adaptations make metabolism per se possible at depth; they do not establish the actual rates of deep-sea metabolism, however. To understand how the metabolic rates of deep-sea organisms are established, it is necessary to broaden our focus beyond pressure and temperature influences and consider additional environmental factors—food quality, distribution of food, darkness, and presence or absence of currents—which have major shaping influences on the design of deep-sea organisms.

1. *Rates of metabolism in typical deep-sea environments.* It is appropriate to begin this analysis with a review of what is currently known about the metabolic rates of deep-sea organisms. During recent years a number of elegant *in situ* studies of the oxygen consumption rates of deep-sea animals have been conducted (Smith and Hessler, 1974; Smith, 1978). Other important contributions have been made by workers who have successfully recovered deep-sea animals and measured their respiratory rates in the laboratory (Childress, 1975, 1977; Quetin and Childress, 1976; Torres et al., 1979). From all of these studies a general conclusion has been established that deep-living animals, both fishes and invertebrates, have markedly lower metabolic rates than shallow-living forms. This relationship between minimal depth of occurrence* and oxygen consumption rate is illustrated for

* Minimal depth of occurrence is generally used as an index of "deepness" of a species' occurrence in the water column. As discussed by Childress and Nygaard (1973), this criterion is preferable to either average depth or maximal depth for the following reason. Many midwater species migrate vertically in the water column, and during the time spent in shallow depths these species have access to a calorically richer and more abundant diet than that experienced by nonmigrators. Since the physiological and biochemical characteristics of marine species are so strongly influenced by food quantity, food distribution, and needs for strong locomotory capabilities, as discussed in this section of the chapter, it would be misleading to group together all migrating and nonmigrating animals which happen to have common average depths of distribution at least during certain times of the day.

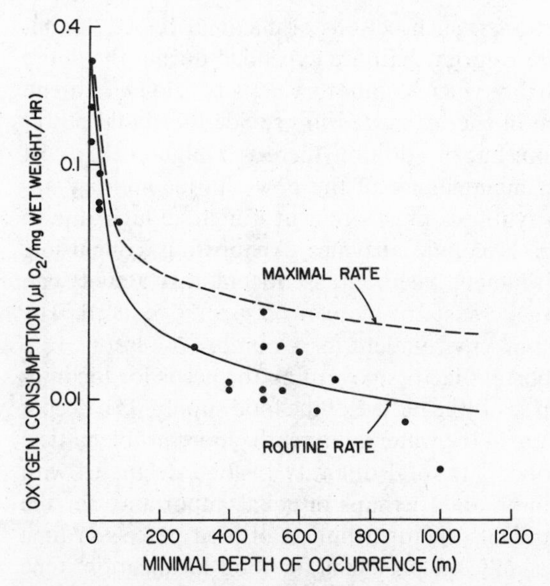

Fig. 12-7. The relationship between routine respiration (solid line), maximal respiration (dashed line), and minimal depth of occurrence for a number of species of midwater teleost fishes. Data of Torres et al. (1979); figure from Somero (1982)

marine fishes in Figure 12-7. It is apparent that deep-living fishes have vastly lower rates of oxygen consumption than shallow-living species. The finding that these low oxygen consumption rates are exhibited both at *in situ* pressures and at 1 atm pressure (see Torres et al., 1979) shows that pressure inhibition of metabolism is not the cause of this low level of metabolic flux at depth. What, then, are the key factors favoring selection for a metabolic apparatus that facilitates such a low rate of energy metabolism?

2. *Factors selecting for low metabolic rates in the deep sea.* To appreciate why deep-sea animals, especially pelagic species, have such low metabolic rates relative to shallow-living forms, it is useful to inquire first, "What factors contribute to the need for *high* levels of energy metabolism?" For those readers who have absorbed the lessons from the sections of this volume devoted to exercise metabolism and muscle biochemistry, one important answer to this question should be obvious. Animals that must engage in frequent vigorous bouts of locomotion live an energetically costly life. To be more quantitative, transitions from a basal metabolic level to peak burst swimming can entail an approximately hundredfold rise in energy expenditure (Brett

and Groves, 1979). Just a few such periods of maximal velocity swimming can burn up more calories than are expended during the entire remaining period of the day, when locomotory activity is low or absent. Not only is a great deal of energy needed to provide the actual power for vigorous locomotion, but in addition there is a high energy cost for the day-in, day-out maintenance of the powerful locomotory apparatus. Thus, protein synthesis must occur at a high enough rate to maintain high titres of glycolytic enzymes. An obvious conclusion, then, is that in an environment where intense locomotory activity can be minimized the energetic costs of life can be greatly reduced. The typical deep sea is such an environment for a number of reasons.

One of the most important factors governing the needs for locomotory activity in the deep sea is the nature of the food supply. The deeper down an organism occurs in the water column, the more problematical is the acquisition of food. The total quantity of food decreases with depth in the water column and, perhaps of equal importance for the design of deep-living animals, the distribution of food parcels in time and space establishes significant problems for food localization. One can ask whether it makes much sense for deep-living pelagic fishes or invertebrates to expend a great deal of energy searching, in the dark, for randomly occurring food parcels settling from the euphotic zone. Whereas certain deep-sea fishes do appear to swim actively in search of food, albeit their locomotory activity is merely a shadow of the swimming performance of fishes like tunas, many deep-sea animals display a "float-and-wait" feeding strategy, thereby reducing the energy costs of locomotion to extremely low levels. The float-and-wait feeders are typically flaccid in texture and have high water contents and reduced skeletal strength. These compositional changes may have a twofold significance for these organisms. First, the feeding strategy they employ may allow a reduction in organic elements and, thereby, a further decrease in the costs of maintenance metabolism. In addition, reduced skeletal elements and high water content facilitate neutral buoyancy, so that the amount of muscular activity required for maintaining position in the water column is lessened. Buoyancy adaptations appear very closely linked to many of the compositional changes that can be rationalized as energy-saving devices and, in fact, energy savings from reduced costs for maintaining neutral buoyancy are a major "spinoff" from the adaptations that appear to be designed primarily to reduce the locomotory apparatus of the organism (see Marshall, 1979, for detailed discussion of the characteristics of deep-sea animals).

Factors in addition to the quality, quantity, and distribution of food also influence the locomotory requirements of deep-sea animals. The absence of light (except for bioluminescence) in the deep sea is a two-edged sword in the case of locomotory performance requirements. Whereas darkness makes it difficult to locate most food items, detection by a visual predator also becomes less likely, thereby reducing needs for bursts of escape swimming. Thus, selection for robust swimming abilities may be absent for two related reasons associated with the darkness of the deep-sea habitat. The lack of strong currents in many deep-sea regions also may play a role in reducing the needs for powerful swimming abilities in deep-living animals. In summary, then, the low metabolic rates of deep-sea animals appear not to be a reflection of any retarding effects of elevated pressure or low temperature (recall the high degree of cold adaptation in metabolic rates seen in polar, shallow-water fishes). Instead, a varied suite of factors in the deep-sea environment selects against a vigorous, fast-moving life-style characterized by high locomotory energy expenditures. It is pertinent now to consider how deep-sea animals effect reduced metabolic rates and whether these metabolic reductions are centered entirely in locomotory organs as the above analysis would seem to suggest, or whether they are characteristic of all tissues of deep-sea animals.

3. *Mechanisms used to achieve low metabolic rates in the deep sea.* The suggested reductions in locomotory energy expenditures by deep-sea animals seem a necessary consequence of the trend displayed in Figure 12-8. The glycolytic enzymes responsible for providing the energy for muscle contraction are grossly reduced in white muscle of most deep-sea fishes. For LDH, the enzyme which is the best indicator enzyme of a muscle's capacity for powering burst swimming, activities in some deep-sea fishes are three orders of magnitude lower than those found in the most powerful swimmers, warm-bodied tunas. PK activity also decreases with increasing minimal depth of occurrence, albeit the decrease is not quite as marked as for LDH.

A noteworthy exception to the downward trends in LDH and PK activities with increasing depth of occurrence is found for an as yet unidentified zoarcid fish found at the 21°N hydrothermal vent site. This fish, which has been found only at the vents, at depths of approximately 2,600 m, has much higher LDH and PK activities than any other deep-living fishes (Table 12-2; note the points corresponding to 2,600 m in Fig. 12-8). As discussed later in this chapter, the presence of a high biomass density (=abundant food) at the vents allows high metabolic rates and, in the case of motile species, may favor high locomotory capacities.

In view of the fact that LDH and PK are glycolytic enzymes associated with anaerobic glycolysis during periods of burst swimming, it

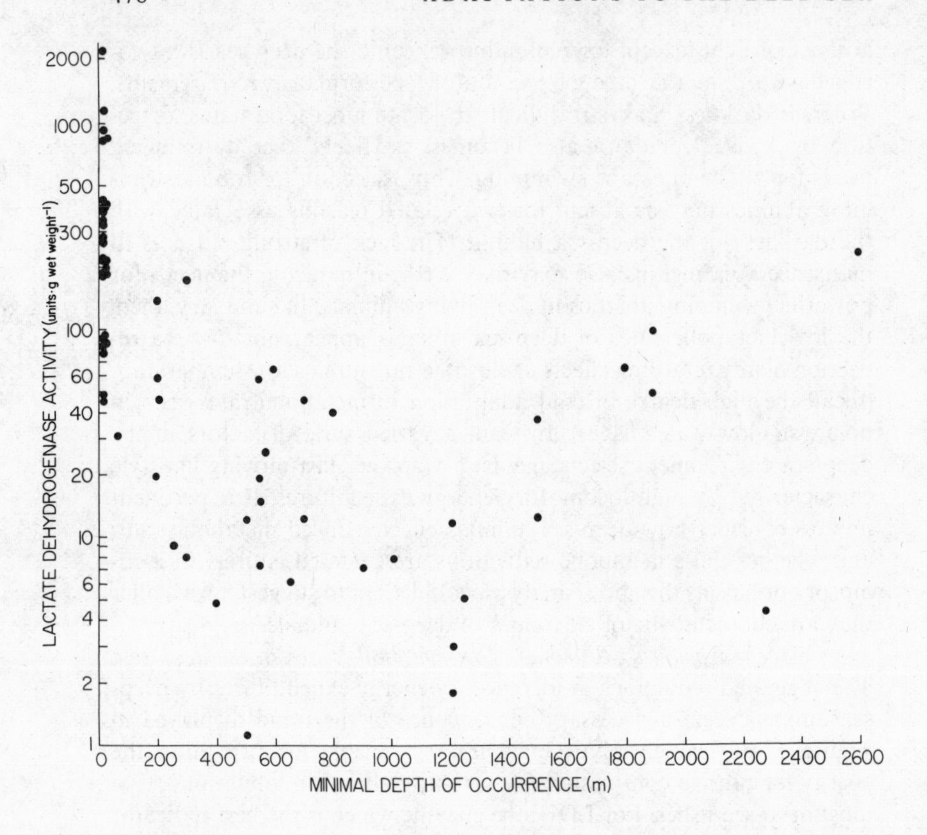

is appropriate to ask whether these enzymes are a proper index for gauging the overall aerobic metabolic rates of fishes. These enzymes can, in fact, serve as such an index, as shown by Figure 12-9. When oxygen consumption rates and LDH and MDH activities of white muscle are obtained for the same species of fishes and plotted as shown in this figure, a strong correlation is noted between enzymic activity and respiration rate. The reason why an anaerobically poised enzyme like LDH can provide an index of aerobic metabolism is that any lactate generated in anaerobic metabolism will lead to a requirement for oxygen consumption at some time and place in the fish (unless the lactate is excreted, which is highly unlikely). Thus, lactate produced during bursts of swimming is typically reconverted to glycogen or re-oxidized to pyruvate, which can then be catabolized via the citric acid cycle. When one views LDH function in a realistic spatial and temporal framework, then, it is not surprising that this enzyme can serve as an indicator of overall aerobic metabolism. It is worth pointing out that

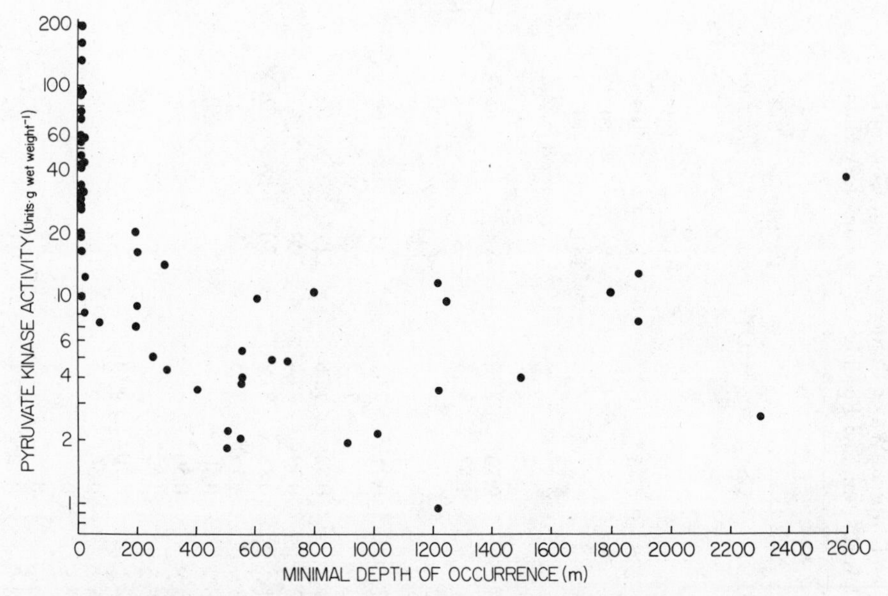

Fig. 12-8. (above and on facing page). The relationships between minimal depth of occurrence and lactate dehydrogenase-LDH (left panel) and pyruvate kinase-PK (right panel) activities of fish white skeletal muscle. Units of activity are in μmoles of substrate converted to product per minute at 10°C per gram wet weight of skeletal muscle. Figure modified after Sullivan and Somero (1980)

the relationship between LDH activity in muscle and whole organism oxygen consumption rate is technically useful for study of deep-sea fishes that cannot be recovered alive (or in good enough condition) for respiration studies. Since LDH activity is highly stable, even during prolonged storage in frozen muscle, assay of muscle LDH activities can provide a useful, indirect method for estimating the oxygen consumption rates of organisms that cannot be used in respirometers.

While deep-sea animals have much lower rates of metabolism than shallow-living species, the data of Figure 12-8 and Table 12-2 reveal considerable heterogeneity in enzymic activity and, by implication, in metabolic rate among deep-sea fishes. These differences appear to be explainable largely in terms of feeding strategy and locomotory habit (Sullivan and Somero, 1980; Siebenaller *et al.*, 1982), so these data lend further support to the hypothesis that locomotory costs are a strong shaping influence on the design of deep-sea species. Note in particular the relatively high—by deep-sea fish standards—enzymic activities for the two rattails, *Coryphaenoides acrolepis* and *C. armatus*. These two species are pelagic forms that cruise through the water column in

Table 12-2. Enzymic Activities of White Skeletal Muscle and Brain and Muscle Water Content, Protein Concentration, and Buffering Capacity of Marine Fishes Having Different Depths of Distribution and Feeding and Locomotory Habits

Muscle	Enzymic activity[a]						
	LDH	PK	MDH	CS	% H_2O[b]	Protein content[c]	Buffering capacity[d]
Group I: Warm-bodied fishes							
Thunnus alalunga	2,100					218	107
Auxis thazard	1,186	192		5.45			109
Thunnus albacares	997					194	105
Euthynnus lineatus	788					184	102
$\bar{X}$	1,268						
Group II: Shallow, ectothermic fishes							
Medialuna californiensis	981	125	23	0.70	75.6	198	63
Engraulis mordax	540	60	61	1.52	80.8		71
Salmo gairdneri	575						60
Sphyraena ensis	512						63
Phanerodon furcatus	414	90	77	0.71	77.3	251	66
Atherinops affinis	412	107	20		77.2	222	
Paralabrax nebulifer	397	71	21	0.52	75.6	211	62
Paralabrax clathratus	389	75	23	0.79	77.4	200	62
Chromis puntipennis	388	92	51	0.85	77.8	230	62
Rhacochilus toxotes	351	42	63	1.15	78.5	211	
Gillichthys mirabilis	321	28	26	0.90			60
Paralabrax maculatofaciatus	287	41		0.50			
Genyonemus lineatus	267	88		0.67			
Caulolatilus princeps	209	32	32	0.50	80.8	213	59
Squalus acanthias	182	58	18	0.56	69.5	166	
Sebastes mystinus	116	72		0.74			
Scorpaena guttata	77	15		0.25			52
$\bar{X}$	378	60	38	0.74	77.1	211	62

Group III: Deep-living fishes							
Histiobranchus bathybius	156	19	7	0.35		103	42
Coryphaenoides acrolepis	154	10	9	0.31	79.8	103	42
Coryphaenoides armatus	153	21	16	0.97	80.3	151	
Anoplopoma fimbria	107	16	16	0.48	71.5	112	
Paraliparis rosaceus	66	10		0.10	93.9		
Antimora rostrata	36	10	7	0.37	82.4	102	44
Coryphaenoides carapinus	15	9	7	1.10	83.8	147	
Halosauropsis macrochir	12	4	3	0.41	83.4	124	48
Nezumia bairdii	9	5	9	0.82	80.5	165	46
Coryphaenoides leptolepis	5	4	4	0.40	82.5	139	46
$\bar{X}$	71	11	9	0.53	82.0	130	45
Anova: II vs III (P)	<.001	<.001	<.001	ns	ns	<.001	<.001
Group IV: Hydrothermal vent fish							
21°N fish (unidentified zoarcid)	216	36	15		76.0		
Brain							
Medialuna californiensis	35	21		2.88			
Paralabrax nebulifer	35	22		1.75			
Paralabrax clathratus	36	20		1.87			
Chromis puntipennis	32	19		2.65			
Genyonemus lineatus	18	14		1.60			
Caulolatilus princeps	31	19		1.88			
Scorpaena guttata	31	17		1.40			
Coryphaenoides acrolepis	36	12		1.46			
Anoplopoma fimbria	40	16		1.65			
$\bar{X}$	33	18		1.90			

[a] Enzymic activities are expressed as international units per gram wet weight of tissue at 10°C. Data are from Sullivan and Somero (1980), Castellini and Somero (1981), and Somero (unpublished)

[b] Water content is expressed as percentage of tissue wet weight. Data from Sullivan and Somero (1980)

[c] Protein concentration is expressed as mg protein per gram wet weight of tissue. Data from Sullivan and Somero (1980) and K. L. Dickson (unpublished)

[d] Buffering capacity is expressed as μmoles of base needed to titrate the pH of a 1-gram sample of muscle by 1 pH unit, between pH values of approximately 6 and 7. Data from Castellini and Somero (1981)

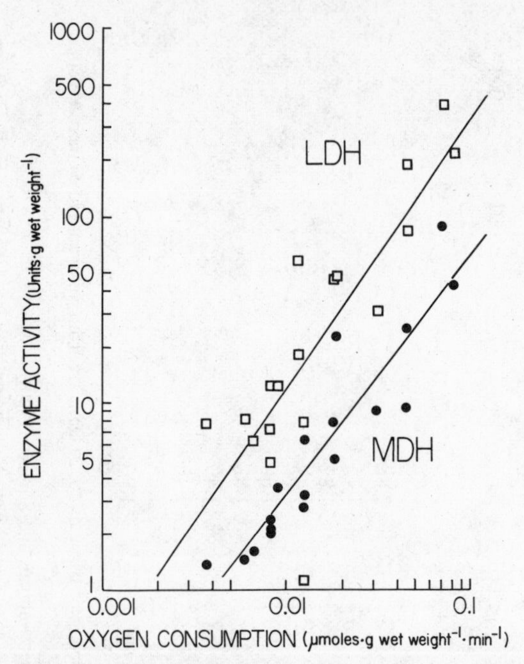

Fig. 12-9. The relationship between routine oxygen consumption rates of midwater fishes and their skeletal muscle lactate dehydrogenase (LDH) and malate dehydrogenase (MDH) activities. LDH: (□); MDH: (●). Figure modified after Childress and Somero (1979)

search of food, and undergo substantial vertical movements through the water column. Unlike more sedentary species, e.g., those deep-sea forms with LDH activities an order of magnitude less than the activities found in *C. acrolepis* and *C. armatus* muscle, these actively cruising rattails may expend considerable energy in their searches for food. The fact that these two rattails have similar enzymic activities in spite of having different depths of occurrence (Fig. 12-4) indicates that depth per se is not a dominant factor in setting enzymic activity levels among deep-sea fishes as a single group. From the data assembled in Table 12-2 and Figure 12-7, we conclude that among all deep-sea fishes studied, where species having minimal depths of occurrence of approximately 200 m or greater are considered as deep-living forms, there is as much variation in metabolic rate as there is among shallow-living fishes, even though the absolute metabolic rates are vastly different for these two broad classes of fishes. The variation in metabolic rate among members of each group is likely to be due principally to differences in feeding strategy and locomotory capacity, much as the large between-group differences are proposed to derive largely from these same factors.

The differences found between the deep- and shallow-living groups of fishes can be discerned in fine-scale comparisons of congeners living at different depths. The muscle enzymic activities of *Sebastolobus altivelis* are only approximately half those found for its shallow-living congener, *Sebastolobus alascanus* (Table 12-3). Comparisons of congeneric species from different depths offers a specially powerful means for discerning depth-related trends. In the case of the two *Sebastolobus* species, two different mechanisms underlie the observed differences in muscle enzymic activity (Siebenaller and Somero, 1982). For LDH, differences in turnover number are adequate to account for the approximately twofold difference in activity per gram of muscle. For PK, the turnover numbers are the same in both species, so the muscle activity differences appear to be due to different enzyme concentrations. The basis for the differences in MDH, CPK, and CS activities is not known. The LDH and PK findings indicate that two distinct regulatory mechanisms (catalytic efficiency and enzyme concentration) are involved in establishing an approximately twofold difference in activities between species, and in conserving a consistent ratio of activities among different enzymes within the muscle of a single species. Thus, the reduced metabolic rates of deep-living forms reflect both the "qualitative" and "quantitative" strategies of biochemical adaptation discussed throughout this volume.

Correlated with the large differences in muscle enzymic activity are smaller, yet significant differences in other muscle properties (Table 12-2). Protein content is lower in deep-sea fishes and water content is higher. Note that protein content is lower by only about 30–40 percent, whereas LDH and PK activities may be up to a thousandfold lower. This indicates that reduction in enzyme content is a specific adaptation and is not merely a reflection of a decrease in all types of muscle proteins. Contractile and structural proteins may decrease in concentration with increasing depth of occurrence much less than enzymes of energy metabolism. In fact, no depth-related changes in skeletal muscle actin content occur (Swezey and Somero, 1928b). Accompanying the decrease in potential for anaerobic glycolysis is a significant decrease in muscle buffering capacity (Table 12-2). As discussed in the context of exercise biochemistry (Fig. 4-3) the capacity to produce lactate (indicated by LDH activity) is closely correlated with a capacity to buffer the muscle cytosol. Thus, the low buffering capacities of the deep-sea fishes are not surprising. In terms of causal relationships, the decrease in buffering capacity is roughly the same as the

Table 12-3. Comparisons of Muscle and Brain Biochemistries of *Sebastolobus alascanus* and *S. altivelis*

	Enzymic activities[a] (units/g wet wt)					Protein content[b]	Water content[b]	Buffering capacity[c]
	CPK	MDH	LDH	PK	CS	(mg/g)	(% wet wt)	(slykes)
Muscle								
S. alascanus	119	5	58	7	0.35	149	81.2	47.4
S. altivelis	75	3	26	4	0.19	152	81.6	43.0
Brain								
S. alascanus			35	23	1.6			
S. altivelis			31	29	1.6			

SOURCE: Data from Siebenaller and Somero (1982)

[a] CPK = creatine phosphokinase. All activities in muscle are significantly different ($p < 0.001$) between species; brain activities do not differ significantly between species.

[b] Protein and water contents do not differ significantly between species.

[c] Buffering capacity differs significantly ($p < 0.01$) between species.

decrease in muscle protein content, suggesting that protein buffering is reduced in the deep-sea forms.

In considering the metabolic rates of deep-sea animals we have focused almost entirely on locomotory costs and the muscle systems associated with swimming. Even though skeletal muscle typically represents 45–60 percent of the body mass of a fish, our treatment of metabolism in the deep sea cannot be comprehensive unless we also consider the metabolism of other tissues. Table 12-2 lists the enzymic activities of brain tissue from a variety of shallow- and deep-living fishes having widely different life-styles and locomotory capacities. It is apparent from these data that the generalizations framed in terms of muscle metabolism and enzyme chemistry are not applicable for brain. Thus, all fishes studied display essentially the same specific activities of enzymes in brain. This discovery does not, of course, refute the locomotory-energy hypothesis developed in this discussion. Brain represents an extremely minute fraction of total fish mass, and while we are not aware of reliable estimates of the contribution made by a fish's brain to total fish metabolism, we feel it is certain to be very small. Viewing brain composition from the standpoint of energy saving, therefore, it does not appear likely that a significant reduction in a fish's total energy costs could be achieved by reducing brain metabolism. Reducing brain mass for buoyancy energy savings also appears to be an unlikely strategy; brain is a lipid-rich (=light) tissue and its small size in fish relative to muscle and skeletal masses makes it a poor candidate for use in buoyancy adaptations.

In summary, the biochemical systems of deep-living animals differ both qualitatively and quantitatively from those of shallow-living animals. Adaptations in proteins leading to reduced perturbation by hydrostatic pressure facilitate survival in the deep sea. Adjustments in the concentrations of enzymes of energy metabolism in locomotory muscle are instrumental in establishing locomotory and metabolic capacities that are suited for life in a dark, food-poor environment. Pelagic animals from the typical deep-sea (benthic species have received very little physiological and biochemical study, so we can say little about their metabolic properties) thus seem admirably designed for a life in slow motion.

The Hydrothermal Vents: Abundant Life in the Deep Sea

Until very recently, the image of the deep sea and its inhabitants that has been sketched in the previous sections of this chapter was considered typical of all regions of the deep sea. Then, during a 1977 expedition to the seafloor spreading center near the Galapagos Islands, oceanographers witnessed for the first time a community of organisms that appears to disobey virtually all of the major rules of deep-sea biology we have discussed. Clustered in exceedingly high densities around cracks in the seafloor, through which warm and chemically modified waters arose, were populations of many types of invertebrates, including clams, mussels, crustaceans, anemones, and, most strikingly, enormous pogonophoran tube worms (*Riftia pachyptila* Jones; Jones, 1981; Corliss et al., 1979). These mouthless and gutless tube worms reach lengths in excess of one meter and at the Galapagos vent site form a large proportion of the total animal biomass.

The discovery at Galapagos, and subsequently at other sites along the spreading center termed the East Pacific Rise (Spiess et al., 1980), of high animal biomass at depths of 2,500 to 2,600 m, and the finding that a dominant member of the vent community was a large worm lacking a digestive system, raised several exciting questions about the nutritional strategies of the vent organisms. What nutritional sources enable such a dense assemblage of animals to exist at these depths? How can an organism like *R. pachyptila* support its metabolic needs in the absence of a means for ingesting large particulate organic matter? Who feeds whom in the trophic web of the vent community? How do the metabolic rates of the vent animals compare with those of animals from typical deep-sea environments, and with animals from shallow water habitats? While investigations of the organisms of the hydrothermal vent communities are still far from answering these and related questions in any detail, enough has been learned about these unique organisms to give us at least a preliminary model for trophic relationships in the vent community. In addition, what we have learned about the vent organisms, which appear to be exceptions to the rules of deep-sea biology, has further substantiated certain of the key rules discussed earlier in this chapter. This is especially true in the case of rules that relate the metabolic and biochemical characteristics of organisms to the properties of the food supply.

Characteristics of the Vent Waters

To understand the biochemical characteristics of the vent animals, especially those features that involve nutritional relationships, we first must consider the physical and chemical properties of the waters that issue through the seafloor at the deep-sea spreading centers. The vent waters are highly modified seawater that originated as typical deep ocean water (see insets to Fig. 12-10). As the bottom water percolates down through the porous seafloor near the spreading centers it comes

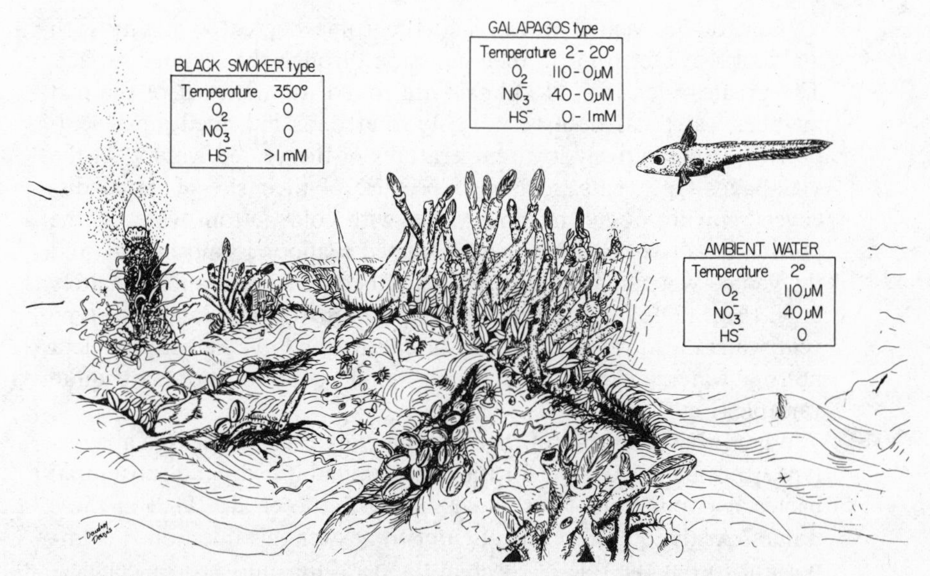

Fig. 12-10. The hydrothermal vent communities. A composite illustration showing major animal species found at the Galapagos and 21°N sites, and the water chemistry of the ambient seawater and the vented waters issuing from the more diffuse "Galapagos-type" vents and the "black smoker" vents. Figure from Somero et al. (1983)

into contact with hot rocks at some distance from the seafloor surface and undergoes extensive chemical modification. These highly modified waters then emerge through the spreading center vents and are mixed with the cold bottom water. The physical and chemical properties of the vent waters are determined partly by the extent to which the vent waters mix with ambient seawater prior to reaching the seafloor surface. The Black Smoker type vent waters (Fig. 12-10), which issue through tall chimneys formed from materials that precipitate from the hot vent waters as these come into contact with the cold bottom water, have temperatures near 350–360°C (Spiess et al., 1980; Edmond et al., 1982). These extremely hot waters represent essentially undiluted or "end member" vent water (Edmond et al., 1982), i.e., the Black Smoker waters have received little, if any, dilution prior to being ejected through the seafloor. The Black Smoker waters are devoid of oxygen and nitrate, but contain hydrogen sulfide (HS^-) at millimolar concentrations. Clearly, these waters seem a hostile environment for life, and at present it is not known if any life forms, e.g., highly thermophilic bacteria, are able to grow in the pure Black Smoker waters.

Most of the vented waters undergo some degree of mixing with ambient seawater prior to their emission through the seafloor surface. The greatest densities of animals are found at vents where the end member water has been very highly diluted, as indicated most obviously by the relatively low temperatures of the vented waters. At the Galapagos-type vents, so named because of their site of initial discovery, mixing of end member water with cold bottom water in the porous basalt rocks of the newly formed seafloor is substantial, and the vented waters have temperatures of between 2 and approximately 20°C (Fig. 12-10). The chemical compositions of the Galapagos-type vent waters also reflect a mixing of end member water with ambient bottom water; oxygen, nitrate, and sulfide concentrations are quite variable (Fig. 12-10).

Which characteristics of the waters arising through the Galapagos-type vents are responsible for the dense animal growth and substantial bacterial populations (Karl et al., 1980; Tuttle et al., 1983) in these waters? At the outset of our attempt to answer this question it is important to put the role of elevated temperatures into proper perspective. Despite the fact that the vent animals and bacteria may experience cell temperatures up to several degrees Celsius higher than those of typical deep-sea organisms, the abundance of life at the vents cannot be attributed to a simple thermal acceleration of metabolic processes. Most of the vent animals have body temperatures within a few degrees of those of typical deep-sea animals, yet there appear to be order of magnitude differences in metabolic flux between animals from typical deep-sea habitats and species from the vents.

Another potential role of high temperature in facilitating the maintenance of high biomass at the vents involves a means for bringing increased supplies of nutrients into the vent regions. As the heated vent waters rise from the site of venting due to their low densities, they are replaced by ambient bottom water. This advective flow of bottom water caused by convection could concentrate organic matter from the surrounding sea floor that has originated in the euphotic zone. In this model, the water circulation pattern created by temperature differences would draw in enough bottom water to enable filter feeders, for example, to obtain much more nutrition than would be possible in deep-sea regions where such current patterns do not exist.

The quantitative importance of this type of nutrient supply to the vent communities is unknown. There is, however, strong evidence suggesting that reduced carbon and nitrogen compounds generated in the euphotic zone play but a minor role in the nutrition of the dominant vent animals and the vent bacteria. Studies of carbon isotopes, which can provide indications of whether organic carbon has been produced photosynthetically or via some other process, have provided evidence that the tube worms and bivalve molluscs of the vents are nourished primarily by reduced carbon of non-photosynthetic origin (Rau, 1981; Williams et al., 1981). Thus, to account for the high biomass at the vents, one must identify a source of energy that is capable of driving a high level of primary production at the vent sites themselves.

Sulfide-based Food Webs at the Hydrothermal Vents

Two key reactants for driving a high rate of primary production have been identified in the Galapagos-type vent waters. As shown in Figure 12-10, these waters contain both the energy-rich molecule, HS^-, and the oxygen needed to burn HS^- to SO_4^{2-} (sulfate). In this oxidative process, which involves a complex series of enzyme catalyzed reactions, the energy released can be trapped in the form of ATP and reducing power, NAD(P)H. These two important energy currency compounds can be spent in driving the reactions of the Calvin-Benson cycle of net CO_2 fixation, and the steps involved in the reduction of nitrate to ammonium ion. Sulfide, then, can be regarded as the major substitute for sunlight in the vent ecosystem. The organisms serving as the primary producers in the vent ecosystem, i.e., the organisms playing the same role as photosynthetic species in sunlight driven ecosystems, are bacteria that possess the enzymes needed to catalyze the oxidation of HS^- and to utilize the energy so released to drive the net fixation of CO_2. These chemolithotrophic sulfide oxidizers are found on the surfaces of vent rocks, on the surfaces of some of the vent animals, e.g., the tubes of *R. pachyptila*, and free-living in the seawater (Karl et al., 1980; Tuttle et al., 1983). These bacteria seem accessible as food sources for motile grazing species, e.g., limpets, and for filter feeders. These primary consumers may, in turn, serve as food for carnivorous species, e.g., the crustaceans and fishes of the community.

While this very simple model of trophic relationships in the vent community is probably accurate as far as it goes, there remains the question of how one of the dominant vent animals, *R. pachyptila*, obtains its nutrition. Lacking a mouth and gut, this animal seems poorly adapted to exploit the reduced carbon and nitrogen compounds synthesized by the sulfide oxidizing chemolithotrophic bacteria. Uptake of dissolved organic matter from the vent waters is one possible way in which *R. pachyptila* could obtain nutrients, yet the unfavorable surface to volume ratio of this large worm (only the tentacle is in direct contact with seawater to serve as a nutrient exchange surface) would seem to make reliance on dissolved organic materials a questionable nutritional strategy.

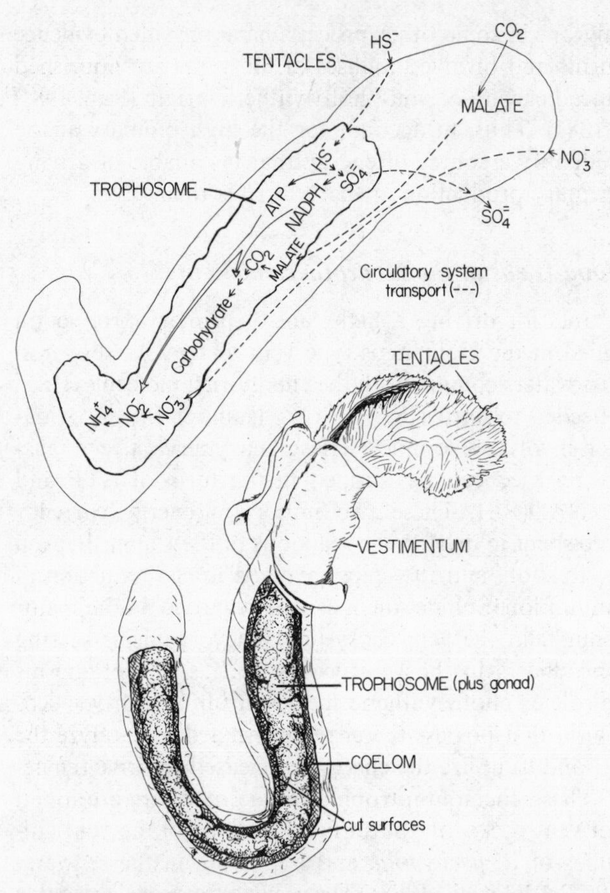

Fig. 12-11. *Riftia pachyptila*, a vestimentiferan tube worm found at the deep-sea hydrothermal vents. Lower panel: anatomy of the worm, showing the trophosome organ in which the endosymbiotic chemolithotrophic bacteria are located. Upper panel: the metabolic transformations and transport processes associated with the symbiotic interactions. Figure from Felbeck and Somero (1982)

Riftia pachyptila does succeed in exploiting the biosynthetic potentials of sulfide oxidizing bacteria, however. The way this is accomplished is shown in Figure 12-11. Rather than relying on the metabolic activities of sulfide oxidizing bacteria living in the surrounding water, *R. pachyptila* contains a massive culture of sulfide-oxidizing bacteria within its body cavity. A large organ termed the trophosome fills most of the body cavity of the worm, and the trophosome is composed partially of bacteria (Cavanaugh et al., 1981; Cavanaugh, 1983). Each gram of trophosome tissue contains up to one billion bacteria. Based on the types and quantities of enzymic activity present in the trophosome (Felbeck, 1981; Felbeck et al., 1981; Felbeck and Somero, 1982), the metabolic model illustrated in the upper portion of Figure 12-11 has been proposed. In this model, HS^- provides the energy for synthesis of ATP and reducing power. (It is not yet known if methane and hydrogen in the vent waters are also exploited by the bacteria.) HS^- is taken up from the seawater through the tentacle and transported by the extensive closed circulatory system of the worm to the bacterial symbionts in the trophosome. The symbionts burn HS^- to SO_4^{2-} or to other partially-oxidized forms of sulfur, and the energy released drives the net fixation of CO_2 and the synthesis of reduced nitrogen compounds. CO_2 obtained from the seawater appears to be fixed into a four-carbon metabolite, malate, for blood transport. Nitrate from the seawater supplies a source of fixed nitrogen for the synthesis of reduced nitrogen compounds.

As in the case of symbioses involving animals and photosynthetic symbionts (cf Felbeck et al., 1983), it is difficult to determine precisely the quantitative significance of nutrients supplied by the symbionts to the host's metabolic needs. Carbon isotope ratio studies again have proven useful. Rau (1981) showed that the different tissues of *R. pachyptila* share a common pool of reduced carbon, and that the ^{13}C to ^{12}C ratio of this pool is radically different from ratios found in other vent animals and in most shallow-living animals. These carbon isotope data argue powerfully that the bacterial symbionts supply a major share of the host's reduced carbon compounds. A further clue to the capacities of the bacterial symbionts to synthesize reduced carbon compounds is provided by the high activities of Calvin-Benson cycle enzymes in the trophosome. Activity of the diagnostic enzyme of the Calvin-Benson cycle, ribulose-1,5-bisphosphate carboxylase, in trophosome is similar to the activity found in fresh spinach leaves (Felbeck, 1981). Thus, a high capacity for net CO_2 fixation is indicated.

The discovery of this new type of symbiosis in *R. pachyptila* led to searches for other examples of animal-sulfur bacteria symbioses. Among the animals from the deep-sea hydrothermal vents, the large clam *Calyptogena magnifica* and the unnamed vent mussel were found to harbor bacterial symbionts in their gill tissues (Cavanaugh, 1983), and these bacteria display capacities for oxidation of sulfide and for net CO_2 fixation via the Calvin-Benson cycle (Felbeck et al., 1981; Felbeck et al., 1983). Thus, like *R. pachyptila*, the dominant bivalve molluscs of the vent community may obtain some of their nutrition from reduced carbon and nitrogen compounds synthesized by bacterial symbionts.

Animals from shallow water habitats in which access to both sulfide and oxygen is possible have also been found to harbor this type of symbiont. For example, in sewage outfall zones the concentrations of sulfide may be several times greater than the sulfide concentrations of the vent waters, and oxygen from the overlying seawater can supply the means for burning this sulfide. In sewage outfall areas off the coast of Los Angeles, high densities of a gutless clam, *Solemya reidi*, are found in sediments in which few other macroscopic forms of animal life are present (Felbeck et al., 1981; Felbeck, 1983). The gills of *S. reidi* are very large and contain numerous bacteria with capacities for sulfide oxidation, for net CO_2 fixation, and for reduction of nitrate. Thus, in many habitats in which animals can acquire sulfide plus the oxygen needed to burn it, the type of symbiosis originally found in the hydrothermal vent tube worm may be an important nutritional mechanism.

Metabolic Rates of Hydrothermal Vent Animals

In view of the relative abundance of food at the deep-sea vent sites compared to typical deep-sea regions it is likely that the metabolic rates of the vent animals are higher than those of animals from the typical deep-sea. To date there have not been extensive studies of oxygen consumption of vent animals; indeed, oxygen consumption may prove to be a deceptive index of metabolic rate in those vent animals that harbor sulfide-oxidizing bacteria since much of the oxygen consumed by the complete symbiosis may reflect sulfide oxidation by the bacteria rather than combustion of foodstuffs by the animal and bacteria.

An alternative means for obtaining at least rough estimates of metabolic rates was discussed earlier in this chapter. By measuring the activities of enzymes of energy metabolism pathways, at least an approximation of the respiration rate of the intact animal is possible, as suggested by the data of Figure 12-9. Enzyme surveys of this type have been done for several vent animals (Hand and Somero, 1983b). Data for the unidentified vent fish from the 21°N vent site off the tip of Baja California are shown in Figure 12-8 (points corresponding to a minimal depth of occurrence of 2,600 m). The LDH and PK activities measured in the white skeletal muscle of this fish fall into the range of values obtained in studies with shallow-living fishes. These enzyme data thus suggest that the vent fish has a metabolic rate similar to many shallow-living fishes, and much higher than rates for all other deep-sea fishes studied to date. Observations of the swimming capacities of the vent fish made from the submersible vehicle DSRV Alvin are consistent with this prediction (Hand and Somero, 1983b).

Enzymic activity data also have been gathered on a number of vent invertebrates, including *R. pachyptila*, *C. magnifica*, and the brachyuran crab, *Bythograea thermydron* (Hand and Somero, 1983b). Comparisons with activities measured in shallow water benthic invertebrates show that the vent invertebrates have apparent metabolic potentials that are very similar to those of shallow-living, related species.

In summary, the animals of the hydrothermal vent community appear to have much higher metabolic rates than the animals that have been studied from typical deep-sea habitats. The high metabolic rates of the vent animals, and the high biomass densities at the Galapagos-type vent sites, show clearly that the low temperatures and high pressures of the deep sea are not, in and of themselves, major determinants of the low metabolic rates found in animals from typical deep-sea habitats. Rather, the metabolic capacities of the hydrothermal vent animals provide a powerful illustration of the close link that exists between metabolic capacity and proximity to a rich food supply.

Avoidance of Poisoning by Hydrogen Sulfide

While a rich energy resource, hydrogen sulfide is also a highly toxic molecule due to its ability to poison hemoglobin and the cytochrome c oxidase system. Organisms that rely on aerobic respiration are typically killed by HS^- concentrations in the micromolar range. Thus few animal species could be expected to survive more than a brief exposure to the sulfide-rich environment of the vent animals. What adaptations allow the vent animals, especially sessile species like *R. pachyptila*, which live continuously in sulfide-rich waters, to avoid poisoning by sulfide?

A variety of mechanisms for allowing metabolism to persist in the face of high concentrations of HS^- can be proposed. First, an animal may rely largely, or entirely, on the types of anaerobic pathways of energy metabolism discussed in Chapter 5. The very low activities of cytochrome c oxidase, and much higher activities of enzymes like malate dehydrogenase which may function in anaerobiosis, in tissues of *Calyptogena magnifica* may indicate that this large clam relies strongly on this strategy for coping with high HS^- concentrations (Hand and Somero, 1983b). A second means for dealing with HS^- involves rapid oxidation of HS^- to less toxic, or nontoxic, forms of sulfur. Some of the meiofaunal species of sulfide-rich sediments may use this strategy (Powell et al., 1980). Third, sulfide-insensitive variants of hemoglobin and cytochrome c oxidase may evolve in animals from sulfide-rich habitats. The extracellular hemoglobin of *R. pachyptila* has been shown to be insensitive to sulfide (Arp and Childress, 1981). However, the cytochrome c oxidase systems of vent animals seem no less sensitive to poisoning by sulfide than the homologous enzymes from other animals (Hand and Somero, 1983b; Powell and Somero, 1983). Fourth, any HS^- that enters the body of an animal that harbors sulfur bacteria symbionts may be tightly bound to some type of transport molecule that, simultaneously, prevents free HS^- from coming into contact with the cytochrome c oxidase system, and protects the HS^- from undergoing spontaneous oxidation prior to the delivery of the HS^- to the symbionts. This latter mechanism for dealing with high concentrations of HS^- appears to be an important part of the sulfide processing mechanism of *R. pachyptila.*

In studies of the blood chemistry of *R. pachyptila*, Arp and Childress (1983) identified an unusual type of protein, termed a sulfide binding protein, that has an extremely high affinity for HS^- and a high carrying capacity for this molecule. This sulfide binding protein performs at least three critical functions for the animal-bacterial symbiosis. First, because the concentration of HS^- in the blood of *R. pachyptila* may exceed 1 mM (Arp and Childress, 1983), even though the HS^- concentration in the water surrounding the tentacle must be very much lower than this, the sulfide binding protein appears to extract HS^- from the environment. Here the sulfide binding protein is functioning analogously to a blood-borne hemoglobin molecule when it extracts oxygen from the external environment. Second, because of its high affinity for HS^- and high sulfide carrying capacity, the sulfide binding protein prevents appreciable amounts of free HS^- from appearing in the blood. Consequently, the potentially toxic free HS^- molecules are held at levels too low to pose a threat to the cytochrome c oxidase system. Indeed, Powell and Somero (1983) showed that minute amounts of blood of *R. pachyptila* could fully reverse HS^- inhibition of the cytochrome c oxidase system of the worm. Third, the sulfide binding protein protects the HS^- from spontaneous oxidation. Arp and Childress (1983) showed that HS^- that is complexed with the sulfide binding protein is extremely stable. The sulfide binding protein thus appears capable of carrying HS^- in a stable state from the site of uptake (the tentacle) to the site of HS^- utilization by symbiotic bacteria (the trophosome). How the HS^- unloading at the trophosome is regulated remains to be established, but it seems likely that, as in the case of O_2 unloading from hemoglobin, some type of effector molecule such as the proton might trigger release of HS^-.

Summary

Our discussion of biochemical adaptations of deep-sea organisms has allowed us to reiterate many of the key strategies of biochemical adaptation that were outlined in the opening chapter of this volume. Comparison of homologous proteins from shallow-living and deep-sea fishes has enabled us to see especially clearly how certain critical traits of proteins are conserved in all species, while other traits of lesser selective significance vary widely. The conservation of ligand binding ability at the apparent expense of reductions in catalytic efficiency is one example of how protein evolution is marked by strong conservative trends and by necessary compromises. The opportunism of biochemical adaptation, that is, the exploitative nature of molecular evolution, is well illustrated by deep-sea species from the hydrothermal vents, notably by those species that have come to exploit the normally toxic substance hydrogen sulfide as a rich energy resource. Deep-sea animals also provide striking illustrations of the ways in which metabolic output is geared to the environmental conditions faced by the organism. The extremely low metabolic rates of pelagic fishes of the deep-sea, which are determined by large-scale reductions in enzyme concentration in muscle tissue, show how radically the rates of a given type of metabolism, e.g., muscle glycolysis, can vary among animals that experience differences in the physical, chemical, and biological attributes of their habitats. Thus, our analysis leads to the same fundamental conclusion that was reached decades ago by Ernest Baldwin, who concluded his pathbreaking volume, *An Introduction to Comparative Biochemistry*, by stating that the study of biochemical variation among organisms adapted to widely different environmental conditions permits us to discern "that all life is based on a fundamental pattern that is common to living organisms of every kind," while teaching us simultaneously how "all the immense diversity of shapes, sizes, habits, habitats and the rest is exclusively due to an agglomeration of secondary and specific adaptational variations on a fundamental theme."

REFERENCES

Abe, H. (1981). Determination of L-histidine-related compounds in fish muscle using high-performance liquid chromatography. *Bull. Jap. Soc. Scient. Fish.* 47:139.

Alexandrov, Y. Ya. (1977). *Cells, Molecules and Temperature.* Berlin: Springer-Verlag.

Alfonso, M., and E. Racker (1979). Components and mechanism of action of ATP-driven proton pumps. *Can. J. Biochem.* 57:1351–1358.

Andersen, M. E., J. S. Olson, and Q. H. Gibson (1973). Studies on ligand binding to hemoglobins from teleosts and elasmobranchs. *J. Biol. Chem.* 248:5544–5555.

Anderson, G. R., V. R. Polonis, J. K. Petell, R. A. Saavedra, K. F. Manly, and L. M. Matovick (1983). LDH_k, a transformation dehydrogenase found in human cancer. In *Isozymes: Current Topics in Biological and Medical Studies*, ed. M. C. Rattazzi, J. G. Scandalios, and G. S. Whitt, 11:155–172. New York: Alan R. Liss.

Argos, P., M. G. Rossmann, U. M. Grau, H. Zuber, G. Frank, and J. D. Tratschin (1979). Thermal stability and protein structure. *Biochemistry* 18: 5698–5703.

Armond, P. A., U. Schreiber, and O. Bjorkman (1978). Photosynthetic acclimation to temperature in the desert shrub, *Larrea divaricata.* II. Light-harvesting efficiency and electron transport, *Plant Physiol.* 61:411–415.

Arnone, A. (1972). X-ray diffraction study of binding of 2,3-diphosphoglycerate to human deoxyhaemoglobin. *Nature* 237:146–149.

Arp, A. J., and J. J. Childress (1981). Blood function in the hydrothermal vent vestimentiferan tube worm. *Science* 213:342–344.

——— (1983). Sulfide binding by the blood of the hydrothermal vent tube worm, *Riftia pachyptila. Science*, 219:295–297.

Ashwood-Smith, M., and J. Farrant (1980). *Low Temperature Preservation in Medicine and Biology.* University Park, Md.: University Park Press.

Assaf, S. A., and J. D. Graves (1969). Structural and catalytic properties of lobster muscle glycogen phosphorylase. *J. Biol. Chem.* 244:5544–5555.

Atkinson, D. E. (1969). Limitation of metabolite concentrations and the conservation of solvent capacity in the living cell. *Current Topics in Cellular Regulation* 1:29–43.

——— (1977). *Cellular Energy Metabolism and Its Regulation.* New York: Academic Press.

———, and M. N. Camien (1982). The role of urea synthesis in the removal of bicarbonate and the regulation of blood pH. *Current Topics in Cellular Regulation* 21:261–302.

Bailey, A. J. (1968). The nature of collagen. In *Comprehensive Biochemistry, 26B*, ed. M. Florkin and E. H. Stotz, 297–423. Amsterdam: Elsevier.

Baldwin, E. (1970). *An Introduction to Comparative Biochemistry.* Cambridge: Cambridge University Press.

Baldwin, J. (1971). Adaptation of enzymes to temperature: acetylcholinesterases in the central nervous system of fishes. *Comp. Biochem. Physiol.* 40:181–187.

———, and P. W. Hochachka (1970). Functional significance of isoenzymes in thermal acclimatization: Acetylcholinesterase from trout brain. *Biochem. J.* 116:883–887.

Ballantyne, J. S., P. W. Hochachka, and T. P. Mommsen (1981). Studies on the metabolism of the migratory squid, *Loligo opalescens*: enzymes of tissues and heart mitochondria. *Marine Biol. Letters* 2:75–85.

Banse, K. (1964). On the vertical distribution of zooplankton in the sea. In *Progress in Oceanography*, ed. M. Sears, 2:53–125. Oxford: Pergamon Press.

Beis, I., and E. A. Newsholme (1975). Contents of adenine nucleotides, phosphagens, and some glycolytic intermediates in resting muscles from vertebrates and invertebrates. *Biochem. J.* 152:23–32.

Benade, A.J.S., and N. Heisler (1978). Comparison of efflux rate of hydrogen and lactate ions from isolated muscle *in vitro. Resp. Physiol.* 32:369–380.

Bennett, A. F. (1972). A comparison of activities of metabolic enzymes in lizards and rats. *Comp. Biochem. Physiol.* 42B:637–647

———, and J. A. Ruben (1979). Endothermy and activity in vertebrates. *Science* 206:649–654.

Bernstam, V. A. (1978). Heat effects on protein biosynthesis. *Ann. Rev. Plant Physiol.* 29:25–46.

Berry, J., and O. Bjorkman (1980). Photosynthetic response and adaptation to temperature in higher plants. *Ann. Rev. Plant Physiol.* 31:491–543.

Bessman, S. P., and P. J. Geiger (1981). Transport of energy in muscle: The phosphorylcreatine shuttle. *Science* 211:448–452.

Beyer, R. F. (1972). Effects of low temperature on cold-sensitive enzymes from mammalian tissues. In *Hibernation-Hypothermia: Perspectives and Challenges*, ed. F. E. South, J. P. Hannon, J. R. Willis, E. T. Pengelley, and N. R. Alpert, 17–54. Amsterdam: Elsevier.

Biesecker, G., J. I. Harris, J. C. Thierry, J. E. Walker, and A. J. Wonacott (1977). Sequence and structure of D-glyceraldehyde 3-phosphate dehydrogenase from *Bacillus stearothermophilus. Nature* 266:328–333.

Bjorkman, O., M. R. Badger, and P. A. Armond (1980). Adaptation to high temperature stress. In *Adaptation of Plants to Water and High Temperature Stress*, ed. N. C. Turner and P. J. Kramer, pp. 231–249. New York: John Wiley and Sons.

Bock, P. E., and C. Frieden (1976). Phosphofructokinase. I. Mechanism of the pH-dependent inactivation and reactivation of the rabbit muscle enzyme. *J. Biol. Chem.* 251:5630–5636.

Bonaventura, C., B. Sullivan, J. Bonaventura, and S. Bourne (1977). Anion modulation of the negative Bohr effect of haemoglobin from a primitive amphibian. *Nature* 265:474–476.

Bonaventura, J., and C. Bonventura (1980). Hemocyanins: relationships in their structure, function and assembly. *Amer. Zool.* 20:7–17.

———, C. Bonaventura, and B. Sullivan (1977). Non-heme oxygen transport proteins. In *Oxygen and Physiological Function*, ed. F. F. Jobsis, 177–220. Dallas, Texas: Professional Information Library.

———, and S. C. Wood (1980). Respiratory pigments: Overview. *Amer. Zool* 20:5–6.

Borgmann, U., and T. W. Moon (1975). A comparison of LDHs from an ectothermic and endothermic animal. *Can. J. Biochem.* 53:998–1004.

Borowitzka, L. J., and A. D. Brown (1974). The salt relations of marine and halophilic species of the unicellular green alga, *Dunaliella.* The role of glycerol as a compatible solute. *Arch. Microbiol.* 96:37–52.

———, S. Demmerle, M. A. Mackay, and R. S. Norton (1980). Carbon-13 nuclear magnetic resonance study of osmoregulation in a blue-green alga. *Science* 210:650–651.

Botelho, L. H., S. H. Friend, J. B. Matthew, L. D. Lehman, G.I.H. Hanania, and F.R.N. Gurd (1978). Proton nuclear magnetic resonance study of histidine ionizations in myoglobins of various species. Comparison of observed and computed pK values. *Biochemistry* 17:5197–5205.

———, and F.R.N. Gurd (1978). Proton nuclear magnetic resonance study of histidine ionizations in myoglobins of various species. Specific assignment of individual resonances. *Biochemistry* 17:5188–5196.

Bowlus, R. D., and G. N. Somero (1979). Solute compatibility with enzyme function and structure: Rationales for the selection of osmotic agents and end-products of anaerobic metabolism in marine invertebrates. *J. Exp. Zool.* 208:137–152.

Brandts, J. F. (1967). Heat effects on proteins and enzymes. In *Thermobiology*, ed. H. Rose, 25–72. New York: Academic Press.

Brauer, R. W., M. Y. Bekman, J. B. Keyser, D. L. Nesbitt, G. N. Sidelev, and S. L. Wright (1980a). Adaptation to high hydrostatic pressures of abyssal gammarids from Lake Baikal in eastern Siberia. *Comp. Biochem. Physiol.* 65A:109–117

———, M. Y. Bekman, J. B. Keyser, D. L. Nesbitt, S. G. Shvetzov, G. N. Sidelev, and S. L. Wright (1980b). Comparative studies of sodium transport and its relation to hydrostatic pressure in deep- and shallow-water gammarid crustaceans from Lake Baikal. *Comp. Biochem. Physiol.* 65A:119–127.

Brett, J. R., and T.D.D. Groves (1979). Physiological energetics. In *Fish Physiology*, ed. W. A. Hoar, D. J. Randall, and J. R. Brett, 7:279–352. New York: Academic Press.

Brooks, S. L., N. J. Rothwell, M. J. Stock, A. E. Goodbody, and P. Trayhurn (1980). Increased proton conductance pathway in brown adipose tissue mitochondria of rats exhibiting diet-induced thermogenesis. *Nature* 286:274–276.

Brown, A. D., and L. J. Borowitzka (1979). Halotolerance of *Dunaliella.* In *Biochemistry and Physiology of Protozoa* ed. M. Levandowsky, and S. H. Hutner (2nd ed.), 139–190. New York: Academic Press.

———, and J. R. Simpson (1972). Water relations of sugar-tolerant yeasts: The role of intracellular polyols. *J. Gen. Microbiol.* 72:589–591.

Brunori, M. (1975). Molecular adaptation to physiological requirements: The hemoglobin system of trout. *Current Topics in Cellular Regulation* 9:1–39.

Bunn, H. F., and P. J. Higgins (1981). Reaction of monosaccharides with proteins: Possible evolutionary significance. *Science* 213:222–224.

Burton, R. F. (1978). Intracellular buffering. *Respir. Physiol.* 33:51–58.

Butler, P. J. and D. R. Jones (1982). The comparative physiology of diving in vertebrates. *Adv. Comp. Physiol. Biochem.* 8:179–364.

Caldwell, R. S. and F. J. Vernberg (1970). The influence of acclimation temperature on the lipid composition of fish gill mitochrondria. *Comp. Biochem. Physiol.* 34:179–191.

Caligiuri, M., E. D. Robin, A. J. Hance, D. A. Robin, N. Lewiston, and J. Theodore (1981). Prolonged diving and recovery in the freshwater turtle, *P. scripta.* II. Magnitude of depression of O_2 requirements and the relation to body O_2 stores. *Comp. Biochem. Physiol.* 70A:365–369.

Carey, F. G., J. M. Teal, J. W. Kanwisher, K. D. Lawson, and J. S. Beckett (1971). Warm-bodied fishes. *Amer. Zool.* 11:135–143.

Castellini, M. A., and Hochachka, P. W. (1982). Glucose and palmitate metabolism during rest and exercise in the gray seal. *The Physiologist* 25 (4): 253.

———, and G. N. Somero (1981). Buffering capacity of vertebrate muscle: Correlations with potentials for anaerobic function. *J. Comp. Physiol.* 143: 191–198.

———, G. N. Somero, and G. L. Kooyman (1981). Glycolytic enzyme activities in tissues of marine and terrestrial mammals. *Physiol. Zool.* 54 (2): 242–252.

Cavanaugh, C. M. (1983). Symbiotic chemoautotrophic bacteria in marine invertebrates. *Nature* 320:58–61.

———, S. Gardiner, M. L. Jones, H. W. Jannasch, and J. B. Waterbury (1981). Prokaryotic cells in the hydrothermal vent tube worm *Riftia pachyptila* Jones: Possible chemoautotrophic symbionts. *Science* 213:340–342.

Chance, B., S. Eleff, J. S. Leigh, D. Sokolow, and A. Sapega (1981). Mitochondrial regulation of phosphocreatine/inorganic phosphate ratios in exercising human muscle: A gated ^{31}P NMR study. *Proc. Natl. Acad Sci. USA* 78:6714–6718.

Childresss, J. J. (1975). The respiratory rates of midwater crustaceans as a function of depth of occurrence and relation to the oxygen minimum layer off southern California. *Comp. Biochem. Physiol.* 50A:787–799.

——— (1977). Effects of pressure, temperature and oxygen on the oxygen consumption rate of the midwater copepod *Gaussia princeps. Mar. Biol.* 39:19–24.

———, and M. H. Nygaard (1973). The chemical composition of midwater fishes as a function of depth of occurrence off southern California. *Deep-Sea Res.* 20:1093–1109.

———, and G. N. Somero (1979). Depth-related enzymic activities in muscle, brain and heart of deep-living pelagic marine teleosts. *Mar. Biol.* 52: 273–283.

Chotia, C. (1975). Structural variants in protein folding. *Nature* 254:304–308.

Clark, M. E., and M. Zounes (1977). The effects of selected cell osmolytes on the activity of lactate dehydrogenase from the euryhaline polychaete, *Nereis succinea. Biol. Bull.* 153:468–484.

Clark, M. G., D. P. Bloxham, P. C. Holland, and H. A. Lardy (1973). Estimation of the fructose diphosphatase-phosphofructokinase substrate cycle in the flight muscle of *Bombus affinisi. Biochem. J.* 134:589–597.

Clegg, J. S. (1962). Free glycrol in dormant cysts of the brine shrimp, *Artemia salina*, and its disappearance during development. *Biol. Bull.* 123:295–301.

——— (1981). Metabolic consequences of the extent and disposition of the aqueous intracellular environment. *J. Exp. Zool.* 215:303–313.

Cohen, P. (1978). The role of cyclic-AMP-dependent protein kinase in the regulation of glycogen metabolism in mammalian skeletal muscle. In *Current Topics in Cellular Regulation* 14:117–196.

Collicutt, J., and P. W. Hochachka (1977). The anaerobic oyster heart. *J. Comp. Physiol.* 115:147–157.

Corliss, J. B., J. Dymond, L. I. Gordon, J. M. Edmond, R. P. von Herzen, R. D. Ballard, K. Green, D. Williams, A. Bainbridge, K. Crane, and T. H. van Andel (1979). Submarine thermal springs on the Galapagos Rift. *Science* 203:1073–1083.

Cossins, A. R., M. J. Friedlander, and C. L. Prosser (1977). Correlations between behavioral temperature adaptations of goldfish and the viscosity and fatty acid composition of their synaptic membranes. *J. Comp. Physiol.* 120: 109–121.

———, and C. L. Prosser (1978). Evolutionary adaptation of membranes to temperature. *Proc. Natl. Acad. Sci. U.S.A.* 75:2040–2043.

Cowey, C. B. (1967). Comparative studies on the activity of D-glyceraldehyde-3-phosphate dehydrogenase from cold- and warm-blooded animals with reference to temperature. *Comp. Biochem. Physiol.* 31:79–93.

Crick, F. (1981). *Life Itself.* Austin, Texas: S and X Press.

Crowe, J. H. (1971). Anhydrobiosis: An unsolved problem. *Amer. Naturalist* 105:563–573.

———, and J. S. Clegg (1978). *Dry Biological Systems.* New York: Academic Press.

Crush, K. G. (1970). Carnosine and related substances in animal tissues. *Comp. Biochem. Physiol.* 34:3–30.

Dann, L. G., and H. G. Britton (1978). Kinetics and mechanism of action of muscle pyruvate kinase. *Biochem. J.* 169:39–54.

Davey, C. L. (1960). The significance of carnosine and anserine in striated skeletal muscle. *Arch. Biochem. Biophys.* 89:303–308.

Davis, B. D. (1958). On the importance of being ionized. *Arch. Biochem. Biophys.* 78:497–509.

Davis, R. W. (1983). Lactate and glucose metabolism in the resting and diving harbor seal (*Phoca vitulina*). *J. Comp. Physiol.* 153:275–288.

Deavers, D. R., and X. J. Musacchia (1980). Water metabolism and renal function during hibernation and hypothermia. *Fed. Proc.* 39:2969–2973.

De Smedt, H., R. Borghgraef, F. Ceuterick, and K. Heremans (1979). Pressure effects on lipid-protein interactions in ($Na^+ + K^+$)-ATPase. *Biophys. Acta* 556:479–489.

DeVries, A. L. (1974). Survival at freezing temperatures. In *Biochemical and Biophysical Perspectives in Marine Biology*, ed. D. C. Malins and J. R. Sargent, 289–330. New York: Academic Press.

——— (1980). Biological antifreezes and survival in freezing environments. In *Animals and Environmental Fitness*, ed. R. Gilles, 583–607. New York. Pergamon Press.

———. (1982). Biological antifreeze agents in coldwater fishes. *Comp. Biochem. Physiol.* 73A:627–640.

De Zwaan, A. (1983). Carbohydrate catabolism in bivalves. In *The Mollusca*, ed. P. W. Hochachka, 1:138–175. New York: Academic Press.

———, and T.C.M. Wijsman (1976). Anaerobic metabolism in Bivalvia (Mollusca). *Comp. Biochem. Physiol.* 54B:313–324.

Dhinsa, D. S., J. Metcalfe, D. W. Blackmore, and R. D. Koler (1981). Postnatal changes in oxygen affinity of rat blood. *Comp. Biochem. Physiol.* 69A:279–283.

Dizon, A. E., and R. W. Brill (1979). Thermoregulation in tunas. *Amer. Zool.* 19:249–265.

Drake, A. (1982). Substrate utilization in the myocardium. *Basic Res. Cardiol.* 77:1–11.

Duman, J. G. (1977). The role of macromolecular antifreeze in the darkling beetle, *Meracantha contracta*. *J. Comp. Physiol.* 115:279–286.

——— (1979). Subzero temperature tolerance in spiders: The role of thermal-hysteresis-factors. *J. Comp. Physiol.* 131:347–352.

——— (1980). Factors involved in overwintering survival of the freeze-tolerant beetle, *Dendroides canadensis*. *J. Comp. Physiol.* 136:53–59.

———, and A. L. DeVries (1973). Freezing behavior of aqueous solutions of glycoproteins from the blood of an Antarctic fish. *Cryobiology* 9:469–472.

———, and A. L. DeVries (1974). The effects of temperature and photoperiod on antifreeze production in cold-water fishes. *J. Exp. Zool.* 190:89–98.

———, and A. L. DeVries (1976). Isolation, characterization, and physical properties of protein antifreezes from the winter flounder, *Pseudopleuronectes americanus*. *Comp. Biochem. Physiol.* 54B:375–380.

———, K. L. Horwath, A. Tomchaney, and J. L. Patterson (1982). Anti-freeze agents of terrestrial arthropods. *Comp. Biochem. Physiol.* 73A:545–555.

Dunn, J. R., W. Davison, G.M.O. Maloiy, P. W. Hochachka, and M. Guppy (1981). An ultrastructural and histochemical study of the axial musculature in the African lungfish. *Cell. Tiss. Res.* 220:599–609.

———, P. W. Hochachka, W. Davison, and M. Guppy (1983). Metabolic adjustments to diving and recovery in the African lungfish. *Amer. J. Physiol.*, 245:R651–R657.

Edmond, J. M., C. Measures, B. Mangum, B. Grant, F. R. Sclater, R. Collier, and A. Hudson (1979). On the formation of metal-rich deposits at ridge crests. *Earth Planet. Sci. Lett.* 46:19–30.

———, K. L. Von Damm, R. E. McDuff, and C. I. Measures (1982). Chemistry of hot springs on the East Pacific Rise and their effluent dispersal. *Nature* 297: 187–191.

Edsall, J. T. (1980). Hemoglobin and the origins of the concept of allosterism. *Fed. Proc.* 39:226–235.

Emmett, B., and P. W. Hochachka (1981). Scaling of oxidative and glycolytic enzymes in the shrew. *Resp. Physiol.* 45:261–267.

Evered, D. F. (1981). Advances in amino acid metabolism in mammals. *Transactions Biochem. Soc.* 9:159–169.

Everse, J., and N. O. Kaplan (1973). Lactate dehydrogenase: Structure and function. *Adv. Enzymol.* 37:61–133.

Felbeck, H. (1981). Chemoautotrophic potential of the hydrothermal vent tube worm, *Riftia pachyptila* Jones (Vestimentifera). *Science* 213:336–338.

——— (1983). Sulfide oxidation and carbon fixation by the gutless clam *Solemya reidi:* An animal-bacteria symbiosis. *J. Comp. Physiol.* 152:3–11.

———, J. J. Childress, and G. N. Somero (1981). Calvin-Benson cycle and sulphide oxidation enzymes in animals from sulphide-rich habitats. *Nature* 293:291–293.

———, and G. N. Somero (1982). Primary production in deep-sea hydrothermal vent organisms: Roles of sulfide-oxidizing bacteria. *Trends in Biochem. Sci.* 7:201–204.

———, G. N. Somero, and J. J. Childress (1983). Biochemical interactions between molluscs and their symbionts. In *The Mollusca*, ed. P. W. Hochachka, 2:331–358. New York: Academic Press.

Fersht, A. R. (1977). *Enzyme Structure and Mechanism.* San Francisco: W. H. Freeman.

——— (1980). Enzymic editing mechanisms in protein synthesis and DNA replication. *Trends in Biochem. Sci.* 5:262–265.

Fields, J.H.A., A. K. Eng, W. D. Ramsden, P. W. Hochachka, and B. Weinstein (1980). Alanopine and strombine are novel imino acids produced by a dehydrogenase found in the adductor muscle of the oyster, *Crassostrea gigas. Arch. Biochem. Biophys.* 201:110–114.

Forster, R. P. and L. Goldstein (1976). Intracellular osmoregulatory role of amino acids and urea in marine elasmobranchs. *Amer. J. Physiol.* 230: 925–931.

French, S. L. (1981). Mechanism of Halotolerance in Some Manganese-oxidizing Bacteria Ph.D. diss., University of California, San Diego.

Fridovich, I. (1978). The biology of oxygen radicals. *Science* 201:875–880.

Fujii, D. K., and Fulco, A. J. (1977). Biosynthesis of unsaturated fatty acids by bacilli. Hyperinduction and modulation of desaturase biosynthesis. *J. Biol. Chem.* 252:3660–3670.

Furuya, E., and K. Uyeda (1980). Regulation of phosphofructokinase by a new mechanism. An "activation factor" binding to the phosphorylated enzyme. *J. Biol. Chem.* 255:11656–11659.

Gadian, D. G., G. K. Radda, T. K. Brown, E. M. Chance, M. J. Dawson, and D. R. Wilkie (1981). The activity of creatine kinase in frog skeletal muscle studied by saturation-transfer nuclear magnetic resonance. *Biochem. J.* 194: 215–228.

Gekko, K., and S. N. Timasheff (1981a). Mechanism of protein stabilization by glycerol: Preferential hydration in glycerol-water mixtures. *Biochemistry* 20:4667–4676.

———, and S. N. Timasheff (1981b). Thermodynamic and kinetic examination of protein stabilization by glycerol. *Biochemistry* 20:4677–4686.

Giles, M. A., and D. J. Randall (1980). Oxygenation characteristics of the polymorphic hemoglobins of coho salmon (*Oncorhynchus kisutch*) at different developmental stages. *Comp. Biochem. Physiol.* 65A:265–271.

Gilles, R., ed. (1979). *Mechanisms of Osmoregulation in Animals. Maintenance of Cell Volume.* New York: Wiley-Interscience.

Gillies, R. J., and D. W. Deamer (1979). Intracellular pH changes during the cell cycle in *Tetrahymena. J. Cell Physiol.* 100:23–32.

Goldberg, A. L., and T. W. Chang (1978). Regulation and significance of amino acid metabolism in skeletal muscle. *Fed. Proc.* 37:2301–2307.

———, and J. F. Dice (1974). Intracellular protein degradation in mammalian and bacterial cells. *Ann. Rev. Biochem.* 43:835–869.

———, and A. C. St John (1976). Intracellular protein degradation in mammalian and bacterial cells: Part 2. *Ann. Rev. Biochem.* 45:747–803.

Goldspink, G. (1977). Mechanics and energetics of muscle in animals of different sizes, with particular reference to the muscle fibre composition of vertebrate muscle. In *Scale Effects in Animal Locomotion*, ed. T. J. Pedley, 37–55. New York: Academic Press.

Gollnick, P. D., B. F. Timson, R. L. Moore, and M. Reidy (1981). Muscular enlargement and number of fibers in skeletal muscles of rats. *J. Appl. Physiol.* 50:936–943.

Gordon, M. S., and V. E. Tucker (1968). Further observations on the physiology and salinity adaptation in the crab-eating frog, *Rana cancrivora. J. Exp. Biol.* 49:185–193.

Gottschalk, G. (1979). *Bacterial Metabolism.* New York: Springer-Verlag.

Graham, J. B., and K. A. Dickson (1982). Physiological thermoregulation in the albacore *Thunnus alalunga. Physiol. Zool.* 54:470–486.

Grainger, J. L., M. M. Winkler, S. S. Shen, and R. A. Steinhardt (1979). Intracellular pH controls protein synthesis rate in the sea urchin egg and early embryo. *Dev. Biol.* 68:396–406

Graves, J. E., R. H. Rosenblatt, and G. N. Somero (1983). Kinetic and electrophoretic differentiation of lactate dehydrogenases of teleost species-pairs from the Atlantic and Pacific Coasts of Panama. *Evolution* 37:30–37.

———, and G. N. Somero (1982). Electrophoretic and functional enzymic evolution in four species of eastern Pacific barracudas from different thermal environments. *Evolution* 36:97–106.

Greaney, G. S., and D. A. Powers (1977). Cellular regulation of an allosteric modifier of fish haemoglobin. *Nature* 270:73–74.

———, and D. A. Powers (1978). Allosteric modifiers of fish hemoglobins: *In vitro* and *in vivo* studies of the effect of ambient oxygen and pH on erythrocyte ATP concentrations. *J. Exp. Zool.* 203:339–350.

———, and G. N. Somero (1979). Effects of anions on the activation thermodynamics and fluorescence emission spectrum of alkaline phosphatase: evidence for enzyme hydration changes during catalysis. *Biochemistry* 18: 5322–5332.

———, and G. N. Somero (1980). Contributions of binding and catalytic rate constants to evolutionary modification in K_m of NADH for muscle-type (M_4) lactate dehydrogenases. *J. Comp. Physiol.* 137:115–121.

Greenwalt, D. E., and S. H. Bishop (1980). Effect of aminotransferase inhibitors on the pattern of free amino acid accumulation in isolated mussel hearts subjected to hyperosmotic stress. *Physiol. Zool.* 53:262–269.

Guppy, M., and P. W. Hochachka (1978). Role of dehydrogenase competition in metabolic regulation: The case of lactate and α-glycerophosphate dehydrogenases. *J. Biol. Chem.* 253:8465–8469.

———, W. C. Hulbert, and P. W. Hochachka (1979). Metabolic sources of heat and power in tuna muscles. II. Enzyme and metabolite profiles. *J. Exp. Biol.* 82:303–320.

Gustafsson, L., and B. Norkrans (1976). On the mechanism of salt tolerance. Production of glycerol and heat during growth of *Debaryomyces hansenii*. *Arch. Microbiol.* 110:177–183.

Guy, P. S., and Snow, D. H. (1977). The effect of training and detraining on lactate dehydrogenase isoenzymes in the horse. *Biochem. Biophys. Res. Comm.* 75:863–869.

Hahn, P. (1982). Nutrition and metabolic development in mammals. In *Nutrition, Pre- and Postnatal Development*, ed. M. Winick, 1–39. New York: Plenum Press.

———, and O. Koldovsky (1966). *Utilization of Nutrients During Postnatal Development.* Oxford: Pergamon Press.

Hand, S. C., and G. N. Somero (1982). Urea and methylamine effects on rabbit muscle phosphofructokinase. Catalytic stability and aggregation state as a function of pH and temperature. *J. Biol. Chem.* 257:734–741.

———, and G. N. Somero (1983a). Phosphofructokinase of the hibernator, *Citellus beecheyi*: Temperature and pH regulation of activity *via* influences on the tetramer-dimer equilibrium. *Physiol. Zool.* 56: 380–388.

——— (1983b). Energy metabolism pathways of hydrothermal vent animals:

adaptation to a food-rich and sulfide-rich deep-sea environment. *Biol. Bull.* (Woods Hole) 165:167–181.

Hazel, J. R. (1972). The effect of temperature acclimation upon succinic dehydrogenase activity from the epaxial muscle of the common goldfish (*Carassius auratus* L). Properties of the enzyme and effect of lipid extraction. *Comp. Biochem. Physiol.* 43B:837–861.

——— (1984). Effects of temperature upon the structure and metabolism of cell membranes in fish. *Amer. J. Physiol.*, in press.

———, W. S. Garlick, and P. A. Sellner (1978). The effects of assay temperature upon the pH optima of enzymes of poikilotherms: A test of the imidazole alphastat hypothesis. *J. Comp. Physiol.* 123:97–104.

———, and C. L. Prosser (1974). Molecular mechanisms of temperature compensation in poikilotherms. *Physiol. Rev.* 54:620–677.

Hazlewood, C. F. (1979). A view of the significance and understanding of the physical properties of cell-associated water. In *Cell-Associated Water*, ed. W. Drost-Hansen and J. Clegg, 165–259. New York: Academic Press.

Hedrick, P. W., M. E. Givevan, and E. P. Ewing (1976). Genetic polymorphism in heterogeneous environments. *Ann. Rev. Ecol. Syst.* 7:1–32.

Heinrich, B. (1971). Temperature regulation of the sphinx moth, *Manduca sexta.* II. Regulation of heat loss by control of blood circulation. *J. Exp. Biol.* 54:153–166.

——— (1974a). Thermoregulation in endothermic insects. *Science* 185:747–756.

——— (1974b). Thermoregulation in bumblebees. I. Brood incubation by *Bombus vosnesenskii* queens. *J. Comp. Physiol.* 88:129–140.

——— (1979). Keeping a cool head: Honeybee thermoregulation. *Science* 205: 1269–1271.

Henderson, L. J. (1913). *The Fitness of the Environment.* Boston: Beacon Press.

Hew, C. L., and G. L. Fletcher (1979). The role of pituitary in regulating antifreeze protein synthesis in the winter flounder. *FEBS Letters* 99:337–339.

Hill, H.A.O. (1981). Oxygen, oxidases, and the essential trace metals. *Phil. Trans. Royal Soc.* London Series B 294:119–128.

Hinkle, P. C. (1981). Coupling ratios of proton transport in mitochondria. In *Chemiosmotic Proton Circuits in Biological Membranes*, ed. V. P. Skulachev and P. C. Hinkle, 49–58. Reading, Mass.: Addison-Wesley.

Hinton, H. E. (1968). Reversible suspension of metabolism and the origin of life. *Proc. Royal Soc.* Ser. B 171:43–57.

Hintz, C. S., M.M.Y. Chi, R. D. Feel, J. L. Ivy, K. K. Kaiser, C. V. Lowry, and O. H. Lowry (1982). Metabolite changes in individual rat muscle fibers during stimulation. *Amer. J. Physiol.* 242:C218–C228.

Hochachka, P. W. (1973). Comparative intermediary metabolism. In *Comparative Animal Physiology*, ed. C. L. Prosser, 212–278. Philadelphia: W. B. Saunders.

——— (1974). Regulation of heat production at the cellular level. *Fed. Proc.* 33:2162–2169.

——— (1975). Fitness of enzyme binding sites for their physical environment: Coenzyme and substrate binding sites for M_4 lactate dehydrogenases. *Comp. Biochem. Physiol.* 52B:25–31.

——— (1979). Cell metabolism, air-breathing, and the origins of endothermy. In *Evolution of Respiratory Processes, A Comparative Approach*, ed. C. Lenfant and S. C. Wood, Monograph 13. 253–288. New York: Marcel Dekker.

——— (1980). *Living Without Oxygen: Closed and Open Systems in Hypoxia Tolerance.* Cambridge, Mass.: Harvard University Press.

——— (1981). Brain, lung, and heart functions during diving and recovery. *Science* 212:509–514.

——— (1982). Anaerobic metabolism: living without oxygen. In *A Companion to Animal Physiology*, ed. C. R. Taylor, K. Johansen, and L. Bolis, 138–150. Cambridge: Cambridge University Press.

———, G. P. Dobson, and T. P. Mommsen (1983). Role of isozymes in metabolic regulation during exercise: Insights from comparative studies. In *Isozymes: Current Topics in Biological and Medical Research*, ed. M. C. Rattazzi, J. G. Scandalios, and G. S. Whitt, 8:91–113. New York: Alan R. Liss.

———, and J. R. Dunn (1984). Metabolic arrest: The most effective means of protecting tissues against hypoxia. In *Third Banff International Hypoxia Symposium*, 297–309. New York: Alan R. Liss.

———, and J.H.A. Fields (1983). Arginine, glutamate, and proline as substrates for oxidation and for glycogenesis in cephalopod tissues. *Pacific Science* 36:325–336.

———, J.H.A. Fields, and T. P. Mommsen (1983). Metabolic and enzyme regulation during rest-to-work transition: A mammal vs. mollusc comparison. In *The Mollusca*, ed. P. W. Hochachka Vol. 1:56–89. New York: Academic Press.

———, and M. Guppy (1977). Variations on a theme by Embden and Meyerhof. In *O_2 in the Organism*, ed. F. Jobsis, 292–310. Dallas, Texas: Prof. Publ. Co.

———, M. Guppy, H. E. Guderley, K. B. Storey, and W. C. Hulbert (1978). Metabolic biochemistry of water- vs. air-breathing fishes: Muscle enzymes and ultrastructure. *Can. J. Zool.* 56:736–750.

———, G. C. Liggins, and W. M. Zapol (1977). Pulmonary metabolism during diving: Conditioning blood for the brain. *Science* 198:831–834.

———, and T. P. Mommsen (1983). Protons and anaerbiosis. *Science*, 219: 1391–1397.

———, and B. Murphy (1979). Metabolic status during diving and recovery in marine mammals. *Intl. Review Physiol.* Vol. 3, *Environmental Physiol.* ed. D. Robertshaw 253–287. Baltimore: University Park Press.

———, C. Norberg, J. Baldwin, and J.H.A. Fields (1975). Enthalpy-entropy compensation of oxamate binding by homologous lactate dehydrogenases. *Nature* 260:648–650.

———, and D. J. Randall (1978). Water-air breathing transition in vertebrates of the Amazon: Alpha Helix Amazon expedition, September-October 1976. *Can. J. Zool.* 56:713–716.

———, and G. N. Somero (1973). *Strategies of Biochemical Adaptation.* Philadelphia: W. B. Saunders.

———, and K. B. Storey (1975). Metabolic consequences of diving in animals and man. *Science* 197:613–621.

Hoh, J. F. Y. (1983). Myosin isoenzymes and muscular contractility. *Proc. Intl. Union Physiol. Sci.* 25, 467.05, 387.

Holbrook, J. J., A. Liljas, S. J. Steindal, and M. G. Rossmann (1975). Lactate dehydrogenase. *The Enzymes* 11:191–292.

Holloszy, J. O. and F. W. Booth (1976). Biochemical adaptations to endurance exercise in muscle. *Ann. Rev. Physiol.* 38:273–291.

———, W. W. Winder, R. H. Fitts, M. J. Rennie, R. C. Hickson, and R. K. Conlee (1978). Energy production during exercise. In *3rd Intl. Symp. on Biochemistry of Exercise*, ed. F. Landry and W.A.R. Orban, 61–74. Miami: Symposia Specialists.

Howald, H., G. von Glutz, and R. Billeter (1978). Energy stores and substrates utilization in muscle during exercise. In *3rd Intl. Symp. on Biochemistry of Exercise*, ed. F. Landry and W.A.R. Orban, 75–86. Miami: Symposia Specialists.

Hultman, E. (1978). Regulation of carbohydrate metabolism in the liver during rest and exercise with special reference to diet. In *3rd Intl. Symp. on Biochemistry of Exercise*, ed. F. Landry and W.A.R. Orban, 99–126. Miami: Symposia Specialists.

Jacobs, H. K., and S. A. Kuby (1980). Studies on muscular dystrophy. A comparison of the steady-state kinetics of the normal human ATP-creatine transphosphorylase isoenzymes (creatine kinases) with those from tissues of Duchenne muscular dystrophy. *J. Biol. Chem.* 255:8477–8482.

Jacobus, W. E., R. W. Moreadith, and K. M. Vandegaer (1982). Mitochondrial respiratory control. Evidence against the regulation of respiration by extramitochondrial phosphorylation potentials or by [ATP]/[ADP] ratios. *J. Biol. Chem.* 257:2397–2402.

Johansen, K. and R. E. Weber (1976). On the adaptability of haemoglobin function to environmental conditions. In *Perspectives in Experimental Biology*, ed. P. Spencer Davies, 1:219–234. Oxford: Pergamon Press.

Johnson, F. H., H. Eyring, and B. J. Stover (1974). *The Theory of Rate Processes in Biology and Medicine.* New York: Wiley.

Johnston, I. A. (1979). Calcium regulatory proteins and temperature acclimation of actomyosin ATPase from a eurythermal teleost (*Carassius auratus* L). *J. Comp. Physiol.* 129:163–167.

——— (1981). Specialization of fish muscle. In *Development and Specialization of Muscle*, ed. G. Goldspink, 123–148. Cambridge: Cambridge University Press.

———, and N. J. Walesby (1977). Molecular mechanisms of temperature adaptation in fish myofibrillar adenosine triphosphatases. *J. Comp. Physiol.* 119: 195–206.

Jones, M. L. (1981). *Riftia pachyptila* Jones: Observations on the vestimentiferan worm from the Galapagos Rift. *Science* 213:333–336.

Jones, R. M. (1980). Metabolic consequences of accelerated urea synthesis during seasonal dormancy of spadefoot toads, *Scaphiopus couchi* and *Scaphiopus multiplicatus. J. Exp. Zool.* 212:255–267.

Kanno, T., K. Sudo, I. Takeuchi, S. Kanda, N. Honda, Y. Nishimura, and K. Oyama (1980). Hereditary deficiency of lactate dehydrogenase M-subunit. *Clinica Chim. Acta* 108:267–276.

Kanwisher, J. (1955). Freezing in intertidal animals. *Biol. Bull.* 109:56–63.

Karl, D. M., C. O. Wirsen, and H. W. Jannasch (1980). Deep-sea primary production at the Galapagos hydrothermal vents. *Science* 207:1345–1347.

Kasai, R., Y. Kitajima, C. E. Martin, Y. Nozawa, L. Skriver, and G. A. Thompson, Jr. (1976). Molecular control of membrane properties during temperature acclimation. Membrane fluidity regulation of fatty acid desaturase action. *Biochemistry* 15:5228–5233.

Katzen, H. M., and D. D. Soderman (1975). The hexokinase isozymes: Sulfydryl considerations in the regulation of the particle-bound and soluble states. In *Isozymes II—Physiological Function*, ed. C. L. Markert, 797–817. New York: Academic Press.

Kauss, H. (1979). Osmotic regulation in algae. In *Progress in Phytochemistry*, ed. L. Reinhold et al., 5:1–27. Oxford: Pergamon Press.

———, K. S. Thomson, M. Thomson, and W. Jeblick (1979). Osmotic regulation. Physiological significance of proteolytic and nonproteolytic activation of isofloridoside-phosphate synthase. *Plant Physiol.* 63:455–459.

Keilin, D. (1959). The problem of anabiosis or latent life: history and current concepts. *Proc. Royal Soc.* Ser. B, 150:149–191.

Kitajima, Y., and G. A. Thompson, Jr. (1977). *Tetrahymena* strives to maintain the fluidity interrelationships of all its membranes constant. Electron microscope evidence. *J. Cell Biol.* 72:744–755.

Kjekshus, J. K., A. S. Blix, R. Elsner, R. Hol, and E. Amundsen (1982). Myocardial blood flow and metabolism in the diving seal. *Amer. J. Physiol.* 242: R97–R104.

Kleiber, M. (1965). Respiratory exchange and metabolic rate. In *Handbook of Respiration*, ed. W. O. Fenn and H. Rahn, Sec. 3, Vol. 2, 927–938. Washington, D. C.: The American Physiological Society.

Koehn, R. K. (1969). Esterase heterogeneity: Dynamics of a polymorphism. *Science* 163:943–944.

Kooyman, G. L., and W. B. Campbell (1972). Heart rates in freely diving Weddell seals, *Leptonychotes weddelli. Comp. Biochem. Physiol.* 43A:31–36.

———, M. A. Castellini, and R. W. Davis (1981). Physiology of diving in marine mammals. *Ann. Rev. Physiol.* 43:343–356.

———, E. A. Wahrenbrock, M. A. Castellini, R. W. Davis, and E. E. Sinnett (1980). Aerobic and anaerobic metabolism during voluntary diving in Weddell seals: Evidence of preferred pathways from blood chemistry and behaviour. *J. Comp. Physiol.* 138:335–346.

Koshland, D. E. (1973). Protein shape and biological control. *Sci. Amer.* 229: 52–64.

Krebs, H. A., H. F. Woods, and K.G.M.M. Alberti (1975). Hyperlactataemia and lactic acidosis. *Essays in Medical Biochemistry* 1:81–103.

Krietsch, W.K.G., and T. Bucher (1970). 3-phosphoglycerate kinase from rabbit skeletal muscle and yeast. *Eur. J. Biochem.* 17:568–580.

Krog, J. O., E. K. Zachariassen, B. Larsen, and O. Smidsrod (1979). Thermal buffering in Afro-alpine plants due to nucleating agent-induced water freezing. *Nature* 282:300–301.

Kuhn, T. S. (1970). *The Structure of Scientific Revolutions.* 2nd ed. Chicago: University of Chicago Press.

Laidler, K. J., and P. S. Bunting (1973). *The Chemical Kinetics of Enzyme Action.* London: Oxford University Press.

Lanyi, J. (1974). Salt dependent properties of proteins from extremely halophilic bacteria. *Bacteriol. Rev.* 38:272–290.

——— (1978). Light energy conversion in *Halobacterium halobium. Microbiol. Rev.* 42:682–706.

Lehninger, A. L. (1975). *Biochemistry*, New York: Worth.

Lenfant, C., R. Elsner, G. L. Kooyman, and C. M. Drabek (1969). Respiratory function of blood of the adult and fetus Weddell seal *Leptonychotes weddelli. Amer. J. Physiol.* 216:1595–1597.

Liggins, G. C., J. Qvist, P. W. Hochachka, B. Murphy, R. Creasy, R. Schneider, M. Snider, and W. M. Zapol (1980). Fetal cardiovascular and metabolic responses to simulated diving in the Weddell seal. *J. Appl. Physiol.* 49:424–430.

Lin, Y. (1979). Environmental regulation of gene expression: *In vitro* translation of winter flounder antifreeze messenger RNA. *J. Biol. Chem.* 254:1422–1426.

———, and D. J. Long (1980). Purification and characterization of winter flounder antifreeze peptide messenger ribonucleic acid. *Biochemistry* 19: 1111–1116.

Ling, G. N. (1979). The polarized multilayer theory of cell water according to the association-induction hypothesis. In *Cell-Associated Water*, ed. W. Drost-Hansen and J. Clegg, 261–269. New York: Academic Press.

Longmuir, I. S., J. A. Knopp, and D. M. Benson (1979). The heterogeneity of intracellular oxygen. *11th Intl. Congress Biochemistry Abstracts*, 428.

Lovelock, J. E. (1979). *Gaia: A New Look At Life On Earth.* Oxford: Oxford University Press.

Low, P. S., and G. N. Somero (1974). Temperature adaptation of enzymes: a proposed molecular basis for the different catalytic efficiencies of enzymes from ectotherms and endotherms. *Comp. Biochem. Physiol.* 49B:307–312.

———, and G. N. Somero (1975). Pressure effects on enzyme structure and function *in vitro* and under simulated *in vivo* conditions. *Comp. Biochem. Physiol.* 52B:67–74.

———, and G. N. Somero (1976). Adaptation of muscle pyruvate kinases to environmental temperatures and pressure. *J. Exp. Zool.* 198:1–12.

Lukton, A., and H. S. Olcott (1958). Content of free imidazole compounds in muscle tissue of aquatic animals. *Food Res.* 23:611–618.

Lumry, R., and R. Biltonen (1969). Thermodynamic and kinetic aspects of protein conformations in relation to physiological function. In *Structure and Stability of Biological Macromolecules*, ed. S. N. Timasheff and G. D. Fasman, 65–212. New York: Marcel Dekker.

Lutz, H., H. Weber, R. Billeter, and E. Jenny (1979). Fast and slow myosin within single skeletal muscle fibres of adult rabbits. *Nature* 281:142–144.

Lutz, P. L., J. C. La Manna, M. R. Adams, and M. Rosenthal (1980). Cerebral resistance to anoxia in the marine turtle. *Resp. Physiol.* 41:241–251.

McClanahan, L., Jr. (1967). Adaptations of the spadefoot toad, *Scaphiopus couchi* to desert environments. *Comp. Biochem. Physiol.* 20:73–99.

———, J. N. Stinner, and V. H. Shoemaker (1978). Skin lipids, water loss, and energy metabolism in a South American tree frog (*Phyllomedusa sauvagei*). *Physiol. Zool.* 51:179–187.

McGilvery, R. W. (1975). The use of fuels for muscular work. In *Metabolic Adaptation to Prolonged Physical Exercise*, ed. H. Howald and J. R. Poortmans, 12–30. Basel: Birkhauser Verlag.

——— (1979). *Biochemistry, A Functional Approach.* Philadelpha: W. B. Saunders.

McLane, J. A., and J. O. Holloszy (1979). Glycogen synthesis from lactate in the three types of skeletal muscle. *J. Biol. Chem.* 254:6548–6553.

Mahler, M. (1980). Kinetics and control of oxygen consumption in skeletal muscle. In *Exercise Bioenergetics and Gas Exchange*, ed. P. Cerretelli and B. J. Whipp, 53–66. Amsterdam: Elsevier/North Holland.

Malan, A. (1978). Intracellular acid-base state at a variable temperature in air-breathing vertebrates and its representation. *Resp. Physiol.* 33:115–119.

Maloiy, G.M.O. ed. (1979). *Comparative Physiology of Osmoregulation in Animals.* 2 vols. New York: Academic Press.

Malpica, J. M., and J. M. Vassallo (1980). A test for the selective origin of environmentally correlated allozyme patterns. *Nature* 286:407–408.

Mangum, C. P., and D. W. Towle (1977). Physiological adaptation to unstable environments. *Amer. Sci.* 65:67–75.

Mansingh, A. (1971). Physiological classification of dormancies in insects. *Can. Entomol.* 103:983–1009.

Manwell, C. (1960). Histological specificity of respiratory pigments I. Comparisons of the coelom and muscle hemoglobins of the polychaete worm, *Travisia pupa* and the echiuroid worm, *Arhynchite pugettensis. Comp. Biochem. Physiol.* 1:267–276.

Margules, D. L. (1979). Beta-endorphin and endoloxone: Hormones of the autonomic nervous system for the conservation or expenditure of bodily resources and energy in anticipation of famine or feast. *Neurosci. Biobehav. Rev.* 3:155–162.

Markert, C. L. (1963). Epigenetic control of specific protein synthesis in differentiating cells. In *Cytodifferentiation and Macromolecular Synthesis*, ed. M. Locke, 65–84. New York: Academic Press.

Marsh, R. L. (1981). Catabolic enzyme activities in relation to premigratory fattening and muscle hypertrophy in the gray catbird (*Dumetella carolinensis*). *J. Comp. Physiol.* 141:417–423.

Marshall, N. B. (1979). *Deep-Sea Biology. Developments and Perspectives.* New York and London: Garland STPM Press.

Matthew, J. B., G.J.H. Hanania, and F.R.N. Gurd (1979a). Electrostatic effects in hemoglobin: Hydrogen ion equilibria in human deoxy- and oxyhemoglobin A. *Biochemistry* 18:1919–1928.

———, G.I.H. Hanania, and F.R.N. Gurd (1979b). Electrostatic effects in hemoglobin: Bohr effect and ionic strength dependence of individual groups. *Biochemistry* 18:1928–1936.

Measures, J. C. (1975). Role of amino acids in osmoregulation of nonhalophilic bacteria. *Nature* 257:398–400.

Merritt, R. B. (1972). Geographic distribution and enzymatic properties of lactate dehydrogenase allozymes in the fathead minnow, *Pimephaes promelas. Amer. Nat.* 196:173–184.

Meschia, G., F. C. Battaglia, W. W. Hay, and J. W. Sparks (1980). Utilization of substrates by the ovine placenta *in vivo. Fed. Proc.* 39:245–249.

Mink, J. W., R. J. Blumenschine, and D. B. Adams (1981). Ratio of central nervous system to body metabolism in vertebrates: Its constancy and functional basis. *Amer. J. Physiol.* 241:R20-3-R212.

Mitchell, P. (1979). Keilin's respiratory chain concept and its chemiosmotic consequences. *Science* 206:1148–1159.

Mommsen, T. P., J. Ballantyne, D. MacDonald, J. Gosline, and P. W. Hochachka (1981). Analogues of red and white muscle in squid mantle. *Proc. Natl. Acad. Sci. USA* 78:3274–3278.

———, C. J. French, and P. W. Hochachka (1980). Sites and patterns of protein and amino acid utilization during the spawning migration of salmon. *Can. J. Zool.* 58:1785–1799.

———, and P. W. Hochachka (1981). Respiratory and enzymatic properties of squid heart mitochondria. *Eur. J. Biochem* 120:345–350.

Moon, T. W., and P. W. Hochachka (1971). Temperature and enzyme activity in poikilotherms: Isocitrate dehydrogenase in rainbow trout liver. *Biochem. J.* 123:695–705.

Moreadith, R. W., and W. E. Jacobus (1982). Creatine kinase of heart mitochondria. Functional coupling of ADP transfer to the adenine nucleotide translocase. *J. Biol. Chem.* 257:899–905.

Mori, M., S. Miura, T. Morita, and M. Tatibana (1982). Transport of ornithine transcarbamylase precursor into mitochondria. *12th Intl. Congress Biochemistry Abstracts*, 359.

Moser, H. G. (1974). Development and distribution of juveniles of *Sebastolobus*

(Pisces; Family Scorpaenidae). *U. S. Nat. Mar. Fish. Ser. Fishery Bull.* 72: 865–884.

Munro, H. N. (1980). Placenta in relation to nutrition. *Fed. Proc.* 39:236–238.

Murphy, B. J., P. W. Hochachka, W. M. Zapol, and G. C. Liggins (1982). Free amino acids in the blood of fetal and maternal Weddell seals. *Amer. J. Physiol.* 242:R85–R88.

———, W. M. Zapol, and P. W. Hochachka (1980). Metabolic activities of heart, lung, and brain during diving and recovery in the Weddell seal. *J. Appl. Physiol.* 48:596–605.

Murphy, D. J., and S. K. Pierce, Jr. (1975). The physiological basis for changes in the freezing tolerance of intertidal molluscs. I. Response to subfreezing temperatures and the influence of salinity and temperature acclimation. *J. Exp. Zool.* 193:313–322.

Musick, W.D.L., and M. G. Rossmann (1979). The structure of mouse testicular lactate dehydrogenase isoenzyme C_4 at 2.9 A resolution. *J. Biol. Chem.* 254: 7611–7620.

Nelson, R. A. (1980). Protein and fat metabolism in hibernating bears. *Fed. Proc.* 39:2955–2958.

Newsholme, E. A. (1978). Control of energy provision and utilization in muscle in relation to sustained exercise. *3rd Intl. Symp. Biochemistry of Exercise*, (ed. F. Landry and W.A.R. Orban), 3–27. Miami: Symposia Specialists.

———, B. Crabtree, S. J. Higgins, S. D. Thornton, and C. Start (1972). The activities of fructose diphosphatase in flight muscles from the bumble-bee and the role of this enzyme in heat generation. *Biochem. J.* 128:89–97.

———, and C. Start (1973). *Regulation in Metabolism.* New York: Wiley-Interscience.

Nicholls, D. G. (1976). The bioenergetics of brown adipose tissue mitochondria. *FEBS Letters* 61:103–110.

——— (1979). Brown adipose tissue mitochondria. *Biochim. Biophys. Acta* 549: 1–29.

Nuccitelli, R., and D. W. Deamer (1982). *Intracellular pH: Its Measurement, Regulation, and Utilization in Cellular Function.* New York: Allen R. Liss.

———, and J. M. Heiple (1982). Summary of the evidence and discussion concerning the involvement of pH_i in the control of cellular functions. In *Intracellular pH: Its Measurement, Regulation, and Utilization in Cellular Function*, ed. R. Nuccitelli and D. W. Deamer, pp. 567–586. New York: Allen R. Liss.

Ohe, M., and A. Kajita (1980). Changes in pK_a values of individual histidine residues of human hemoglobin upon reaction with carbon monoxide. *Biochemistry* 19:4443–4450.

Ohno, S. (1970). *Evolution by Gene Duplication.* Berlin: Springer-Verlag.

Oshima, T. (1979). Molecular basis for unusual thermostabilities of cell constituents from an extreme thermophile, *Thermus thermophilus.* In *Strategies of Microbial Life in Extreme Environments*, ed. M. Shilo, 455–469. Berlin: Dahlem Konferenzen 1979.

Parkhouse, W. S., D. C. McKenzie, P. W. Hochachka, T. P. Mommsen, W. K. Ovalle, S. L. Shinn, and E. C. Rhodes (1982). The relationship between carnosine levels, buffering capacity, fiber type, and anaerobic capacity in elite athletes. *5th Intl. Symp. Biochemistry of Exercise Abstracts*, 2.

Patterson, J. L., and J. G. Duman (1979). Composition of a protein antifreeze from larvae of the beetle, *Tenebrio molitor. J. Exp. Zool.* 210:361–367.

———, T. J. Kelley, and J. G. Duman (1981). Purification and composition of a thermal hysteresis producing protein from the milkweed bug, *Oncopeltus fasciatus. J. Comp. Physiol.* 142:539–542.

Patton, J. S. (1975). The effect of pressure and temperature on phospholipid and triglyceride fatty acids of fish white muscle: A comparison of deepwater and surface marine species. *Comp. Biochem. Physiol.* 52B:105–110.

Penniston, J. T. (1971). High hydrostatic pressure and enzymatic activity: Inhibition of multimeric enzymes by dissociation. *Arch. Biochem. Biophys.* 142: 322–332.

Perutz, M. (1970). Stereochemistry of cooperative effects in haemoglobin. *Nature* 228:726–734.

———, C. Bauer, G. Gros, F. Leclercq, C. Vandecasserie, A. G. Schnek, G. Braunitzer, A. E. Friday, and K. A. Joysey (1981). Allosteric regulation of crocodilian haemoglobin. *Nature* 291:682–684.

Pettigrew, D. W., and C. Frieden (1978). Rabbit muscle phosphofructokinase. *J. Biol. Chem.* 253:3623–3627.

Pfeiler, E. (1978). Effects of hydrostatic pressure on ($Na^+ + K^+$)-ATPase and Mg^{2+}-ATPase in gills of marine teleost fish. *J. Exp. Zool.* 205:393–402.

Phelps, M. E., D. E. Kuhl, and J. C. Mazziotta (1981). Metabolic mapping of the brain's response to visual stimulation: studies on humans. *Science* 211: 1445–1448.

Pike, C. S., and J. A. Berry (1980). Membrane phospholipid phase separations in plants adapted to or acclimated to different thermal regimes. *Plant Physiol.* 66:238–241.

Place, A. R., and D. A. Powers (1979). Genetic variation and relative catalytic efficiencies: Lactate dehydrogenase B allozymes of *Fundulus heteroclitus. Proc. Natl. Acad. Sci. USA* 76:2354–2358.

Podesta, R. B., T. Mustafa, T. W. Moon, W. C. Hulbert, and D. F. Mettrick (1976). Anaerobes in a aerobic environment: Role of CO_2 in energy metabolism of *Hymenolepis diminuta.* In *Biochemistry of Parasites and Host-Parasite Relationships*, ed. Van den Bossche, 81–88. New York: North-Holland Press.

Pollard, A., and R. G. Wyn Jones (1979). Enzyme activities in concentrated solutions of glycinebetaine and other solutes. *Planta* 144:291–298.

Powell, E. N., M. A. Crenshaw, and R. M. Rieger (1980). Adaptations to sulfide in sulfide-system meiofauna. Endproducts of sulfide detoxification in three Turbellarians and a Gastrotrich. *Mar. Ecol. Prog.* Ser 2:169–177.

Powell, M. A., and G. N. Somero (1983). Blood components prevent sulfide poisoning of respiration of the hydrothermal vent tube worm, *Riftia pachyptila. Science*, 219:297–299.

Powers, D. A. (1972). Hemoglobin adaptation for fast and slow water habitats in sympatric catostomid fishes. *Science* 177:360–362.

———, G. S. Greaney, and A. R. Place (1979). Physical correlation between lactate dehydrogenase genotype and haemoglobin function in killifish. *Nature* 277:240–241.

Prosser, C. L. (1973). *Comparative Animal Physiology*. Philadelphia: W. B. Saunders.

Prusiner, S., and M. Poe (1968). Thermodynamic considerations of mammalian thermogenesis. *Nature* 220:235–237.

Quetin, L. B., and J. J. Childress (1976). Respiratory adaptations of *Pleuroncodes planipes* to its environment off Baja California. *Mar. Biol.* 38:327–334.

Rahn, H. (1982). Comparison of embryonic development in birds and mammals: Birth weight, time, and cost. In *A Companion to Animal Physiology*, ed. C. R. Taylor, K. Johansen, and L. Bolis, 124–137. Cambridge: Cambridge University Press.

———, R. B. Reeves, and B. J. Howell (1975). Hydrogen ion regulation, temperature and evolution. *Amer Rev. Respir. Dis.* 112:165–172.

Raison, J. K., J. A. Berry, P. A. Armond, and C. S. Pike (1980). Membrane properties in relation to the adaptation of plants to temperature stress. In *Adaptation of Plants to Water and High Temperature Stress*, ed. N. C. Turner and P. J. Kramer, 261–273. New York: Wiley.

———, J.K.M. Roberts, and J. A. Berry (1982). Correlations between the thermal stability of chloroplasts (thylakoid membranes) and the composition and fluidity of the polar lipids upon acclimation of the higher plant, *Nerium oleander*, to growth temperatures. *Biochim. Biophys. Acta* 688:218–228.

Ramsay, J. A. (1964). The rectal complex of the mealworm, *Tenebrio molitor* L. (Coleoptera, Tenebrionidae). *Phil. Trans. Royal Soc.* London Series B 248:279–314.

Rau, G. H. (1981). Hydrothermal vent clam and tube worm $^{13}C/^{12}C$: Further evidence of nonphotosynthetic food sources. *Science* 213:338–340.

Raymond, J. A., and A. L. DeVries (1977). Adsorption inhibition as a mechanism of freezing resistance in polar fishes. *Proc. Nat. Acad. Sci. USA* 74:2589–2593.

———, Y. Lin, and A. L. DeVries (1975). Glycoprotein and protein antifreezes in two Alaskan fishes. *J. Exp. Zool.* 193:125–130.

Reeves, R. B. (1972). An imidazole alphastat hypothesis for vertebrate acid-base regulation: Tissue carbon dioxide content and body temperature in bullfrogs. *Resp. Physiol.* 14:219–236.

———, (1977). The interaction of body temperature and acid-base balance in ectothermic vertebrates. *Ann. Rev. Physiol.* 39:559–586.

Riddiford, L. M., and J. W. Truman (1978). Biochemistry of insect hormones and insect growth regulators. In *Biochemistry of Insects*, ed. M. Rockstein, 308–357. New York: Academic Press.

Riedesel, M. L., and J. M. Steffen (1980). Protein metabolism and urea recycling in rodent hibernators. *Fed. Proc.* 39:2959–2963.

Rigby, B. J. (1968). Temperature relationships of poikilotherms and the melting temperature of molecular collagen. *Biol. Bull.* 135:223–229.

Riggs, A. (1960). The nature and significance of the Bohr effect in mammalian hemoglobins. *J. Gen. Physiol.* 43:737–752.

Robin, E. D., J. Ensinck, A. J. Hance, M. D. Newman, N. Lewiston, L. Cornell, R. W. Davis, and J. Theodore (1981). Glucoregulation and prolonged diving in the harbor seal, *Phoca vitulina*. *Amer. J. Physiol.* 241:R293–R300.

———, N. Lewiston, A. Newman, L. M. Simon, and J. Theodore (1979). Bioenergetic patterns of turtle brain and resistance to profound loss of mitochondrial ATP generation. *Proc. Nat. Acad. Sci. U.S.A.* 76:3922–3926.

Rolph, T. P., and C. T. Jones (1981). Glucose metabolism in the perfused heart of the foetal guinea pig. *Biochem. Soc. Transactions* 9:65.

Rossi-Fanelli, A., and E. Antonini (1960). Oxygen equilibrium of hemoglobin from *Thunnus thynnus*. *Nature* 186:895–896.

Rosso, P., and C. Cramoy (1982). Nutrition and pregnancy. In *Nutrition, Pre- and Postnatal Development*, ed. M. Winick, 133–210. New York: Plenum Press.

Rutledge, P. S. (1981). Effects of temperature acclimation on crayfish hemocyanin oxygen binding. *Amer. J. Physiol.* 240:R93–R98.

Sacktor, B. (1976). Biochemical adaptations for flight in the insect. *Biochem. Soc. Symp.* 41:111–131.

Saks, V. A., G. B. Chernousova, D. E. Gukovsky, V. N. Smirnov, and E. I. Chazov (1975). Studies of energy transport in heart cells. Mitochondrial isoenzyme of creatine phosphokinase: Kinetic properties and regulatory action of Mg^{2+} ions. *Eur. J. Biochem.* 57:273–290.

Salt, R. W. (1961). Principles of insect cold hardiness. *Ann. Rev. Entomol.* 6:55–76.

Saz, H. J. (1981). Energy metabolism of parasitic helminths: Adaptations to parasitism. *Ann. Rev. Physiol.* 43:323–341.

Schade, B. C., R. Rudolph, H.-D Ludemann, and R. Jaenicke (1980). Reversible high-pressure dissociation of lactic dehydrogenase from pig muscle. *Biochemistry* 19:1121–1126.

Schmidt-Nielsen, K. (1979). *Animal Physiology: Adaptation and Environment.* Cambridge: Cambridge University Press.

Schneiderman, H. A., and C. M. Williams (1953). The physiology of insect diapause. VII. The respiratory metabolism of cecropia silkworm during diapause and development. *Biol. Bull.* 105:320–334.

Schneppenheim, R. and H. Theede (1979). Abstracts of the First European Society of Comparative Physiology and Biochemistry Conference. In *Animal and Environmental Fitness*, p. 97. Oxford: Pergamon Press.

———, and H. Theede (1980). Isolation and characterization of freezing-point depressing peptides from larvae of *Tenebrio molitor*. *Comp. Biochem. Physiol.* 67B:561–568.

Schoffeniels, E. (1976). Adaptations with respect to salinity. *Biochem. Soc. Symp.* 41:179–204.

Scholander, P. F. (1940). Experimental investigations in diving mammals and birds. *Hvalrad. Skr.* 22:1–131.

———, W. Flagg, R. J. Hoch, and L. Irving (1953). Studies on the physiology of frozen plants and animals in the Arctic. *J. Cell. Comp. Physiol. Suppl.* 1:1–56.

Schöttler, U. (1977). The energy-yielding oxidation of NADH by fumarate in anaerobic mitochondria of *Tubifex* sp. *Comp. Biochem. Physiol.* 58B:151–156.

Setlow, B., and P. Setlow (1980). Measurements of the pH within dormant and germinated bacterial spores. *Proc. Nat. Acad. Sci. USA* 77:2472–2476.

Shaklee, J. B., J. A. Christiansen, B. D. Sidell, C. L. Prosser, and G. S. Whitt (1977). Molecular aspects of temperature acclimation in fish: Contributions of changes in enzyme activities and isozyme patterns to metabolic reorganization in the green sunfish. *J. Exp. Zool.* 201:1–20.

Shoubridge, E. A., and P. W. Hochachka (1981). The origin and significance of metabolic carbon dioxide production in the anoxic goldfish. *Molec. Physiol.* 1:315–338.

Sidell, B. D. (1977). Turnover of cytochrome C in skeletal muscle of green sunfish (*Lepomis cyanellus*, R.) during thermal acclimation. *J. exp. Zool.* 199:233–250.

———, F. R. Wilson, J. Hazel, and C. L. Prosser (1973). Time course of thermal acclimation in goldfish. *J. Comp. Physiol.* 84:119–127.

Siebenaller, J. F., and G. N. Somero (1978). Pressure-adaptive differences in lactate dehydrogenases of congeneric fishes living at different depths. *Science* 210:255–257.

———, and G. N. Somero (1979). Pressure-adaptive differences in the binding and catalytic properties of muscle-type (M_4) lactate dehydrogenases of shallow- and deep-living marine fishes. *J. Comp. Physiol.* 129:295–300.

———, and G. N. Somero (1982). The maintenance of different enzyme activity levels in congeneric fishes living at different depths. *Physiol. Zool.* 55:171–179.

———, G. N. Somero, and R. L. Haedrich (1982). Biochemical characteristics of macrourid fishes differing in their depths of distribution. *Biol. Bull.* (Woods Hole) 163:240–249.

Sinensky, M. (1974). Homeoviscous adaptation—a homeostatic process that regulates the viscosity of membrane lipids in *Escherichia coli*. *Proc. Natl. Acad. Sci. USA* 71:522–525.

Singer, S. J., and G. L. Nicolson (1972). The fluid mosaic model of the structure of cell membranes. *Science* 175:720–731.

Singleton, R., Jr., and R. E. Amelunxen (1973). Proteins from thermophilic microorganisms. *Bacteriol. Rev.* 37:320–342.

———, C. R. Middaugh, and R. D. MacElroy (1977). Comparison of proteins from thermophilic and nonthermophilic sources in terms of structural parameters inferred from amino acid composition. *Int. J. Peptide Protein Res.* 10:39–50.

Slaughter, D., G. L. Fletcher, V. S. Ananthanarayanan, and C. L. Hew (1981). Antifreeze proteins from the sea raven, *Hemitripterus americanus*. *J. Biol. Chem.* 256:2022–2026.

Smith, C., and S. F. Velick (1972). The glyceraldehyde-3-phosphate dehydrogenase of liver and muscle. *J. Biol. Chem.* 247:273–284.

Smith, H. W. (1930). Metabolism of the lungfish *Protopterus aethiopicus*. *J. Biol. Chem.* 88:97–130.

Smith, K. L., Jr. (1978). Metabolism of the abyssopelagic rattail *Coryphaenoides armatus* measured *in situ*. *Nature* 274:362–364.

———, and R. R. Hessler (1974). Respiration of benthopelagic fishes: *in situ*-measurements at 1230 meters. *Science* 184:72—73.

Snapp, B. D., and H. C. Heller (1981). Suppression of metabolism during hibernation in ground squirrels (*Citellus lateralis*). *Physiol. Zool.* 54:297–307.

Snell, K. (1980). Muscle alanine synthesis and hepatic gluconeogenesis. *Transaction Biochem. Soc.* 8:205—213.

Somero, G. N. (1969). Pyruvate kinase variants of the Alaskan king-carb. *Biochem. J.* 114:237—241.

——— (1975). The role of isozymes in adaptation to varying temperatures. In *Isozymes II: Physiological Function*, ed. by C. L. Markert, pp. 221–234. New York: Academic Press.

——— (1978). Temperature adaptation of enzymes: Biological optimization through structure-function compromises. *Ann. Rev. Ecol. Syst.* 9:1–29.

——— (1981). pH-temperature interactions on proteins: Principles of optimal pH and buffer system design. *Marine Biol. Letters* 2:163–178.

——— (1982). Physiological and biochemical adaptations of deep-sea fishes: Adaptive responses to the physical and biological characteristics of the abyss. In *The Environment of the Deep Sea*, ed. W. G. Ernst and J. G. Morin, pp. 256–278. Englewood Cliffs, N. J.: Prentice-Hall, Inc.

———, and R. D. Bowlus (1983). Solute compatibility with enzyme structure and function. In *The Mollusca*, ed. P. W. Hochachka, 2. 77–100. New York: Academic Press.

———, and J. J. Childress (1980). A violation of the metabolism-size scaling paradigm: activities of glycolytic enzymes in muscle increase in larger size fishes *Physiol. Zool.* 53:322–337.

———, and D. Doyle (1973). Temperature and rates of protein degradation in the fish *Gillichthys mirabilis*. *Comp. Biochem. Physiol.* 46B:463–474.

———, and P. S. Low (1976). Temperature: A "shaping force" in protein evolution. *Biochem. Soc. Symp.* 41:33–42.

———, and P. S. Low (1977). Eurytolerant proteins: mechanisms for extending the environmental tolerance range of enzyme-ligand interactions. *Amer. Nat.* 111:527–538.

———, and J. F. Siebenaller (1979). Inefficient lactate dehydrogenases of deep-sea fishes. *Nature* 282:100–102.

———, J. F. Siebenaller, and P. W. Hochachka (1983). Biochemical and physiological adaptations of deep-sea animals. In *The Sea*, ed. G. T. Rowe, 8:261–330. New York: Wiley.

———, and P. H. Yancey (1978). Evolutionary adaptation of K_m and k_{cat} values: Fitting the enzyme to its environment through modifications in amino acid sequences and changes in the solute composition of the cytosol. *Symp. Biol. Hungarica* 21:249–276.

Sonoda, T., S. Kawamoto, M. Mori, and M. Tatibana (1982). *12th Intl. Congress of Biochemistry Abstracts*, 312.

Spiess, F. N., K. C. Macdonald, T. Atwater, R. Ballard, A. Carranza, D. Cordoba, C. Cox, V. M. Diaz Garcia, J. Francheteau, J. Guerrero, J. Hawkins, R. Haymon, R. Hessler, T. Juteau, M. Kastner, R. Larson, B. Luyendyk, J. D. Macdougall, S. Miller, W. Normark, J. Orcutt, and C. Rangin (1980). East Pacific rise: hot springs and geophysical experiments. *Science* 207:1421–1433.

Steinberg, D., and J. C. Khoo (1977). Hormone-sensitive lipase of adipose tissues. *Fed. Proc.* 36:1986–1990.

Stevens, E. D., and F.E.J. Fry (1971). Brain and muscle temperature in ocean caught and captive skipjack tuna. *Comp. Biochem. Physiol.* 38A:203–211.

Stoeckenius, W., R. H. Lozier, and R. A. Bogomolni (1979). Bacteriorhodopsin and the purple membrane of Halobacteria. *Biochim. Biophys. Acta Bioenergetics* 505:215–278.

Storey, K. B., J. G. Baust, and J. M. Storey (1981). Intermediary metabolism during low temperature acclimation in the overwintering gall fly larva, *Eurosta solidaginis*. *J. Comp. Physiol.* 144:183–190.

———, and P. W. Hochachka (1974a). Enzymes of energy metabolism from a vertebrate facultative anaerobe, *Pseudemys scripta*. Turtle heart phosphofructokinase. *J. Biol. Chem.* 249:1417–1422.

———, and P. W. Hochachka (1974b). Enzymes of energy metabolism in a vertebrate facultative anaerobe, *Pseudemys scripta*. Turtle heart pyruvate kinase. *J. Biol. Chem.* 249:1423–1427.

———, and J. M. Storey (1983). Carbohydrate metabolism in cephalopods. In *The Mollusca*, ed. P. W. Hochachka, 92–136. New York: Academic Press.

Sugiyama, T., M. R. Schmitt, S. B. Ku, and G. E. Edwards (1979). Differences in cold lability of pyruvate, P_i dikinase among C_4 species. *Plant Cell Physiol.* 20:965–971.

Sullivan, K. M., and G. N. Somero (1980). Enzyme activities of fish skeletal muscle and brain as influenced by depth of occurrence and habits of feeding and locomotion. *Mar. Biol.* 60:91–99.

Swan, H., and C. L. Schatte (1977). Antimetabolic extract from the brain of the hibernating ground squirrel. *Citellus tridecemlineatus*. *Science* 195: 84–85.

Swezey, R. R., and G. N. Somero (1982a). Polymerization thermodynamics and structural stabilities of skeletal muscle actins from vertebrates adapted to different temperatures and hydrostatic pressures. *Biochemistry* 21:4496–4503.

———, and G. N. Somero (1982b). Skeletal muscle actin content is strongly conserved in fishes having different depths of distribution and capacities of locomotion. *Marine Biol. Letters* 3:307–315.

Theede, H., R. Schneppenheim, and Beress. (1976). Frostschutz-Glycoproteine bei *Mytilus edulis*? *Mar. Biol* 36:183–189.

Thomas, D. P., and G. F. Fregin (1981). Cardiorespiratory and metabolic responses to treadmill exercise in the horse. *J. Appl. Physiol.* 50:864–868.

Torres, J. J., B. W. Belman, and J. J. Childress (1979). Oxygen consumption rates of midwater fishes off California. *Deep-Sea Res.* 26A:185–197.

Tracy, C. R. (1977). Minimum size of mammalian homeotherms: Role of the thermal environment. *Science* 198:1034–1035.

Trivedi, B., and W. H. Danforth (1966). Effects of pH on the kinetics of frog muscle phosphofructokinase, *J. Biol. Chem.* 241:4110–4112.

Tsai, M. Y., F. Gonzalez, and R. G. Kemp (1975). Physiological significance of phosphofructokinase isozymes. In *Isozymes II. Physiological Function*, ed. C. L. Markert, 819–835. New York: Academic Press.

Tuttle, J. H., C. O. Wirsen, and H. W. Jannasch (1983). Microbial activities in the emitted hydrothermal waters of the Galapagos rift vents. *Marine Biol.* 73:293–299.

Ultsch, G. R., and D. C. Jackson (1982). Long-term submergence at 3°C of the turtle, *Chrysemys picta bellii*, in normoxic and severely hypoxic water. *J. Exp. Biol.* 96:11–28.

Uyeda, K., E. Furuya, and L. J. Luby (1981). The effect of natural and synthetic D-fructose 2,6,-bisphosphate on the regulatory kinetic properties of liver and muscle phosphofructokinase. *J. Biol. Chem.* 256:8394–8399.

———, and E. Racker (1965). Regulatory mechanisms in carbohydrate metabolism. *J. Biol. Chem.* 240:4682–4688.

Vaghy, P. L. (1979). Role of mitochondrial oxidative phosphorylation in the maintenance of intracellular pH. *J. Mol. Cell. Cardiol.* 11:933–940.

Van Voorhies, W. V., J. A. Raymond, and A. L. DeVries (1978). Glycoproteins as biological antifreeze agents in the cod, *Gadus ogac* (Richardson). *Physiol. Zool.* 51:347–353.

Van den Thillart, G. (1982). Adaptations of fish energy metabolism to hypoxia and anoxia. *Molec. Physiol.* 2:49–62.

Vary, T. C., D. K. Reibel, and J. R. Neely (1981). Control of energy metabolism of heart muscle. *Ann. Rev. Physiol.* 43:419–430.

Vik, S. B., and Capaldi, R. A. (1977). Lipid requirements for cytochrome c oxidase activity. *Biochemistry* 16:5755–5759.

Von Hippel, P. and T. Schleich (1969). The effects of neutral salts on the structure and conformational stability of macromolecules in solution. In *Structure and Stability of Biological Macromolecules*, ed. S. N. Timascheff and G. D. Fasman, 417–574. New York: Marcel Dekker.

Wald, G. (1964). The origins of life. *Proc. Nat. Acad. Sci. U.S.A.* 52:595–611.

Walsh, P. J. (1981). Purification and characterization of glutamate dehydrogenase from three species of sea anemones: Adaptation to temperature within and among species from different thermal environments. *Marine Biol. Letters* 2:289–299.

———, and G. N. Somero (1982). Interactions among pyruvate concentration, pH, and K_m of pyruvate in determining *in vivo* Q_{10} values of the lactate dehydrogenase reaction. *Can. J. Zool.* 60:1293–1299.

Wang, L.C.H. (1978). Energetic and field aspects of mammalian torpor: The Richardson's ground squirrel. In *Strategies in Cold—Natural Torpidity and Thermogenesis*, ed. L.C.H. Wang and J. W. Hudson, 109–145. New York: Academic Press.

Wann, K. T., and A. G. Macdonald (1980). The effects of pressure on excitable cells. *Comp. Biochem. Physiol.* 66A:1–12.

Watt, W. B. (1977). Adaptation at specific loci. I. Natural selection on phosphoglucose isomerase of *Colias* butterflies: biochemical and population aspects. *Genetics* 87:177–194.

Weber, R. E., J. Bonaventura, B. Sullivan, and C. Bonaventura (1978). Oxygen equilibrium and ligand-binding kinetics of erythrocruorins from two burrowing polychaetes of different modes of life, *Marphysa sanguinea* and *Diopatra cuprea. J. Comp. Physiol.* 123:177–184.

Weeda, E., A. B. Koopmanschap, C.A.D. de Kort, and A.M.Th. Beenakkers (1980). Proline synthesis in fat body of *Leptinotarsa decemlineata. Insect Biochem.* 10:631–636.

Weibel, E. R., and C. R. Taylor (1981). Design of the mammalian respiratory system. *Resp. Physiol.* 44:1–164.

Whalen, R. G., S. M. Sell, G. S. Butler-Browne, K. Schwartz, P. Bouveret, and I. Pinset-Harstrom (1981). Three myosin heavy-chain isozymes appear sequentially in rat muscle development. *Nature* 292:805–809.

White, F. N., and G. N. Somero (1982). Acid-base regulation and phospholipid adaptations to temperature: Time courses and physiological significance of modifying the milieu for protein function. *Physiol. Rev.* 62:40–90.

Wiggins, P. M. (1971). Water structure as a determinant of ion distribution in living tissue. *J. Theor. Biol.* 32:131–146.

———, (1979). Metabolic control of the properties of intracellular water as a universal driving force for active transport. In *Cell-Associated Water*, ed. W. Drost-Hansen and J. Clegg, 69–114. New York: Academic Press.

Williams, P. M., K. L. Smith, E. M. Druffel, and T. W. Linick (1981). Dietary carbon sources of mussels and tubeworms from Galapagos hydrothermal vents determined from tissue ^{14}C activity. *Nature* 292:448–449.

Williamson, J. R., B. Safer, K. F. LaNoue, C. M. Smith, and E. Walajtys (1973). Mitochondrial-cytosolic interactions in cardiac tissue: role of the malate-aspartate cycle in the removal of glycolytic NADH from the cytosol. *Soc. Exp. Biol. Symp.* 27:241–281.

Wilson, D. F., M. Erecinska, C. Drown, and I. A. Silver (1979). The oxygen dependence of cellular energy metabolism. *Arch. Biochem. Biophys.* 195:485–493.

Wilson, F. R., G. N. Somero, and C. L. Prosser (1974). Temperature-metabolism relations of two species of *Sebastes* from different thermal environments. *Comp. Biochem. Physiol.* 47B:485–491.

Wilson, J. E. (1980). Brain hexokinase, the prototype ambiquitous enzyme. *Curr. Top. Cell. Reg.* 16:1–44.

Wolfe, R. R., P. W. Hochachka, R. L. Trelstad, and J. F. Burke (1979). Lactate oxidation in perfused rat lung. *Amer. J. Physiol.* 236:E276–282.

Wood, S. C., and K. Johansen (1972). Adaptation to hypoxia by increased HbO_2 affinity and decreased red cell ATP concentration. *Nature New Biol.* 237:278–279.

———, and K. Johansen (1973). Organic phosphate metabolism in nucleated red cells. Influence of hypoxia on eel HbO_2 affinity. *Neth. J. Sea Res.* 7: 328–338.

———, and K. Johansen (1974). Oxygen uptake and cardiac output in eel adapted to hypoxia. *Physiologist* 17:362.

Wu, T.F.L, and E. J. Davis (1981). Regulation of glycolytic flux in an energetically controlled cell-free system: the effects of adenine nucleotide ratios, inorganic phosphate, pH, and citrate. *Arch. Biochem. Biophys.* 209:85–99.

Yancey, P. H., M. E. Clark, S. C. Hand, R. D. Bowlus, and G. N. Somero (1982). Living with water stress: Evolution of osmolyte systems. *Science* 217:1214–1222.

———, and G. N. Somero (1978a). Temperature dependence of intracellular pH: its role in the conservation of pyruvate apparent K_m values of vertebrate lactate dehydrogenases. *J. Comp. Physiol.* 125:129–134.

———, and G. N. Somero (1978b). Urea-requiring lactate dehydrogenases of marine elasmobranch fishes. *J. Comp. Physiol.* 125:135–141.

———, and G. N. Somero (1979). Counteraction of urea destabilization of protein structure by methylamine osmoregulatory compounds of elasmobranch fishes. *Biochem. J.* 183:317–323.

———, and G. N. Somero (1980). Methylamine osmoregulatory solutes of elasmobranch fishes counteract urea inhibition of enzymes. *J. Exp. Zool.* 212:205–213.

Zachariassen, K. E. (1980). The role of polyols and nucleating agents in cold-hardy beetles. *J. Comp. Physiol.* 140:227–234.

Zapol, W. M., G. C. Liggins, R. C. Schneider, J. Qvist, M. T. Snider, R. K. Creasy, and P. W Hochachka (1979). Regional blood flow during simulated diving in the conscious Weddell seal. *J. Appl. Physiol.* 47:968–973.

Zuber, H. (1979). Structure and function of enzymes from thermophilic microorganisms. In *Strategies of Microbial Life in Extreme Environments*, ed. M. Shilo, 393–415. Berlin: Dahlem Konferenzen 1979.

INDEX

Library of Congress Cataloging in Publication Data

Hochachka, Peter W.
Biochemical adaptation.

Bibliography: p.
Includes index.
1. Adaptation (Physiology) 2. Biological chemistry.
I. Somero, George N. II. Title.
QP82.H63 1984 574.5 83-43076

ISBN 0-691-08343-6
ISBN 0-691-08344-4 (pbk.)

Peter W. Hochachka is Professor of Biology at the University of British Columbia in Vancouver. George N. Somero is Professor of Biology at the Scripps Institution of Oceanography at the University of California at San Diego.